THE
V·I·S·U·A·L
HANDBOOK OF
BUILDING
AND REMODELING

Completely Updated and Expanded!

THE

V·I·S·U·A·L

HANDBOOK OF

BUILDING
AND REMODELING

■ PROFESSIONAL EDITION ■

THE ONLY GUIDE TO CHOOSING
THE RIGHT MATERIALS AND SYSTEMS
FOR EVERY PART OF YOUR HOME

CHARLIE WING

Rodale Press, Inc.
Emmaus, Pennsylvania

OUR PURPOSE

*"We inspire and enable people to improve
their lives and the world around them."*

The author and editors who compiled this book have
tried to make all of the contents as accurate and as correct as possible. Plans, illustrations, photographs, and
text have all been carefully checked and cross-checked.
However, due to the variability of local conditions, construction materials, personal skill, and so on, neither
the author nor Rodale Press assumes any responsibility
for any injuries suffered or for damages or other losses
incurred that result from the material presented herein.
All instructions and plans should be carefully studied
and clearly understood before beginning construction.

Printed in the United States of America on
acid-free ∞, recycled ♲ paper

Editor: David Schiff
Interior Book Designer: Glen Burris
Cover Designer: Jerry O'Brien
Copy Editor: Patty Sinnott
Production Editor: Barbara McIntosh Webb
Manufacturing Coordinator: Patrick T. Smith
Indexer: Nan Badgett

Distributed in the book trade by St. Martin's Press

2 4 6 8 10 9 7 5 3 1   hardcover
2 4 6 8 10 9 7 5 3 1   paperback

Rodale Home and Garden Books
Vice President and Editorial Director:
    Margaret J. Lydic
Managing Editor, Woodworking Books:
    Kevin Ireland
Director of Design and Production: Michael Ward
Associate Art Director: Carol Angstadt
Studio Manager: Leslie M. Keefe
Copy Director: Dolores Plikaitis
Book Manufacturing Director: Helen Clogston
Office Manager: Karen Earl-Braymer

We're always happy to hear from you. For questions or comments concerning the editorial content of this book, please write to:
    Rodale Press, Inc.
    Book Readers' Service
    33 East Minor Street
    Emmaus, PA 18098
Look for other Rodale books wherever books are
sold. Or call us at (800) 848-4735.

For more information about Rodale Press and the
books and magazines we publish, visit our World
Wide Web site at:
    http://www.rodalepress.com

**Library of Congress
Cataloging-in-Publication Data**

Wing, Charlie, 1939
    The visual handbook of building and remod-
eling : the only guide to choosing the right
materials and systems for every part of your
home / Charlie Wing.—Updated and expanded,
2nd ed.
            p.   cm.
    Includes index.
    ISBN 0-87596-808-2   hardcover
    ISBN 0-87596-981-X   paperback
    1. House construction—Handbooks, manuals,
etc.   2. Dwellings—Remodeling—Handbooks,
manuals, etc.   3. Building materials—Handbooks, manuals, etc.   I. Title.
TH4813.W56   1998
690'.837—dc21
                                    97-53115

Contents

Acknowledgments

This book is a dream come true. For as long as I have been writing about house building, I have dreamed of collecting and organizing the bits and scraps of information I knew every builder would find truly useful.

For many years the dream remained just that. No publisher shared my enthusiasm for the project, no publisher until the wonderful people at Rodale Press.

The first and most important credit must go to Ray Wolf for selling the concept to the unknown (to me) powers that be.

The most trying task, no doubt, fell upon the shoulders of Maggie Balitas, who patiently tempered my dream with the realities of publishing.

Ensuring accuracy and practicality became the personal obsession of David Schiff, who, thank God, has a sense of humor.

Finally, an unexpected reward was the opportunity to apprentice under master graphic artist and designer Glen Burris.

To all on my team, a sincere thank you.

Introduction

Eight years ago I proposed to Rodale Press a "visual handbook for small builders and do-it-yourselfers." I stated that such a book could sell 50,000 copies and might, with periodic revision and addition, enjoy a long life.

"Let's just see how the first edition sells" was their dubious but predictable reply.

Well, I was right. In fact, I was so right I was almost wrong! The first edition sold not 50,000 but 200,000 copies and, at age seven, was showing little sign of slowing.

When I proposed a second edition, their (again predictable) response was, "Why fix it if it ain't broke?" But when they saw what I proposed to add, they agreed we should wait no longer. Of course, all of the original material in the first edition has been brought up-to-date, but the 100 new pages cover elements I had always felt to be either weak or missing entirely in the original:

- span tables for US and Canadian lumber
- checklists of building-code requirements
- a veritable catalog of metal framing aids
- a whole new chapter on decks and fences
- twice the information on design for access
- three times as many framing details

There you have the reason for a second edition. But let me tell you why I felt compelled to do the first.

"Building a house requires thousands of decisions based on a million bits of information." This was the opening line of my talk to thousands of potential owner/builders who attended my three-week courses at Shelter Institute and, later, at Cornerstones.

Teaching that course, and an earlier house-building course at Bowdoin College, taught me how to convey technical design information to people who are not professional builders. I put that training to work in my five previous books, which covered every aspect of building, from retrofitting insulation in a drafty old house in *House Warming* to siting, planning, and constructing a house in *Breaking New Ground*.

But still, something was missing. I'd written all I'd wanted to about designing and constructing a house, but I hadn't thoroughly covered the topic of what to construct a house with. What's more, it didn't seem to me that anyone else had, either. Certainly there were thorough technical references for architects and structural engineers. But there was no thorough guide for people without that kind of formal technical training. The fact is, the majority of people who actually lay their hands on materials—tradesmen, owner/builders, and do-it-yourselfers—are not trained architects or engineers. These people needed a book that was just as thorough as the ones the architects were using but that also offered explanations, formulas, and charts that would make the information accessible. *The Visual Handbook of Building and Remodeling*, I believe, is that book.

Naturally, in my previous books, I discussed materials in the context of how to use them, as have many other excellent how-to writers. But the *Visual Handbook* focuses on the materials. Its purpose is to enable you to decide how much of which material of what size should be used for any given house on any given site. You will find some how-to information in these pages because the way materials are used is often relevant in deciding which material to use. The point is, the how-to is incidental to the what-to, instead of the other way around. For example, the siding chapter discusses the pros and cons of each type of siding, from clapboards to vinyl, so you can make an informed decision about which is best for your particular climate and site. But it also illustrates construction details, to help you decide which will work best on your particular house.

I could see right away that this had to be a visual handbook. There's no way words could describe, for example, every commercially available molding profile. But it's possible to illustrate each one. Two things made it possible for me to produce this visual book. One is the wonders of personal computing technologies that enabled me to produce illustrations even though I can't draw. The other is that I am an information pack rat, a looter of lists, a burglar of booklets, a swiper of spec sheets.

In this, I am insatiable. No builder, no ancient hardware clerk, no sawyer in the backwoods, not even architects were spared my quest. I raided their files, their bookshelves, and their minds. Over the years I have accumulated the best of what they found useful in the actual building of houses: tables, lists, government pamphlets, manufacturers' literature, building-trade association publications, even the instructions from a package of asphalt shingles.

The result, I think, is a book that should be useful to anyone who puts his or her hands on building materials or who hires others who do. If you hire a builder, you won't be limited to his preferences but will be able to take a more active role in deciding what materials to use. If you are an owner/builder, this book should complement the how-to books I'm sure you are accumulating. And if you are a tradesman, I hope that you'll toss a copy behind the seat of your truck for easy reference.

By the way, the opening line of my owner/builder course proved to be a bit off the mark. According to my computer, this book actually contains 40 million bytes of information.

Abbreviations

AASHO: American Association of State Highway Officials

ABS: acrylonitrile-butadiene-styrene

AC: armored cable

ACA: ammoniacal copper arsenate

ACZA: ammoniacal copper zinc arsenate

ADA: airtight drywall approach

AFUE: annual fuel utilization efficiency

ag: above grade

amp: ampere

ANSI: American National Standards Institute

APA: The Engineered Wood Association (formerly known as the American Plywood Association)

ARM: adjustable-rate mortgage

avg: average

AWPB: American Wood Preservers Bureau

b: breadth

bg: below grade

Btu: British thermal unit

Btuh: British thermal unit per hour

BUR: built-up roof

BX: armored cable (used interchangeably with AC)

C: centigrade; corrosion-resistant

CABO: Council of American Building Officials

CCA: chromated copper arsenate

CCF: hundred cubic feet

C_D: duration of load factor

CDD: cooling degree-day

C_F: size factor

cfm: cubic feet per minute

CFR: Code of Federal Regulations

C_H: horizontal shear adjustment

CMU: concrete masonry unit

CPSC: Consumer Product Safety Commission

CPVC: chlorinated polyvinyl chloride

C_r: repetitive member factor

CRI: color-rendering index

cu ft: cubic foot

d: depth; pennyweight

D: deciduous

db: decibel

dbl: double

DD_{65}: base 65°F degree-day

dia: diagonal; diameter

DMT: design minimum temperature

Abbreviations

DS: double-strength

DWV: drain, waste, and vent

E: east; modulus of elasticity; evergreen; excellent

EER: energy efficiency ratio

EMT: thin-wall meta conduit

EPA: US Environmental Protection Agency

EPDM: ethylene propylene polymer membrane

EWS: Engineered Wood Society

Exp 1: exposure 1

EXT: exterior

F: Fahrenheit; fair; feeder; female

Fb: extreme fiber stress in bending

Fc: compression parallel to grain

FHA: Federal Housing Administration

fipt: female iron pipe thread

fnpt: female national pipe thread tapered

FRM: fixed-rate mortgage

Ft: fiber stress in tension

Fv: horizontal shear stress

G: good

ga: gauge

gal: gallon

GEM: growing-equity mortgage

GFCI: ground fault circuit interrupter

GFI: ground fault interrupter

GPM: graduated-payment mortgage; gallon per minute

H: heat-resistant; height; run

HDD: heating degree-day

HDO: high-density overlay

hp: horsepower

hr: hour

HUD: US Department of Housing and Urban Development

HVAC: heating, ventilating, and air-conditioning

Hz: hertz

ID: internal diameter

IIC: impact insulation class

IMC: intermediate metal conduit

ins: insert thread

INT: interior

IRMA: insulated roof membrane assembly

K: Kelvin

kwhr: kilowatt-hour

L: left; length

lb: pound

lin ft: linear feet

M: male

max: maximum

MC: moisture content

MDO: medium-density overlay

min: minimum; minute

mipt: male iron pipe thread

mnpt: male national pipe thread tapered

mph: miles per hour

MW: moderate weather

N: north

NM: nonmetallic

NRC: noise-reduction coefficient

NWWDA: National Wood Window and Door Association

oc: on-center

OD: outside diameter

OSB: oriented-strand board

oz: ounce

P: perennial; plastic-weld; poor

PB: polybutylene

pcf: pounds per cubic foot

PE: polyethylene

perm: measure of permeability

PS: product standard

psf: pounds per square foot

psi: pounds per square inch

PVC: polyvinyl chloride

R: right; riser; rubber

rec: recommended

Ref: reference point

RMC: rigid metal conduit

RTV: room temperature vulcanized

R-value: thermal resistance

S: south

SAM: shared-appreciation mortgage

SE: service entrance

sel str: select structural

sgl: single(s)

SJ: slip joint

slip: slip fitting

spig: spigot fitting

sq ft: square feet

sq in: square inch

STC: sound transmission class

std: standard

SW: severe weather

S1S: surfaced one side

S4S: surfaced four sides

T: texture; thermo-plastic; tread

T&G: tongue-and-groove

TCE: trichloroethylene

THM: trihalomethane

U: underground

USDA: US Department of Agriculture

V: rise

VB: vapor barrier

W: water-resistant; west; width

WM: Wood Moulding and Millwork Producers Association

WWPA: Western Wood Products Association

yr: year

1

Design

Houses are designed for people. Residential dimensions must, therefore, be tied to *human dimensions*.

In the sections on *kitchen dimensions, bathroom dimensions,* and *bedroom dimensions,* you will find minimum and standard heights, widths, and clearances for the respective rooms and their furnishings, as well as for *cabinetry*. Similar minimum and standard criteria for *window, closet, and passageway dimensions* will make your house a better home.

Use the 1/4-inch-scale *furnishing templates* and *fixture and appliance templates* as tools for scaling your floor plans.

Begin your kitchen design by considering the wide range of practical *kitchen plans* illustrated. A similarly wide range of *bathroom plans* is illustrated.

Stair design is complex and is subject to many code regulations. You will find the critical elements listed, along with a simple table for adjusting stair treads to any ceiling height.

Nearly one-quarter of us have some form of disability. Use the section on *access* to make the house work for all.

Blueprints are the visual language of building. To understand the ideas of building-trade professionals and to make yourself understood, you must learn *how to read blueprints*.

Much as we dislike design restraints, the various building codes have evolved primarily to keep us from hurting ourselves. The *checklist of code requirements* lists the most important provisions of the codes.

HUMAN DIMENSIONS

Why do we need typical human dimensions to design buildings? Because buildings are built for people. People need to reach shelves, move furniture around, get by each other, work side by side, and see out windows. The special requirements of the handicapped mean that spaces sometimes have to be designed for two sets of human dimensions.

The *typical human dimensions* shown below are for an adult male of height 5 feet 10 inches. To determine typical dimensions for other heights, simply multiply the dimensions shown by the ratio of heights.

Example: What is the standing eye level of a 5-foot (60-inch) person?

The standing height of the figure in the illustration is 70 inches, and the eye height is 66 inches. The standing eye level of a 60-inch person is therefore $60 \div 70 \times 66 = 57$ (inches).

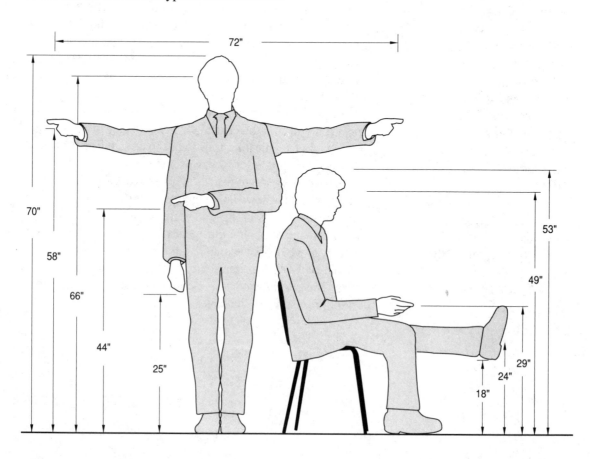

KITCHEN DIMENSIONS

Unless you special-order or build your own, chances are your cabinetry will be of standard dimensions. The only choice open to you is the height of the wall cabinets above the counter.

Low wall cabinets make wall shelving more accessible. On the other hand, if you plan to hang a microwave oven and a cookbook shelf under the wall cabinets, you will need greater height for a useful counter surface.

The countertop frontages shown are the minimum. Cooks who use lots of pots and pans or who entertain a lot will appreciate 36 inches to the left of the sink and a separate 24 × 48-inch work surface, such as the island shown.

Recessed counter surfaces on inside corners do not count toward either frontage or surface area.

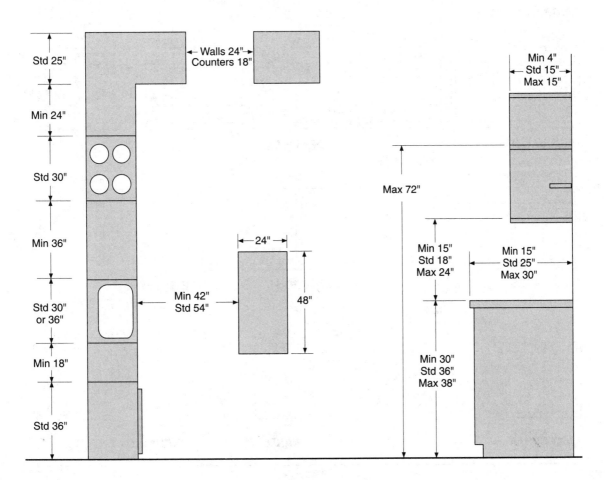

KITCHEN CABINETRY

Standard modular kitchen cabinets are available in a wide range of styles and prices.

Standard base cabinet height is 34-1/2 inches, assuming a 1-1/2-inch countertop and 36-inch counter surface height. Cabinet height can be raised or lowered by adding to or removing from the kick space.

Ideal counter height is considered to be the bottom of a bent elbow less 3 inches, but modifying counter heights should be taken seriously because it decreases the home's resale value.

Cabinet widths range from 9 to 36 inches in 3-inch increments, and 36 to 48 inches in 6-inch increments. Exact fit is made by site-trimming standard 3- and 6-inch filler strips.

Typical modular kitchen cabinets are shown below and on the opposite page.

Base Cabinets

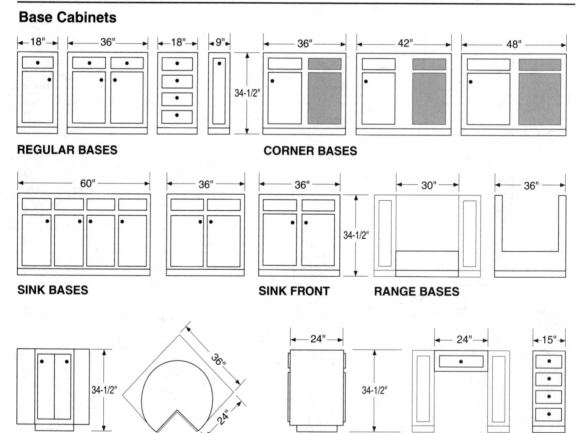

REGULAR BASES   CORNER BASES

SINK BASES   SINK FRONT   RANGE BASES

LAZY SUSAN   ISLAND   PLANNING CENTERS

Wall Cabinets

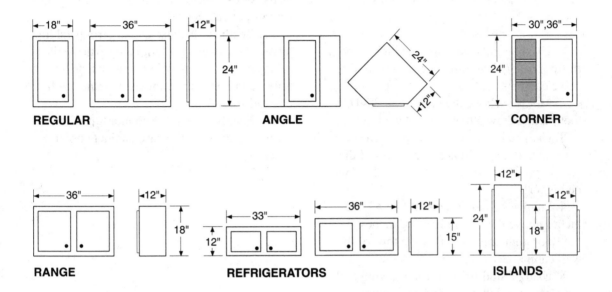

REGULAR   **ANGLE**   **CORNER**

RANGE   **REFRIGERATORS**   **ISLANDS**

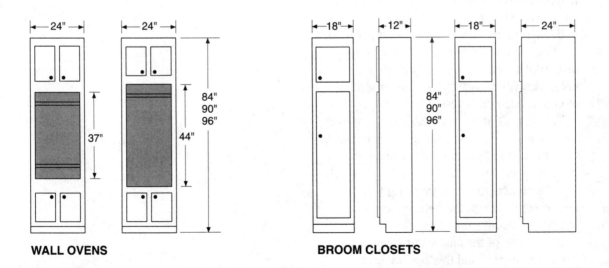

WALL OVENS   **BROOM CLOSETS**

Installing Kitchen Cabinets

Installation of kitchen cabinets will go smoothly in either new construction or remodeling if the standard procedure below is followed:

1. Using a line level, find the highest point of floor along the entire length of wall to receive cabinets.

2. Measure above the high point 34-1/2 inches, and draw a mark on the wall.

3. Using a line level and a chalk line, snap a horizontal line to indicate the level of the base cabinet tops (not countertop, which will be higher).

4. Snap a second horizontal line 49-1/2 inches above the first line, or 84 inches above the floor high point, to mark the tops of the wall cabinets.

5. Using a stud finder (or a finishing nail where holes will be covered by cabinets), mark all locations of studs. Be sure the stud marks are visible above both the 34-1/2-inch and 84-inch heights.

6. Starting in a corner, have a helper hold the first wall cabinet up to the 84-inch line, drill pilot holes, and fasten through the cabinet back with two #10 x 2-1/2-inch screws. Do not tighten the screws all the way.

7. When all of the wall cabinets have been loosely hung, shim if necessary to bring all of the fronts into alignment, then tighten the mounting screws.

8. Starting at the same corner, place the corner base cabinet, the flanking cabinets, then the remaining cabinets.

9. Shim all of the base cabinets until the fronts are aligned and the tops are level.

10. Drill 1/8-inch holes through the top rails of the cabinet backs until the drill reaches the studs. Fasten with two #10 x 2-1/2-inch screws per cabinet.

11. Readjust the base cabinet fronts (shim again, if necessary). Clamp the vertical stiles of adjacent cabinets with C-clamps, drill pilot holes, and fasten the cabinets together with drywall screws.

12. Finally, attach the countertop with screws up through the base cabinet corner blocks.

BATHROOM DIMENSIONS

Except in custom and luxurious homes, bathrooms tend to be of minimal dimensions. This is probably a holdover from the early days when the new "bath" room was stuffed into an existing closet.

Both clearance and fixture dimensions shown below are the minimums allowed by most building codes. Where applicable, standard fixture sizes are noted as well. Standard fixture sizes are used in the section "Bathroom Plans."

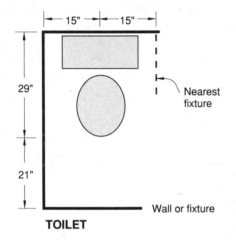

TOILET

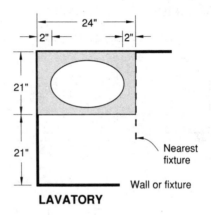

LAVATORY

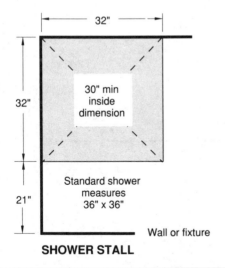

SHOWER STALL

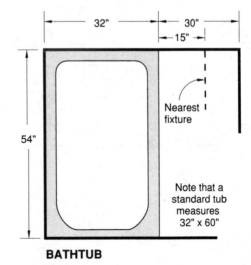

BATHTUB

BATHROOM VANITY CABINETRY

Most manufacturers of kitchen cabinets also offer a standardized line of bathroom vanity cabinets. Standard bathroom cabinet height is 29 inches; standard depth is 21 inches. Standard widths are 12 to 48 inches in 3-inch increments and 60 inches.

Vanity Cabinets

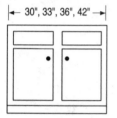
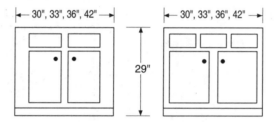
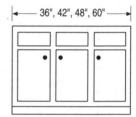

SINK BASES

BASE

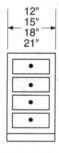

DRAWER BASE

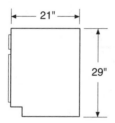

BEDROOM DIMENSIONS

The illustration below shows the most common sizes of beds and minimum recommended bedroom sizes, based on those bed sizes and minimum recommended side and foot clearances of 24 inches. Dressers add 18 inches in room width or depth. Other bed sizes include twin (39 x 80 inches), three-quarter (48 x 75 inches), and king (72 x 84 inches).

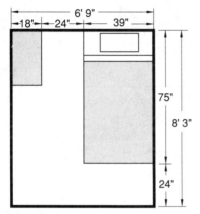

TWIN

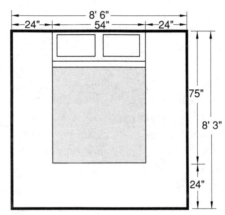

FULL

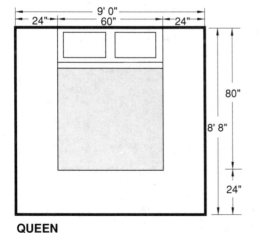

QUEEN

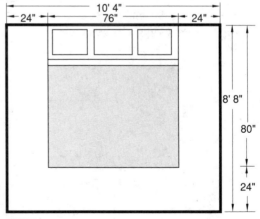

KING

WINDOW, CLOSET, AND PASSAGEWAY DIMENSIONS

Windows

Much thought should be given to window geometries. For exterior symmetry the tops of all windows should align with the tops of exterior doors (80 inches). Sill heights of windows adjacent to counters or furniture should be at least 42 inches; view or picture window sills should not exceed 38 inches; no window sill except that of a sliding glass door should be less than 10 inches high.

Closets

Allow at least 36 inches of closet pole per occupant, with a hanger depth of 24 inches minimum. Provide at least one closet per bedroom, closets near front and rear entrances for coats, a linen closet, and at least one generous walk-in closet.

Passageways

Minimum widths of residential passageways are dictated by the need to move large furnishings. The table at upper right lists both minimum and recommended widths of passageways.

WIDTHS OF PASSAGEWAYS

Passageway	Minimum	Recommended
Stairs	36"	40"
Landings	36"	40"
Main hall	36"	48"
Minor hall	30"	36"
Interior door	28"	32"
Exterior door	36"	36"

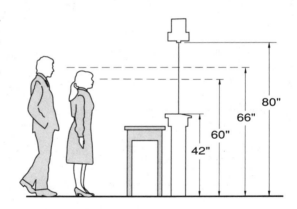

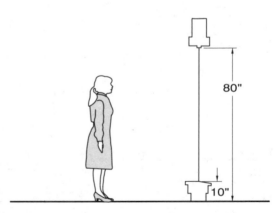

TEMPLATES FOR FURNISHINGS

Many designers have difficulty judging the relative sizes of spaces and furnishings. The furnishings below and the fixtures and appliances on the following page are drawn to a scale of 1/4 inch = 1 foot. A useful trick is to photocopy, cut out, and paste these objects into your floor plans. The sizes of many other furnishings can often be found in mail-order catalogs. Better yet, take a tape measure and notepad to your favorite furniture store.

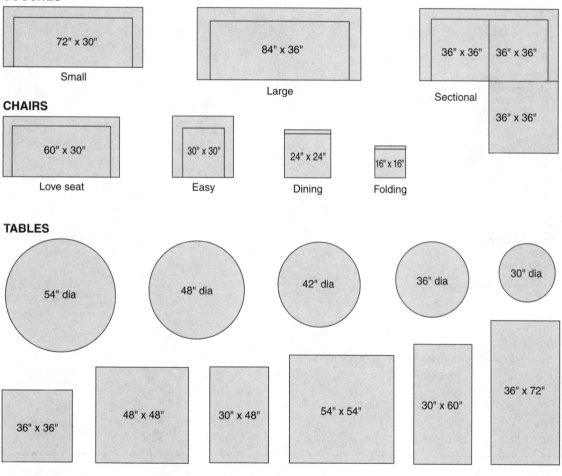

COUCHES

72" x 30" — Small

84" x 36" — Large

36" x 36" / 36" x 36" / 36" x 36" — Sectional

CHAIRS

60" x 30" — Love seat

30" x 30" — Easy

24" x 24" — Dining

16" x 16" — Folding

TABLES

54" dia

48" dia

42" dia

36" dia

30" dia

36" x 36"

48" x 48"

30" x 48"

54" x 54"

30" x 60"

36" x 72"

TEMPLATES FOR FIXTURES AND APPLIANCES

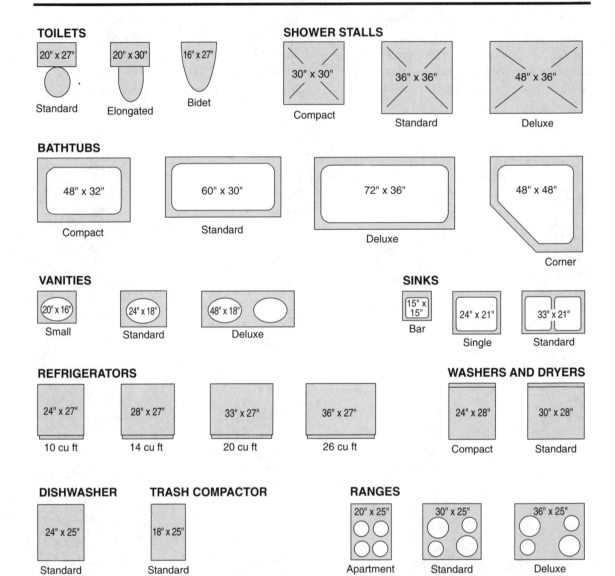

TOILETS

20" x 27"
Standard

20" x 30"
Elongated

16" x 27"
Bidet

SHOWER STALLS

30" x 30"
Compact

36" x 36"
Standard

48" x 36"
Deluxe

BATHTUBS

48" x 32"
Compact

60" x 30"
Standard

72" x 36"
Deluxe

48" x 48"
Corner

VANITIES

20" x 16"
Small

24" x 18"
Standard

48" x 18"
Deluxe

SINKS

15" x 15"
Bar

24" x 21"
Single

33" x 21"
Standard

REFRIGERATORS

24" x 27"
10 cu ft

28" x 27"
14 cu ft

33" x 27"
20 cu ft

36" x 27"
26 cu ft

WASHERS AND DRYERS

24" x 28"
Compact

30" x 28"
Standard

DISHWASHER

24" x 25"
Standard

TRASH COMPACTOR

18" x 25"
Standard

RANGES

20" x 25"
Apartment

30" x 25"
Standard

36" x 25"
Deluxe

KITCHEN PLANS

Kitchen design involves four considerations:

- appliances appropriate to cooking needs
- work surfaces for preparation, serving, and cleaning up
- storage for food, dishes, and utensils
- arrangement of the above for efficiency

Work Centers

Kitchen tasks can be assigned to one of five work centers. The table below shows the normal order of flow and the approximate percentage of trips made to each.

Work Center	Trips
Sink and cleanup	35%
Cooking	25%
Preparation	20%
Serving	10%
Refrigerator and storage	10%

Lay out work centers this way:

- Provide each center with enough counter and storage space.
- Locate the refrigerator and food storage center near the entrance door.
- Locate the serving center near the usual eating area.
- Locate the sink and cleanup center on an outside wall where natural light and an outdoor view can be enjoyed.

Work Triangle

A layout that follows the normal order of flow will maximize the efficiency of the kitchen. The total length of the work triangle (paths between the refrigerator, sink, and range) is a measure of efficiency. Less than 11 feet indicates too little counter space; more than 26 feet indicates an inefficient layout.

The kitchen plans that follow show the lengths of work triangles.

Basic Kitchen Plans

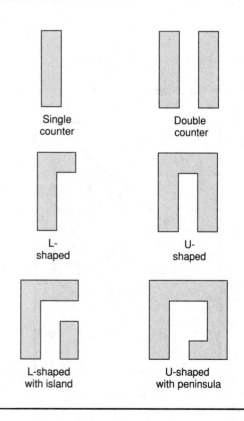

Single counter

Double counter

L-shaped

U-shaped

L-shaped with island

U-shaped with peninsula

Single Counter

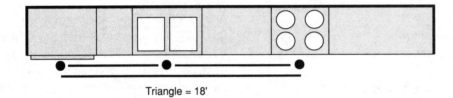

Triangle = 18'

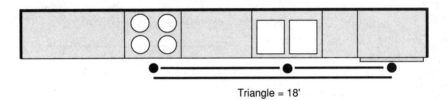

Triangle = 18'

Double Counter

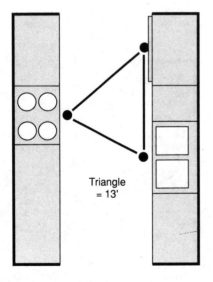

Triangle = 13'

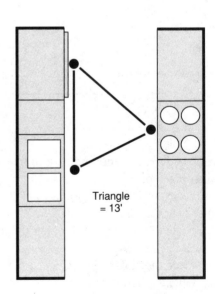

Triangle = 13'

L-Shaped

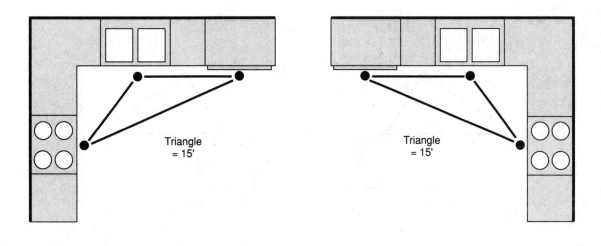

U-Shaped

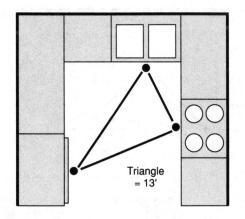 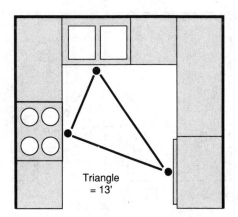

L-Shaped with Island

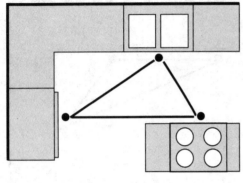 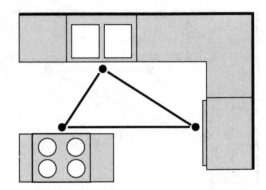

Triangle = 14'   Triangle = 14'

U-Shaped with Peninsula

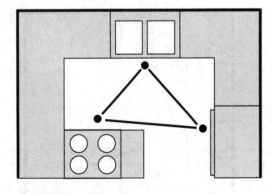 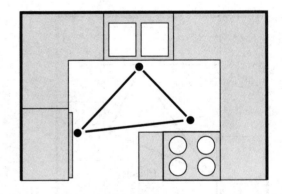

Triangle = 13'   Triangle = 13'

BATHROOM PLANS

Bathroom design involves five decisions:

Who
Who will use this bathroom: children, parents, both, or guests? Are any of them handicapped?

What
What will the user do in this bathroom: wash up, shower, soak, dress for work, develop film, wash clothes, or read philosophy books? The bathroom has come a long way since the outhouse. Consider all of the possibilities before settling for the bare minimum.

Privacy
People vary widely in how they regard their bodies and natural functions. Do you need absolute visual and acoustic privacy, attainable only in the remotest corner of the third floor, or do you enjoy bathing with your children and/or friends?

Location
Is it handy to all who want to use it when they want to use it? A toilet near the children's bedrooms allows parents to sleep through the night. A master bath/dressing room is luxurious. An outside wall admits natural, healthful sunlight. On the other hand, a head-of-the-stairs location places all users on center stage.

Fixtures
Standard fixtures are the best buy. There is a limitless variety of nonstandard fixtures available, but at *much higher* cost. The plans that follow assume the use of standard fixtures and minimum code-required clearances. If you plan to use nonstandard fixtures, get the dimensions from your supplier and adjust the plans accordingly.

Half Baths

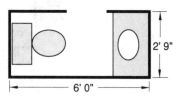
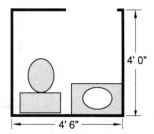
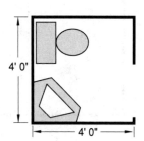

Half Baths with Laundry

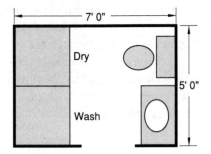

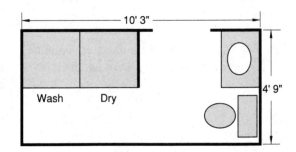

Full Baths

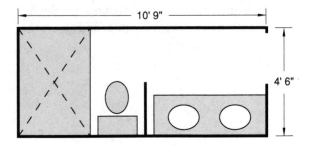

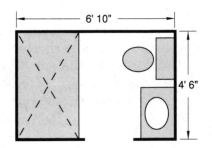

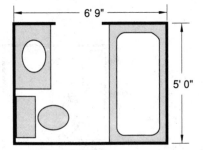

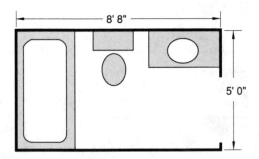

Full Baths

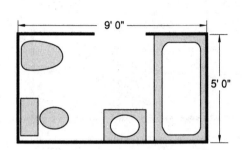

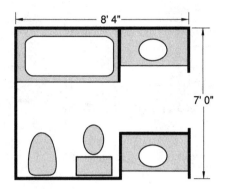

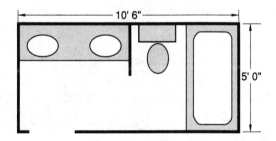

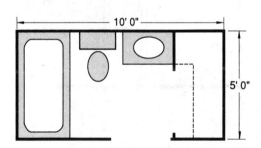

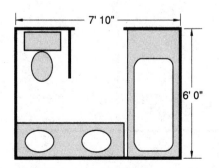

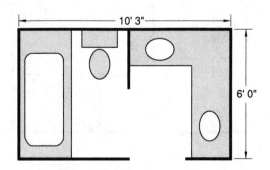

Multi-user Baths

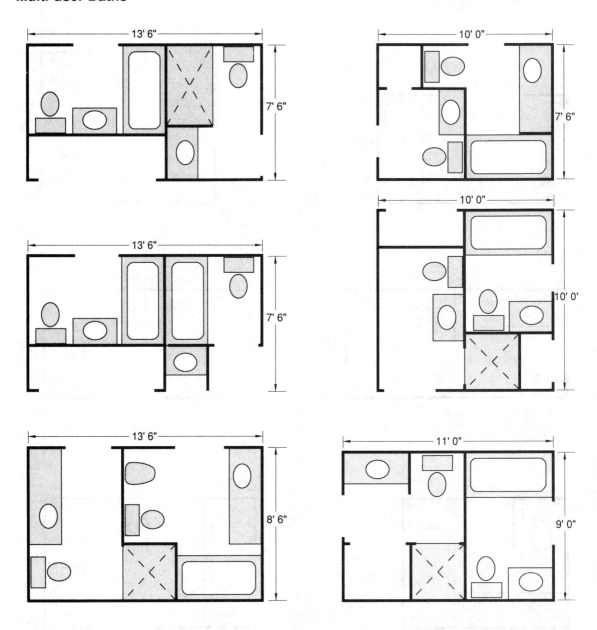

STAIR DESIGN

Riser and Tread Formulas

Risers and treads are designed together to be neither too shallow nor too steep. If too shallow, the stairs occupy too much space; if too steep, they are dangerous and difficult to climb.

Rules for proportioning risers (R) and treads (T) in inches are

$$R + T = 17" \text{ to } 17\text{-}1/2"$$
$$R \times T = 72" \text{ to } 75"$$
$$2R + T = 24" \text{ to } 25"$$

Total rise (V) is the height between finished floors. It is simply the number of steps × R.

Total run (H) is T × (the number of steps − 1).

Stair Dimensions

Given the total rise between finished floors, what should the riser and tread of each step be? In the table below, find the approximate total rise (V) in the left column and read across: number of steps, riser (R), tread (T), and total run (H). Small differences in V are made up by adjusting R and T by 1/8" for the required number of steps. Variations of 1/8" will not be noticeable when the stairs are used.

Example: The total rise between two finished floors is 8' 1". In the table below we find V = 8' 1-1/2": 13 steps of R = 7-1/2", T = 10", and H = 10'. We need to "lose" 1/2" in total rise. For 4 steps we will decrease R by 1/8" and increase T by 1/8". The result is a total rise of 8' 1" and a total run of 10' 1/2".

STAIR DIMENSIONS

Rise (V)	Steps	Riser (R)	Tread (T)	Run (H)	Rise (V)	Steps	Riser (R)	Tread (T)	Run (H)
7'1-1/2"	12	7-1/8"	10-3/8"	9'6-1/8"	7'11-7/8"	13	7-3/8"	10-1/8"	10'1-1/2"
7'2-5/8"	11	7-7/8"	9-5/8"	8'1/4"	8'0"	12	8"	9-1/2"	8'8-1/2"
7'3"	12	7-1/4"	10-1/4"	9'4-3/4"	8'1-1/2"	13	7-1/2"	10"	10'0"
7'4"	11	8"	9-1/2"	7'11"	8'2"	14	7"	10-1/2"	11'4-1/2"
7'4-1/2"	12	7-3/8"	10-1/8"	9'3-3/8"	8'3-1/8"	13	7-5/8"	9-7/8"	9'10-1/2"
7'6"	12	7-1/2"	10"	9'2"	8'3-3/4"	14	7-1/8"	10-3/8"	11'2-7/8"
7'7"	13	7"	10-1/2"	10'6"	8'4-3/4"	13	7-3/4"	9-3/4"	9'9"
7'7-1/2"	12	7-5/8"	9-7/8"	9'5/8"	8'5-1/2"	14	7-1/4"	10-1/4"	11'1-1/4"
7'8-5/8"	13	7-1/8"	10-3/8"	10'4-1/2"	8'6-3/8"	13	7-7/8"	9-5/8"	9'7-1/2"
7'9"	12	7-3/4"	9-3/4"	8'11-1/4"	8'7-1/4"	14	7-3/8"	10-1/8"	10'11-5/8"
7'10-1/4"	13	7-1/4"	10-1/4"	10'3"	8'8"	13	8"	9-1/2"	9'6"
7'10-1/2"	12	7-7/8"	9-5/8"	8'9-7/8"	8'9"	14	7-1/2"	10"	10'10"

ACCESS

The regulations implementing the 1990 Americans with Disabilities Act are contained on 28 CFR Part 36. The illustrations in the following 9 pages are adaptations of those in Appendix A of Part 36 and should prove useful to those designing for handicap access in private residential as well as public and commercial buildings.

Reach Limits

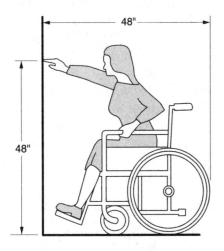

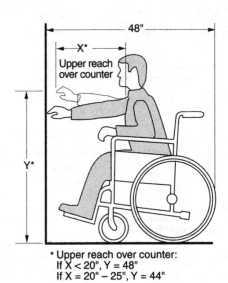

* Upper reach over counter:
If X < 20", Y = 48"
If X = 20" – 25", Y = 44"

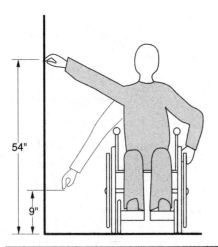

Clear Space and Turning

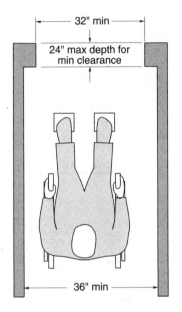

32" min

24" max depth for min clearance

36" min

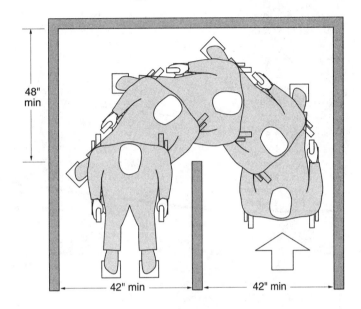

48" min

42" min

42" min

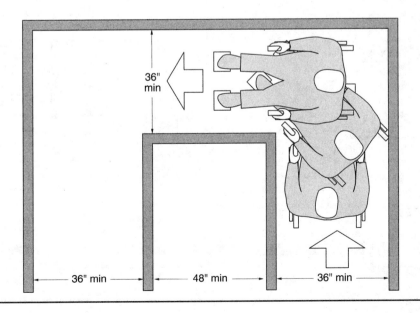

36" min

36" min

48" min

36" min

Walks, Ramps, and Level Changes

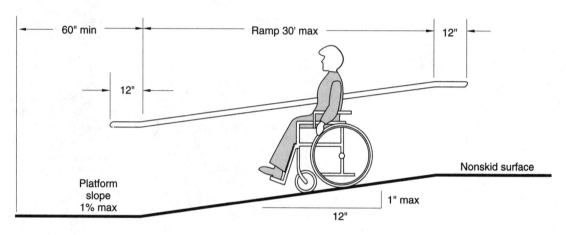

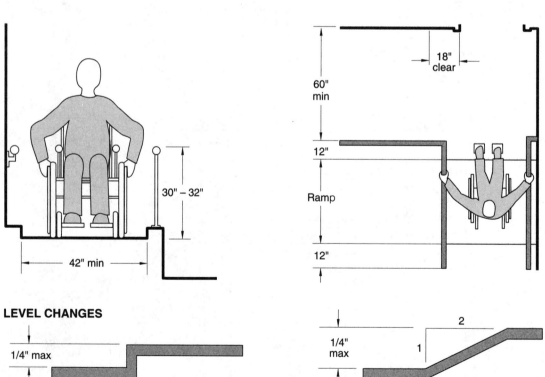

LEVEL CHANGES

Doorways

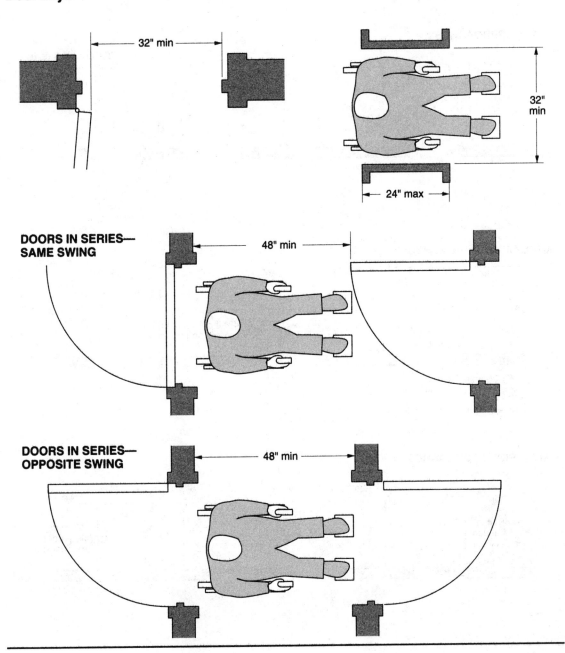

DOORS IN SERIES— SAME SWING

DOORS IN SERIES— OPPOSITE SWING

Doorways

FRONT APPROACHES

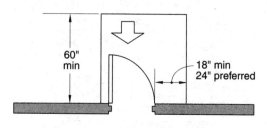

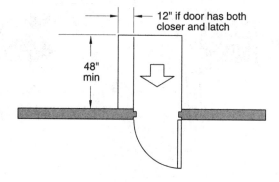

HINGE-SIDE APPROACHES

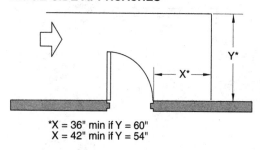

*X = 36" min if Y = 60"
 X = 42" min if Y = 54"

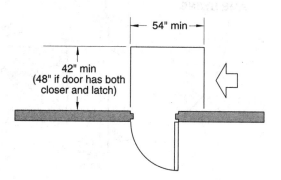

LATCH-SIDE APPROACHES

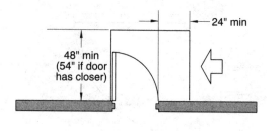

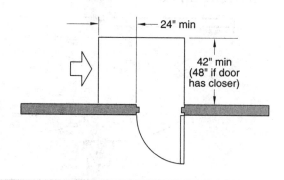

Kitchen and Dining Room

COUNTER AND RANGE

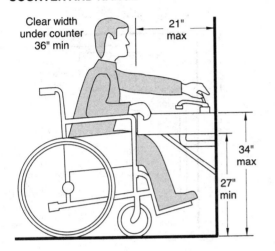

Clear width under counter 36" min

21" max

34" max

27" min

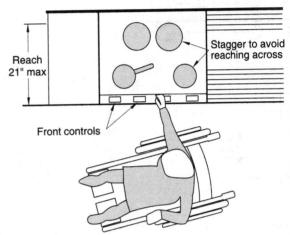

Reach 21" max

Stagger to avoid reaching across

Front controls

SEATING AT TABLES

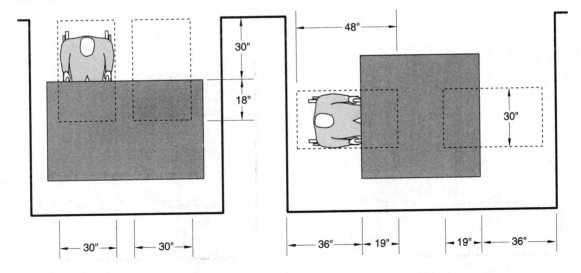

30"

18"

30"

30"

48"

30"

36"

19"

19"

36"

Water Closets

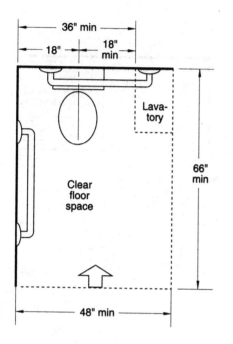

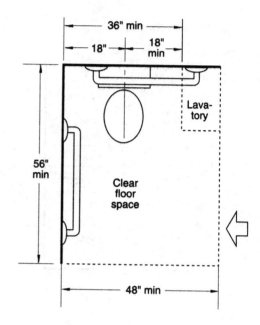

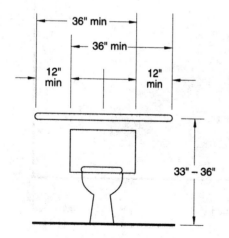

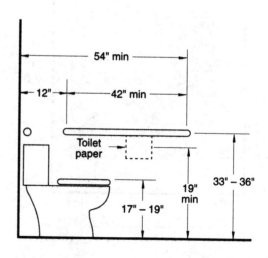

Tub and Shower Floor Space

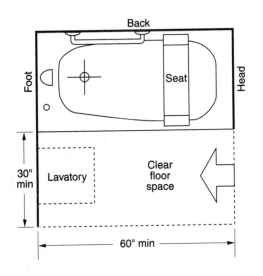

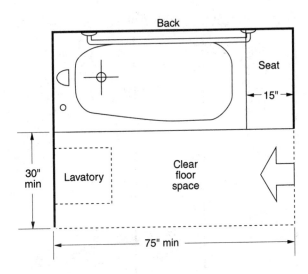

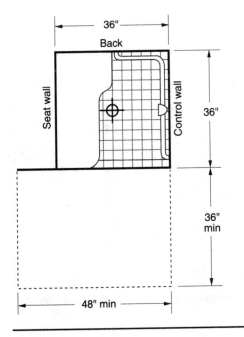

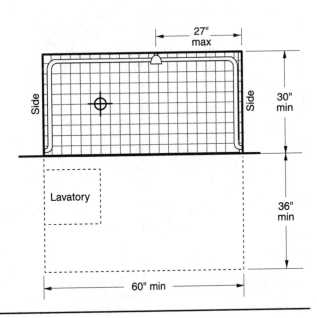

Tub Grab Bars

WITH PORTABLE SEAT IN TUB

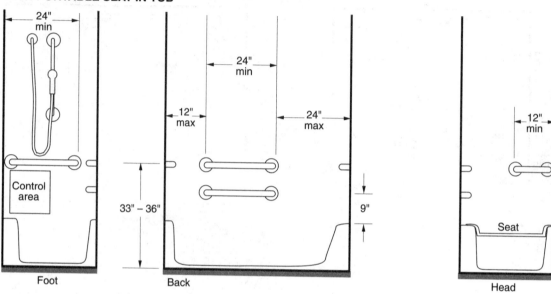

Foot   Back   Head

WITH BUILT-IN SEAT AT HEAD OF TUB

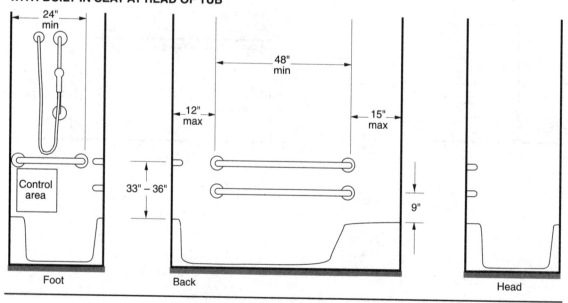

Foot   Back   Head

Shower Grab Bars

36" X 36" STALL

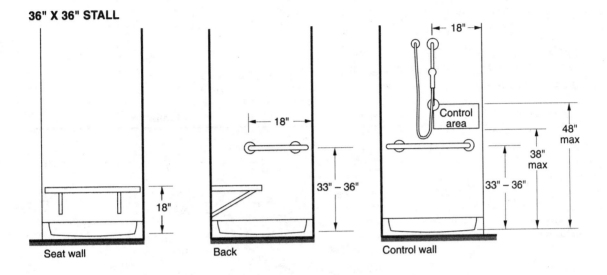

Seat wall   Back   Control wall

30" X 60" STALL

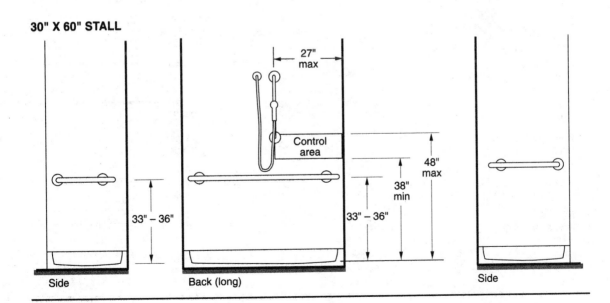

Side   Back (long)   Side

HOW TO READ BLUEPRINTS

A picture is worth a thousand words. Never is this statement more true than in the blueprints for a building. The symbols below are the language between the designer and the builder. Learn this language if you'd like to get in on the conversation!

Materials

Earth	
Gravel	
Concrete	
Clay tile	
Brick	
Metal	
Plaster	
Finish wood	
Framing lumber	
Plywood	
Fibrous insulation	
Rigid insulation	

Wiring

Surface light	Wall / Ceiling
Recessed light	R — R
Duplex receptacle	
Weatherproof receptacle	WP
Quadruple receptacle	
Range outlet	R
Thermostat	T
Telephone jack	
Switch	S
Three-way switch	S3
Four-way switch	S4
Switch & receptacle	S

Plumbing

Cold water	
Hot water	
Vent pipe	
Waste pipe	
Gas pipe	— G — G —
90° elbow	
Tee	
Clean-out	
Hose bibb	
Gate valve	
Meter	M
Floor drain, vent	

CHECKLIST OF CODE REQUIREMENTS

The following is a partial list of requirements from the 1995 Council of American Building Officials (CABO) *One and Two Family Dwelling Code*. Consult the publication for the full text and additional provisions.

Habitable Rooms
(304.1) Floor areas:
- one ≥ 150 sq. ft.
- others ≥ 70 sq. ft.
- kitchen ≥ 50 sq. ft.
- minimum dimension 7', except kitchen

(305.1) Ceiling heights:
- general 50% of area ≥ 7'6"
- kitchen, bathrooms, and halls ≥ 7'0"
- basement (uninhabitable) ≥ 6'8"
- basement beams ≥ 6'4"

Garages
(309.1) Doors to dwelling:
- to sleeping room prohibited
- to others solid wood or 20-minute rated

(309.2) Dwelling wall 1/2" gypsum board
(309.3) Floor noncombustible and sloped

Smoke Detectors
(316.1) Required locations (interconnected):
- in each sleeping room
- halls outside sleeping areas
- every floor, including basement

(316.1.1) Alterations, repairs, and additions:
- add detectors as in 316.1
- interconnection not required

(316.2) Power source:
- primary power not to be switched
- backup power from battery

Light and Ventilation
(303.1) Habitable rooms:
- glazing ≥ 8% of floor area, unless not required for egress, mechanical ventilation provided, and lighting provides ≥ 6 ft.-candles
- 50% glazing openable, unless not required for egress and mechanical ventilation provided

Egress
(310.2) Sleeping room egress windows:
- window sill ≤ 44" above floor
- net area ≥ 5.7 sq.ft. (grade-floor ≥ 5 sq.ft.)
- obstructions removable without key

(311.1) Egress doors:
- operable from inside without key
- side-hinged
- min. width 3'
- min. height 6'8"

(311.1) Exit hall min. width 3'

Landings and Ramps
(312.1) Landings:
- min. 3' × 3' landing each side of egress door
- landing ≤ 1-1/2" lower than threshold

Exceptions:
- top of stairs where door does not swing over stairs
- exterior door landing where landing is ≤ 8-1/4" below threshold and where door does not swing over landing

(313) Ramps:
- max. slope 1/8 (12.5%)
- handrail required for slope > 1/12 (8.3%)
- min. 3' × 3' landing each end of ramp

Stairways

(314.1) Width:
- min. width above handrail 36"
- min. width below one handrail 32"
- min. width below two handrails 28"

(314.2) Risers and treads:
- max. riser 7-3/4"
- max. riser variation 3/8"
- min. tread (nose to nose) 10"
- max. tread variation 3/8"
- max. tread slope 1/48 (2%)

(314.2.1) Profile:
- nosing 3/4" min. to 1-1/4" max.
- nosing not required if tread ≥ 11"
- riser sloped 0 to 30° from vertical

(314.3) Headroom 6'8" min.

(314.4) Winders:
- tread ≥ 10" at 12" from narrower side
- tread ≥ 6" at all points
- handrail required on narrow-tread side

(314.5) Spiral stairs:
- min. width 26"
- tread ≥ 7-1/2" at 12" from narrower side
- all treads identical
- riser ≤ 7-3/4"
- headroom 6'6" min.

Handrails and Guardrails

(315.1) Handrails:
- 30" to 38" above tread nosing
- at least one side if three or more risers
- on outside of spiral stairway
- ends return or terminate at newel posts
- spaced from wall ≥ 1-1/2"

(315.3) Guardrails:
- if floor > 30" above floor or grade below
- min. height of rail 36"
- min. height of rail above stair nosing 34"
- not allow passage of 4" diameter sphere

2

Site

Architects have a favorite saying: "You can't build a good house on a bad site." Which is to say, no matter how clever or aesthetic the design, a building cannot overcome the limitations of its site.

The most basic descriptors of a site are its size, shape, location, and orientation. *Plot plans* display all of these specifications and the location of the house on the site.

To develop a plot plan, you first need orientation, or the direction of true north. Surveyors usually *orient by the noon sun,* using expensive instruments. You can orient by the noon sun within a degree or two with only a stick and a watch. Or you can *orient by magnetic compass* at any time.

Next, you'll want to find the relative heights (elevations) of each point of your site. Surveyors obtain *elevations with a level.* You can rent a level and do the same.

After you pick the spot for a house, you'll want to *lay out the driveway.* As you'll see, there are guidelines for driveway slopes, widths, and turning radii.

Site grading is most economical when done at one time. If you know you want a tennis court or other *playing field,* save money by laying it out now.

When it comes time to construct the foundation, you'll need to know your site's *soil properties* and whether you should take precautions against *radon* in the soil.

Finally, you'll want to make the most of a site and enhance the home's setting by *landscaping.* Trees can form *windbreaks,* protecting a house from cold winter winds and lowering fuel bills. You may want to keep some of the mature trees on the site and add others. If so, you'll need to know how to both *plant and protect trees.*

PLOT PLANS

Plots of original and final site contours (lines of equal elevation) are useful for setting foundation heights, grading drainage swales, laying out leach fields, and estimating excavation costs. Creating a plot plan with site contours requires three steps:

1. Lay out a grid of measuring points using a straight baseline (such as the 250-foot western boundary in the illustration below), a transit (to establish secondary reference lines at right angles to the baseline), and a 100-foot tape measure (to establish points along the line). Mark the grid points with dots of spray paint or small stakes.

2. Starting at a permanent reference point (such as a pipe or stone referenced in the property deed), measure the relative elevation of each grid point, using the technique shown in "Elevations with a Level."

3. Draw a plot plan to scale (usually 10, 20, 50, or 100 feet per inch), interpolating curves of equal elevation (contour lines) between the grid points. If grading is planned, draw the proposed final contours, as well. Draw on vellum if you want blueprint reproductions.

Typical Plot Plan

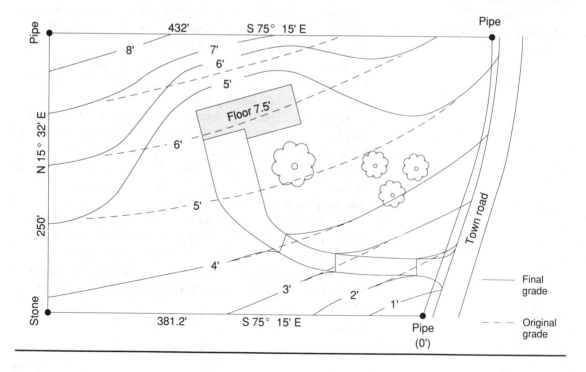

ORIENTATION BY NOON SUN

Finding true north by the sun is very simple. The sun rises in the east and sets in the west. Precisely halfway between sunrise and sunset (solar noon), the sun reaches its highest elevation and lies in the direction of true, or geographic, south. Find true south at your site this way:

1. Obtain the local times of sunrise and sunset from the newspaper or weatherman.

2. Average the times of sunrise and sunset (24-hour system) to get the precise time of local solar noon.

3. Erect a vertical stick or plumb bob where it will be struck by the noon sun.

4. At solar noon, mark the direction of the shadow. This is the direction of true north.

Example (using almanac below):

1. Sunrise = 6:29 AM (06:29)
 Sunset = 6:37 PM (18:37)
2. Solar noon = (06:29 + 18:37)/2
  = 24:66/2
  = 12:33 PM

Today's Almanac

Sun and Moon

Sunrise...............	6:29 AM
Sunset................	6:37 PM
Daylight..............	12hr 8m
Moonrise.............	6:53 AM
Moonset..............	6:50 PM

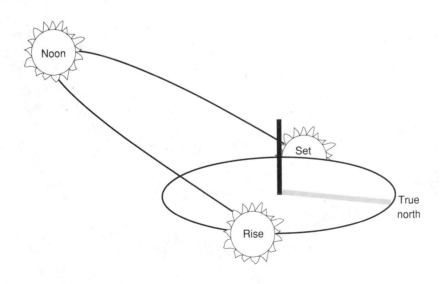

ORIENTATION BY MAGNETIC COMPASS

The earth has two sets of poles. The *geographic poles* are the ends of the axis upon which the earth revolves. True north is the direction toward the geographic north pole. The second set are the *magnetic poles.* The earth is magnetic, and a floating compass needle points toward the north magnetic pole, located at 76°N latitude and 101°W longitude in northern Canada.

In general, then, a magnetic compass needle will point to either the east (east variation) or the west (west variation) of true north.

Finding the direction of true north is simple:

1. Select a reference point (a corner of the foundation, for example).

2. Find the local variation (for example, 12°W for New York City) from the map below.

3. Hold a magnetic compass to your eye and turn your body until the sight line reads the opposite of the local variation (12°E for the example). You are now sighting along the true north-south line. Have an assistant mark a second reference sighting point along the line.

Magnetic Compass Variation

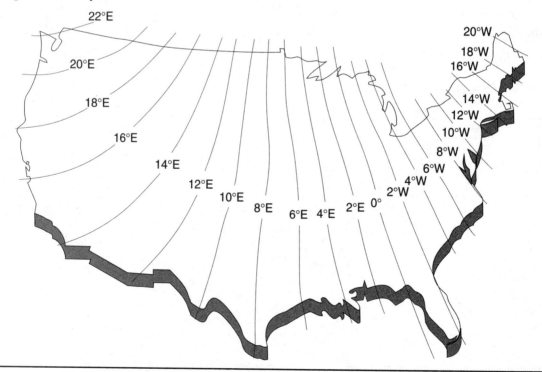

ELEVATIONS WITH A LEVEL

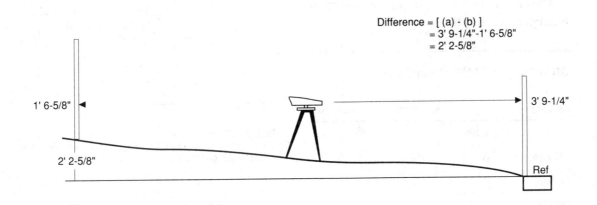

Difference = [(a) - (b)]
= 3' 9-1/4"-1' 6-5/8"
= 2' 2-5/8"

1' 6-5/8"

3' 9-1/4"

2' 2-5/8"

Ref

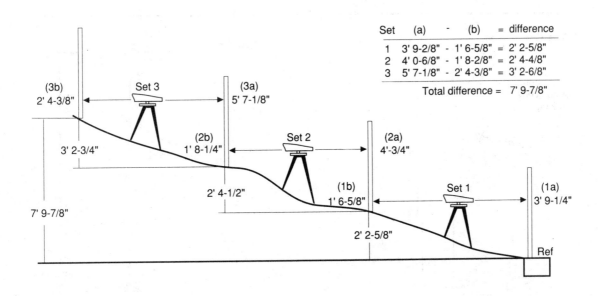

Set	(a)	-	(b)	= difference
1	3' 9-2/8"	-	1' 6-5/8"	= 2' 2-5/8"
2	4' 0-6/8"	-	1' 8-2/8"	= 2' 4-4/8"
3	5' 7-1/8"	-	2' 4-3/8"	= 3' 2-6/8"

Total difference = 7' 9-7/8"

(3b)
2' 4-3/8"

Set 3

(3a)
5' 7-1/8"

3' 2-3/4"

(2b)
1' 8-1/4"

Set 2

(2a)
4'-3/4"

2' 4-1/2"

(1b)
1' 6-5/8"

Set 1

(1a)
3' 9-1/4"

7' 9-7/8"

2' 2-5/8"

Ref

LAYING OUT DRIVEWAYS

The *slope* of a surface is the ratio of horizontal run to vertical rise. The *grade* is the percentage ratio of vertical rise to horizontal run.

The table below shows minimum and maximum grades for traffic surfaces.

GRADES FOR TRAFFIC SURFACES

Surface	Minimum	Maximum
Driveways in the North	1%	10%
Driveways in the South	1%	15%
Walks	1%	4%
Ramps	—	15%
Wheelchair ramps	—	8%
Patios	1%	2%

DEFINITION OF SLOPE AND GRADE

Example

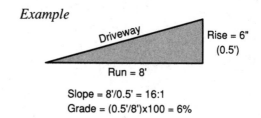

Slope = 8'/0.5' = 16:1
Grade = (0.5'/8')x100 = 6%

MEASURING TRICK
(Use a common 2x4 stud (96" long) and a carpenter's level to estimate grades.)

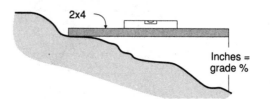

One-Car

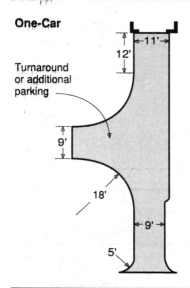

Two-Car

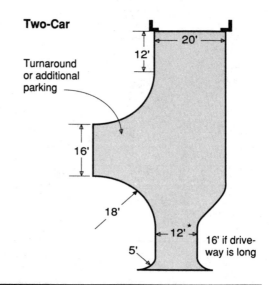

PLAYING FIELDS

Badminton

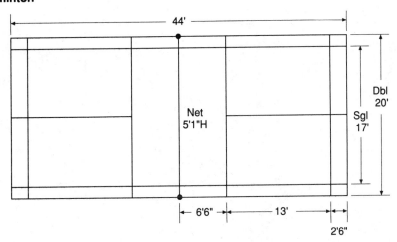

Tennis

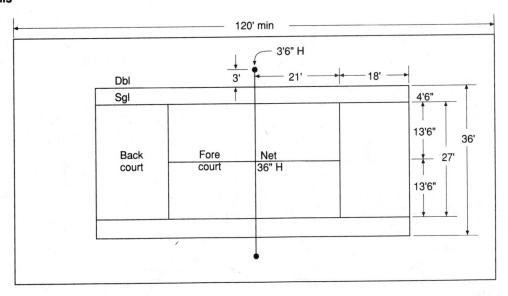

Volleyball

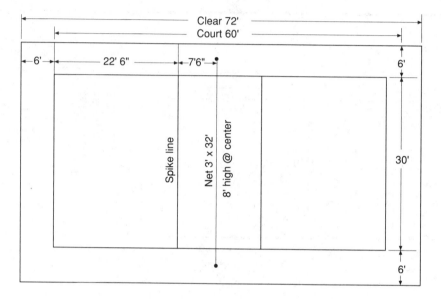

Croquet

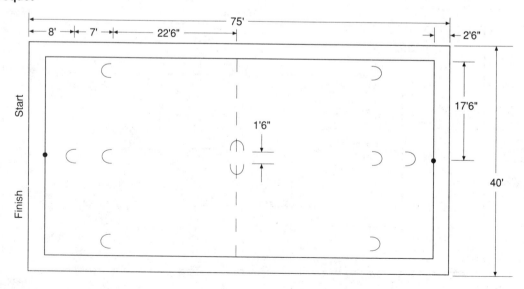

SOIL PROPERTIES

Soil Classifications
Soils may be classified with reference to one or more of the following systems:

• AASHO (American Association of State Highway Officials)
• Unified (Army Corps of Engineers)
• USDA (US Department of Agriculture Soil Conservation Service)

USDA Soil Surveys
Homeowners, developers, and local governments should avail themselves of the *Soil Surveys* published by the USDA Soil Conservation Service and available free through the local soil and water conservation district offices in every state. These surveys show the boundaries between soil types on aerial photographs (see example below), as well as

• definition of local soil types (abbreviations on map below) and their equivalent AASHO and Unified classifications
  • depth to bedrock and seasonal high water
  • grain size distribution
  • permeability and expansion potential
  • engineering and agriculture suitability
  • soil pH (acidity) and natural vegetation

Typical Soil Conservation Service Soil Map

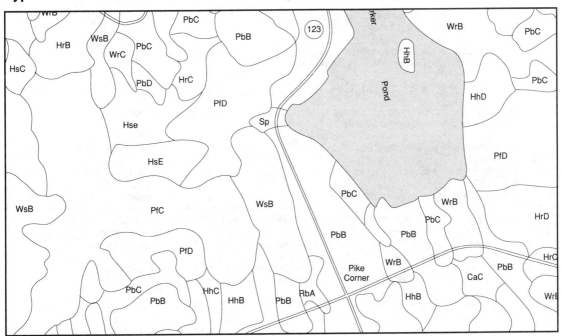

UNIFIED SOIL CLASSIFICATION SYSTEM

Soil Type	Description	Allowable Bearing[1] lb/sq ft	Drain-age[2]	Frost Heave Potential	Expan-sion Potential[3]
BR[4]	Bedrock	30,000	Poor	Low	Low
GW	Well-graded gravels, gravel-sand mixtures, little or no fines	8,000	Good	Low	Low
GP	Poorly graded gravels or gravel-sand mixtures, little or no fines	8,000	Good	Low	Low
SW	Well-graded sands, gravelly sands, little or no fines	6,000	Good	Low	Low
SP	Poorly graded sands or gravelly sands, little or no fines	5,000	Good	Low	Low
GM	Silty gravels, gravel-sand-silt mixtures	4,000	Good	Med	Low
SM	Silty sand, sand-silt mixtures	4,000	Good	Med	Low
GC	Clayey gravels, gravel-clay-sand mixtures	4,000	Med	Med	Low
SC	Clayey sands, sand-clay mixture	4,000	Med	Med	Low
ML	Inorganic silts and very fine sands, rock flour, silty or clayey fine sands with slight plasticity	2,000	Med	High	Low
CL	Inorganic clays of low to medium plasticity, gravelly clays, sandy clays, silty clays, lean clays	2,000	Med	Med	Med[5]
CH	Inorganic clays of high plasticity, fat clays	2,000	Poor	Med	High[5]
MH	Inorganic silts, micaceous or diatomaceous fine sandy or silty soils, elastic silts	2,000	Poor	High	High
OL	Organic silts and organic silty clays	400	Poor	Med	Med
OH	Organic clays of medium to high plasticity	0	Unsat	Med	High
PT	Peat and other highly organic soils	0	Unsat	Med	High

Source: *Permanent Wood Foundations* (Tacoma, Wash: American Plywood Association, 1985).
[1] Allowable bearing value may be increased 25% for very compact, coarse-grained, gravelly or sandy soils or very stiff, fine-grained, clayey or silty soils. Allowable bearing value shall be decreased 25% for loose, coarse-grained, gravelly or sandy soils or soft, fine-grained, clayey or silty soils.
[2] Percolation rate for good drainage is over 4"/hr, medium drainage is 2–4"/hr, and poor is less than 2"/hr.
[3] For expansive soils, contact local soils engineer for verification of design assumptions.
[4] Added by author.
[5] Dangerous expansion might occur if these soil types are dry but subject to future wetting.

RADON

Radon is a naturally occurring, invisible, odorless gas resulting from the radioactive breakdown of uranium in rocks and soils. It is estimated to cause between 5,000 and 20,000 lung cancer deaths per year in the United States. (By comparison, cigarette smoking is thought to cause 110,000 deaths.)

The map below shows areas that often have high radon concentrations. Homes and building sites in these areas should be tested for radon. For your guidance, two booklets are available free from state radiation protection offices and regional US Environmental Protection Agency (EPA) offices:

• *A Citizen's Guide to Radon* (what it is, how to test for it, how to interpret results)
• *Radon Reduction Methods* (methods that a contractor or skilled homeowner might apply to lower radon levels in the home)

See chapter 4 for more information.

Radon-Producing Areas

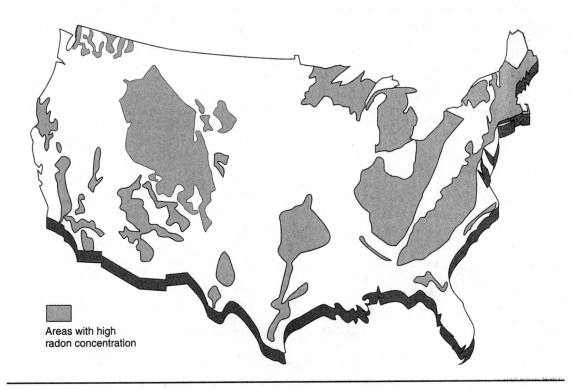

Areas with high
radon concentration

LANDSCAPING

Plant Hardiness

The climatic suitability of landscaping plants is usually defined by reference to the US Department of Agriculture's "Plant Hardiness Zone Map" below. The zones refer to the range of annual minimum temperatures experienced. Planting tables often refer to the hardiness zones, but you should remember that local (microclimatic) temperatures can vary widely and that plants respond to variables other than temperature, such as soil pH, degree of shade, exposure to wind, and rainfall. Consult a local professional nurseryman before selecting plants for your site.

Approximate Range of Average Minimum Temperatures for Plant Hardiness Zones, °F	
Zone 2	-50 to -40
Zone 3	-40 to -30
Zone 4	-30 to -20
Zone 5	-20 to -10
Zone 6	-10 to 0
Zone 7	0 to 10
Zone 8	10 to 20
Zone 9	20 to 30
Zone 10	30 to 40

Plant Hardiness Zones

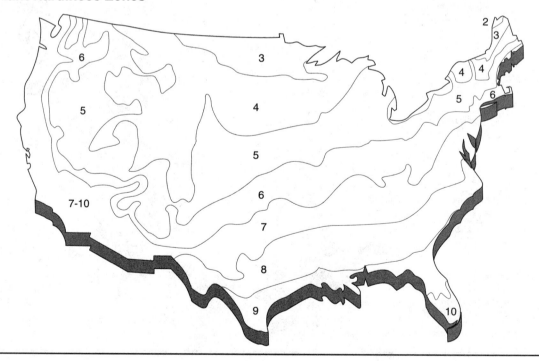

TREES FOR LANDSCAPING

Common Name	Zones	Type[1]	Height feet	Width feet	Spacing feet	Features
1. Quaking aspen	2-5	D	35	5	7	Excellent visual screen
2. Paper birch	2-5	D	45	20	15	White bark, very hardy
3. White spruce	2-5	E	45	20	10	Good for windbreak
4. Bur oak	2-5	D	45	20	15	Requires good soil for full size
5. Eastern red cedar	2-8	E	50	10	7	Good screen, tolerates dry soil
6. Norway maple	3-6	D	50	30	25	Grows in city, grows fast
7. Sugar maple	3-6	D	80	50	40	Beautiful foliage, sugar sap
8. Norway spruce	3-5	E	60	25	14	Grows fast, prefers sun
9. Red maple	3-6	D	40	30	25	Brilliant foliage, grows fast
10. Green ash	3-6	D	50	30	25	Grows fast, most soils
11. Eastern white pine	3-6	E	70	40	12	Grows very fast, most soils
12. Eastern hemlock	3-7	E	60	30	8	Good screen, grows in shade
13. White poplar	3-5	D	50	12	10	Grows very fast, short life
14. Pin oak	4-7	D	80	50	30	Keeps leaves in winter
15. Japanese cryptomeria	6-8	E	70	20	10	Good screen, grows fast
16. Oriental arborvitae	7-9	E	16	6	3	Grows fast, most soils
17. Rocky Mountain juniper	4-7	E	25	10	6	In West only, dry soils
18. Black haw	4-7	D	15	15	5	White flowers, red berries
19. American holly	6-9	E	20	8	8	Spined leaves, red berries
20. Lombardy poplar	6-8	D	40	6	4	Grows fast, all soils
21. Weeping willow	6-8	D	30	30	30	Drooping branches, wet soils
22. Sea grape	10	E	20	8	4	Very decorative
23. Northern white cedar	3-6	E	30	12	8	Good screen, loamy soil
24. Southern magnolia	8-9	E	30	10	5	Large white flowers
25. Douglas fir	4-6	E	60	25	12	Grows fast, up to 200'

Source: W. R. Nelson, Jr., *Landscaping Your Home* (Urbana-Champaign: University of Illinois, 1975).
[1] D = deciduous, E = evergreen.

Trees for Landscaping (typical mature heights compared with two-story house)

WINDBREAKS

A strategically located group of trees can reduce winter wind speeds around a home by 50 percent or more, reducing fuel bills. The chart below shows measured wind speed reductions upwind and downwind of a shelterbelt of mixed deciduous and hardwood trees. The maximum downwind reduction is 50 to 60 percent, occurring at a distance of five times the average tree height. Upwind the reduction is negligible at a distance of five tree heights.

In areas where prevailing winter and summer winds are in opposite directions, shelterbelts can be used to block winter winds without diminishing summer breezes.

The maps on the following page show the directions of the predominant winter and summer winds at weather stations across the United States. The map should be used only as a crude guide, however, since topography can alter wind directions markedly.

The illustrations on the page following the maps show overall strategies for dealing with both wind and sun. The bottom plan shows how both winter wind and summer sun can be blocked without losing cooling summer breezes or the warming winter sun. The winter and summer wind directions in the example are typical for most of the eastern United States.

Wind Speed Reduction Upwind and Downwind of a Shelterbelt

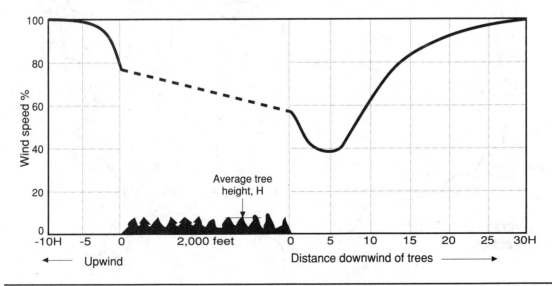

Site

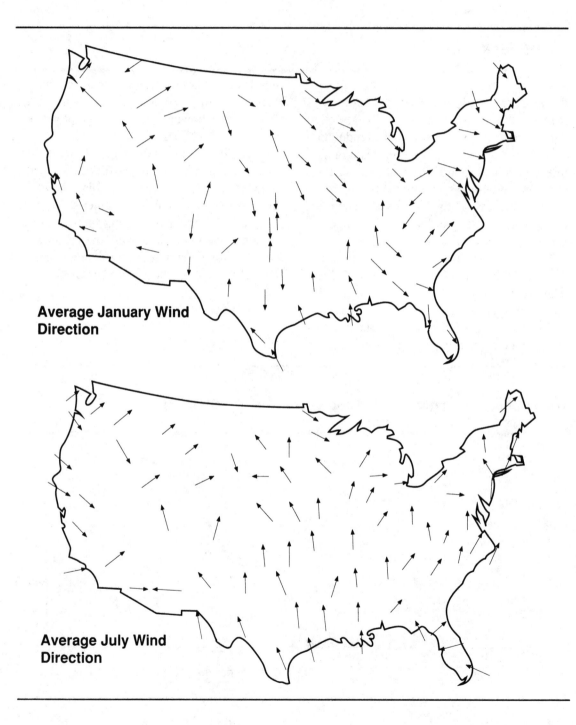

**Average January Wind
Direction**

**Average July Wind
Direction**

Siting for Wind and Sun

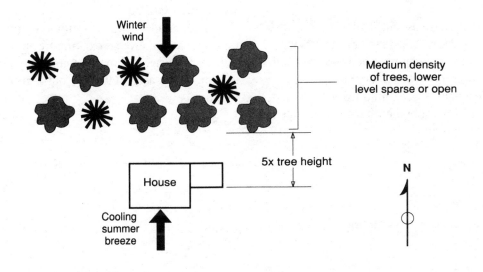

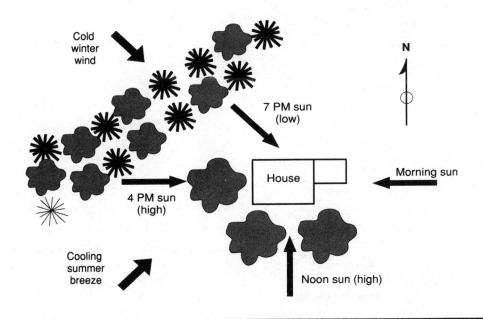

PLANTING AND PROTECTING TREES

Planting

Make a hole twice as wide and as deep as the
root ball, break up the sides, and refill with
the original soil. Plant trees at the same depth
as in the nursery. Water every day the first
week, then once a week until frost. Paint the
trunk to the lowest branches with white latex
to prevent sunscald.

Changing Grade

When you grade a site, it is important to maintain aerated soil around existing trees. Short
of transplanting, that means maintaining the
existing grade around the tree out to its canopy line.

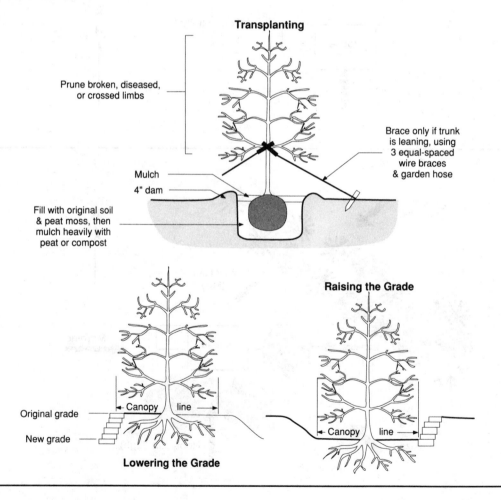

Transplanting

Prune broken, diseased,
or crossed limbs

Brace only if trunk
is leaning, using
3 equal-spaced
wire braces
& garden hose

Mulch
4" dam

Fill with original soil
& peat moss, then
mulch heavily with
peat or compost

Original grade

New grade

Canopy line

Lowering the Grade

Raising the Grade

Canopy line

3 Masonry

Masonry is the art of assembling stone, brick, and other mineral units, with or without the use of mortar. All masonry materials share the properties of massiveness, strength in compression, and immunity to decay.

This chapter begins by describing the properties of *concrete* and *mortar* and the procedures for estimating their required volumes.

The section on *brick walls* illustrates the time-honored terminology of brickwork and the classic wall bonds (patterns). You will also find illustrated four modern methods of constructing residential brick walls.

You will see there are many "standard" *brick sizes*. Included are *coursing* tables, which make it simple to lay brick walls to any height. Brick size and coursing are key considerations when *estimating brick and mortar* quantities.

Bricks may also be used to construct attractive and durable walks and driveways. You will see how to construct *brick pavement* in a variety of patterns.

Concrete masonry is construction using concrete masonry units (CMUs, or concrete blocks). This section shows how a steel-reinforced concrete masonry wall is constructed. It also includes a listing of *block sizes* and a *coursing* table for laying out walls.

Walks and drives may be constructed of *concrete pavers*, using the same techniques as for brick. Several methods of building *stone veneer* walls and the six most common patterns are also shown.

Finally, we provide you with a *checklist of code requirements* relating to masonry.

CONCRETE

Concrete is man-made rock. It consists of gravel aggregate, sand, and portland cement. Made properly, it is as strong as most naturally occurring rock and will last as long. Made improperly, it cracks and crumbles and may have to be replaced within a few years.

Properly made concrete has four characteristics:

• It has the correct ratio of clean and well-sized materials. The proper ratios of materials ensure that all sand grains are surrounded by cement and that all gravel pebbles are surrounded by sand. Clean materials ensure that all of the materials will bond, resulting in maximum strength.

• The amount of water is correct. During curing, the portland cement chemically combines with the water (the cement hydrates). Too little water results in unhydrated cement; too much water leaves voids when the water eventually escapes. As a practical matter, the proper amount of water is the minimum that results in a cohesive mixture.

• It has a long curing time. Complete hydration requires many days. To prevent evaporation of the water, the surface of the concrete should be kept damp. This may require dampening the soil or placing polyethylene or wetted burlap over exposed surfaces for up to a week. Freezing poses a similar problem: The formation of ice crystals withholds water from the hydration process. When the ice later melts, it leaves air pockets.

• There is air entrainment. When concrete is exposed to freezing, water within the cured concrete can freeze and expand and crack the concrete. Concrete that will be exposed to freezing should contain an air-entraining admixture, which forms billions of microscopic air bubbles. The tiny air pockets act as pressure relief valves for freezing water.

Availability

For a very small job, such as a fence post or two, purchase premixed 1-cubic-foot bags containing all of the ingredients except water. Mix in a wheelbarrow.

For up to 1 cubic yard (27 cubic feet) of concrete, purchase sand and gravel by the fractional yard and portland cement in 1-cubic-foot bags. Mix in a portable mixer, which you can rent.

For more than a cubic yard, order ready-mix concrete delivered by truck.

Specifying Ready-Mix

In addition to simplicity, a big advantage of ready-mix concrete is the opportunity to specify exactly what you need. Unless you are a concrete professional, discuss your application with the ready-mix salesman. Together you can decide the cement content (in bags per yard), aggregate size (should not exceed one-third of concrete thickness), percentage of entrained air, and the need for accelerators or retarders. You alone will have to specify the amount of concrete (allow an extra 10 percent for spreading of the forms and spillage) and the time of delivery.

The following page contains tables for mixing concrete and estimating required volumes.

RECOMMENDED CONCRETE MIXES

Application	Cement Sacks[1]	Sand cu ft	Gravel cu ft	Gal. Water If Sand Dry	Wet	Makes cu ft
Severe weather, heavy wear, slabs < 3" thick	1.0	2.0	2.2	5.0	4.0	3.5
Slabs > 3" thick, sidewalks, driveways, patios	1.0	2.2	3.0	6.0	5.0	4.1
Footings, foundation walls, retaining walls	1.0	3.0	4.0	7.0	5.5	5.0

[1]One 94-lb sack contains 1 cu ft.

ESTIMATING CUBIC YARDS OF CONCRETE FOR SLABS, WALKS, AND DRIVES

Slab Area sq ft	Slab Thickness, inches							
	2	3	4	5	6	8	10	12
10	0.1	0.1	0.1	0.2	0.2	0.3	0.3	0.4
50	0.3	0.5	0.6	0.7	0.9	1.2	1.4	1.9
100	0.6	0.9	1.2	1.5	1.9	2.5	3.0	3.7
300	1.9	2.8	3.7	4.7	5.6	7.4	9.4	11.1
500	3.1	4.7	6.2	7.2	9.3	12.4	14.4	18.6

Note: Amount does not include extra 10% for form spreading and waste.

ESTIMATING CONCRETE FOR FOOTINGS AND WALLS (cubic yards per linear foot)

Depth or Height, inches	Width or Thickness, inches				
	6	8	12	18	24
8	0.012	0.017	0.025	0.037	0.049
12	0.019	0.025	0.037	0.056	0.074
24	0.037	0.049	0.074	0.111	0.148
72	0.111	0.148	0.222	0.333	0.444
96	0.148	0.198	0.296	0.444	0.593

MORTAR

Mortar is a mixture of cement, lime, and sand for bonding unit masonry (bricks, blocks, or stone) together. It differs from concrete in two ways: 1) because it is used in thin layers, it contains no gravel aggregate, and 2) because brick and block laying is a slow process, it contains hydrated lime to retain water and retard set-up.

Availability

For small jobs requiring up to 1 cubic foot of mortar (about 30 8 x 8 x 16-inch blocks or 70 2 x 4 x 8-inch bricks), purchase a bag of mortar mix that requires only the addition of water. For larger jobs buy sacks of masonry cement.To each sack add 3 cubic feet of clean, sharp sand and enough water to make the mix plastic. Precolored mortar mixes are also available through masonry supply outlets.

Mixing

Unless you are a professional, or more than one bricklayer is at work, mix small batches of mortar by hand, because it must be used within approximately 1 hour. Do not lay masonry when there is any danger of freezing. Calcium chloride accelerator is sometimes used to speed set-up and lessen the chances of freezing, but it can later cause efflorescence (a white, powdery deposit) on wall surfaces.

Use the table at right to estimate quantities of mortar for various masonry wall constructions.

MATERIALS REQUIRED PER 100 SQUARE FEET OF MASONRY WALLS

Unit Type	Thickness of Wall	Masonry Units	Cement Sacks	Sand cu ft
Standard brick	4"	616	3	9
Standard brick	8"	1,232	7	21
Standard block	8"	112	1	3

Note: Assume 20% mortar waste.

BRICK WALL CONSTRUCTION

BRICK POSITIONS

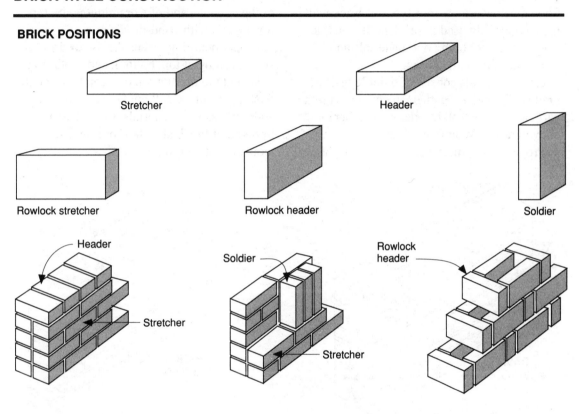

Stretcher

Header

Rowlock stretcher

Rowlock header

Soldier

Header

Stretcher

Soldier

Stretcher

Rowlock header

MORTAR JOINTS

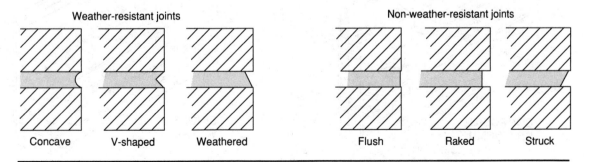

Weather-resistant joints

Non-weather-resistant joints

Concave   V-shaped   Weathered   Flush   Raked   Struck

Brick Wall Types

Shown below are the four types of brick wall most commonly used in residential construction. All require both good materials and proper design and detailing.

Good materials consist of durable brick: grade SW (severe weather) in contact with the earth and grade MW (moderate weather) above grade (SW in freeze areas).

Proper design means choice of the right type for the climate: insulated for most areas of the country, and of a draining-cavity type for regions with wind-driven rain.

Proper detailing means the use of drips at sills; flashing at top, bottom, and heads and sills; and the installation of weep holes at all flashing points. Caulking between masonry and nonmasonry materials is mandatory. Sprayed or brushed waterproofing, such as silicone, is not recommended.

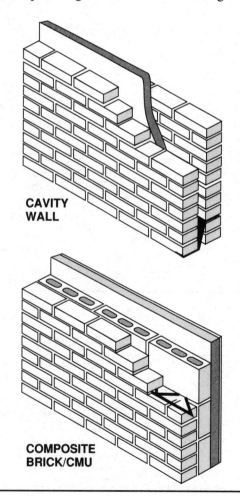

CAVITY WALL

COMPOSITE BRICK/CMU

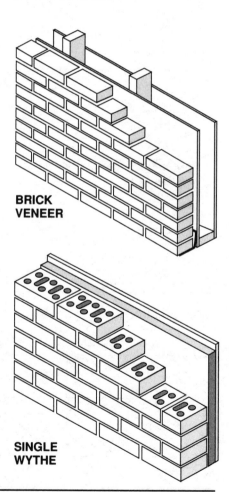

BRICK VENEER

SINGLE WYTHE

Brick Wall Bonds (Patterns)

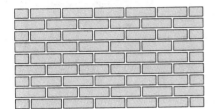

RUNNING

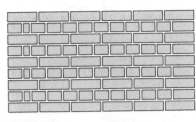

ENGLISH

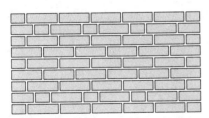

COMMON

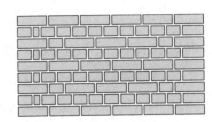

DUTCH

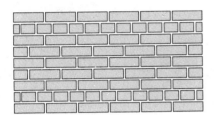

**COMMON WITH
FLEMISH HEADERS**

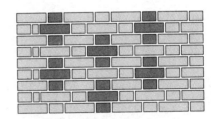

FLEMISH CROSS

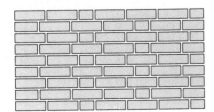

GARDEN WALL

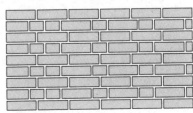

FLEMISH

BRICK SIZES AND COURSING

NONMODULAR BRICK (dimensions actual W x L x H)

Standard
3-5/8" x 8" x 2-1/4"

Engineer standard
3-3/4" x 8" x 2-3/4"

Closure standard
3-5/8" x 8" x 3-5/8"

King size
3-5/8" x 9-5/8" x 2-3/4"

Queen size
3" x 9-5/8" x 2-5/8"

MODULAR BRICK (dimensions nominal W x L x H)

Modular
4" x 8" x 2-2/3 "

Engineer modular
4" x 8" x 3-1/5"

Closure modular
4" x 8" x 4"

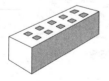

Utility
4" x 12" x 4"

Roman
4" x 12" x 2"

Norman
4" x 12" x 2-2/3"

Engineer norman
4" x 12" x 3-1/5"

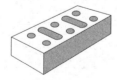

6" Norwegian
6" x 12" x 3-1/5"

8" jumbo
8" x 12" x 4"

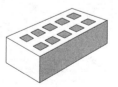

6" jumbo
6" x 12" x 4"

HEIGHTS OF NONMODULAR BRICK WALLS

Number of Courses	2-1/4" Thick Bricks 3/8" Joint	1/2" Joint	2-5/8" Thick Bricks 3/8" Joint	1/2" Joint	2-3/4" Thick Bricks 3/8" Joint	1/2" Joint
1	0' 2-5/8"	0' 2-3/4"	0' 3"	0' 3-1/8"	0' 3-1/8"	0' 3-1/4"
2	0' 5-1/4"	0' 5-1/2"	0' 6"	0' 6-1/4"	0' 6-1/4"	0' 6-1/2"
3	0' 7-7/8"	0' 8-1/4"	0' 9"	0' 9-3/8"	0' 9-3/8"	0' 9-3/4"
4	0' 10-1/2"	0' 11"	1' 0"	1' 0-1/2"	1' 0-1/2"	1' 1"
5	1' 1-1/8"	1' 1-3/4"	1' 3"	1' 3-5/8"	1' 3-5/8"	1' 4-1/4"
6	1' 3-3/4"	1' 4-1/2"	1' 6"	1' 6-3/4"	1' 6-3/4"	1' 7-1/2"
7	1' 6-3/8"	1' 7-1/4"	1' 9"	1' 9-7/8"	1' 9-7/8"	1' 10-3/4"
8	1' 9"	1' 10"	2' 0"	2' 1"	2' 1"	2' 2"
9	1' 11-5/8"	2' 0-3/4"	2' 3"	2' 4-1/8"	2' 4-1/8"	2' 5-1/4"
10	2' 2-1/4"	2' 3-1/2"	2' 6"	2' 7-1/4"	2' 7-1/4"	2' 8-1/2"
11	2' 4-7/8"	2' 6-1/4"	2' 9"	2' 10-3/8"	2' 10-3/8"	2' 11-3/4"
12	2' 7-1/2"	2' 9"	3' 0"	3' 1-1/2"	3' 1-1/2"	3' 3"
13	2' 10-1/8"	2' 11-3/4"	3' 3"	3' 4-5/8"	3' 4-5/8"	3' 6-1/4"
14	3' 0-3/4"	3' 2-1/2"	3' 6"	3' 7-3/4"	3' 7-3/4"	3' 9-1/2"
15	3' 3-3/8"	3' 5-1/4"	3' 9"	3' 10-7/8"	3' 10-7/8"	4' 0-3/4"
16	3' 6"	3' 8"	4' 0"	4' 2"	4' 2"	4' 4"
17	3' 8-5/8"	3' 10-3/4"	4' 3"	4' 5-1/8"	4' 5-1/8"	4' 7-1/4"
18	3' 11-1/4"	4' 1-1/2"	4' 6"	4' 8-1/4"	4' 8-1/4"	4' 10-1/2"
19	4' 1-7/8"	4' 4-1/4"	4' 9"	4' 11-3/8"	4' 11-3/8"	5' 1-3/4"
20	4' 4-1/2"	4' 7"	5' 0"	5' 2-1/2"	5' 2-1/2"	5' 5"
21	4' 7-1/8"	4' 9-3/4"	5' 3"	5' 5-5/8"	5' 5-5/8"	5' 8-1/4"
22	4' 9-3/4"	5' 0-1/2"	5' 6"	5' 8-3/4"	5' 8-3/4"	5' 11-1/2"
23	5' 0-3/8"	5' 3-1/4"	5' 9"	5' 11-7/8"	5' 11-7/8"	6' 2-3/4"
24	5' 3"	5' 6"	6' 0"	6' 3"	6' 3"	6' 6"

Source: *Brick Sizes and Related Information* (McLean, Va: Brick Institute of America, 1986).

HEIGHTS OF MODULAR BRICK WALLS

Number of Courses	Nominal Height (thickness) of Brick [1]				
	2"	2-2/3"	3-1/5"	4"	5-1/3"
1	0' 2"	0' 2-11/16"	0' 3-3/16"	0' 4"	0' 5-5/16"
2	0' 4"	0' 5-5/16"	0' 6-3/8"	0' 8"	0' 10-11/16"
3	0' 6"	0' 8"	0' 9-5/8"	1' 0"	1' 4"
4	0' 8"	0' 10-11/16"	1' 0-13/16"	1' 4"	1' 9-5/16"
5	0' 10"	1' 1-5/16"	1' 4"	1' 8"	2' 2-11/16"
6	1' 0"	1' 4"	1' 7-3/16"	2' 0"	2' 8"
7	1' 2"	1' 6-11/16"	1' 10-3/8"	2' 4"	3' 1-5/16"
8	1' 4"	1' 9-5/16"	2' 1-5/8"	2' 8"	3' 6-11/16"
9	1' 6"	2' 0"	2' 4-13/16"	3' 0"	4' 0"
10	1' 8"	2' 2-11/16"	2' 8"	3' 4"	4' 5-5/16"
11	1' 10"	2' 5-5/16"	2' 11-3/16"	3' 8"	4' 10-11/16"
12	2' 0"	2' 8"	3' 2-3/8"	4' 0"	5' 4"
13	2' 2"	2' 10-11/16"	3' 5-5/8"	4' 4"	5' 9-5/16"
14	2' 4"	3' 1-5/16"	3' 8-13/16"	4' 8"	6' 2-11/16"
15	2' 6"	3' 4"	4' 0"	5' 0"	6' 8"
16	2' 8"	3' 6-11/16"	4' 4-3/16"	5' 4"	7' 1-5/16"
17	2' 10"	3' 9-5/16"	4' 6-3/8"	5' 8"	7' 6-11/16"
18	3' 0"	4' 0"	4' 9-5/8"	6' 0"	8' 0"
19	3' 2"	4' 2-11/16"	5' 0-13/16"	6' 4"	8' 5-5/16"
20	3' 4"	4' 5-5/16"	5' 4"	6' 8"	8' 10-11/16"
21	3' 6"	4' 8"	5' 7-3/16"	7' 0"	9' 4"
22	3' 8"	4' 10-11/16"	5' 10-3/8"	7' 4"	9' 9-5/16"
23	3' 10"	5' 1-5/16"	6' 1-5/8"	7' 8"	10' 2-11/16"
24	4' 0"	5' 4"	6' 4-13/16"	8' 0"	10' 8"

Source: *Brick Sizes and Related Information* (McLean, Va: Brick Institute of America, 1986).
[1] Mortar joint thickness approximately equals difference between actual and nominal brick height.

ESTIMATING BRICK AND MORTAR

NONMODULAR BRICK[1] AND MORTAR[2] REQUIRED FOR SINGLE-THICKNESS WALLS IN RUNNING BOND

Size of Brick, Inches T x H x L	With 3/8" Joint			With 1/2" Joint		
	Number of Brick/ 100 Sq Ft	Cubic Feet of Mortar/ 100 Sq Ft	Cubic Feet of Mortar/ 1,000 Brick	Number of Brick/ 100 Sq Ft	Cubic Feet of Mortar/ 100 Sq Ft	Cubic Feet of Mortar/ 1,000 Brick
2-3/4 x 2-3/4 x 9-3/4	455	3.2	7.1	432	4.5	10.4
2-5/8 x 2-3/4 x 8-3/4	504	3.4	6.8	470	4.1	8.7
3-3/4 x 2-1/4 x 8	655	5.8	8.8	616	7.2	11.7
3-3/4 x 2-3/4 x 8	551	5.0	9.1	522	6.4	12.2

[1] Add at least 5% for breakage.
[2] Add 10 to 25% for waste.

MODULAR BRICK AND MORTAR REQUIRED FOR SINGLE-THICKNESS WALLS IN RUNNING BOND

Nominal Size of Brick, Inches T x H x L	Number of Brick/ 100 Sq Ft	Cubic Feet of Mortar			
		Per 100 Sq Ft		Per 1,000 Brick	
		3/8" Joint	1/2" Joint	3/8" Joint	1/2" Joint
4 x 2-2/3 x 8	675	5.5	7.0	8.1	10.3
4 x 3-1/5 x 8	563	4.8	6.1	8.6	10.9
4 x 4 x 8	450	4.2	5.3	9.2	11.7
4 x 5-1/3 x 8	338	3.5	4.4	10.2	12.9
4 x 2 x 12	600	6.5	8.2	10.8	13.7
4 x 2-2/3 x 12	450	5.1	6.5	11.3	14.4
4 x 3-1/5 x 12	375	4.4	5.6	11.7	14.9
4 x 4 x 12	300	3.7	4.8	12.3	15.7
4 x 5-1/3 x 12	225	3.0	3.9	13.4	17.1
6 x 2-2/3 x 12	450	7.9	10.2	17.5	22.6
6 x 3-1/5 x 12	375	6.8	8.8	18.1	23.4
6 x 4 x 12	300	5.6	7.4	19.1	24.7

[1] Add at least 5% for breakage.
[2] Add 10 to 25% for waste.

Source: *Estimating Brick Masonry* (Reston, Va: Brick Institute of America, 1997).

BRICK PAVEMENT

Advantages

Compared with asphalt or concrete, brick paving is more attractive, will not crack, and is easy to repair.

Details

Below are designs for a walk, a driveway, and two patios. The use of grade SW brick is recommended for all. All should be sloped for drainage. Walks and drives should slope at least 1/4 inch per foot (2 percent), but not more than 1-3/4 inches per foot (15 percent).

Patios should slope from 1/8 to 1/4 inch per foot (1 to 2 percent). The slopes can be from one edge to the other or from the center to the edges.

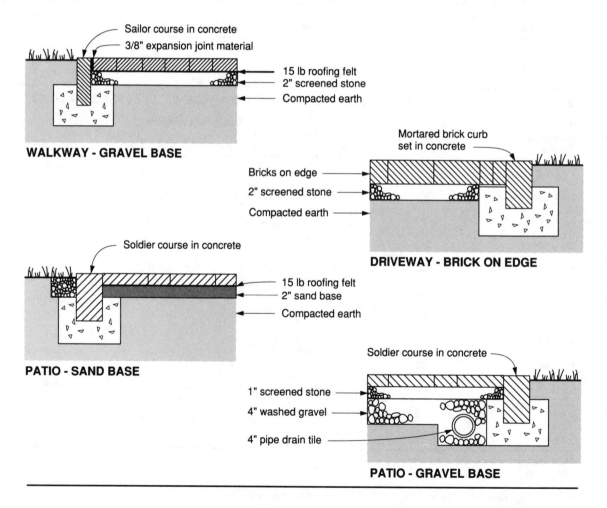

Sailor course in concrete
3/8" expansion joint material
15 lb roofing felt
2" screened stone
Compacted earth

WALKWAY - GRAVEL BASE

Mortared brick curb set in concrete
Bricks on edge
2" screened stone
Compacted earth

DRIVEWAY - BRICK ON EDGE

Soldier course in concrete
15 lb roofing felt
2" sand base
Compacted earth

PATIO - SAND BASE

Soldier course in concrete
1" screened stone
4" washed gravel
4" pipe drain tile

PATIO - GRAVEL BASE

Brick Pavement Bonds (Patterns)

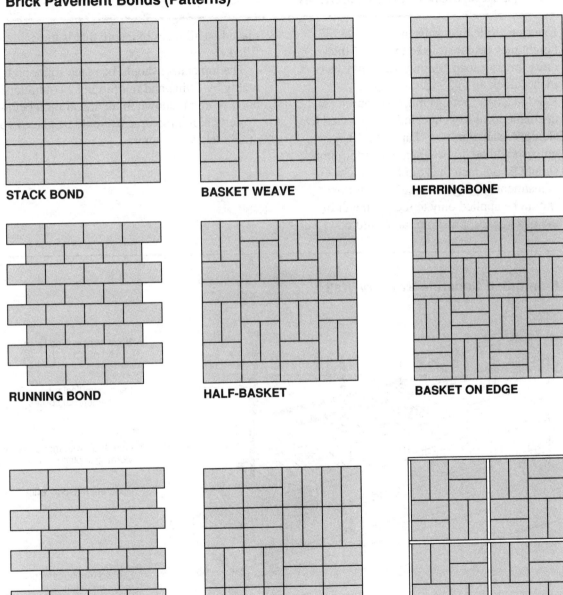

STACK BOND

BASKET WEAVE

HERRINGBONE

RUNNING BOND

HALF-BASKET

BASKET ON EDGE

OFFSET BOND

DOUBLE-BASKET

BASKET/WOOD GRID

CONCRETE MASONRY CONSTRUCTION

Exterior residential walls up to 35 feet in height may be constructed of 8-inch-thick units. Interior load-bearing walls may be 6 inches thick up to 20 feet in height.

Foundation footings must be poured beneath the local frost line, unless protected by shallow insulation (see chapter 4). Masonry units should be dry when laid. Mortar joints should be 3/8 inch thick, unless a slight adjustment of height is required. Mortar should be applied only to the perimeter of the block, except that mortar should be placed on all faces of a core that is to be filled.

The top course should be made solid, preferably by filling and reinforcing a continuous bond beam, as shown in the illustration below.

Vertical reinforcement should be placed 16 or 24 inches on-center.

Reinforced Concrete Masonry Wall

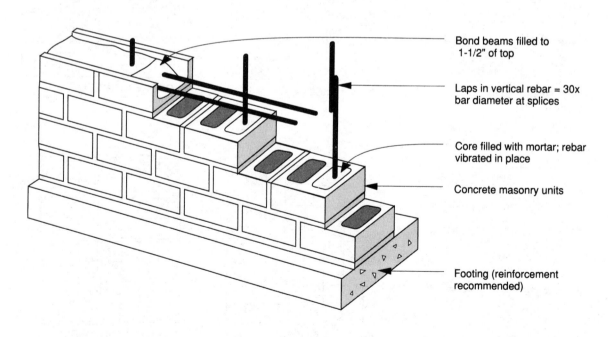

Bond beams filled to 1-1/2" of top

Laps in vertical rebar = 30x bar diameter at splices

Core filled with mortar; rebar vibrated in place

Concrete masonry units

Footing (reinforcement recommended)

BLOCK SIZES AND COURSING

(dimensions actual W x L x H)

Partition
3-5/8"x7-5/8"x3-5/8"

1/2 partition
3-5/8"x15-5/8"x3-5/8"

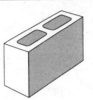

Partition
3-5/8"x15-5/8"x7-5/8"

Double ends
3-5/8"x7-5/8"x7-5/8"

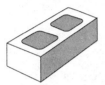

Double ends
5-5/8"x15-5/8"x3-5/8"

Double ends
5-5/8"x11-5/8"x3-5/8"

Double ends
5-5/8"x7-5/8"x7-5/8"

Half jamb
5-5/8"x7-5/8"x7-5/8"

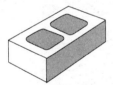

Half-hi double end
7-5/8"x15-5/8"x3-5/8"

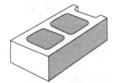

Half-hi
7-5/8"x15-5/8"x3-5/8"

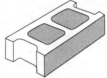

Half-hi stretcher
7-5/8"x15-5/8"x3-5/8"

Half-hi
7-5/8"x7-5/8"x3-5/8"

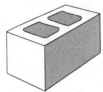

Double end
7-5/8"x15-5/8"x7-5/8"

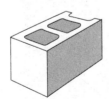

Regular
7-5/8"x15-5/8"x7-5/8"

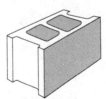

Stretcher
7-5/8"x15-5/8"x7-5/8"

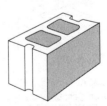

Steel sash jamb
7-5/8"x15-5/8"x7-5/8"

Plain ends
7-5/8"x7-5/8"x7-5/8"

Half block
7-5/8"x7-5/8"x7-5/8"

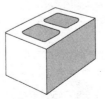

Double ends
7-5/8"x11-5/8"x7-5/8"

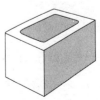

Plain ends
7-5/8"x11-5/8"x7-5/8"

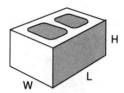

Double ends
11-5/8"x15-5/8"x7-5/8"

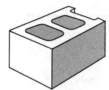

Regular
11-5/8"x15-5/8"x7-5/8"

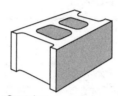

Stretcher
11-5/8"x15-5/8"x7-5/8"

Single bull nose
11-5/8"x15-5/8"x7-5/8"

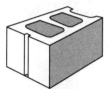

Steel sash jamb
11-5/8"x15-5/8"x7-5/8"

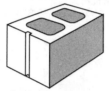

Jamb
11-5/8"x15-5/8"x7-5/8"

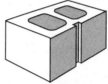

Twin
1-5/8"x15-5/8"x7-5/8"

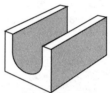

Solid-bottom U-block
11-5/8"x15-5/8"x7-5/8"

Beam
7-5/8"x7-5/8"x11-5/8"

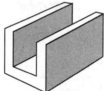

U-lintel
7-5/8"x15-5/8"x7-5/8"

U-lintel
7-5/8"x7-5/8"x7-5/8"

U-lintel
7-5/8"x3-5/8"x7-5/8"

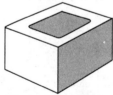

Pilaster & chimney block
15-5/8"x15-5/8"x7-5/8"

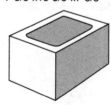

Pilaster
11-5/8"x15-5/8"x7-5/8"

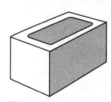

Pilaster
7-5/8"x15-5/8"x7-5/8"

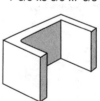

Pilaster
7-5/8"x15-5/8"x7-5/8"

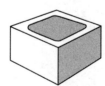

Chimney block
14"x14"x7-5/8"

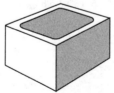

Column style #1
15-5/8"x15-5/8"x7-5/8"

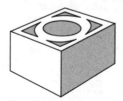

Flue block
15-5/8"x15-5/8"x7-5/8"

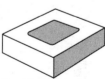

Flue block
15-5/8"x15-5/8"x3-5/8"

Heights of Concrete Block Walls

Number of Courses	Actual Height of Block [1] 3-5/8"	7-5/8"	Number of Courses	Actual Height of Block [1] 3-5/8"	7-5/8"
1	0' 4"	0' 8"	25	8' 4"	16' 8"
2	0' 8"	1' 4"	26	8' 8"	17' 4"
3	1' 0"	2' 0"	27	9' 0"	18' 0"
4	1' 4"	2' 8"	28	9' 4"	18' 8"
5	1' 8"	3' 4"	29	9' 8"	19' 4"
6	2' 0"	4' 0"	30	10' 0"	20' 0"
7	2' 4"	4' 8"	31	10' 4"	20' 8"
8	2' 8"	5' 4"	32	10' 8"	21' 4"
9	3' 0"	6' 0"	33	11' 0"	22' 0"
10	3' 4"	6' 8"	34	11' 4"	22' 8"
11	3' 8"	7' 4"	35	11' 8"	23' 4"
12	4' 0"	8' 0"	36	12' 0"	24' 0"
13	4' 4"	8' 8"	37	12' 4"	24' 8"
14	4' 8"	9' 4"	38	12' 8"	25' 4"
15	5' 0"	10' 0"	39	13' 0"	26' 0"
16	5' 4"	10' 8"	40	13' 4"	26' 8"
17	5' 8"	11' 4"	41	13' 8"	27' 4"
18	6' 0"	12' 0"	42	14' 0"	28' 0"
19	6' 4"	12' 8"	43	14' 4"	28' 8"
20	6' 8"	13' 4"	44	14' 8"	29' 4"
21	7' 0"	14' 0"	45	15' 0"	30' 0"
22	7' 4"	14' 8"	46	15' 4"	30' 8"
23	7' 8"	15' 4"	47	15' 8"	31' 4"
24	8' 0"	16' 0"	48	16' 0"	32' 0"

Source: *Passive Solar Construction Handbook* (Herndon, Va: National Concrete Masonry Association, 1984).
[1] With 3/8-inch mortar joint.

CONCRETE PAVERS

The details for paving with CMU pavers are, for the most part, the same as for brick pavers (see "Brick Pavement"). There is an almost infinite variety of sizes and shapes of CMU pavers. Some of the most common are shown below. Many are available in colors. Check with local suppliers before designing your project.

(dimensions actual W x L x H)

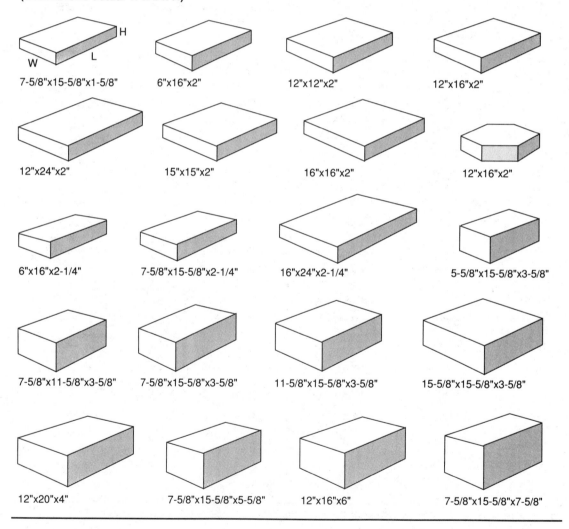

7-5/8"x15-5/8"x1-5/8"   6"x16"x2"   12"x12"x2"   12"x16"x2"

12"x24"x2"   15"x15"x2"   16"x16"x2"   12"x16"x2"

6"x16"x2-1/4"   7-5/8"x15-5/8"x2-1/4"   16"x24"x2-1/4"   5-5/8"x15-5/8"x3-5/8"

7-5/8"x11-5/8"x3-5/8"   7-5/8"x15-5/8"x3-5/8"   11-5/8"x15-5/8"x3-5/8"   15-5/8"x15-5/8"x3-5/8"

12"x20"x4"   7-5/8"x15-5/8"x5-5/8"   12"x16"x6"   7-5/8"x15-5/8"x7-5/8"

STONE VENEER CONSTRUCTION

Stone walls may be built entirely of stone or attached to poured concrete or concrete masonry walls, but the majority of residential walls are of one of the stone veneer types shown below. The wood frame simplifies insulating and finishing and eliminates the need for bracing of the stone wall.

Mortar joints should be 1/2 to 1 inch thick for rubble patterns and 3/8 to 3/4 inch thick for ashlar (cut stone) patterns.

To prevent staining over time, all fasteners should be stainless steel; flashing metal should be stainless or hot-dipped galvanized steel; and mortar should be nonstaining.

Solid Veneer Wall

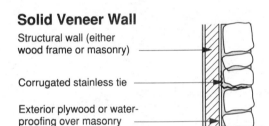

Structural wall (either wood frame or masonry)

Corrugated stainless tie

Exterior plywood or water-proofing over masonry

Thin Veneer Wall

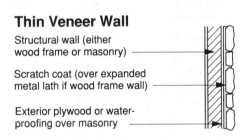

Structural wall (either wood frame or masonry)

Scratch coat (over expanded metal lath if wood frame wall)

Exterior plywood or water-proofing over masonry

Stone Wall Patterns

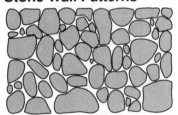

Random rubble or fieldstone

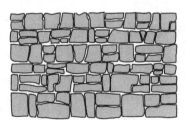

Coursed rubble

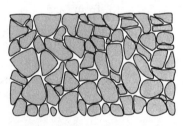

Web or mosaic

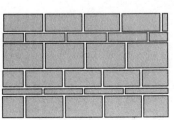

Coursed ashlar

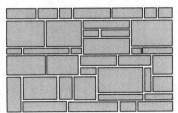

Random ashlar

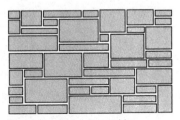

Three-height random ashlar

CHECKLIST OF CODE REQUIREMENTS

The following is a partial list of requirements from the 1995 Council of American Building Officials (CABO) *One and Two Family Dwelling Code*. Consult the publication for the full text and additional provisions.

Masonry Fireplaces

(1003.7) Hearth extension material:
- masonry or concrete ≥ 2" thick (3/8" for firebox raised ≥ 8")
- supported by noncombustible materials
- distinguishable from surrounding materials

(1003.8) Hearth extension dimensions:

Opening < 6 sq ft
- ≥ 36" from back of firebox
- ≥ 16" from face of opening
- ≥ 8" from sides of opening

Opening ≥ 6 sq ft
- ≥ 36" from back of firebox
- ≥ 20" from face of opening
- ≥ 12" from sides of opening

(1003.9) Combustible clearances:
- ≥ 2" to outside face of opening
- ≥ 2" to back surface of fireplace
- ≥ 6" to inside surface of flue lining

REQUIREMENTS FOR MASONRY FIREPLACES AND CHIMNEYS

Item	Requirement
Hearth thickness	4"
Hearth side extension	
Opening < 6 sq ft	8"
Opening ≥ 6 sq ft	12"
Hearth front extension	
Opening < 6 sq ft	16"
Opening ≥ 6 sq ft	20"
Firebox wall thickness	
Solid brick	10"
Firebrick lining	8"
Top of opening to throat	8"
Smoke chamber	
Rear wall thickness	6"
Front wall thickness	8"
Side wall thickness	8"
Wall with unlined flue	8" solid
Wall with flue lining	Brick with grout around lining or 1/2" airspace
Chimney footing	
Thickness	≥ 12"
Width beyond wall	6"

4

Foundations

Many homeowners think the words *foundation* and *basement* are synonymous. In northern states this is understandable, since about 90 percent of homes have basements. But as the *foundation design* section shows, there are many other options. Your choice should be determined by the functions you expect the foundation to perform.

After you choose the style (*basement, crawl space,* or *slab-on-grade*), you must choose the material (poured concrete, masonry, or all-weather wood).

In this age of dwindling energy supplies, foundations should be insulated against heat loss. Tables for each style of foundation show how much insulation you should install in 10 regions of the country.

Additional sections cover the details of effective *moisture control, termite control,* and *radon control.*

In all, 18 specific foundation designs are detailed. You should have no problem following the large illustrations of each.

Finally, we provide you with a *checklist of code requirements* relating to foundations.

FOUNDATION DESIGN

Functions of Foundations

In selecting or designing an appropriate building foundation, one must keep in mind *all* of its possible functions. It may be used to

- transfer building loads to the ground
- anchor the building against wind
- isolate the building from frost heaving
- isolate the building from expansive soil
- hold the building above ground moisture
- retard heat loss from conditioned space
- provide storage space
- provide living space
- house the mechanical systems

The Design Process

The design process (or decision-making process) should involve these steps:

1. Select the type (basement, crawl space, or slab).

2. Select the material (poured concrete, masonry, or wood).

3. Decide if it is to be conditioned (heated and cooled).

4. If it is to be conditioned, select the type and R-value of insulation.

5. Detail structure, insulation, and finish.

6. Detail moisture control.

7. Detail termite control.

8. Detail radon control.

This chapter generally follows the order of the above design process. It starts by defining foundation types, at right.

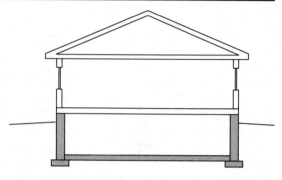

BASEMENT

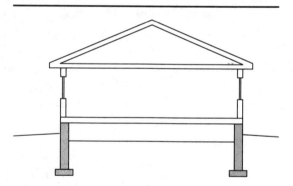

CRAWL SPACE

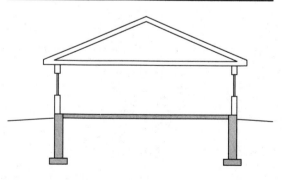

SLAB-ON-GRADE

BASEMENTS

Recommended Insulation

The recommended configurations and R-values (for more on R-value, see chapter 13) of foundation insulation are given below for combinations of conditioned and unconditioned basements, full and half depth of burial, and masonry and wood construction. *Recommended* means that configuration having the lowest 30-year life-cycle cost assuming fuel inflation of 7 percent, general inflation of 5 percent, and the present fuel prices: natural

gas – $0.56/ therm, fuel oil – $0.79/ gallon, electric heat – $0.028/ kwhr, and electric cooling – $0.076/ kwhr.

Most of the recommended configurations place the insulation on the outside of the foundation, where it is most effective. However, optimum R-values and savings for interior insulation placement are nearly identical.

Basement Configurations

CONDITIONED

4' of exterior foam

8' of exterior foam

8' of interior blanket

CONDITIONED

Interior blanket above grade;
4' of exterior foam below grade

8' of interior blanket

UNCONDITIONED

4' of exterior foam

Blanket between joists

8' of interior blanket

Foundations

RECOMMENDED INSULATION FOR BASEMENTS

Location	HDD_{65}[1]	CDD_{75}[2]	Conditioned[3] - Full Masonry[4]	Wood	Conditioned - Half Masonry	Wood	Unconditioned Masonry	Wood	Ceiling
Minneapolis	8,007	98	8' R15	R19	R19/15	R19	4' R5	R19	R30
Chicago	6,177	181	8' R10	R19	R19/10	R19	4' R5	R11	R30
Denver	6,014	83	8' R10	R19	R19/15	R19	4' R5	R11	R30
Boston	5,593	74	8' R10	R19	R19/10	R19	4' R5	R11	R30
Washington	4,122	299	8' R10	R19	R19/10	R19	none	R11	R11
Atlanta	3,021	415	8' R5	R19	R19/10	R19	none	none	none
Fort Worth	2,407	1,139	8' R10	R19	R19/10	R19	none	none	none
Phoenix	1,442	1,856	8' R10	R19	R19/5	R19	none	none	none
Los Angeles	1,595	36	none	none	R19/0	R11	none	none	none
Miami	199	1,257	none	none	R19/0	R11	none	none	none

[1] Heating degree-days, base 65°F, is the accumulated difference between the average daily temperature and 65°F through the heating season.

[2] Cooling degree-days, base 75°F, is the accumulated difference between the average daily temperature and 75°F through the cooling season.

[3] *Conditioned* means heated to 65°F and cooled to 75°F.

[4] *Masonry* here means masonry block or poured concrete below grade (bg) and wood above grade (ag). Insulation values are presented as R (ag) / R (bg).

Concrete Basement Wall with 4' Exterior Insulation

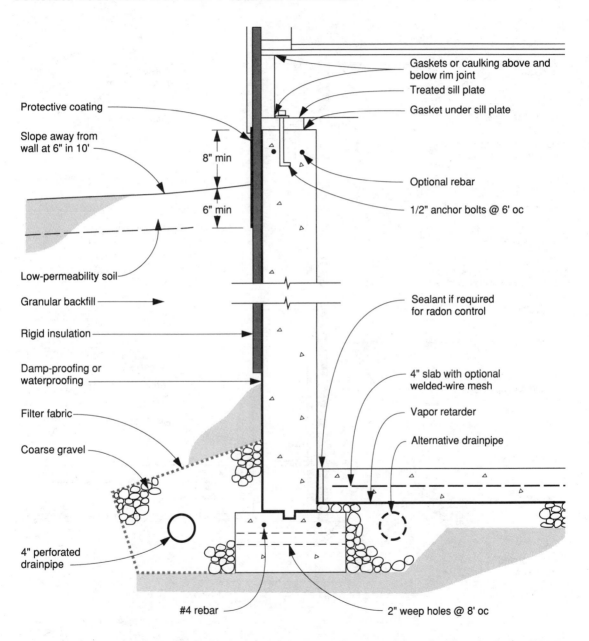

Protective coating

Slope away from wall at 6" in 10'

8" min

6" min

Low-permeability soil

Granular backfill

Rigid insulation

Damp-proofing or waterproofing

Filter fabric

Coarse gravel

4" perforated drainpipe

Gaskets or caulking above and below rim joint

Treated sill plate

Gasket under sill plate

Optional rebar

1/2" anchor bolts @ 6' oc

Sealant if required for radon control

4" slab with optional welded-wire mesh

Vapor retarder

Alternative drainpipe

#4 rebar

2" weep holes @ 8' oc

Masonry Basement Wall with 8' Exterior Insulation

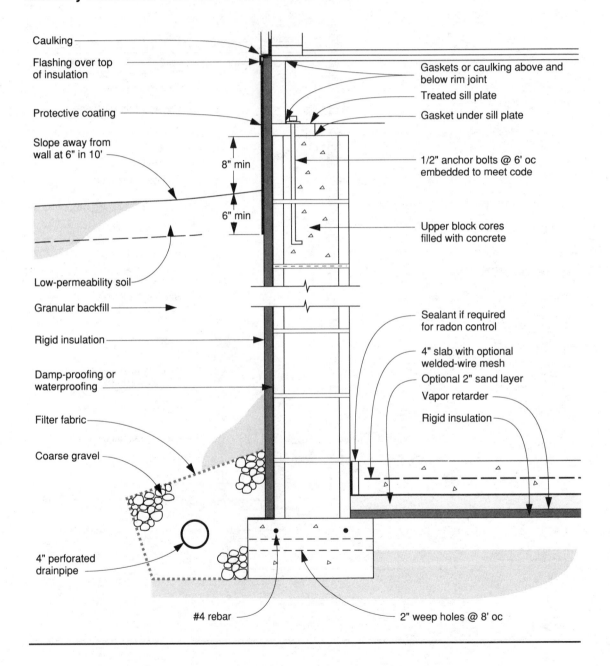

Caulking

Flashing over top of insulation

Protective coating

Slope away from wall at 6" in 10'

8" min

6" min

Low-permeability soil

Granular backfill

Rigid insulation

Damp-proofing or waterproofing

Filter fabric

Coarse gravel

4" perforated drainpipe

Gaskets or caulking above and below rim joint

Treated sill plate

Gasket under sill plate

1/2" anchor bolts @ 6' oc embedded to meet code

Upper block cores filled with concrete

Sealant if required for radon control

4" slab with optional welded-wire mesh

Optional 2" sand layer

Vapor retarder

Rigid insulation

#4 rebar

2" weep holes @ 8' oc

Concrete Basement Wall with Interior Insulation

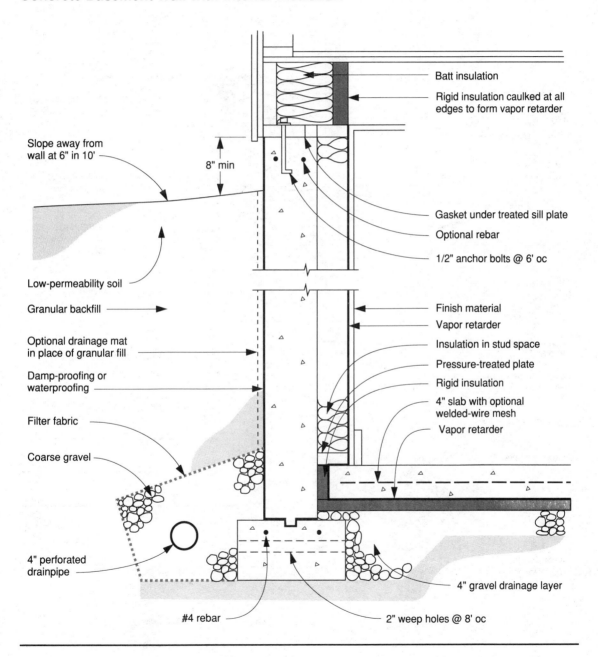

Slope away from wall at 6" in 10'

8" min

Low-permeability soil

Granular backfill

Optional drainage mat in place of granular fill

Damp-proofing or waterproofing

Filter fabric

Coarse gravel

4" perforated drainpipe

#4 rebar

2" weep holes @ 8' oc

Batt insulation

Rigid insulation caulked at all edges to form vapor retarder

Gasket under treated sill plate

Optional rebar

1/2" anchor bolts @ 6' oc

Finish material

Vapor retarder

Insulation in stud space

Pressure-treated plate

Rigid insulation

4" slab with optional welded-wire mesh

Vapor retarder

4" gravel drainage layer

Masonry Basement Wall with Interior Insulation

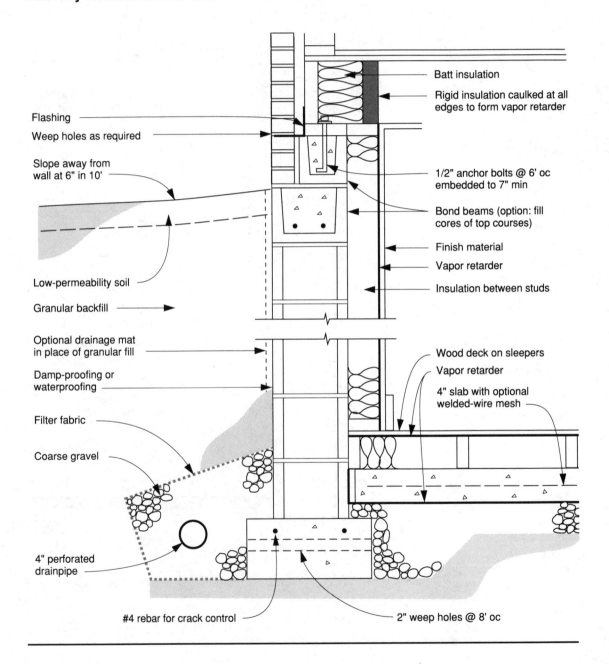

Flashing

Weep holes as required

Slope away from
wall at 6" in 10'

Low-permeability soil

Granular backfill

Optional drainage mat
in place of granular fill

Damp-proofing or
waterproofing

Filter fabric

Coarse gravel

4" perforated
drainpipe

#4 rebar for crack control

Batt insulation

Rigid insulation caulked at all
edges to form vapor retarder

1/2" anchor bolts @ 6' oc
embedded to 7" min

Bond beams (option: fill
cores of top courses)

Finish material

Vapor retarder

Insulation between studs

Wood deck on sleepers

Vapor retarder

4" slab with optional
welded-wire mesh

2" weep holes @ 8' oc

Pressure-Treated Wood Basement Wall

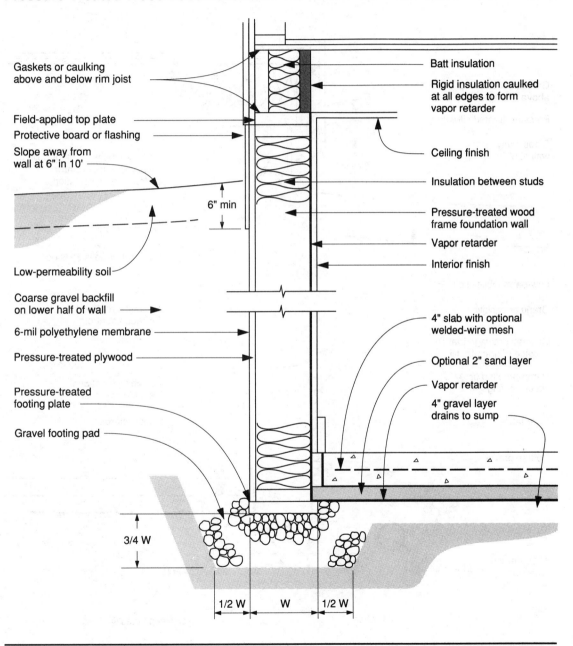

Gaskets or caulking above and below rim joist

Field-applied top plate

Protective board or flashing

Slope away from wall at 6" in 10'

6" min

Low-permeability soil

Coarse gravel backfill on lower half of wall

6-mil polyethylene membrane

Pressure-treated plywood

Pressure-treated footing plate

Gravel footing pad

3/4 W

1/2 W   W   1/2 W

Batt insulation

Rigid insulation caulked at all edges to form vapor retarder

Ceiling finish

Insulation between studs

Pressure-treated wood frame foundation wall

Vapor retarder

Interior finish

4" slab with optional welded-wire mesh

Optional 2" sand layer

Vapor retarder

4" gravel layer drains to sump

Concrete Basement Wall with Ceiling Insulation

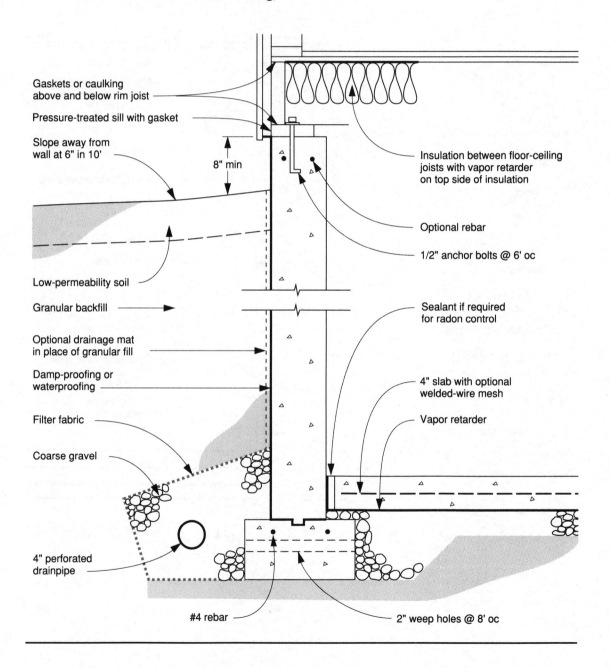

Gaskets or caulking
above and below rim joist

Pressure-treated sill with gasket

Slope away from
wall at 6" in 10'

8" min

Low-permeability soil

Granular backfill

Optional drainage mat
in place of granular fill

Damp-proofing or
waterproofing

Filter fabric

Coarse gravel

4" perforated
drainpipe

Insulation between floor-ceiling
joists with vapor retarder
on top side of insulation

Optional rebar

1/2" anchor bolts @ 6' oc

Sealant if required
for radon control

4" slab with optional
welded-wire mesh

Vapor retarder

#4 rebar

2" weep holes @ 8' oc

CRAWL SPACES

Recommended Insulation

The recommended configurations and R-values of foundation insulation are given below for combinations of vented and unvented, 2 and 4-foot masonry and wood crawl spaces. *Recommended* means the configuration having the lowest 30-year life-cycle cost assuming fuel inflation of 7 percent, general inflation of 5 percent, and present fuel prices: natural gas – $0.56/ therm, fuel oil – $0.79/ gallon, electric heat – $0.028/ kwhr, and electric cooling – $0.076/ kwhr.

Most of the recommended configurations place the insulation on the outside of the foun-

dation, where it is most effective, both at stopping heat flow and protecting the foundation. However, in many situations, such as retrofits and masonry walls, interior placement is more practical; optimum R-values and savings for interior insulation placement are nearly the same. The illustration below shows interior application.

Crawl Space Configurations

UNVENTED 4' WALL

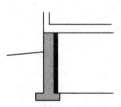

Masonry foundation with interior foam insulation

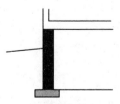

Wood foundation with blanket between studs

UNVENTED 2' WALL

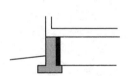

Masonry foundation with interior foam insulation

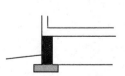

Wood foundation with blanket between studs

VENTED 2' WALL

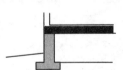

Masonry foundation with blanket in floor

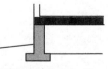

Wood foundation with blanket in floor

Foundations

RECOMMENDED INSULATION FOR CRAWL SPACES

Location	HDD_{65}[1]	CDD_{75}[2]	Unvented 4' Wall of Masonry	Wood	Unvented 2' Wall of Masonry	Wood	Vented 2' Wall of[3] Masonry	Wood
Minneapolis	8,007	98	4' R5	R19	2' R10	R19	R30	R30
Chicago	6,177	181	4' R5	R11	2' R10	R19	R19	R30
Denver	6,014	83	4' R5	R11	2' R10	R11	R19	R30
Boston	5,593	74	4' R5	R11	2' R10	R19	R19	R30
Washington	4,122	299	4' R5	none	2' R5	R11	R19	R30
Atlanta	3,021	415	4' R5	none	2' R5	none	R11	R11
Fort Worth	2,407	1,139	4' R5	none	2' R5	none	R11	R11
Phoenix	1,442	1,856	4' R5	none	2' R5	none	R11	R11
Los Angeles	1,595	36	none	none	none	none	R11	R11
Miami	199	1,257	none	none	none	none	none	none

[1] Heating degree-days, base 65°F, is the accumulated difference between the average daily temperature and 65°F through the heating season.

[2] Cooling degree-days, base 75°F, is the accumulated difference between the average daily temperature and 75°F through the cooling season.

[3] Values are for ceiling insulation.

Concrete Crawl Space Wall with Interior Insulation

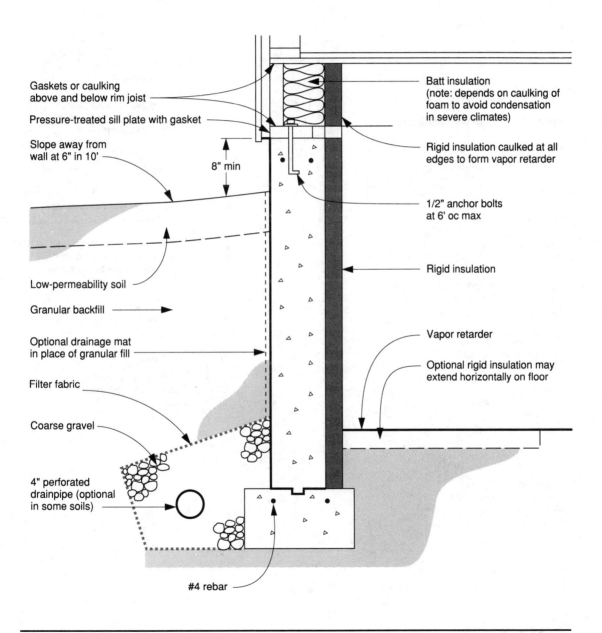

Gaskets or caulking above and below rim joist

Pressure-treated sill plate with gasket

Slope away from wall at 6" in 10'

8" min

Low-permeability soil

Granular backfill

Optional drainage mat in place of granular fill

Filter fabric

Coarse gravel

4" perforated drainpipe (optional in some soils)

#4 rebar

Batt insulation (note: depends on caulking of foam to avoid condensation in severe climates)

Rigid insulation caulked at all edges to form vapor retarder

1/2" anchor bolts at 6' oc max

Rigid insulation

Vapor retarder

Optional rigid insulation may extend horizontally on floor

Concrete Crawl Space Wall with Interior Insulation

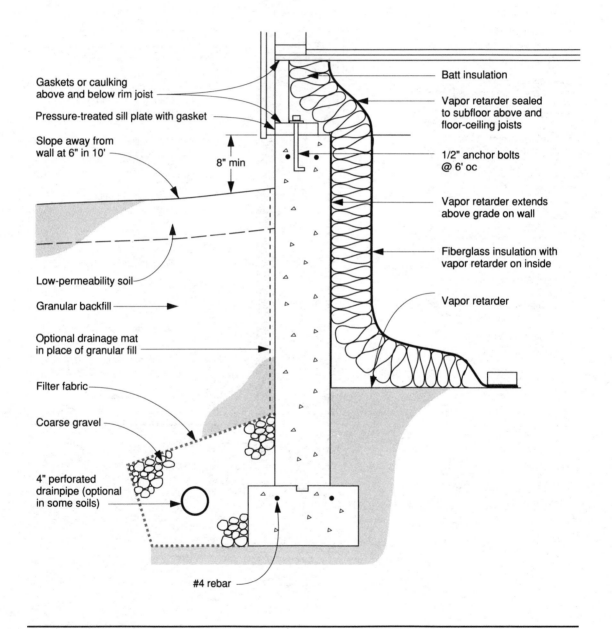

Gaskets or caulking above and below rim joist

Pressure-treated sill plate with gasket

Slope away from wall at 6" in 10'

8" min

Low-permeability soil

Granular backfill

Optional drainage mat in place of granular fill

Filter fabric

Coarse gravel

4" perforated drainpipe (optional in some soils)

#4 rebar

Batt insulation

Vapor retarder sealed to subfloor above and floor-ceiling joists

1/2" anchor bolts @ 6' oc

Vapor retarder extends above grade on wall

Fiberglass insulation with vapor retarder on inside

Vapor retarder

Masonry Crawl Space Wall with Exterior Insulation

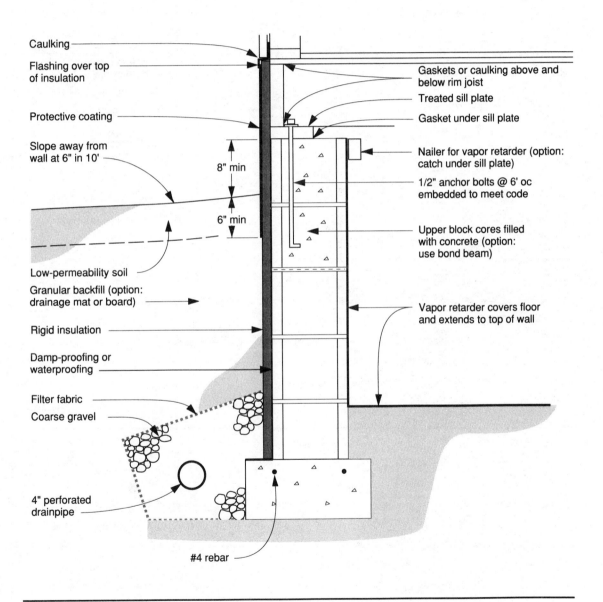

Caulking

Flashing over top of insulation

Protective coating

Slope away from wall at 6" in 10'

8" min

6" min

Low-permeability soil

Granular backfill (option: drainage mat or board)

Rigid insulation

Damp-proofing or waterproofing

Filter fabric

Coarse gravel

4" perforated drainpipe

#4 rebar

Gaskets or caulking above and below rim joist

Treated sill plate

Gasket under sill plate

Nailer for vapor retarder (option: catch under sill plate)

1/2" anchor bolts @ 6' oc embedded to meet code

Upper block cores filled with concrete (option: use bond beam)

Vapor retarder covers floor and extends to top of wall

Concrete Crawl Space Wall with Exterior Insulation

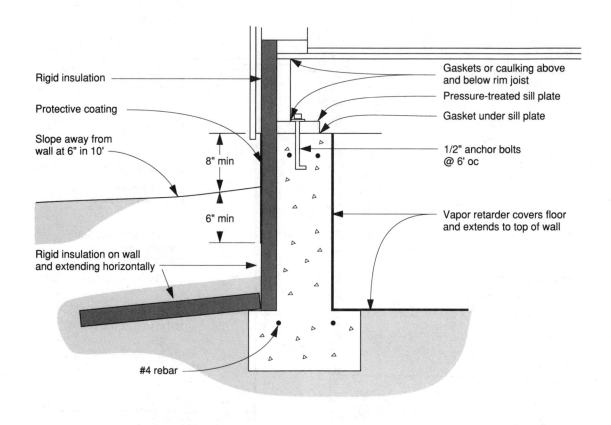

Rigid insulation

Protective coating

Slope away from
wall at 6" in 10'

8" min

6" min

Rigid insulation on wall
and extending horizontally

#4 rebar

Gaskets or caulking above
and below rim joist

Pressure-treated sill plate

Gasket under sill plate

1/2" anchor bolts
@ 6' oc

Vapor retarder covers floor
and extends to top of wall

Concrete Crawl Space Wall with Ceiling Insulation

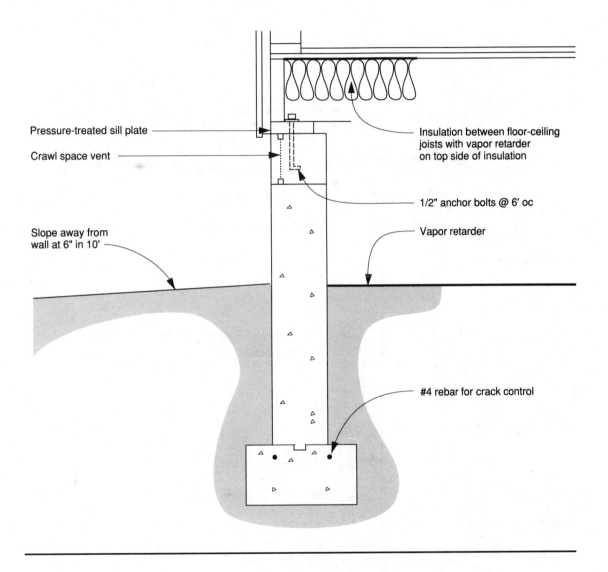

Pressure-treated sill plate

Crawl space vent

Slope away from wall at 6" in 10'

Insulation between floor-ceiling joists with vapor retarder on top side of insulation

1/2" anchor bolts @ 6' oc

Vapor retarder

#4 rebar for crack control

Piers with Floor Insulation

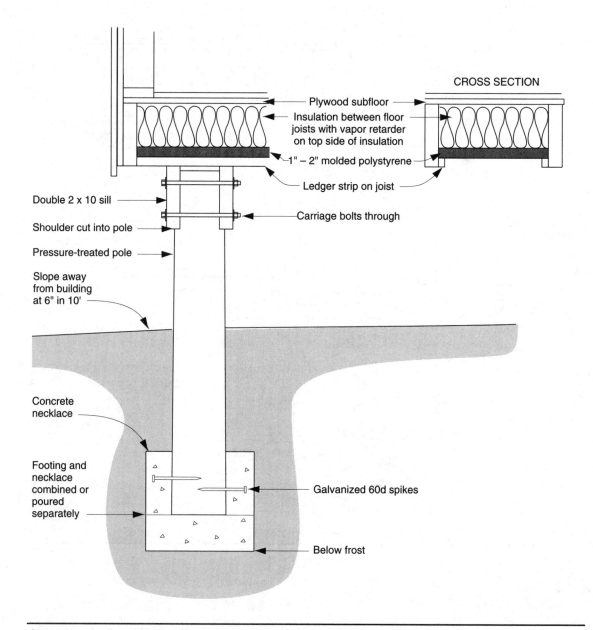

CROSS SECTION

Plywood subfloor

Insulation between floor joists with vapor retarder on top side of insulation

1" – 2" molded polystyrene

Ledger strip on joist

Double 2 x 10 sill

Carriage bolts through

Shoulder cut into pole

Pressure-treated pole

Slope away from building at 6" in 10'

Concrete necklace

Footing and necklace combined or poured separately

Galvanized 60d spikes

Below frost

Source: Adapted from *Low-Cost Wood Homes for Rural America* (Washington, DC: US Department of Agriculture, 1969).

SLABS-ON-GRADE

Recommended Insulation

The recommended configurations and R-values of foundation insulation are given below for combinations of exterior and interior, vertical and horizontal insulation. *Recommended* means that configuration having the lowest 30-year life-cycle cost assuming fuel inflation of 7 percent, general inflation of 5 percent, and present fuel prices: natural gas – $0.56/ therm, fuel oil – $0.79/ gallon, electric heat – $0.028/ kwhr, and electric cooling – $0.076/ kwhr.

Slab-on-Grade Configurations

VERTICAL INSULATION

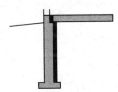

4' of interior foam insulation

2' of interior foam insulation

HORIZONTAL INSULATION

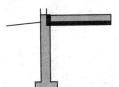

4' of interior foam insulation

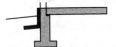

2' of exterior foam insulation

RECOMMENDED INSULATION FOR SLABS-ON-GRADE

Location	HDD$_{65}$[1]	CDD$_{75}$[2]	Interior Insulation	Exterior Insulation
Minneapolis	8,007	98	4' R5 vertical	2' R5 horizontal
Chicago	6,177	181	4' R5 horizontal	2' R5 horizontal
Denver	6,014	83	4' R5 horizontal	2' R5 horizontal
Boston	5,593	74	4' R5 horizontal	2' R5 horizontal
Washington	4,122	299	4' R5 vertical	2' R5 horizontal
Atlanta	3,021	415	4' R5 vertical	2' R5 horizontal
Fort Worth	2,407	1,139	4' R5 vertical	2' R5 horizontal
Phoenix	1,442	1,856	4' R5 vertical	2' R5 horizontal
Los Angeles	1,595	36	none	none
Miami	199	1,257	none	none

[1] Heating degree-days, base 65°F, is the accumulated difference between the average daily temperature and 65°F through the heating season.

[2] Cooling degree-days, base 75°F, is the accumulated difference between the average daily temperature and 75°F through the cooling season.

Slab-on-Grade and Integral Grade Beam with Exterior Insulation

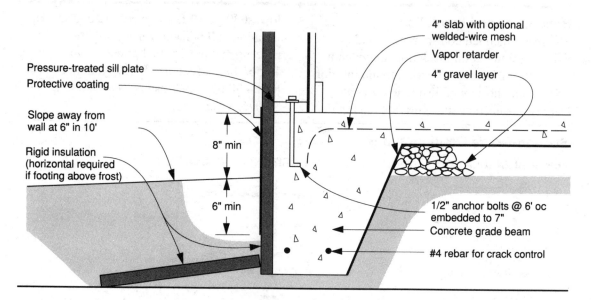

4" slab with optional welded-wire mesh

Vapor retarder

4" gravel layer

Pressure-treated sill plate
Protective coating

Slope away from wall at 6" in 10'

Rigid insulation (horizontal required if footing above frost)

8" min

6" min

1/2" anchor bolts @ 6' oc embedded to 7"
Concrete grade beam
#4 rebar for crack control

Slab-on-Grade and Concrete Foundation Wall with Exterior Insulation

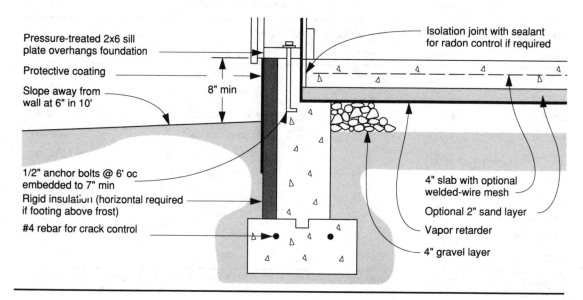

Isolation joint with sealant for radon control if required

Pressure-treated 2x6 sill plate overhangs foundation

Protective coating

Slope away from wall at 6' in 10'

8" min

1/2" anchor bolts @ 6' oc embedded to 7" min
Rigid insulation (horizontal required if footing above frost)
#4 rebar for crack control

4" slab with optional welded-wire mesh

Optional 2" sand layer

Vapor retarder

4" gravel layer

Slab-on-Grade and Concrete Foundation Wall with Interior Insulation

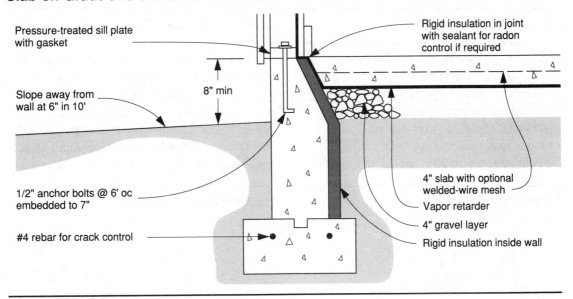

Pressure-treated sill plate with gasket

Rigid insulation in joint with sealant for radon control if required

8" min

Slope away from wall at 6" in 10'

1/2" anchor bolts @ 6' oc embedded to 7"

#4 rebar for crack control

4" slab with optional welded-wire mesh

Vapor retarder

4" gravel layer

Rigid insulation inside wall

Slab-on-Grade and Masonry Foundation Wall with Interior Insulation

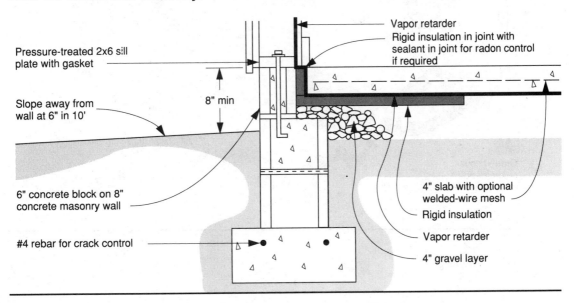

Pressure-treated 2x6 sill plate with gasket

Vapor retarder

Rigid insulation in joint with sealant in joint for radon control if required

8" min

Slope away from wall at 6" in 10'

6" concrete block on 8" concrete masonry wall

#4 rebar for crack control

4" slab with optional welded-wire mesh

Rigid insulation

Vapor retarder

4" gravel layer

MOISTURE CONTROL

Ground Covers

A ground cover membrane that restricts evaporation of soil moisture is the most effective way to prevent condensation and wood decay problems in a crawl space, as well as in an unheated full basement. However, ground covers aren't required everywhere; a rule of thumb is to use ground cover membranes in cool climates where total annual precipitation exceeds 20 inches of water (the shaded area in the map below).

The ground cover should have a permeability rating of 1.0 perm maximum and should be rugged enough to withstand foot traffic. Recommended materials include 6-mil polyethylene and 45-mil ethylene polymer membranes (EPDM). Membrane edges should overlap by 6 inches but need not be sealed.

Areas Where Vapor-Retarding Ground Covers Are Recommended

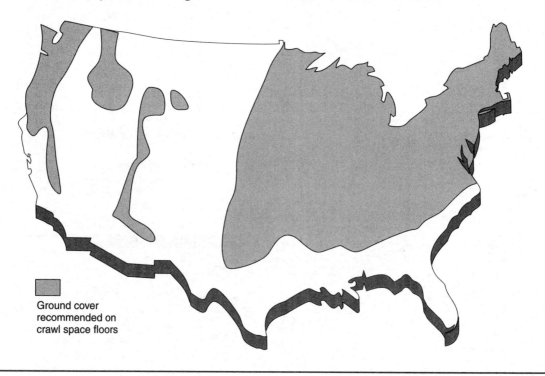

Ground cover recommended on crawl space floors

Vent Requirements

A rectangular crawl space requires a minimum of four vents, one on each wall, located no farther than 3 feet from each corner. The vents should be as high on the wall as possible to capture breezes, and landscaping should be planned to prevent obstruction. The free (open) area of all vents should total no less than 1/1,500 of the floor area. The gross area of vents depends on the type of vent cover. The gross area can be found by multiplying the vents' free area by one of the factors in the table below. In the absence of a ground cover (if recommended), the vent area should be increased tenfold, to 1/150 of the floor area.

Unvented Crawl Spaces

It is not necessary to vent a crawl space for moisture control if it is continuous with an adjacent basement. Venting is also incompatible with crawl spaces that are used as heat distribution plenums (in which the space acts as a giant warm-air duct). In such a case, duct insulation is not necessary, since the foundation is insulated around the perimeter. However, high radon levels may require modifications, including venting of the crawl space.

Example: What gross vent area is required for a 1,000-square-foot dirt-floored basement without soil cover membrane if the vents are covered with 1/8-inch mesh screen?

From the table below, the multiplier is 12.5. The gross vent area is therefore 1,000 square feet ÷ 1,500 × 12.5 = 8.33 square feet.

CRAWL SPACE GROSS VENT AREA REQUIREMENTS

Vent Cover Material	Multiply Free Vent Area By:	
	With Soil Cover	No Soil Cover
1/4" mesh hardware cloth	1.0	10
1/8" mesh screen	1.25	12.5
16-mesh insect screen	2.0	20
Louvers + 1/4" hardware cloth	2.0	20
Louvers + 1/8" mesh screen	2.25	22.5
Louvers + 16-mesh screen	3.0	30

Note: Data also apply to unconditioned full basements with dirt floors.

Drainage

Most drainage system failures are due to improper position, slope, size, or damage during construction. Footing drains (see below) have two basic functions whose relative importance depends on local conditions. They draw down the groundwater level to below the basement walls and floor, and they collect and drain away water that seeps down through the backfill from rain and melting snow. They are aided by vertical wall and underfloor drainage blankets and, especially in the case of exterior footing drains, by weep holes through the footing.

A footing drain must be placed so that the top of the pipe is beneath the bottom of the floor slab, and the bottom of the pipe and its bedding must be above the bottom of the footing. Drains can be located outside the foundation, either next to or on top of the footing, or next to the footing beneath the floor slab. When the drain line is inside, a gravel fill should still be provided at the outside, and weep holes should be provided through the footing every 8 feet.

Exterior placement beside the footing is more effective at drawing down the water table and coupling with optional wall drainage blankets. Interior drains are more effective for collecting soil gas in radon control systems.

Because groundwater enters the drain line from beneath, the pipe is placed with the two rows of holes facing down. A foundation fabric filter keeps coarse sediment from washing into the drain line.

Drain lines are also sized and sloped to wash out fine sediment that may infiltrate. The common 4-inch pipe should be sloped at least 1 inch in 20 feet, although 1 inch in 10 feet is desirable to compensate for settling after construction.

Perimeter Drain Placement

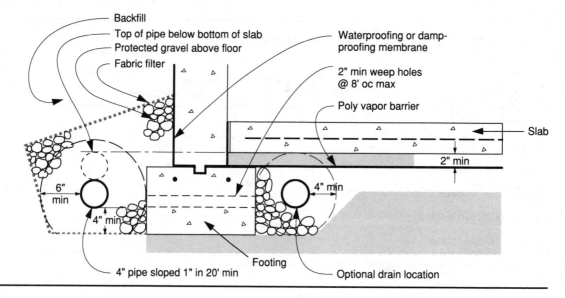

Backfill
Top of pipe below bottom of slab
Protected gravel above floor
Fabric filter
Waterproofing or damp-proofing membrane
2" min weep holes @ 8' oc max
Poly vapor barrier
Slab
2" min
6" min
4" min
4" min
Optional drain location
Footing
4" pipe sloped 1" in 20' min

TERMITE CONTROL

Termites in the United States

Termites occur naturally in woodlands, where they help break down dead plant material and play an important part in the nutrient cycle. The problem lies in the fact that termites don't distinguish between the wood in dead trees and the wood in houses. Subterranean termites, which account for 95 percent of all termite damage, are found throughout the United States wherever the average annual air temperature is 50°F or above and the ground is sufficiently moist (see map below).

Subterranean Termite

Termite Hazard Distribution

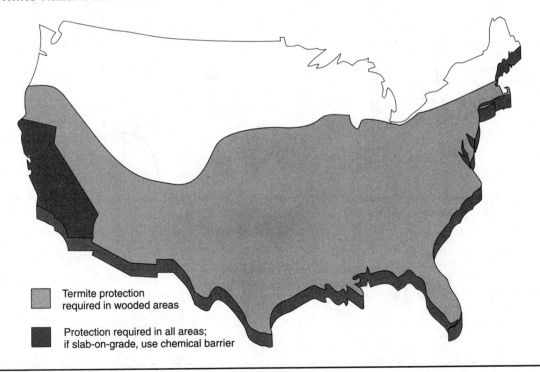

Termite protection required in wooded areas

Protection required in all areas; if slab-on-grade, use chemical barrier

Points of Entry

Since termites occur naturally in forests and brushlands, clearing wooded sites robs termites of their food supply. They adapt by feeding on the wood in structures (wouldn't you?). Termite control begins by blocking easy routes of entry from the soil to wood in the structure. The most common points of entry are shown below.

Common Points of Termite Entry

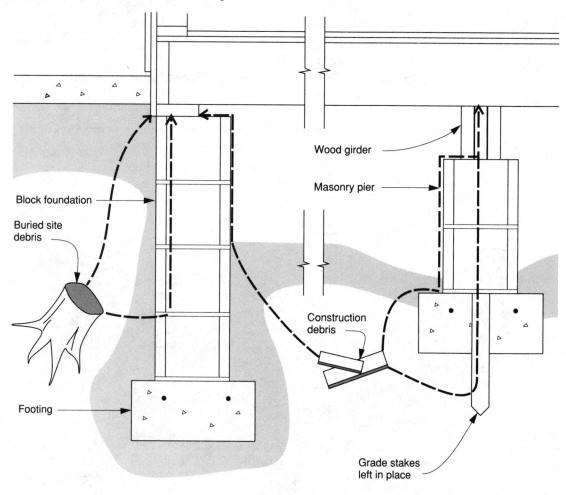

Wood girder

Masonry pier

Block foundation

Buried site debris

Construction debris

Footing

Grade stakes left in place

Termite Control Measures

Termite control consists of keeping them in the wild, rather than exterminating them within the building. This is simple in concept, but difficult in practice, since termites can pass through cracks as narrow as 1/32 inch. The major strategies of termite control include

- separating wood structures from the soil
- minimizing cracks in slabs and walls
- installing barriers to force termites into the open
- treating the soil with chemicals
- keeping the soil and foundation dry

Separation of Wood from Soil

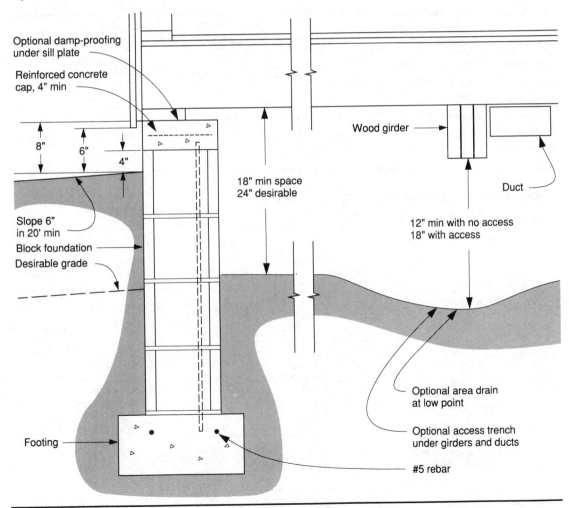

Optional damp-proofing under sill plate

Reinforced concrete cap, 4" min

8"

6"

4"

18" min space
24" desirable

Wood girder

Duct

Slope 6"
in 20' min

12" min with no access
18" with access

Block foundation

Desirable grade

Footing

Optional area drain
at low point

Optional access trench
under girders and ducts

#5 rebar

Soil Treatments

The purpose of soil treatment is to create a continuous barrier in the soil surrounding the foundation. The chemical must be applied thoroughly and uniformly to block all routes of termite entry. Special consideration must be given to joints, pipes, and utility conduits that pass through the wall and floor.

Only approved termiticides should be used. Chemicals for pest management are controlled by the Environmental Protection Agency (EPA) and by state and local governments. The most familiar termiticides and their recommended concentrations (by weight) are shown in the table at right.

Chlordane and heptachlor were suspended in August 1987 by the EPA. At the time of the suspension, about two-thirds of all residential termite treatments used these chemicals, which are now suspected of being carcinogenic.

In anticipation of restrictions on some commonly used termiticides, chemical companies have registered alternatives whose active ingredient is a pyrethroid—a man-made substance based on a pesticide produced naturally in certain plants. A disadvantage is that it is expected to be effective for a shorter time.

RECOMMENDED TERMITICIDE CONCENTRATIONS

Chemical	% by Weight
Aldrin	0.5
Chlorpyrifos	1.0
Permethrin	0.5
Isofenphos	1.0

Application of Termiticide to Slabs-on-Grade

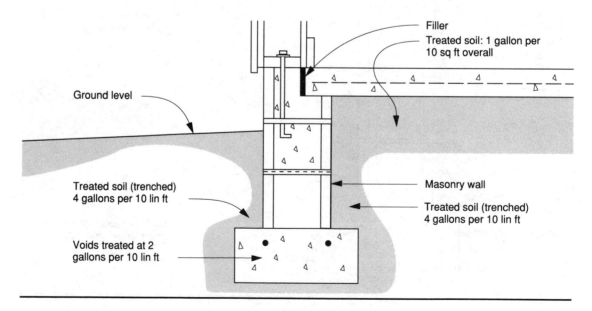

Filler
Treated soil: 1 gallon per 10 sq ft overall
Ground level
Treated soil (trenched) 4 gallons per 10 lin ft
Voids treated at 2 gallons per 10 lin ft
Masonry wall
Treated soil (trenched) 4 gallons per 10 lin ft

Application of Termiticide to Crawl Spaces

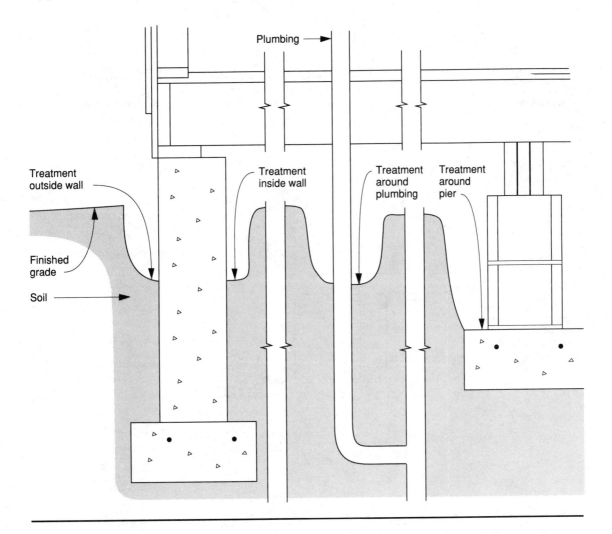

Application of Termiticide to Basement Foundations

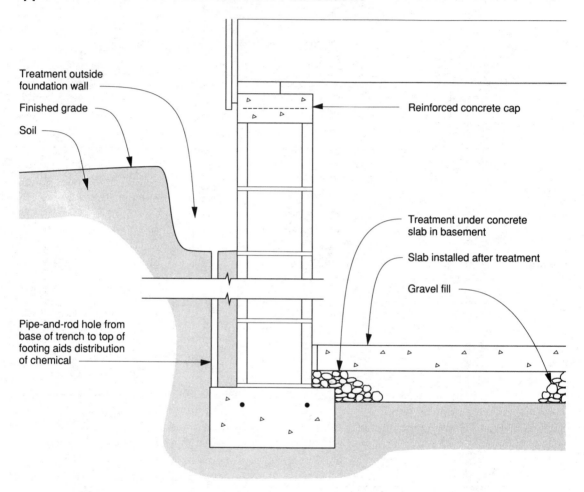

Treatment outside foundation wall

Finished grade

Soil

Reinforced concrete cap

Treatment under concrete slab in basement

Slab installed after treatment

Gravel fill

Pipe-and-rod hole from base of trench to top of footing aids distribution of chemical

RADON CONTROL

What is Radon?

Radon is a colorless, odorless, radioactive gas found in soils and underground water. An element with atomic weight 222, radon is produced in the natural decay of radium and exists at varying concentrations throughout the United States (see map in chapter 2). Radon is emitted from the ground to the outdoor air, where it is diluted to an insignificant level by the atmosphere. Because radon is a gas, it can travel through the soil and into a building through cracks, joints, and other openings in the foundation floor and walls. Earth-based building materials such as poured concrete, concrete blocks, brick, and adobe ordinarily are not significant sources of indoor radon. Radon from well water sometimes contributes in a minor way to radon levels in indoor air.

How Does Radon Enter Buildings?

For the most part, radon gas is drawn from the soil through the foundation when the indoor pressure is less than the pressure outside in the soil. Radon levels in buildings are usually much higher in winter due to the suction caused by

- the buoyancy of indoor air in winter
- wind on the outside
- furnaces, heaters, and fireplaces
- power exhaust fans

There are two highly effective approaches to radon management in buildings. Neither requires substantial investment, so they should become standard practice in new construction.

The Barrier Approach

The barrier approach keeps radon out by making it difficult for it to get in. An analogy is building a watertight hull to keep water out of a boat. Thus it should not be surprising that what works to keep a basement dry also works toward keeping radon out.

Since radon is a gas, the barrier approach relies on infiltration control measures such as minimizing cracks, joints, and other openings through the foundation to the soil. Waterproofing and damp-proofing membranes outside the wall and under the slab are excellent barriers that, if performing properly, cover cracks and joints.

The Suction Approach

Suction systems collect gas outside the foundation and vent it to the outdoor air. They keep radon out by creating a stronger suction than that of the building itself. They are recommended by the EPA for new construction where high potential for radon problems exists.

The suction systems consist of two parts: the collection system and the discharge system. Collection systems add little to the cost of new construction, and the discharge system can be deferred until proven necessary by tests.

The radon collection system may be independent of the moisture drainage system, or it may use it. Individual suction taps may be installed instead of underfloor pipe. One tap should be provided for every 500 square feet of floor area. A single tap is adequate for a slab poured over a 4-inch layer of clean, coarse gravel.

Radon Barrier and Suction Design

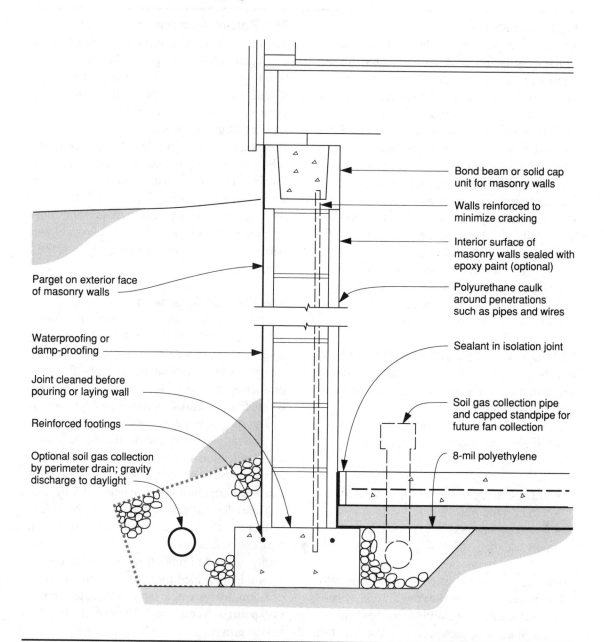

Bond beam or solid cap
unit for masonry walls

Walls reinforced to
minimize cracking

Interior surface of
masonry walls sealed with
epoxy paint (optional)

Polyurethane caulk
around penetrations
such as pipes and wires

Sealant in isolation joint

Soil gas collection pipe
and capped standpipe for
future fan collection

8-mil polyethylene

Parget on exterior face
of masonry walls

Waterproofing or
damp-proofing

Joint cleaned before
pouring or laying wall

Reinforced footings

Optional soil gas collection
by perimeter drain; gravity
discharge to daylight

CHECKLIST OF CODE REQUIREMENTS

The following is a partial list of requirements from the 1995 Council of American Building Officials (CABO) *One and Two Family Dwelling Code*. Consult the publication for the full text and additional provisions.

Drainage and Grading
(401.3) Drainage:
- surface water diverted to not cause hazard
- grade away from foundation 6" in 10' min.

(406.3.5) Final grading:
- away from foundation on all sides
- slope 1/24 (4%) min. for 6' min.

Footings
(403.1.1) Slope of footing:
- top surface level
- bottom surface sloped 1/10 (10%) max.
- footings stepped where slope >1/10 (10%)

(3103.5) Pipes through or under footing arched or in sleeves two pipe sizes larger

Concrete Floors on Ground
(505.2) All vegetation, topsoil, and foreign material removed

(505.2.1) Fill clean and compacted:
- earth 8" max.
- sand or gravel 24" max.

(505.2.1) Below-grade base course, except on well-drained or sand-gravel mixture:
- clean sand, gravel, or crushed stone or slag
- 4" thick

(505.2.3) Vapor barrier under slab. Exceptions:
- detached unheated structures
- drives, walks, and patios never to be heated
- when excepted by code official

Protection against Decay
(322.1) Pressure-treated wood required:
- joists < 18" above grade within foundation
- girders < 12" above grade within foundation
- sills or plates on concrete < 8" above grade
- siding or sheathing < 6" above grade
- basement wall furring strips without barrier
- sills and sleepers on slab without barrier
- wood on ground supporting permanent structures, except entirely below water
- posts and columns in direct contact or embedded in concrete exposed to weather

Details
(Fig. 403.1a):
- slab 3-1/2" thick min.
- 1/2" anchor bolts 6' on-center
- anchor bolts ≤ 12" from corners
- bolts 15" into masonry or 7" into concrete

Crawl Spaces
(409.1) Ventilation openings:
- net area 1/150 floor area without VB
- net area 1/1,500 floor area with VB
- opening within 3' of each corner
- corrosion-resistant 1/8" mesh wire screens

Access
(409.2) 18" × 24" access hole to crawl space

Damp-proofing
(406.1) Walls must be damp-proofed from grade to top of footing:
- masonry construction 3/8" min. parging and approved coating
- concrete approved coating

5　　　　　Wood

Wood is nature's must wonderful building material. Its combination of strength and beauty has never been surpassed in the laboratory. The first section of this chapter explores the *nature of wood*.

The beauty of wood, however, is partly due to its imperfections. So next, the chapter looks at how *lumber defects* affect its grading and how *lumber grade stamps* are interpreted.

Wood is categorized as being either softwood (from evergreens) or hardwood (from deciduous trees). The *"Properties of Wood"* table lists the qualities of 40 species.

A second table compares *moisture and shrinkage* of the same 40 wood species.

Wood will last a long time if kept dry. The building codes recognize that many outdoor and underground applications lead to decay, however. This chapter lists and illustrates the applications for which the codes call for *pressure-treated wood*.

Finally, when is a 2 × 4 really a 2 × 4? Building projects often require that we know the exact dimensions of the lumber. The table of *standard lumber sizes* lists both nominal and actual dimensions of all standard categories and sizes of lumber.

THE NATURE OF WOOD

Bark is a thick layer of dead cells, similar in function to the outer layers of human skin, that protects the living parts of the tree from insects and fire. A tree is very resistant to insects as long as its bark forms a complete barrier.

The *phloem* is the inner bark, consisting of live cells that transmit nutrients, as do the cells of the sapwood.

The *cambium* is a single layer of cells where, remarkably, all tree growth occurs. The cells of the cambium continually divide, first adding a cell to the phloem outside and then a cell to the sapwood inside. As a result, a tree limb that first appears at a height of 5 feet aboveground will remain 5 feet high, even though the tree grows taller.

Sapwood consists of the most recently formed layers of wood and, as its name implies, it carries sap up and down the tree. When the rate of growth varies throughout the year, or even ceases during cold winters, the sapwood shows annual growth rings. Wide rings are due to rapid growth in wet summers; narrow rings indicate dry summers.

Heartwood is formed of dead sapwood cells. Chemicals and minerals are deposited in and between the heartwood cells, making the wood more dense, strong, dark, and resistant to decay than the sapwood.

The *pith,* at the very center of the tree, is the overgrown remnant of the original shoot.

Rays are at right angles to the circular rings. Not defects or cracks, as they appear, rays are bundles of cells that transport and store food across the annual rings.

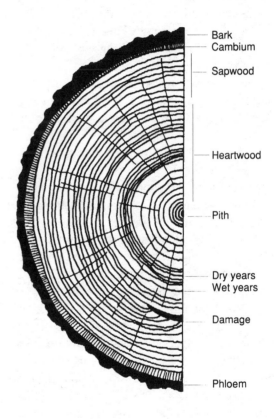

LUMBER DEFECTS

Bow is deviation from a flat plane of the wide face, end to end. It is caused by a change in moisture content after sawing and by fibers not being exactly parallel to the surfaces. It has no effect on strength. Therefore, feel free to use a piece wherever nailing will constrain it to a flat plane.

Cup is deviation from a flat plane of the narrow face, edge to edge. It is caused by a change in moisture content after surfacing. It tends to loosen fasteners.

Crook is deviation from a flat plane of the narrow face, end to end. It is caused by change of moisture content after sawing and by fibers not parallel to the surfaces. It makes wood unsuitable for framing.

Twist is deviation from a flat plane of all faces, end-to-end. It results from spiral wood grain and changes in moisture content. Twist also makes lumber unsuitable for framing.

Check is the lumber version of a stretch mark, a rift in the surface caused when the surface of a timber dries more rapidly than the interior. End checks weaken a timber in shear; other checks are mostly cosmetic. The development of check can be very dramatic and unsightly in exposed beams that are dried rapidly in a warm house the first winter. Solutions include air drying for several years before use, or treatment of timber surfaces with oil to retard the drying process.

Split passes clear through the wood and is often the result of rough handling. It constitutes a serious structural weakness. Lumber with splits should not be used in bending (joists and rafters) or in compression as a post.

Shake is a separation of growth rings. Lumber with shake should not be used to support bending loads (beams, joists, and rafters), since it must be presumed that the zone of weakness extends the entire length of the piece.

Wane is the presence of bark or lack of wood at an edge. It results from a slight miscalculation on the part of the sawyer. It has very little effect on strength. The main drawback is the lack of a full-width nailing surface.

Knots are the high-density roots of the limbs. Knots are very strong but not well connected to the surrounding wood. The rules for use in joists and rafters are 1) tight knots are allowed in the top third, 2) loose or missing knots are allowed in the middle third, and 3) no knots at all over 1 inch are allowed in the bottom third.

Cross grain occurs when a board is sawn from a crooked log. Since wood is ten times stronger in the direction of grain than across the grain, a cross-grain angle greater than one part in ten seriously weakens the wood in bending (beams, joists, and rafters).

Decay is destruction of the wood structure by fungi or insects. It prohibits structural use of the wood but may enhance its decorative value, provided the decay process has been halted.

Pitch pockets are accumulations of natural resins. They have little effect on strength but will bleed through paint and should not be allowed in lumber that will be painted.

Lumber Defects

DEFECT	END VIEW	LONG VIEW
BOW		
CUP		
CROOK		
TWIST		
CHECK		
SPLIT		
SHAKE		
WANE		
KNOT		
CROSS GRAIN		
DECAY		
PITCH POCKET		

LUMBER GRADE STAMPS

The grade stamp assures the buyer of uniform standards. There is a bewildering number of grade stamps, but all stamps from certified agencies conform to the guidelines set by the American Lumber Standards Committee.

Reading grade stamps is quite simple, as demonstrated by the example below.

Mill number where the lumber was manufactured

Certified agency under whose rules the lumber was graded

Species may be a single species or a group of species having similar characteristics

Grade:
SEL STR = select structural
1 = No. 1
2 = No. 2
3 = No. 3
CONST = construction
STAND = standard
UTIL = utility
STUD = stud

Moisture content (MC) at the time of surfacing (planing):
S-GRN = MC 20% or more
S-DRY = MC 19% or less
MC 15 = MC 15% or less

PROPERTIES OF WOOD SPECIES

Below is a table of properties of North American wood species. For all of the numerical ratings, 1 is best and 5 is worst.

Density is an indirect measure of strength. Refer to chapter 20 for tables of density versus the holding power of nails and screws.

Paintability refers to the relative ease of maintaining exterior painted surfaces such as clapboards, decking, and trim.

Cupping, or deviation of the narrow face from a flat plane, tends to loosen nails. Only species with cupping ratings of 1 or 2 should be used for exterior siding and trim.

Checking is cracking of the surface from drying too quickly. It does not usually affect strength but is unsightly.

Species	Density lb/cu ft	Paint-ability	Resistance to Cupping	Resistance to Checking	Color of Heartwood
Hardwoods					
Alder, red	25.6	3	–	–	Pale brown
Ash, white	37.4	–	–	–	Gray-brown
Aspen, quaking	24.3	3	2	1	Pale brown
Basswood, American	23.1	3	2	2	Cream
Beech, American	40.0	4	4	2	Pale brown
Birch, yellow	38.7	4	4	2	Light brown
Butternut	23.7	3-5	–	–	Light brown
Cherry, black	31.2	4	–	–	Brown
Chestnut	26.8	3-5	3	2	Light brown
Cottonwood, black	21.8	3	4	2	White
Elm, American	31.2	4-5	4	2	Brown
Hickory, pecan	41.2	4-5	4	2	Light brown
Locust, black	43.1	–	–	–	Golden brown
Magnolia, southern	31.2	3	2	–	Pale brown
Maple, sugar	39.3	4	4	2	Light brown
Oak, red	39.3	4-5	4	2	Brown

(continued)

Species	Density lb/cu ft	Paint- ability	Resistance to Cupping	Resistance to Checking	Color of Heartwood
Hardwoods—*Continued*					
Oak, white	42.4	4-5	4	2	Brown
Poplar, yellow	26.2	3	2	1	Pale brown
Sycamore, American	30.6	4	–	–	Pale brown
Walnut, black	34.3	3-5	3	2	Dark brown
Softwoods					
Cedar, Alaska	27.5	1	1	1	Yellow
Cedar, incense	23.1	1	–	–	Brown
Cedar, Port-Orford	26.8	1	–	1	Cream
Cedar, western red	20.0	1	1	1	Brown
Cedar, white	19.3	1	1	–	Light brown
Cypress	28.7	1	1	1	Light brown
Fir, Douglas	30.0	4	2	2	Pale red
Fir, white	24.3	3	2	2	White
Hemlock, eastern	25.0	3	2	2	Pale brown
Hemlock, western	28.1	3	2	2	Pale brown
Larch	32.4	4	2	2	Brown
Pine, eastern white	21.8	2	2	2	Cream
Pine, Norway	–	4	2	2	Light brown
Pine, ponderosa	25.0	3	2	2	Cream
Pine, southern	34.3	4	2	2	Light brown
Pine, sugar	22.5	2	2	2	Cream
Pine, western white	23.7	2	2	2	Cream
Redwood	22.5	1	1	1	Dark brown
Spruce, white	21.8	3	2	2	White
Tamarack	33.1	4	2	2	Brown

Source: *Wood Handbook* (Washington, DC: US Department of Agriculture, 1987).

MOISTURE AND SHRINKAGE OF WOOD

The amount of water in wood is expressed as a percentage of its oven-dry (dry as possible) weight. For example, 1 cubic foot of oven-dry red oak weighs 39.3 pounds. In drying from the just-cut, or green, stage, the sapwood loses 69 percent of 39.3, or 27.1 pounds of water.

As wood dries, it first loses moisture from within its cells without shrinking; after reaching the fiber saturation point (cells dry), further drying results in shrinkage. Eventually wood comes to dynamic equilibrium with the relative humidity of the surrounding air — interior wood typically shrinking in winter and swelling in summer. Average equilibrium moisture content ranges from 6 to 11 percent, depending on climatic region.

In the table below, the terms *radial* and *tangential* refer to orientation relative to the growth rings.

| Species | Moisture Content Green, %[1] | | Shrinkage, from Green to Oven-Dry, % | | |
	Heartwood	Sapwood	Radial	Tangential	Volume
Hardwoods					
Alder, red	–	97	4.4	7.3	12.6
Ash, white	46	44	4.9	7.8	13.3
Aspen, quaking	95	113	3.5	6.7	11.5
Basswood, American	81	133	6.6	9.3	15.8
Beech, American	55	72	5.5	11.9	17.2
Birch, yellow	74	72	7.3	9.5	16.8
Cherry, black	58	–	3.7	7.1	11.5
Chestnut, American	120	–	3.4	6.7	11.6
Cottonwood, black	162	146	3.6	8.6	12.4
Elm, American	95	92	4.2	9.5	14.6
Hickory, pecan	80	54	4.9	8.9	13.6
Locust, black	–	–	4.6	7.2	10.2
Magnolia, southern	80	104	5.4	6.6	12.3
Maple, sugar	65	72	4.8	9.9	14.7
Oak, red	80	69	4.0	8.6	13.7

(continued)

Species	Moisture Content Green, %[1]		Shrinkage, Green to Oven-Dry, %		
	Heartwood	Sapwood	Radial	Tangential	Volume
Hardwoods—Continued					
Oak, white	64	78	5.6	10.5	16.3
Poplar, yellow	83	106	4.6	8.2	12.7
Sycamore, American	114	130	5.0	8.4	14.1
Walnut, black	90	73	5.5	7.8	12.8
Softwoods					
Cedar, Alaska	32	166	2.8	6.0	9.2
Cedar, incense	40	213	3.3	5.2	7.7
Cedar, Port-Orford	50	98	4.6	6.9	10.1
Cedar, western red	58	249	2.4	5.0	6.8
Cedar, white	–	–	2.2	4.9	7.2
Cypress	121	171	3.8	6.2	10.5
Fir, Douglas	37	115	4.8	7.6	12.4
Fir, white	98	160	3.3	7.0	9.8
Hemlock	97	119	3.0	6.8	9.7
Hemlock, western	85	170	4.2	7.8	12.4
Larch, western	54	110	4.5	9.1	14.0
Pine, eastern white	–	–	2.1	6.1	8.2
Pine, lodgepole	41	120	4.3	6.7	11.1
Pine, longleaf	31	106	5.1	7.5	12.2
Pine, sugar	98	219	2.9	5.6	7.9
Pine, western white	62	148	4.1	7.4	11.8
Redwood	86	210	2.6	4.4	6.8
Spruce, Sitka	41	142	4.3	7.5	11.5
Tamarack	49	–	3.7	7.4	13.6

Source: *Wood Handbook* (Washington, DC: US Department of Agriculture, 1987).

[1]Moisture content is expressed as percentage of oven-dry weight. When the moisture in the wood weighs more than oven-dry wood, this percentage will be more than 100%.

PRESSURE-TREATED WOOD

Pressure-treated lumber is softwood lumber which has been treated by forcing chemicals into the cells of the wood. The result is a material that is immune to decay. The table below shows the retention levels required for various applications.

REQUIRED RETENTION LEVELS

Retention, pcf	Application
0.25	Above ground
0.40	Ground contact
0.60	Permanent wood foundation
2.50	Salt water

Building codes generally require the use of pressure-treated wood for the applications below and illustrated on page 117:

1. wood embedded in, or in direct contact with, earth for support of permanent structures

2. floor joists less than 18 inches and girders less than 12 inches from the ground

3. foundation plates, sills, or sleepers on a masonry slab or foundation in earth contact

4. posts or columns placed directly on masonry exposed to weather, or in basements

5. ends of girders entering concrete or masonry walls without a 1/2-inch air space

6. wood in permanent structures and located less than 6 inches from earth

7. wood structural members supporting moisture-permeable floors or roofs exposed to the weather unless separated by an impervious moisture barrier

8. retaining walls (not shown)

9. all-weather wood foundations (see details in chapter 4)

10. in hot and humid areas, structural supports of buildings, balconies, porches, etc., when exposed to the weather without protection from a roof, overhang, or other covering to prevent moisture accumulation

Typical Quality Stamp for Pressure-Treated Lumber

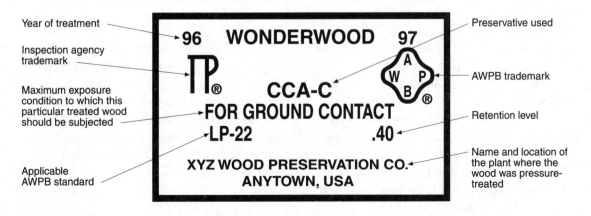

Inorganic Arsenical Pressure-Treated Wood (Including Chromated Copper Arsenate [CCA], Ammoniacal Copper Quat [ACQ], and Ammoniacal Copper Zinc Arsenate [ACZA])

Consumer Information

This wood has been preserved by pressure-treatment with an Environmental Protection Agency (EPA)-registered pesticide containing inorganic arsenic to protect it from insect attack and decay. Wood treated with inorganic arsenic should be used only where such protection is important.

Inorganic arsenic penetrates deeply into the pressure-treated wood and remains there for a long time. Exposure to inorganic arsenic may present certain hazards. Therefore, the following precautions should be taken both when handling the treated wood and in determining where to use or dispose of the treated wood.

Use Site Precautions

Wood that is pressure-treated with waterborne arsenical preservatives may be used inside residences as long as all sawdust and construction debris are cleaned up and disposed of after construction.

Do not use treated wood under circumstances where the preservative may become a component of food or animal feed. Examples of such sites would be structures or containers for storing silage or food.

Do not use treated wood for cutting boards or countertops.

Only treated wood that is visibly clean and free of surface residue should be used for patios, decks, and walkways.

Do not use treated wood for construction of those portions of beehives that may come into contact with the honey.

Treated wood should not be used where it may come into direct or indirect contact with public drinking water, except for uses involving incidental contact such as docks and bridges.

Handling Precautions

Dispose of treated wood by ordinary trash collection or burial. Treated wood should not be burned in open fires or in stoves, fireplaces, or residential boilers because toxic chemicals may be produced as part of the smoke and ashes. Treated wood from commercial or industrial use (e.g., construction sites) may be burned only in commercial or industrial incinerators or boilers in accordance with state and federal regulations.

Avoid frequent or prolonged inhalation of sawdust from treated wood. When sawing and machining treated wood, wear a dust mask. Whenever possible, these operations should be performed outdoors to avoid indoor accumulations of airborne sawdust from treated wood.

When power-sawing and machining, wear goggles to protect eyes from flying particles.

After working with the wood, and before eating, drinking, or using tobacco products, wash exposed areas thoroughly.

If preservative or sawdust accumulates on clothes, launder before reuse. Wash work clothes separately from other household clothing.

Source: US Environmental Protection Agency, 1987.

Pressure-Treated Wood Applications
(numbers refer to page 115)

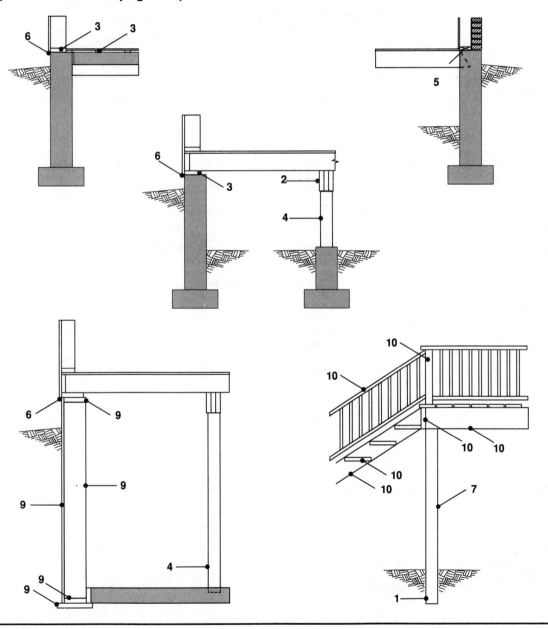

STANDARD LUMBER SIZES

STANDARD LUMBER SIZES BASED ON WESTERN WOOD PRODUCTS ASSOCIATION RULES

Product	Nominal Dimensions inches		Dressed Dimensions When Surfaced, inches	
	Thickness	**Width**	**Dry**	**Green**
Dimension	2	2	1-1/2	1-9/16
	3	3	2-1/2	2-9/16
	4	4	3-1/2	3-9/16
		5	4-1/2	4-5/8
		6	5-1/2	5-5/8
		8	7-1/4	7-1/2
		10	9-1/4	9-1/2
		12	11-1/4	11-1/2
		>12	Off 3/4	Off 1/2
Scaffold plank	1-1/4 and thicker	8 and wider	Same as for "dimension"	
Timbers	5 and larger	5 and larger	Nominal less 1/2"	
			Thickness	**Width**
Decking	2	5	1-1/2	4
		6		5
		8		6-3/4
		10		8-3/4
		12		10-3/4
	3	6	2-1/2	5-1/4
	4		3-1/2	
Flooring	3/8	2	5/16	1-1/8
	1/2	3	7/16	2-1/8
	5/8	4	9/16	3-1/8
	1	5	3/4	4-1/8
	1-1/4	6	1	5-1/8
	1-1/2		1-1/4	
Ceiling and partition	3/8	3	5/16	2-1/8
	1/2	4	7/16	3-1/8
	5/8	5	9/16	4-1/8
	3/4	6	11/16	5-1/8
Factory and shop lumber (mostly hardwoods used in woodworking)	1 (4/4)	5	3/4	Usually
	1-1/4 (5/4)	and	1-5/32	sold
	1-1/2 (6/4)	wider	1-13/32	random
	1-3/4 (7/4)		1-19/32	width
	2 (8/4)		1-13/16	
	2-1/2 (10/4)		2-3/8	
	3 (12/4)		2-3/4	
	4 (16/4)		3-3/4	

STANDARD LUMBER SIZES —*Continued*

Product	Nominal Dimensions inches		Dry Dressed Dimensions inches	
	Thickness	Width	Thickness	Width
Selects and commons	4/4	2	3/4	1-1/2
	5/4	3	1-5/32	2-1/2
	6/4	4	1-13/32	3-1/2
	7/4	5	1-19/32	4-1/2
	8/4	6	1-13/16	5-1/2
	9/4	7	2-3/32	6-1/2
	10/4	8 and	2-3/8	3/4 off
	11/4	wider	2-9/16	nominal
	12/4		2-3/4	
	16/4		3-3/4	
Finish and boards	3/8	2	5/16	1-1/2
	1/2	3	7/16	2-1/2
	5/8	4	9/16	3-1/2
	3/4	5	5/8	4-1/2
	1	6	3/4	5-1/2
	1-1/4	7	1	6-1/2
	1-1/2	8 and	1-1/4	3/4 off
	1-3/4	wider	1-3/8	nominal
	2		1-1/2	
	2-1/2		2	
	3		2-1/2	
	3-1/2		3	
	4		3-1/2	
Rustic and drop siding	1	6	23/32	5-3/8
		8		7-1/8
		10		9-1/8
		12		11-1/8
Paneling and siding	1	6	23/32	5-7/16
		8		7-1/8
		10		9-1/8
		12		11-1/8
Ceiling and partition	5/8	4	9/16	3-3/8
	1	6	23/32	5-3/8
Bevel siding	1/2	4	15/32 butt,	3-1/2
		5	3/16 tip	4-1/2
		6		5-1/2
	3/4	8	3/4 butt,	7-1/4
		10	3/16 tip	9-1/4
		12		11-1/4

Source: *Product Use Manual* (Portland, Oreg: Western Wood Products Association, 1986).

6

Framing

To make a human analogy, the frame of a building is its skeleton. To assure that this skeleton is strong enough, we need to specify how much weight it will support—the *building loads*. Knowing the loads and the *design values* (maximum allowable loads) *for structural lumber*, we can consult the *span tables for rough lumber* or for *surfaced-four-sides (S4S) lumber* to select the proper size joists and rafters.

Builders are increasingly turning to *trusses* and *prefabricated-wood I-joists*, so span tables for both are included here.

Normal sheathing materials are covered in chapter 7. However, *stressed-skin panels*, a modern development structurally combining the strengths of joists or rafters and sheathing, are discussed in this chapter.

Span tables are given for *plank floors and roofs*, often used with the greater-than-usual joist and rafter spacing that results from post-and-beam framing.

When loads become too large for ordinary timbers, you can turn to *glued laminated beams* and even *steel beams*. This chapter includes simple span tables for both.

The section on *post-and-beam framing* traces the origins of this revitalized art and illustrates the joinery that many find so beautiful.

If you are remodeling an older home, you may be dealing with *balloon framing*. If your home is less than 50 years old, you will probably recognize its *platform framing*.

Regardless of your framing system, you should find the section on *framing details* helpful—especially the section on framing for superinsulation.

Finally, we provide you with a *checklist of code requirements* relating to framing.

BUILDING LOADS

Types of Load

Building loads are the weights a building frame is required to support or resist. There are four types of load:

- dead load – the weight of the building materials alone
- live load – the additional weight due to occupancy; i.e., the weights of people, furnishings, and stored materials
- snow load – the weight of accumulated snow on the roof of the building
- wind load – the force of wind pressure against the exterior building surfaces

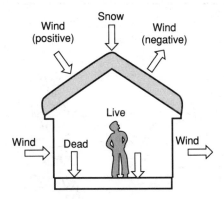

Dead Loads

For floors and roofs, add the weights of building components, using the table at right to find the total dead load in pounds per square foot (psf).

Example: A first floor framed with 2x8 joists, spaced 16 inches on-center, and covered with 1/2-inch plywood and wall-to-wall carpet weighs 3 psf + 1.5 psf + 0.5 psf = 5 psf.

WEIGHTS OF BUILDING MATERIALS

Component	Material	Load, psf
Framing (16" oc)	2x4 and 2x6	2
	2x8 and 2x10	3
Floor-ceiling	Softwood, per inch	3
	Hardwood, per inch	4
	Plywood, per inch	3
	Concrete, per inch	12
	Stone, per inch	13
	Carpet	0.5
	Drywall, per inch	5
Roofing	Softwood, per inch	3
	Plywood, per inch	3
	Foam insulation, per inch	0.2
	Asphalt shingle	3
	Asphalt roll roofing	1
	Asphalt, built-up	6
	Wood shingle	3
	Copper	1
	Steel	2
	Slate, 3/8"	12
	Roman tile	12
	Spanish tile	19

Live Loads

Live loads are specified by building code. The table on the following page is representative of the major national codes, but verify particulars with your local building inspector.

Snow Load

Snow load is the greater of the following:
- *minimum roof live load,* including the effects of snow and roof construction/repair activities (found in the table at right)
- *design snow load,* 0.8 times the 50-year ground snow load (from the map below)

Example: What is the snow load for a house in Boston with a roof slope of 7/12?

From the table at right, the minimum roof live load for slope = 7/12 is 16 psf.

From the map below, the 50-year ground snow load for Boston is 40 psf. The product of 0.8 and 40 psf is 32 psf, so the greater snow load (and the answer) is 32 psf.

RESIDENTIAL LIVE LOADS

Area/Activity	Live Load, psf
First floor	40
Second floor and habitable attics	30
Balconies, fire escapes, and stairs	100
Garages	50

MINIMUM ROOF LIVE LOADS

Roof Slope	Load, psf
Less than 4/12	20
4/12 to less than 12/12	16
12/12 or greater	12

50-Year Ground Snow Load (psf)

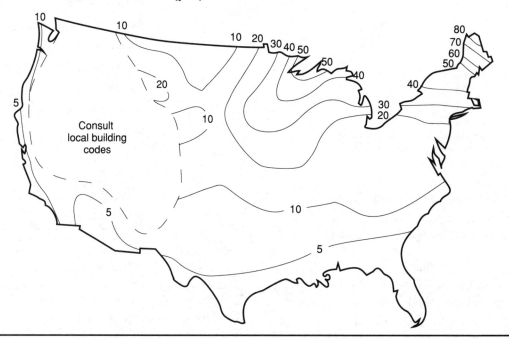

DESIGN VALUES FOR STRUCTURAL LUMBER

Below are the four types of stress in wood structures. The following pages list the maximum allowable stresses (design values) for joists, rafters, and planks.

Fiber Stress in Bending, Fb
Loads at right angles to structural members cause the members to bend, resulting in tension in the fibers on the outside of the bend. Fb is one of the two common criteria used to size joists, rafters, and planks. Single-member Fb values are used when a load is carried by one or two members. Repetitive-member Fb values may be used when at least three members share the load, spacing is 24 inches or less, and members are connected by load-distributing elements.

Modulus of Elasticity, E
The modulus of elasticity predicts how much a beam will deflect under a bending load. It is the second common joist criterion, determining the "bounciness" of a floor.

Horizontal Shear, Fv
Horizontal shear tends to slide wood fibers over each other horizontally. Short beams carrying heavy loads should be checked for shear.

Compression Parallel to Grain, Fc
Studs, posts, and columns are compressed parallel to the grain of the wood. Loads on wall studs are usually very low, but posts and columns carrying heavy loads should be checked.

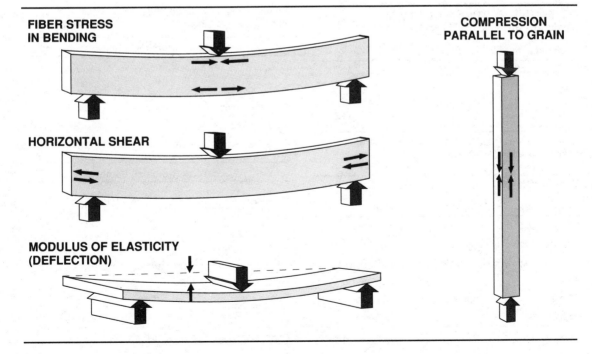

FIBER STRESS
IN BENDING

HORIZONTAL SHEAR

MODULUS OF ELASTICITY
(DEFLECTION)

COMPRESSION
PARALLEL TO GRAIN

Stress Value Adjustment Factors

Since different sizes of visually graded lumber have different values, the design values are tabulated as *base values* (see table on opposite page).

The designer selects a base value for a given species group and grade. Then the designer multiplies the base value by a size factor, C_F, to determine the value for a given size.

Example: For DF-L No. 2 grade 2×4 size

$$F_b = F_b(\text{base value}) \times C_F$$
$$= 875 \text{ psi} \times 1.5$$
$$= 1,312 \text{ psi}$$

Similar adjustments are made for repetitive members (C_r), duration of load (C_D), and horizontal shear (C_H). The tables for each adjustment factor follow.

REPETITIVE MEMBER FACTOR, C_r
(applies to tabulated design values for Fb)

Type of Member	Factor
Joists and rafters	1.15
Headers*	
2 plies	1.10
≥ 3 plies	1.15

*Header factors based on American Forest & Paper Association's header tables.

DURATION OF LOAD ADJUSTMENT, C_D

Load Duration	Factor
Permanent	0.9
10 years (normal load)	1.0
2 months (snow load)	1.15
7 days	1.25
10 minutes (wind, earthquake)	1.6
Impact	2.0

SIZE FACTORS, C_F

Grade	Nom. Width	Fb<4" Thick	Fb 4" Thick	Ft	Fc
Sel str,	≤ 4"	1.5	1.5	1.5	1.15
No.1,	5"	1.4	1.4	1.4	1.1
No. 2,	6"	1.3	1.3	1.3	1.1
and No. 3	8"	1.2	1.3	1.2	1.05
	10"	1.1	1.2	1.1	1.0
	12"	1.0	1.1	1.0	1.0
	≥ 14"	0.9	1.0	0.9	0.9
Const and std	≤ 4"	1.0	1.0	1.0	1.0
Utility	4"	1.0	1.0	1.0	1.0
Stud	≤ 4"	1.1	1.1	1.1	1.05
	5" and 6"	1.0	1.0	1.0	1.0

HORIZONTAL SHEAR ADJUSTMENT, C_H

When Length of Split on Wide Face Does Not Exceed	Multiply Fv Value by
2" Thick (Nominal) Lumber	
No split	2.00
1/2 x wide face	1.67
3/4 x wide face	1.50
1 x wide face	1.33
1-1/2 x wide face or more	1.00
3" Thick (Nominal) Lumber	
No split	2.00
1/2 x wide face	1.67
1 x wide face	1.33
1-1/2 x wide face or more	1.00

Source: *The U.S. Span Book for Major Lumber Species* (Ottawa, Ont: Canadian Wood Council, 1996).

BASE DESIGN VALUES FOR US DIMENSION LUMBER
(2"–4" thick x 2" and wider[1])

Species Group	Grade	Extreme Fiber Stress in Bending Fb	Tension Parallel to Grain Ft	Horizontal Shear Fv	Compression Perpendicular to Grain Fc⊥	Parallel to Grain Fc	Modulus of Elasticity (million psi) E
Douglas fir-larch	Sel str	1,450	1,000	95	625	1,700	1.9
	No. 1	1,000	675	95	625	1,450	1.7
	No. 2	875	575	95	625	1,300	1.6
	No. 3	500	325	95	625	750	1.4
	Const	1,000	650	95	625	1,600	1.5
	Std	550	375	95	625	1,350	1.4
	Utility	275	175	95	625	875	1.3
	Stud	675	450	95	625	825	1.4
Hem-fir	Sel str	1,400	900	75	405	1,500	1.6
	No. 1	950	600	75	405	1,300	1.5
	No. 2	850	500	75	405	1,250	1.3
	No. 3	500	300	75	405	725	1.2
	Const	975	575	75	405	1,500	1.3
	Std	550	325	75	405	1,300	1.2
	Utility	250	150	75	405	850	1.1
	Stud	675	400	75	405	800	1.2

ALLOWABLE DESIGN VALUES FOR SOUTHERN PINE[2]

Size Classification	Grade	Extreme Fiber Stress in Bending Fb	Tension Parallel to Grain Ft	Horizontal Shear Fv	Compression Perpendicular to Grain Fc⊥	Parallel to Grain Fc	Modulus of Elasticity (million psi) E
2"–4" thick, 2"–4" wide	Sel str	2,850	1,600	100	565	2,100	1.8
	No.1	1,850	1,050	100	565	1,850	1.7
	No. 2	1,500	825	90	565	1,650	1.6
	No. 3	850	475	90	565	975	1.4
	Stud	875	500	90	565	975	1.4
2"–4" thick, 4" wide	Const	1,100	625	100	565	1,800	1.5
	Std	625	350	90	565	1,500	1.3
	Utility	300	175	90	565	975	1.3
2"–4" thick, 5"–6" wide	Sel str	2,550	1,400	90	565	2,000	1.8
	No.1	1,650	900	90	565	1,750	1.7
	No. 2	1,250	725	90	565	1,600	1.6
	No. 3	750	425	90	565	925	1.4
	Stud	775	425	90	565	925	1.4
2"–4" thick, 8" wide	Sel str	2,300	1,300	90	565	1,900	1.8
	No.1	1,500	825	90	565	1,650	1.7
	No. 2	1,200	650	90	565	1,550	1.6
	No. 3	700	400	90	565	875	1.4

[1] Use with adjustment factors on previous page.
[2] Use with adjustment factors on previous page except for size factor (CF).

SPAN TABLES FOR ROUGH LUMBER

On the following pages are span tables for rough-sawn joists and rafters, assuming uniform loading over a single span. Use the tables as follows:

1. Determine the sum of dead and live loads.
2. Estimate design stress values for the species and quality of lumber you plan to use.
3. Multiply the design stress values by any applicable stress adjustment factors.
4. Find in the appropriate rough-sawn span table the size of framing lumber required.
5. If you are designing floor joists, consult the deflection table below to see whether the floor will be stiff enough.

Example: What size rough-sawn hemlock joists are required for a first floor with a clear span of 14 feet 3 inches, on-center spacing of 16 inches, and a structural subfloor if the quality is equivalent to No. 2 hem-fir?

To be conservative, allow 10 psf for the weight of the floor (dead load). The code specifies a live load of 40 psf, so the total load is 50 psf. Repetitive-member $F_b = 1,150$ psi and $E = 1.4 \times 10^6$ psi. No multipliers apply. From the span table for rough-sawn lumber, you can find that 2x8 joists will span nearly 16 feet, and in the table below you can find that the 2x8 deflection ratio is less than 1/360 up to a span of 14 feet 8 inches.

ROUGH-SAWN LUMBER: SPAN LIMITED BY 1/360 DEFLECTION AT 40 PSF LIVE LOAD

Spacing inches oc	Actual Size b x d	$E, 10^6$ psi					
		1.0	1.2	1.4	1.6	1.8	2.0
12	2 x 6	11 - 00	11 - 08	12 - 04	12 - 11	13 - 05	13 - 10
	2 x 8	14 - 08	15 - 07	16 - 04	17 - 02	17 - 09	18 - 06
	2 x 10	18 - 04	19 - 06	20 - 06	21 - 05	22 - 04	23 - 01
	2 x 12	22 - 00	23 - 04	24 - 08	25 - 06	26 - 09	27 - 09
16	2 x 6	10 - 00	10 - 07	11 - 03	11 - 08	12 - 03	12 - 07
	2 x 8	13 - 04	14 - 02	14 - 08	15 - 07	16 - 02	16 - 10
	2 x 10	16 - 08	17 - 09	18 - 08	19 - 06	20 - 03	21 - 00
	2 x 12	19 - 11	21 - 03	22 - 04	23 - 04	24 - 04	25 - 01
24	2 x 6	8 - 07	9 - 04	9 - 09	10 - 03	10 - 07	10 - 10
	2 x 8	11 - 08	12 - 04	13 - 04	13 - 08	14 - 02	14 - 08
	2 x 10	14 - 06	15 - 05	16 - 01	17 - 01	17 - 09	18 - 03
	2 x 12	17 - 06	18 - 07	19 - 07	20 - 05	21 - 03	22 - 01

Note: Spans are given as feet - inches.

ROUGH-SAWN LUMBER: SPANS FOR TOTAL UNIFORM DEAD PLUS LIVE LOAD 30 PSF

Spacing inches oc	Actual Size b x d	Fb, psi					
		1,000	1,200	1,400	1,600	1,800	2,000
16	2 x 4	9 - 05	10 - 04	11 - 02	11 - 11	12 - 08	13 - 04
	2 x 6	14 - 02	15 - 06	16 - 09	17 - 11	19 - 00	20 - 00
	2 x 8	18 - 10	20 - 08	22 - 04	23 - 10	25 - 03	26 - 08
	2 x 10	23 - 07	25 - 10	27 - 10	29 - 10	31 - 08	33 - 04
	2 x 12	28 - 03	31 - 00	33 - 06	35 - 09	37 - 11	39 - 11
24	2 x 4	7 - 08	8 - 05	9 - 01	9 - 08	10 - 03	10 - 10
	2 x 6	11 - 07	12 - 08	13 - 08	14 - 08	15 - 06	16 - 05
	2 x 8	15 - 05	16 - 10	18 - 02	19 - 06	20 - 08	21 - 10
	2 x 10	19 - 03	21 - 01	22 - 09	24 - 04	25 - 10	27 - 03
	2 x 12	23 - 01	25 - 04	27 - 04	29 - 02	31 - 00	32 - 08
48	3 x 4	6 - 08	7 - 04	7 - 11	8 - 05	8 - 11	9 - 05
	3 x 6	10 - 00	10 - 11	11 - 10	12 - 08	13 - 05	14 - 02
	3 x 8	13 - 04	14 - 07	15 - 09	16 - 10	17 - 11	18 - 10
	3 x 10	16 - 08	18 - 03	19 - 09	21 - 01	22 - 04	23 - 07
	3 x 12	20 - 00	21 - 11	23 - 08	25 - 04	26 - 10	28 - 03
	4 x 4	7 - 08	8 - 05	9 - 01	8 - 08	10 - 04	10 - 10
	4 x 6	11 - 07	12 - 08	13 - 08	14 - 08	15 - 06	16 - 05
	4 x 8	15 - 05	16 - 10	18 - 03	19 - 06	20 - 08	21 - 10
	4 x 10	19 - 03	21 - 01	22 - 09	24 - 04	25 - 10	27 - 03
	4 x 12	23 - 01	25 - 04	27 - 04	29 - 02	31 - 00	32 - 08
96	6 x 6	10 - 00	10 - 11	11 - 10	12 - 09	13 - 06	14 - 03
	6 x 8	13 - 04	14 - 07	15 - 09	16 - 10	17 - 11	18 - 10
	6 x 10	16 - 08	18 - 03	19 - 09	21 - 01	22 - 04	23 - 07
	6 x 12	20 - 00	21 - 11	23 - 08	25 - 04	26 - 10	28 - 03
	8 x 8	15 - 05	16 - 10	18 - 03	19 - 06	20 - 08	21 - 10
	8 x 10	19 - 03	21 - 01	22 - 09	24 - 04	25 - 10	27 - 03
	8 x 12	23 - 01	25 - 04	27 - 04	29 - 02	31 - 00	32 - 08

Note: Spans are given as feet - inches.

ROUGH-SAWN LUMBER: SPANS FOR TOTAL UNIFORM DEAD PLUS LIVE LOAD 40 PSF

Spacing inches oc	Actual Size b x d	Fb, psi					
		1,000	1,200	1,400	1,600	1,800	2,000
16	2 x 4	8 - 02	8 - 11	9 - 08	10 - 04	10 - 11	11 - 07
	2 x 6	12 - 03	13 - 05	14 - 06	15 - 06	16 - 05	17 - 04
	2 x 8	16 - 04	17 - 10	19 - 04	20 - 08	21 - 11	23 - 01
	2 x 10	20 - 05	22 - 04	24 - 02	25 - 10	27 - 05	28 - 10
	2 x 12	24 - 06	26 - 10	29 - 00	31 - 00	32 - 10	34 - 08
24	2 x 4	6 - 08	7 - 04	7 - 11	8 - 05	8 - 11	9 - 05
	2 x 6	10 - 00	10 - 11	11 - 10	12 - 08	13 - 05	14 - 02
	2 x 8	13 - 04	14 - 07	15 - 09	16 - 10	17 - 11	18 - 10
	2 x 10	16 - 08	18 - 03	19 - 09	21 - 01	22 - 04	23 - 07
	2 x 12	20 - 00	21 - 11	23 - 08	25 - 04	26 - 10	28 - 03
48	3 x 4	5 - 09	6 - 04	6 - 10	7 - 03	7 - 09	8 - 02
	3 x 6	8 - 08	9 - 06	10 - 03	11 - 00	11 - 07	12 - 03
	3 x 8	11 - 07	12 - 08	13 - 08	14 - 08	15 - 06	16 - 04
	3 x 10	14 - 05	15 - 10	17 - 01	18 - 03	19 - 04	20 - 05
	3 x 12	17 - 04	19 - 00	20 - 06	21 - 11	23 - 03	24 - 06
	4 x 4	6 - 08	7 - 04	7 - 11	8 - 05	8 - 11	9 - 05
	4 x 6	10 - 00	10 - 11	11 - 10	12 - 08	13 - 05	14 - 02
	4 x 8	13 - 04	14 - 07	15 - 09	16 - 10	17 - 11	18 - 10
	4 x 10	16 - 08	18 - 03	19 - 09	21 - 01	22 - 04	23 - 07
	4 x 12	20 - 00	21 - 11	23 - 08	25 - 04	26 - 10	28 - 03
96	6 x 6	8 - 08	9 - 06	10 - 03	11 - 00	11 - 07	12 - 03
	6 x 8	11 - 07	12 - 08	13 - 08	14 - 08	15 - 06	16 - 04
	6 x 10	14 - 05	15 - 10	17 - 01	18 - 03	19 - 04	20 - 05
	6 x 12	17 - 04	19 - 00	20 - 06	21 - 11	23 - 03	24 - 06
	8 x 8	13 - 04	14 - 07	15 - 09	16 - 10	17 - 11	18 - 10
	8 x 10	16 - 08	18 - 03	19 - 09	21 - 01	22 - 04	23 - 07
	8 x 12	20 - 00	21 - 11	23 - 08	25 - 04	26 - 10	28 - 03

Note: Spans are given as feet - inches.

ROUGH-SAWN LUMBER: SPANS FOR TOTAL UNIFORM DEAD PLUS LIVE LOAD 50 PSF

Spacing inches oc	Actual Size b x d	Fb, psi					
		1,000	1,200	1,400	1,600	1,800	2,000
16	2 x 4	7 - 04	8 - 00	8 - 08	9 - 03	9 - 10	10 - 04
	2 x 6	10 - 11	12 - 00	13 - 00	13 - 10	14 - 08	15 - 05
	2 x 8	14 - 07	16 - 00	17 - 03	18 - 05	19 - 07	20 - 07
	2 x 10	18 - 03	20 - 00	21 - 07	23 - 01	24 - 06	25 - 10
	2 x 12	21 - 11	24 - 00	25 - 11	27 - 09	29 - 05	31 - 00
24	2 x 4	6 - 00	6 - 06	7 - 01	7 - 07	8 - 01	8 - 06
	2 x 6	8 - 11	9 - 10	10 - 07	11 - 03	12 - 00	12 - 07
	2 x 8	11 - 11	13 - 01	14 - 01	15 - 01	16 - 00	16 - 10
	2 x 10	14 - 11	16 - 04	17 - 08	18 - 10	20 - 00	21 - 01
	2 x 12	17 - 11	19 - 07	21 - 02	22 - 08	24 - 00	25 - 04
48	3 x 4	5 - 02	5 - 08	6 - 01	6 - 06	6 - 11	7 - 04
	3 x 6	7 - 09	8 - 06	9 - 02	9 - 10	10 - 05	10 - 11
	3 x 8	10 - 04	11 - 04	12 - 03	13 - 01	13 - 10	14 - 07
	3 x 10	12 - 11	14 - 02	15 - 03	16 - 04	17 - 04	18 - 03
	3 x 12	15 - 06	17 - 00	18 - 04	19 - 07	20 - 09	21 - 11
	4 x 4	6 - 00	6 - 06	7 - 01	7 - 07	8 - 01	8 - 06
	4 x 6	8 - 11	9 - 10	10 - 07	11 - 03	12 - 00	12 - 07
	4 x 8	11 - 11	13 - 01	14 - 01	15 - 01	16 - 00	16 - 10
	4 x 10	14 - 11	16 - 04	17 - 08	18 - 10	20 - 00	21 - 01
	4 x 12	17 - 11	19 - 07	21 - 02	22 - 08	24 - 00	25 - 04
96	6 x 6	7 - 09	8 - 06	9 - 02	9 - 10	10 - 05	10 - 11
	6 x 8	10 - 04	11 - 04	12 - 03	13 - 01	13 - 10	14 - 07
	6 x 10	12 - 11	14 - 02	15 - 03	16 - 04	17 - 04	18 - 03
	6 x 12	15 - 06	17 - 00	18 - 04	19 - 07	20 - 09	21 - 11
	8 x 8	11 - 11	13 - 01	14 - 01	15 - 01	16 - 00	16 - 10
	8 x 10	14 - 11	16 - 04	17 - 08	18 - 10	20 - 00	21 - 01
	8 x 12	17 - 11	19 - 07	21 - 02	22 - 08	24 - 00	25 - 04

Note: Spans are given as feet - inches.

ROUGH-SAWN LUMBER: SPANS FOR TOTAL UNIFORM DEAD PLUS LIVE LOAD 60 PSF

Spacing inches oc	Actual Size b x d	Fb, psi					
		1,000	1,200	1,400	1,600	1,800	2,000
16	2 x 4	6 - 08	7 - 04	7 - 11	8 - 05	8 - 11	9 - 05
	2 x 6	10 - 00	10 - 11	11 - 10	12 - 08	13 - 05	14 - 02
	2 x 8	13 - 04	14 - 07	15 - 09	16 - 10	17 - 11	18 - 10
	2 x 10	16 - 08	18 - 03	19 - 09	21 - 01	22 - 04	23 - 07
	2 x 12	20 - 00	21 - 11	23 - 08	25 - 04	26 - 10	28 - 03
24	2 x 4	5 - 05	6 - 00	6 - 05	6 - 10	7 - 03	7 - 11
	2 x 6	8 - 02	8 - 11	9 - 08	10 - 04	10 - 11	11 - 07
	2 x 8	10 - 11	11 - 11	12 - 10	13 - 10	14 - 08	15 - 05
	2 x 10	13 - 07	14 - 11	16 - 01	15 - 01	16 - 00	16 - 10
	2 x 12	16 - 04	17 - 11	19 - 04	20 - 08	21 - 11	23 - 01
48	3 x 4	4 - 09	5 - 02	5 - 07	6 - 00	6 - 04	6 - 09
	3 x 6	7 - 01	7 - 09	8 - 04	9 - 00	9 - 06	10 - 00
	3 x 8	9 - 05	10 - 04	11 - 02	11 - 11	12 - 08	13 - 04
	3 x 10	11 - 09	12 - 11	13 - 11	14 - 10	15 - 09	16 - 07
	3 x 12	14 - 02	15 - 06	16 - 09	17 - 11	19 - 00	20 - 00
	4 x 4	5 - 05	6 - 00	6 - 05	6 - 10	7 - 03	7 - 11
	4 x 6	8 - 02	8 - 11	9 - 08	10 - 04	10 - 11	11 - 07
	4 x 8	10 - 11	11 - 11	12 - 10	13 - 10	14 - 08	15 - 05
	4 x 10	13 - 07	14 - 11	16 - 01	15 - 01	16 - 00	16 - 10
	4 x 12	16 - 04	17 - 11	19 - 04	20 - 08	21 - 11	23 - 01
96	6 x 6	7 - 01	7 - 09	8 - 04	9 - 00	9 - 06	10 - 00
	6 x 8	9 - 05	10 - 04	11 - 02	11 - 11	12 - 08	13 - 04
	6 x 10	11 - 09	12 - 11	13 - 11	14 - 10	15 - 09	16 - 07
	6 x 12	14 - 02	15 - 06	16 - 09	17 - 11	19 - 00	20 - 00
	8 x 8	10 - 11	11 - 11	12 - 10	13 - 10	14 - 08	15 - 05
	8 x 10	13 - 07	14 - 11	16 - 01	15 - 01	16 - 00	16 - 10
	8 x 12	16 - 04	17 - 11	19 - 04	20 - 08	21 - 11	23 - 01

Note: Spans are given as feet - inches.

SPAN TABLES FOR SURFACED-FOUR-SIDES (S4S) LUMBER

The span tables on pages 133–164 are adapted from *The U.S. Span Book for Major Lumber Species.* The book, available from the Canadian Wood Council, was created in conformance with the procedures of US grading and building authorities and with the design methods of the American Forest and Paper Association's *National Design Specification for Wood Construction.*

The book provides a convenient reference for spans for common species of US and Canadian dimension lumber, fully in accord with US building codes and Federal Housing Authority (FHA) requirements.

Allowable Span

The allowable span is the maximum clear horizontal distance between supports. For horizontal members such as joists, the clear span is the actual length between supports (see illustration at right). For sloping members such as roof rafters, a factor must be applied to the horizontal rafter span to determine the actual clear length or sloping distance. The table at right provides a method for converting horizontal span to sloping distance, and vice versa.

Spans are based on the use of lumber in dry conditions, as in most covered structures. They have been calculated using strength and stiffness values adjusted for size, repetitive member use, and appropriate load duration.

Maximum allowable spans in some cases (e.g., rafters or ceiling joists) may exceed available lengths. Availability of all grades and sizes should be confirmed before specifying for a project.

Definition of Span

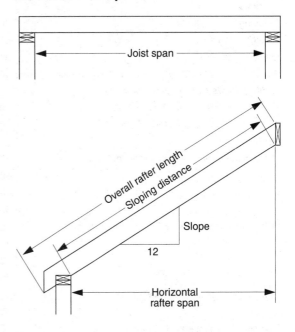

CONVERSION FACTORS FOR RAFTERS

Slope in 12	Slope Factor	Slope in 12	Slope Factor
3	1.031	12	1.414
4	1.054	13	1.474
5	1.083	14	1.537
6	1.118	15	1.601
7	1.158	16	1.667
8	1.202	17	1.734
9	1.250	18	1.803
10	1.302	19	1.873
11	1.357	20	1.944

To convert rafter span to sloping distance:
1. Find slope factor for given slope in table above.
2. Multiply horizontal span by slope factor.

Source: *The U.S. Span Book for Major Lumber Species* (Ottawa, Ont: Canadian Wood Council, 1996).

DESIGN ASSUMPTIONS FOR S4S SPAN TABLES

Application	Live Load psf	Dead Load psf	Deflection Limit under Live Load[1]	Page
Floor Joists				
Sleeping rooms and attics	30	10	Span/360	133, 149
All other rooms	40	10	Span/360	133, 149
Office space	50	10	Span/360	134, 150
All rooms with ≤ 1-1/2" lightweight concrete topping	40	20	Span/360	134, 150
Ceiling Joists (with Drywall Ceiling)				
No future rooms and no attic storage	10	5	Span/240	135, 151
No future rooms and limited storage	20	10	Span/240	135, 151
Rafters				
With drywall ceiling and light roof covering	20	10	Span/240	136, 152
	30	10	Span/240	136, 152
	40	10	Span/240	137, 153
	50	10	Span/240	137, 153
With drywall ceiling and medium roof covering	20	15	Span/240	138, 154
	30	15	Span/240	138, 154
	40	15	Span/240	139, 155
	50	15	Span/240	139, 155
With drywall ceiling and heavy roof covering	20	20	Span/240	140, 156
	30	20	Span/240	140, 156
No finished ceiling and light roof covering	20	10	Span/180	141, 157
	30	10	Span/180	141, 157
	40	10	Span/180	142, 158
	50	10	Span/180	142, 158
No finished ceiling and medium roof covering	20	15	Span/180	143, 159
	30	15	Span/180	143, 159
	40	15	Span/180	144, 160
	50	15	Span/180	144, 160
No finished ceiling and heavy roof covering	20	20	Span/180	145, 161
	30	20	Span/180	145, 161
Headers				
Supported roofs	20	20	Span/360	146, 162
	30	20	Span/360	↓ ↓
	40	20	Span/360	
	50	20	Span/360	
Supported floors	40	10	Span/360	
Supported walls (up to 9' height)	–	11	Span/360	148, 164

[1] Span/360 means a vertical deflection of 1/360 of the clear span.

Floor Joists

SLEEPING ROOMS AND ATTICS: 30 PSF LIVE, 10 PSF DEAD
Maximum Allowable Span, feet - inches

Species Group	Spacing inches oc	2 × 6 Sel Str	No.1	No.2	No.3	2 × 8 Sel Str	No.1	No.2	No.3	2 × 10 Sel Str	No.1	No.2	No.3	2 × 12 Sel Str	No.1	No.2	No.3
Douglas fir-larch	12	12-6	12-0	11-10	9-8	16-6	15-10	15-7	12-4	21-0	20-3	19-10	15-0	25-7	24-8	23-0	17-5
	16	11-4	10-11	10-9	8-5	15-0	14-5	14-1	10-8	19-1	18-5	17-2	13-0	23-3	21-4	19-11	15-1
	19.2	10-8	10-4	10-1	7-8	14-1	13-7	12-10	9-9	18-0	16-9	15-8	11-10	21-10	19-6	18-3	13-9
	24	9-11	9-7	9-1	6-10	13-1	12-4	11-6	8-8	16-8	15-0	14-1	10-7	20-3	17-5	16-3	12-4
Hem-fir	12	11-10	11-7	11-0	9-8	15-7	15-3	14-6	12-4	19-10	19-5	18-6	15-0	24-2	23-7	22-6	17-5
	16	10-9	10-6	10-0	8-5	14-2	13-10	13-2	10-8	18-0	17-8	16-10	13-0	21-11	20-9	19-8	15-1
	19.2	10-1	9-10	9-5	7-8	13-4	13-0	12-5	9-9	17-0	16-4	15-6	11-10	20-8	19-0	17-11	13-9
	24	9-4	9-2	8-9	6-10	12-4	12-0	11-4	8-8	15-9	14-8	13-10	10-7	19-2	17-0	16-1	12-4
Southern pine	12	12-3	12-0	11-10	10-5	16-2	15-10	15-7	13-3	20-8	20-3	19-10	15-8	25-1	24-8	24-2	18-8
	16	11-2	10-11	10-9	9-0	14-8	14-5	14-2	11-6	18-9	18-5	18-0	13-7	22-10	22-5	21-1	16-2
	19.2	10-6	10-4	10-1	8-3	13-10	13-7	13-4	10-6	17-8	17-4	16-5	12-5	21-6	21-1	19-3	14-9
	24	9-9	9-7	9-4	7-4	12-10	12-7	12-4	9-5	16-5	16-1	14-8	11-1	19-11	19-6	17-2	13-2

ALL ROOMS EXCEPT SLEEPING ROOMS AND ATTICS: 40 PSF LIVE, 10 PSF DEAD
Maximum Allowable Span, feet - inches

Species Group	Spacing inches oc	2 × 6 Sel Str	No.1	No.2	No.3	2 × 8 Sel Str	No.1	No.2	No.3	2 × 10 Sel Str	No.1	No.2	No.3	2 × 12 Sel Str	No.1	No.2	No.3
Douglas fir-larch	12	11-4	10-11	10-9	8-8	15-0	14-5	14-2	11-0	19-1	18-5	17-9	13-5	23-3	22-0	20-7	15-7
	16	10-4	9-11	9-9	7-6	13-7	13-1	12-7	9-6	17-4	16-5	15-5	11-8	21-1	19-1	17-10	13-6
	19.2	9-8	9-4	9-1	6-10	12-10	12-4	11-6	8-8	16-4	15-0	14-1	10-7	19-10	17-5	16-3	12-4
	24	9-0	8-8	8-1	6-2	11-11	11-0	10-3	7-9	15-2	13-5	12-7	9-6	18-5	15-7	14-7	11-0
Hem-fir	12	10-9	10-6	10-0	8-8	14-2	13-10	13-2	11-0	18-0	17-8	16-10	13-5	21-11	21-6	20-4	15-7
	16	9-9	9-6	9-1	7-6	12-10	12-7	12-0	9-6	16-5	16-0	15-2	11-8	19-11	18-7	17-7	13-6
	19.2	9-2	9-0	8-7	6-10	12-1	11-10	11-3	8-8	15-5	14-8	13-10	10-7	18-9	17-0	16-1	12-4
	24	8-6	8-4	7-11	6-2	11-3	10-9	10-2	7-9	14-4	13-1	12-5	9-6	17-5	15-2	14-4	11-0
Southern pine	12	11-2	10-11	10-9	9-4	14-8	14-5	14-2	11-11	18-9	18-5	18-0	14-0	22-10	22-5	21-9	16-8
	16	10-2	9-11	9-9	8-1	13-4	13-1	12-10	10-3	17-0	16-9	16-1	12-2	20-9	20-4	18-10	14-6
	19.2	9-6	9-4	9-2	7-4	12-7	12-4	12-1	9-5	16-0	15-9	14-8	11-1	19-6	19-2	17-2	13-2
	24	8-10	8-8	8-6	6-7	11-8	11-5	11-0	8-5	14-11	14-7	13-1	9-11	18-1	17-5	15-5	11-10

Floor Joists

OFFICE SPACE: 50 PSF LIVE, 10 PSF DEAD

Maximum Allowable Span, feet - inches

Species Group	Spacing inches oc	2 × 6				2 × 8				2 × 10				2 × 12			
		Sel Str	No.1	No.2	No.3	Sel Str	No.1	No.2	No.3	Sel Str	No.1	No.2	No.3	Sel Str	No.1	No.2	No.3
Douglas fir-larch	12	10-6	10-2	9-11	7-11	13-11	13-5	13-1	10-0	17-9	17-1	16-3	12-3	21-7	20-1	18-10	14-3
	16	9-7	9-3	9-1	6-10	12-7	12-2	11-6	8-8	16-1	15-0	14-1	10-7	19-7	17-5	16-3	12-4
	19.2	9-0	8-8	8-3	6-3	11-11	11-3	10-6	7-11	15-2	13-8	12-10	9-8	18-5	15-11	14-10	11-3
	24	8-4	7-11	7-5	5-7	11-0	10-0	9-5	7-1	14-1	12-3	11-6	8-8	17-1	14-3	13-4	10-1
Hem-fir	12	9-11	9-9	9-3	7-11	13-1	12-10	12-3	10-0	16-9	16-5	15-7	12-3	20-4	19-7	18-6	14-3
	16	9-1	8-10	8-5	6-10	11-11	11-8	11-1	8-8	15-2	14-8	13-10	10-7	18-6	17-0	16-1	12-4
	19.2	8-6	8-4	7-11	6-3	11-3	10-11	10-4	7-11	14-4	13-4	12-8	9-8	17-5	15-6	14-8	11-3
	24	7-11	7-9	7-4	5-7	10-5	9-9	9-3	7-1	13-3	11-11	11-4	8-8	16-2	13-10	13-1	10-1
Southern pine	12	10-4	10-2	9-11	8-6	13-8	13-5	13-1	10-10	17-5	17-1	16-9	12-10	21-2	20-9	19-10	15-3
	16	9-5	9-3	9-1	7-4	12-5	12-2	11-11	9-5	15-10	15-6	14-8	11-1	19-3	18-10	17-2	13-2
	19.2	8-10	8-8	8-6	6-9	11-8	11-5	11-3	8-7	14-11	14-7	13-5	10-1	18-1	17-9	15-8	12-1
	24	8-3	8-1	7-9	6-0	10-10	10-8	10-0	7-8	13-10	13-4	12-0	9-1	16-10	15-11	14-0	10-9

(ALL ROOMS) 1.5" OR LESS CONCRETE FLOOR FILL: 40 PSF LIVE, 20 PSF DEAD

Maximum Allowable Span, feet - inches

Species Group	Spacing inches oc	2 × 6				2 × 8				2 × 10				2 × 12			
		Sel Str	No.1	No.2	No.3	Sel Str	No.1	No.2	No.3	Sel Str	No.1	No.2	No.3	Sel Str	No.1	No.2	No.3
Douglas fir-larch	12	11-4	10-11	10-6	7-11	15-0	14-2	13-3	10-0	19-1	17-4	16-3	12-3	23-3	20-1	18-10	14-3
	16	10-4	9-8	9-1	6-10	13-7	12-4	11-6	8-8	17-4	15-0	14-1	10-7	21-0	17-5	16-3	12-4
	19.2	9-8	8-10	8-3	6-3	12-10	11-3	10-6	7-11	16-4	13-8	12-10	9-8	19-2	15-11	14-10	11-3
	24	9-0	7-11	7-5	5-7	11-11	10-0	9-5	7-1	14-9	12-3	11-6	8-8	17-1	14-3	13-4	10-1
Hem-fir	12	10-9	10-6	10-0	7-11	14-2	13-10	13-1	10-0	18-0	16-11	16-0	12-3	21-11	19-7	18-6	14-3
	16	9-9	9-6	8-11	6-10	12-10	12-0	11-4	8-8	16-5	14-8	13-10	10-7	19-11	17-0	16-1	12-4
	19.2	9-2	8-8	8-2	6-3	12-1	10-11	10-4	7-11	15-5	13-4	12-8	9-8	18-9	15-6	14-8	11-3
	24	8-6	7-9	7-4	5-7	11-3	9-9	9-3	7-1	14-4	11-11	11-4	8-8	16-10	13-10	13-1	10-1
Southern pine	12	11-2	10-11	10-9	8-6	14-8	14-5	14-2	10-10	18-9	18-5	16-11	12-10	22-10	22-5	19-10	15-3
	16	10-2	9-11	9-6	7-4	13-4	13-1	12-4	9-5	17-0	16-4	14-8	11-1	20-9	19-6	17-2	13-2
	19.2	9-6	9-4	8-8	6-9	12-7	12-4	11-3	8-7	16-0	14-11	13-5	10-1	19-6	17-9	15-8	12-1
	24	8-10	8-8	7-9	6-0	11-8	11-3	10-0	7-8	14-11	13-4	12-0	9-1	18-1	15-11	14-0	10-9

Ceiling Joists

DRYWALL–NO FUTURE ROOMS AND NO ATTIC STORAGE: 10 PSF LIVE, 5 PSF DEAD
Maximum Allowable Span, feet - inches

Species Group	Spacing inches oc	2 × 6				2 × 8				2 × 10				2 × 12			
		Sel Str	No.1	No.2	No.3	Sel Str	No.1	No.2	No.3	Sel Str	No.1	No.2	No.3	Sel Str	No.1	No.2	No.3
Douglas fir- larch	12	13-2	12-8	12-5	10-10	20-8	19-11	19-6	15-10	27-2	26-2	25-8	20-1	34-8	33-5	32-5	24-6
	16	11-11	11-6	11-3	9-5	18-9	18-1	17-8	13-9	24-8	23-10	23-0	17-5	31-6	30-0	28-1	21-3
	19.2	11-3	10-10	10-7	8-7	17-8	17-0	16-7	12-6	23-3	22-5	21-0	15-10	29-8	27-5	25-8	19-5
	24	10-5	10-0	9-10	7-8	16-4	15-9	14-10	11-2	21-7	20-1	18-9	14-2	27-6	24-6	22-11	17-4
Hem-fir	12	12-5	12-2	11-7	10-10	19-6	19-1	18-2	15-10	25-8	25-2	24-0	20-1	32-9	32-1	30-7	24-6
	16	11-3	11-0	10-6	9-5	17-8	17-4	16-6	13-9	23-4	22-10	21-9	17-5	29-9	29-2	27-8	21-3
	19.2	10-7	10-4	9-11	8-7	16-8	16-4	15-7	12-6	21-11	21-6	20-6	15-10	28-0	26-9	25-3	19-5
	24	9-10	9-8	9-2	7-8	15-6	15-2	14-5	11-2	20-5	19-7	18-6	14-2	26-0	23-11	22-7	17-4
Southern pine	12	12-11	12-8	12-5	11-6	20-3	19-11	19-6	17-0	26-9	26-2	25-8	21-8	34-1	33-5	32-9	25-7
	16	11-9	11-6	11-3	10-0	18-5	18-1	17-8	14-9	24-3	23-10	23-4	18-9	31-0	30-5	29-4	22-2
	19.2	11-10	10-10	10-7	9-1	17-4	17-0	16-8	13-6	22-10	22-5	21-11	17-2	29-2	28-7	26-9	20-3
	24	10-3	10-0	9-10	8-2	16-1	15-9	15-6	12-0	21-2	20-10	20-1	15-4	27-1	26-6	23-11	18-1

DRYWALL–NO FUTURE ROOMS AND LIMITED ATTIC STORAGE: 20 PSF LIVE, 10 PSF DEAD
Maximum Allowable Span, feet - inches

Species Group	Spacing inches oc	2 × 6				2 × 8				2 × 10				2 × 12			
		Sel Str	No.1	No.2	No.3	Sel Str	No.1	No.2	No.3	Sel Str	No.1	No.2	No.3	Sel Str	No.1	No.2	No.3
Douglas fir- larch	12	10-5	10-0	9-10	7-8	16-4	15-9	14-10	11-2	21-7	20-1	18-9	14-2	27-6	24-6	22-11	17-4
	16	9-6	9-1	8-9	6-8	14-11	13-9	12-10	9-8	19-7	17-5	16-3	12-4	25-0	21-3	19-10	15-0
	19.2	8-11	8-7	8-0	6-1	14-0	12-6	11-9	8-10	18-5	15-10	14-10	11-3	23-4	19-5	18-2	13-8
	24	8-3	7-8	7-2	5-5	13-0	11-2	10-6	7-11	17-1	14-2	13-3	10-0	20-11	17-4	16-3	12-3
Hem-fir	12	9-10	9-8	9-2	7-8	15-6	15-2	14-5	11-2	20-5	19-7	18-6	14-2	26-0	23-11	22-7	17-4
	16	8-11	8-9	8-4	6-8	14-1	13-5	12-8	9-8	18-6	16-11	16-0	12-4	23-8	20-8	19-7	15-0
	19.2	8-5	8-3	7-10	6-1	13-3	12-3	11-7	8-10	17-5	15-6	14-8	11-3	22-3	18-11	17-10	13-8
	24	7-10	7-6	7-1	5-5	12-3	10-11	10-4	7-11	16-2	13-10	13-1	10-0	20-6	16-11	16-0	12-3
Southern pine	12	10-3	10-0	9-10	8-2	16-1	15-9	15-6	12-0	21-2	20-10	20-1	15-4	27-1	26-6	23-11	18-1
	16	9-4	9-1	8-11	7-1	14-7	14-4	13-6	10-5	19-3	18-11	17-5	13-3	24-7	23-1	20-9	15-8
	19.2	8-9	8-7	8-5	6-5	13-9	13-6	12-3	9-6	18-2	17-9	15-10	12-1	23-2	21-1	18-11	14-4
	24	8-1	8-0	7-8	5-9	12-9	12-6	11-0	8-6	16-10	15-10	14-2	10-10	21-6	18-10	16-11	12-10

Rafters

LIGHT ROOF COVERING, DRYWALL, NO ATTIC SPACE: 20 PSF LIVE, 10 PSF DEAD
Maximum Allowable Span, feet - inches

Species Group	Spacing inches oc	2 × 6				2 × 8				2 × 10				2 × 12			
		Sel Str	No.1	No.2	No.3	Sel Str	No.1	No.2	No.3	Sel Str	No.1	No.2	No.3	Sel Str	No.1	No.2	No.3
Douglas fir-larch	12	16-4	15-9	15-6	12-0	21-7	20-10	20-2	15-3	27-6	26-4	24-7	18-7	33-6	30-6	28-6	21-7
	16	14-11	14-4	13-9	10-5	19-7	18-8	17-5	13-2	25-0	22-9	21-4	16-1	30-5	26-5	24-8	18-8
	19.2	14-0	13-5	12-7	9-6	18-5	17-0	15-11	12-0	23-7	20-9	19-5	14-8	28-8	24-1	22-7	17-1
	24	13-0	12-0	11-3	8-6	17-2	15-3	14-3	10-9	21-10	18-7	17-5	13-2	26-0	21-7	20-2	15-3
Hem-fir	12	15-6	15-2	14-5	12-0	20-5	19-11	19-0	15-3	26-0	25-5	24-3	18-7	31-8	29-9	28-1	21-7
	16	14-1	13-9	13-1	10-5	18-6	18-2	17-2	13-2	23-8	22-2	21-0	16-1	28-9	25-9	24-4	18-8
	19.2	13-3	12-11	12-4	9-6	17-5	16-7	15-8	12-0	22-3	20-3	19-2	14-8	27-1	23-6	22-3	17-1
	24	12-3	11-9	11-1	8-6	16-2	14-10	14-0	10-9	20-8	18-1	17-2	13-2	25-1	21-0	19-11	15-3
Southern pine	12	16-1	15-9	15-6	12-11	21-2	20-10	20-5	16-5	27-1	26-6	25-8	19-5	32-11	32-3	30-1	23-1
	16	14-7	14-4	14-1	11-2	19-3	18-11	18-6	14-3	24-7	24-1	22-3	16-10	29-11	29-4	26-1	20-0
	19.2	13-9	13-6	13-2	10-2	18-2	17-9	17-0	13-0	23-2	22-7	20-4	15-4	28-1	26-11	23-10	18-3
	24	12-9	12-6	11-9	9-2	16-10	16-6	15-3	11-8	21-6	20-3	18-2	13-9	26-1	24-1	21-3	16-4

LIGHT ROOF COVERING, DRYWALL, NO ATTIC SPACE: 30 PSF LIVE, 10 PSF DEAD
Maximum Allowable Span, feet - inches

Species Group	Spacing inches oc	2 × 6				2 × 8				2 × 10				2 × 12			
		Sel Str	No.1	No.2	No.3	Sel Str	No.1	No.2	No.3	Sel Str	No.1	No.2	No.3	Sel Str	No.1	No.2	No.3
Douglas fir-larch	12	14-4	13-9	13-6	10-5	18-10	18-2	17-5	13-2	24-1	22-9	21-4	16-1	29-3	26-5	24-8	18-8
	16	13-0	12-6	11-11	9-0	17-2	16-2	15-1	11-5	21-10	19-9	18-5	13-11	26-7	22-10	21-5	16-2
	19.2	12-3	11-8	10-11	8-3	16-1	14-9	13-9	10-5	20-7	18-0	16-10	12-9	25-0	20-11	19-6	14-9
	24	11-4	10-5	9-9	7-4	15-0	13-2	12-4	9-4	19-1	16-1	15-1	11-5	22-6	18-8	17-6	13-2
Hem-fir	12	13-6	13-3	12-7	10-5	17-10	17-5	16-7	13-2	22-9	22-2	21-0	16-1	27-8	25-9	24-4	18-8
	16	12-3	12-0	11-5	9-0	16-2	15-9	14-11	11-5	20-8	19-3	18-2	13-11	25-1	22-3	21-1	16-2
	19.2	11-7	11-4	10-9	8-3	15-3	14-4	13-7	10-5	19-5	17-7	16-7	12-9	23-7	20-4	19-3	14-9
	24	10-9	10-2	9-7	7-4	14-2	12-10	12-2	9-4	18-0	15-8	14-10	11-5	21-11	18-2	17-3	13-2
Southern pine	12	14-1	13-9	13-6	11-2	18-6	18-2	17-10	14-3	23-8	23-2	22-3	16-10	28-9	28-2	26-1	20-0
	16	12-9	12-6	12-3	9-8	16-10	16-6	16-2	12-4	21-6	21-1	19-3	14-7	26-1	25-7	22-7	17-4
	19.2	12-0	11-9	11-5	8-10	15-10	15-6	14-9	11-3	20-2	19-7	17-7	13-4	24-7	23-4	20-7	15-10
	24	11-2	10-11	10-2	7-11	14-8	14-5	13-2	10-1	18-9	17-6	15-9	11-11	22-10	20-11	18-5	14-2

Rafters

LIGHT ROOF COVERING, DRYWALL, NO ATTIC SPACE: 40 PSF LIVE, 10 PSF DEAD
Maximum Allowable Span, feet - inches

Species Group	Spacing inches oc	2 × 6 Sel Str	No.1	No.2	No.3	2 × 8 Sel Str	No.1	No.2	No.3	2 × 10 Sel Str	No.1	No.2	No.3	2 × 12 Sel Str	No.1	No.2	No.3
Douglas fir-	12	13-0	12-6	12-3	9-4	17-2	16-6	15-7	11-9	21-10	20-4	19-1	14-5	26-7	23-7	22-1	16-8
larch	16	11-10	11-5	10-8	8-1	15-7	14-5	13-6	10-3	19-10	17-8	16-6	12-6	24-2	20-5	19-2	14-6
	19.2	11-1	10-5	9-9	7-4	14-8	13-2	12-4	9-4	18-8	16-1	15-1	11-5	22-6	18-8	17-6	13-2
	24	10-4	9-4	8-9	6-7	13-7	11-9	11-0	8-4	17-4	14-5	13-6	10-2	20-1	16-8	15-7	11-10
Hem-fir	12	12-3	12-0	11-5	9-4	16-2	15-10	15-1	11-9	20-8	19-10	18-9	14-5	25-1	23-0	21-9	16-8
	16	11-2	10-11	10-5	8-1	14-8	14-1	13-4	10-3	18-9	17-2	16-3	12-6	22-10	19-11	18-10	14-6
	19.2	10-6	10-2	9-7	7-4	13-10	12-10	12-2	9-4	17-8	15-8	14-10	11-5	21-6	18-2	17-3	13-2
	24	9-9	9-1	8-7	6-7	12-10	11-6	10-10	8-4	16-5	14-0	13-3	10-2	19-9	16-3	15-5	11-10
Southern	12	12-9	12-6	12-3	10-0	16-10	16-6	16-2	12-9	21-6	21-1	19-11	15-1	26-1	25-7	23-4	17-11
pine	16	11-7	11-5	11-2	8-8	15-3	15-0	14-5	11-0	19-6	19-2	17-3	13-0	23-9	22-10	20-2	15-6
	19.2	10-11	10-8	10-2	7-11	14-5	14-1	13-2	10-1	18-4	17-6	15-9	11-11	22-4	20-11	18-5	14-2
	24	10-2	9-11	9-2	7-1	13-4	13-1	11-9	9-0	17-0	15-8	14-1	10-8	20-9	18-8	16-6	12-8

LIGHT ROOF COVERING, DRYWALL, NO ATTIC SPACE: 50 PSF LIVE, 10 PSF DEAD
Maximum Allowable Span, feet - inches

Species Group	Spacing inches oc	2 × 6 Sel Str	No.1	No.2	No.3	2 × 8 Sel Str	No.1	No.2	No.3	2 × 10 Sel Str	No.1	No.2	No.3	2 × 12 Sel Str	No.1	No.2	No.3
Douglas fir-	12	12-1	11-8	11-3	8-6	15-11	15-3	14-3	10-9	20-3	18-7	17-5	13-2	24-8	21-7	20-2	15-3
larch	16	11-0	10-5	9-9	7-4	14-5	13-2	12-4	9-4	18-5	16-1	15-1	11-5	22-5	18-8	17-6	13-2
	19.2	10-4	9-6	8-11	6-9	13-7	12-0	11-3	8-6	17-4	14-8	13-9	10-5	20-6	17-1	15-11	12-1
	24	9-7	8-6	7-11	6-0	12-7	10-9	10-1	7-7	15-10	13-2	12-4	9-4	18-4	15-3	14-3	10-9
Hem-fir	12	11-5	11-2	10-8	8-6	15-0	14-8	14-0	10-9	19-2	18-1	17-2	13-2	23-4	21-0	19-11	15-3
	16	10-4	10-2	9-7	7-4	13-8	12-10	12-2	9-4	17-5	15-8	14-10	11-5	21-2	18-2	17-3	13-2
	19.2	9-9	9-3	8-9	6-9	12-10	11-9	11-1	8-6	16-5	14-4	13-7	10-5	19-11	16-7	15-9	12-1
	24	9-1	8-3	7-10	6-0	11-11	10-6	9-11	7-7	15-2	12-10	12-1	9-4	18-0	14-10	14-1	10-9
Southern	12	11-10	11-8	11-5	9-2	15-7	15-4	15-0	11-8	19-11	19-7	18-2	13-9	24-3	23-9	21-3	16-4
pine	16	10-9	10-7	10-2	7-11	14-2	13-11	13-2	10-1	18-1	17-6	15-9	11-11	22-0	20-11	18-5	14-2
	19.2	10-2	9-11	9-4	7-3	13-4	13-1	12-0	9-2	17-0	16-0	14-4	10-10	20-9	19-1	16-10	12-11
	24	9-5	9-3	8-4	6-5	12-5	12-0	10-9	8-3	15-10	14-4	12-10	9-9	19-3	17-1	15-1	11-7

Rafters

MEDIUM ROOF COVERING, DRYWALL, NO ATTIC SPACE: 20 PSF LIVE, 15 PSF DEAD

Maximum Allowable Span, feet - inches

Species Group	Spacing inches oc	2 × 6 Sel Str	No.1	No.2	No.3	2 × 8 Sel Str	No.1	No.2	No.3	2 × 10 Sel Str	No.1	No.2	No.3	2 × 12 Sel Str	No.1	No.2	No.3
Douglas fir-larch	12	16-4	15-9	14-9	11-2	21-7	19-11	18-8	14-1	27-6	24-4	22-9	17-3	33-6	28-3	26-5	20-0
	16	14-11	13-8	12-9	9-8	19-7	17-3	16-2	12-2	25-0	21-1	19-9	14-11	29-5	24-5	22-10	17-3
	19.2	14-0	12-5	11-8	8-10	18-5	15-9	14-9	11-2	23-2	19-3	18-0	13-7	26-11	22-4	20-11	15-9
	24	13-0	11-2	10-5	7-10	17-0	14-1	13-2	10-0	20-9	17-3	16-1	12-2	24-0	20-0	18-8	14-1
Hem-fir	12	15-6	15-2	14-5	11-2	20-5	19-5	18-4	14-1	26-0	23-9	22-5	17-3	31-8	27-6	26-0	20-0
	16	14-1	13-3	12-7	9-8	18-6	16-10	15-11	12-2	23-8	20-7	19-5	14-11	28-9	23-10	22-6	17-3
	19.2	13-3	12-2	11-6	8-10	17-5	15-4	14-6	11-2	22-3	18-9	17-9	13-7	26-5	21-9	20-7	15-9
	24	12-3	10-10	10-3	7-10	16-2	13-9	13-0	10-0	20-4	16-9	15-10	12-2	23-7	19-5	18-5	14-1
Southern pine	12	16-1	15-9	15-5	11-11	21-2	20-10	19-11	15-3	27-1	26-6	23-9	18-0	32-11	31-7	27-11	21-5
	16	14-7	14-4	13-4	10-4	19-3	18-11	17-3	13-2	24-7	22-11	20-7	15-7	29-11	27-4	24-2	18-6
	19.2	13-9	13-6	12-2	9-5	18-2	17-7	15-9	12-0	23-2	20-11	18-10	14-3	28-1	24-11	22-0	16-11
	24	12-9	12-6	10-11	8-5	16-10	15-9	14-1	10-9	21-6	18-9	16-10	12-9	26-1	22-4	19-9	15-2

MEDIUM ROOF COVERING, DRYWALL, NO ATTIC SPACE: 30 PSF LIVE, 15 PSF DEAD

Maximum Allowable Span, feet - inches

Species Group	Spacing inches oc	2 × 6 Sel Str	No.1	No.2	No.3	2 × 8 Sel Str	No.1	No.2	No.3	2 × 10 Sel Str	No.1	No.2	No.3	2 × 12 Sel Str	No.1	No.2	No.3
Douglas fir-larch	12	14-4	13-9	13-0	9-10	18-10	17-7	16-5	12-5	24-1	21-6	20-1	15-2	29-3	24-11	23-3	17-7
	16	13-0	12-0	11-3	8-6	17-2	15-3	14-3	10-9	21-10	18-7	17-5	13-2	26-0	21-7	20-2	15-3
	19.2	12-3	11-0	10-3	7-9	16-1	13-11	13-0	9-10	20-5	17-0	15-11	12-0	23-8	19-8	18-5	13-11
	24	11-4	9-10	9-2	6-11	15-0	12-5	11-8	8-9	18-3	15-2	14-2	10-9	21-2	17-7	16-6	12-5
Hem-fir	12	13-6	13-3	12-7	9-10	17-10	17-2	16-2	12-5	22-9	20-11	19-10	15-2	27-8	24-3	22-11	17-7
	16	12-3	11-9	11-1	8-6	16-2	14-10	14-0	10-9	20-8	18-1	17-2	13-2	25-1	21-0	19-11	15-3
	19.2	11-7	10-8	10-1	7-9	15-3	13-7	12-10	9-10	19-5	16-7	15-8	12-0	23-3	19-2	18-2	13-11
	24	10-9	9-7	9-1	6-11	14-2	12-1	11-6	8-9	18-0	14-10	14-0	10-9	20-10	17-2	16-3	12-5
Southern pine	12	14-1	13-9	13-6	10-6	18-6	18-2	17-7	13-5	23-8	23-2	21-0	15-10	28-9	27-10	24-7	18-11
	16	12-9	12-6	11-9	9-2	16-10	16-6	15-3	11-8	21-6	20-3	18-2	13-9	26-1	24-1	21-3	16-4
	19.2	12-0	11-9	10-9	8-4	15-10	15-6	13-11	10-7	20-2	18-5	16-7	12-6	24-7	22-0	19-5	14-11
	24	11-2	10-11	9-7	7-5	14-8	13-11	12-5	9-6	18-9	16-6	14-10	11-3	22-10	19-8	17-5	13-4

Rafters

MEDIUM ROOF COVERING, DRYWALL, NO ATTIC SPACE: 40 PSF LIVE, 15 PSF DEAD
Maximum Allowable Span, feet - inches

Species Group	Spacing inches oc	2 × 6 Sel Str	No.1	No.2	No.3	2 × 8 Sel Str	No.1	No.2	No.3	2 × 10 Sel Str	No.1	No.2	No.3	2 × 12 Sel Str	No.1	No.2	No.3
Douglas fir-larch	12	13-0	12-6	11-9	8-11	17-2	15-11	14-10	11-3	21-10	19-5	18-2	13-9	26-7	22-6	21-1	15-11
	16	11-10	10-10	10-2	7-8	15-7	13-9	12-11	9-9	19-10	16-10	15-9	11-11	23-6	19-6	18-3	13-9
	19.2	11-1	9-11	9-3	7-0	14-8	12-7	11-9	8-11	18-6	15-4	14-4	10-10	21-5	17-10	16-8	12-7
	24	10-4	8-11	8-4	6-3	13-6	11-3	10-6	7-11	16-6	13-9	12-10	9-9	19-2	15-11	14-11	11-3
Hem-fir	12	12-3	12-0	11-5	8-11	16-2	15-6	14-8	11-3	20-8	18-11	17-11	13-9	25-1	21-11	20-9	15-11
	16	11-2	10-7	10-0	7-8	14-8	13-5	12-8	9-9	18-9	16-5	15-6	11-11	22-10	19-0	18-0	13-9
	19.2	10-6	9-8	9-2	7-0	13-10	12-3	11-7	8-11	17-8	15-0	14-2	10-10	21-1	17-4	16-5	12-7
	24	9-9	8-8	8-2	6-3	12-10	10-11	10-4	7-11	16-3	13-5	12-8	9-9	18-10	15-6	14-8	11-3
Southern pine	12	12-9	12-6	12-3	9-6	16-10	16-6	15-11	12-2	21-6	21-1	19-0	14-4	26-1	25-2	22-3	17-1
	16	11-7	11-5	10-8	8-3	15-3	15-0	13-9	10-6	19-6	18-3	16-5	12-5	23-9	21-10	19-3	14-9
	19.2	10-11	10-8	9-9	7-6	14-5	14-1	12-7	9-7	18-4	16-8	15-0	11-4	22-4	19-11	17-7	13-6
	24	10-2	9-11	8-8	6-9	13-4	12-7	11-3	8-7	17-0	14-11	13-5	10-2	20-9	17-10	15-9	12-1

MEDIUM ROOF COVERING, DRYWALL, NO ATTIC SPACE: 50 PSF LIVE, 15 PSF DEAD
Maximum Allowable Span, feet - inches

Species Group	Spacing inches oc	2 × 6 Sel Str	No.1	No.2	No.3	2 × 8 Sel Str	No.1	No.2	No.3	2 × 10 Sel Str	No.1	No.2	No.3	2 × 12 Sel Str	No.1	No.2	No.3
Douglas fir-larch	12	12-1	11-7	10-10	8-2	15-11	14-7	13-8	10-4	20-3	17-10	16-9	12-8	24-8	20-9	19-5	14-8
	16	11-0	10-0	9-4	7-1	14-5	12-8	11-10	8-11	18-5	15-6	14-6	10-11	21-7	17-11	16-9	12-8
	19.2	10-4	9-2	8-6	6-5	13-7	11-7	10-10	8-2	17-0	14-1	13-3	10-0	19-9	16-5	15-4	11-7
	24	9-7	8-2	7-8	5-9	12-5	10-4	9-8	7-4	15-3	12-8	11-10	8-11	17-8	14-8	13-8	10-4
Hem-fir	12	11-5	11-2	10-8	8-2	15-0	14-3	13-6	10-4	19-2	17-5	16-6	12-8	23-4	20-2	19-1	14-8
	16	10-4	9-9	9-3	7-1	13-8	12-4	11-8	8-11	17-5	15-1	14-3	10-11	21-2	17-6	16-6	12-8
	19.2	9-9	8-11	8-5	6-5	12-10	11-3	10-8	8-2	16-5	13-9	13-0	10-0	19-5	16-0	15-1	11-7
	24	9-1	8-0	7-6	5-9	11-11	10-1	9-6	7-4	14-11	12-4	11-8	8-11	17-4	14-3	13-6	10-4
Southern pine	12	11-10	11-8	11-4	8-9	15-7	15-4	14-7	11-2	19-11	19-5	17-5	13-2	24-3	23-2	20-5	15-9
	16	10-9	10-7	9-10	7-7	14-2	13-11	12-8	9-8	18-1	16-10	15-1	11-5	22-0	20-1	17-9	13-7
	19.2	10-2	9-11	8-11	6-11	13-4	12-11	11-7	8-10	17-0	15-4	13-10	10-5	20-9	18-4	16-2	12-5
	24	9-5	9-2	8-0	6-2	12-5	11-7	10-4	7-11	15-10	13-9	12-4	9-4	19-3	16-5	14-6	11-1

Rafters

HEAVY ROOF COVERING, DRYWALL, NO ATTIC SPACE: 20 PSF LIVE, 20 PSF DEAD

Maximum Allowable Span, feet - inches

Species Group	Spacing inches oc	2 × 6 Sel Str	No.1	No.2	No.3	2 × 8 Sel Str	No.1	No.2	No.3	2 × 10 Sel Str	No.1	No.2	No.3	2 × 12 Sel Str	No.1	No.2	No.3
Douglas fir-larch	12	16-4	14-9	13-9	10-5	21-7	18-8	17-5	13-2	27-5	22-9	21-4	16-1	31-10	26-5	24-8	18-8
	16	14-11	12-9	11-11	9-0	19-5	16-2	15-1	11-5	23-9	19-9	18-5	13-11	27-6	22-10	21-5	16-2
	19.2	14-0	11-8	10-11	8-3	17-9	14-9	13-9	10-5	21-8	18-0	16-10	12-9	25-2	20-11	19-6	14-9
	24	12-6	10-5	9-9	7-4	15-10	13-2	12-4	9-4	19-5	16-1	15-1	11-5	22-6	18-8	17-6	13-2
Hem-fir	12	15-6	14-4	13-7	10-5	20-5	18-2	17-2	13-2	26-0	22-2	21-0	16-1	31-3	25-9	24-4	18-8
	16	14-1	12-5	11-9	9-0	18-6	15-9	14-11	11-5	23-4	19-3	18-2	13-11	27-1	22-3	21-1	16-2
	19.2	13-3	11-4	10-9	8-3	17-5	14-4	13-7	10-5	21-4	17-7	16-7	12-9	24-8	20-4	19-3	14-9
	24	12-3	10-2	9-7	7-4	15-7	12-10	12-2	9-4	19-1	15-8	14-10	11-5	22-1	18-2	17-3	13-2
Southern pine	12	16-1	15-9	14-5	11-2	21-2	20-10	18-8	14-3	27-1	24-9	22-3	16-10	32-11	29-6	26-1	20-0
	16	14-7	14-4	12-6	9-8	19-3	18-1	16-2	12-4	24-7	21-5	19-3	14-7	29-11	25-7	22-7	17-4
	19.2	13-9	13-1	11-5	8-10	18-2	16-6	14-9	11-3	23-2	19-7	17-7	13-4	28-1	23-4	20-7	15-10
	24	12-9	11-9	10-2	7-11	16-10	14-9	13-2	10-1	21-6	17-6	15-9	11-11	25-9	20-11	18-5	14-2

HEAVY ROOF COVERING, DRYWALL, NO ATTIC SPACE: 30 PSF LIVE, 20 PSF DEAD

Maximum Allowable Span, feet - inches

Species Group	Spacing inches oc	2 × 6 Sel Str	No.1	No.2	No.3	2 × 8 Sel Str	No.1	No.2	No.3	2 × 10 Sel Str	No.1	No.2	No.3	2 × 12 Sel Str	No.1	No.2	No.3
Douglas fir-larch	12	14-4	13-2	12-4	9-4	18-10	16-8	15-7	11-9	24-1	20-4	19-1	14-5	28-5	23-7	22-1	16-8
	16	13-0	11-5	10-8	8-1	17-2	14-5	13-6	10-3	21-3	17-8	16-6	12-6	24-8	20-5	19-2	14-6
	19.2	12-3	10-5	9-9	7-4	15-10	13-2	12-4	9-4	19-5	16-1	15-1	11-5	22-6	18-8	17-6	13-2
	24	11-3	9-4	8-9	6-7	14-2	11-9	11-0	8-4	17-4	14-5	13-6	10-2	20-1	16-8	15-7	11-10
Hem-fir	12	13-6	12-10	12-2	9-4	17-10	16-3	15-4	11-9	22-9	19-10	18-9	14-5	27-8	23-0	21-9	16-8
	16	12-3	11-1	10-6	8-1	16-2	14-1	13-4	10-3	20-8	17-2	16-3	12-6	24-2	19-11	18-10	14-6
	19.2	11-7	10-2	9-7	7-4	15-3	12-10	12-2	9-4	19-1	15-8	14-10	11-5	22-1	18-2	17-3	13-2
	24	10-9	9-1	8-7	6-7	13-11	11-6	10-10	8-4	17-1	14-0	13-3	10-2	19-9	16-3	15-5	11-10
Southern pine	12	14-1	13-9	12-11	10-0	18-6	18-2	16-8	12-9	23-8	22-2	19-11	15-1	28-9	26-5	23-4	17-11
	16	12-9	12-6	11-2	8-8	16-10	16-2	14-5	11-0	21-6	19-2	17-3	13-0	26-1	22-10	20-2	15-6
	19.2	12-0	11-9	10-2	7-11	15-10	14-9	13-2	10-1	20-2	17-6	15-9	11-11	24-7	20-11	18-5	14-2
	24	11-2	10-6	9-2	7-1	14-8	13-2	11-9	9-0	18-9	15-8	14-1	10-8	22-10	18-8	16-6	12-8

Rafters

LIGHT ROOF COVERING, NO CEILING: 20 PSF LIVE, 10 PSF DEAD
Maximum Allowable Span, feet - inches

Species Group	Spacing inches oc	2 × 6 Sel Str	No.1	No.2	No.3	2 × 8 Sel Str	No.1	No.2	No.3	2 × 10 Sel Str	No.1	No.2	No.3	2 × 12 Sel Str	No.1	No.2	No.3
Douglas fir-larch	12	11-6	11-1	10-10	8-3	18-0	17-0	15-11	12-0	23-9	21-6	20-2	15-3	30-4	26-4	24-7	18-7
	16	10-5	10-0	9-5	7-1	16-4	14-9	13-9	10-5	21-7	18-8	17-5	13-2	27-5	22-9	21-4	16-1
	19.2	9-10	9-2	8-7	6-6	15-5	13-5	12-7	9-6	20-4	17-0	15-11	12-0	25-0	20-9	19-5	14-8
	24	9-1	8-3	7-8	5-10	14-4	12-0	11-3	8-6	18-4	15-3	14-3	10-9	22-5	18-7	17-5	13-2
Hem-fir	12	10-10	10-7	10-1	8-3	17-0	16-7	15-8	12-0	22-5	21-0	19-10	15-3	28-7	25-8	24-3	18-7
	16	9-10	9-8	9-2	7-1	15-6	14-4	13-7	10-5	20-5	18-2	17-2	13-2	26-0	22-2	21-0	16-1
	19.2	9-3	8-11	8-6	6-6	14-7	13-1	12-5	9-6	19-2	16-7	15-8	12-0	24-6	20-3	19-2	14-8
	24	8-7	8-0	7-7	5-10	13-6	11-9	11-1	8-6	17-10	14-10	14-0	10-9	22-0	18-1	17-2	13-2
Southern pine	12	11-3	11-1	10-10	8-9	17-8	17-4	16-8	12-11	23-4	22-11	21-6	16-5	29-9	28-7	25-8	19-5
	16	10-3	10-0	9-10	7-7	16-1	15-9	14-5	11-2	21-2	20-10	18-8	14-3	27-1	24-9	22-3	16-10
	19.2	9-8	9-5	9-2	6-11	15-2	14-10	13-2	10-2	19-11	19-0	17-0	13-0	25-5	22-7	20-4	15-4
	24	8-11	8-9	8-3	6-2	14-1	13-6	11-9	9-2	18-6	17-0	15-3	11-8	23-8	20-3	18-2	13-9

LIGHT ROOF COVERING, NO CEILING: 30 PSF LIVE, 10 PSF DEAD
Maximum Allowable Span, feet - inches

Species Group	Spacing inches oc	2 × 4 Sel Str	No.1	No.2	No.3	2 × 6 Sel Str	No.1	No.2	No.3	2 × 8 Sel Str	No.1	No.2	No.3	2 × 10 Sel Str	No.1	No.2	No.3
Douglas fir-larch	12	10-0	9-8	9-5	7-1	15-9	14-9	13-9	10-5	20-9	18-8	17-5	13-2	26-6	22-9	21-4	16-1
	16	9-1	8-9	8-2	6-2	14-4	12-9	11-11	9-0	18-10	16-2	15-1	11-5	23-9	19-9	18-5	13-11
	19.2	8-7	7-11	7-5	5-8	13-6	11-8	10-11	8-3	17-9	14-9	13-9	10-5	21-8	18-0	16-10	12-9
	24	7-11	7-1	6-8	5-0	12-6	10-5	9-9	7-4	15-10	13-2	12-4	9-4	19-5	16-1	15-1	11-5
Hem-fir	12	9-6	9-3	8-10	7-1	14-10	14-4	13-7	10-5	19-7	18-2	17-2	13-2	25-0	22-2	21-0	16-1
	16	8-7	8-5	8-0	6-2	13-6	12-5	11-9	9-0	17-10	15-9	14-11	11-5	22-9	19-3	18-2	13-11
	19.2	8-1	7-9	7-4	5-8	12-9	11-4	10-9	8-3	16-9	14-4	13-7	10-5	21-4	17-7	16-7	12-9
	24	7-6	6-11	6-7	5-0	11-10	10-2	9-7	7-4	15-7	12-10	12-2	9-4	19-1	15-8	14-10	11-5
Southern pine	12	9-10	9-8	9-6	7-7	15-6	15-2	14-5	11-2	20-5	20-0	18-8	14-3	26-0	24-9	22-3	16-10
	16	8-11	8-9	8-7	6-7	14-1	13-9	12-6	9-8	18-6	18-1	16-2	12-4	23-8	21-5	19-3	14-7
	19.2	8-5	8-3	7-11	6-0	13-3	13-0	11-5	8-10	17-5	16-6	14-9	11-3	22-3	19-7	17-7	13-4
	24	7-10	7-8	7-1	5-4	12-3	11-9	10-2	7-11	16-2	14-9	13-2	10-1	20-8	17-6	15-9	11-11

Rafters

LIGHT ROOF COVERING, NO CEILING: 40 PSF LIVE, 10 PSF DEAD
Maximum Allowable Span, feet - inches

Species Group	Spacing inches oc	2×6 Sel Str	No.1	No.2	No.3	2×8 Sel Str	No.1	No.2	No.3	2×10 Sel Str	No.1	No.2	No.3	2×12 Sel Str	No.1	No.2	No.3
Douglas fir-larch	12	9-1	8-9	8-5	6-4	14-4	13-2	12-4	9-4	18-10	16-8	15-7	11-9	24-1	20-4	19-1	14-5
	16	8-3	7-10	7-3	5-6	13-0	11-5	10-8	8-1	17-2	14-5	13-6	10-3	21-3	17-8	16-6	12-6
	19.2	7-9	7-1	6-8	5-0	12-3	10-5	9-9	7-4	15-10	13-2	12-4	9-4	19-5	16-1	15-1	11-5
	24	7-3	6-4	5-11	4-6	11-3	9-4	8-9	6-7	14-2	11-9	11-0	8-4	17-4	14-5	13-6	10-2
Hem-fir	12	8-7	8-5	8-0	6-4	13-6	12-10	12-2	9-4	17-10	16-3	15-4	11-9	22-9	19-10	18-9	14-5
	16	7-10	7-7	7-2	5-6	12-3	11-1	10-6	8-1	16-2	14-1	13-4	10-3	20-8	17-2	16-3	12-6
	19.2	7-4	6-11	6-7	5-0	11-7	10-2	9-7	7-4	15-3	12-10	12-2	9-4	19-1	15-8	14-10	11-5
	24	6-10	6-2	5-10	4-6	10-9	9-1	8-7	6-7	13-11	11-6	10-10	8-4	17-1	14-0	13-3	10-2
Southern pine	12	8-11	8-9	8-7	6-9	14-1	13-9	12-11	10-0	18-6	18-2	16-8	12-9	23-8	22-2	19-11	15-1
	16	8-1	8-0	7-10	5-10	12-9	12-6	11-2	8-8	16-10	16-2	14-5	11-0	21-6	19-2	17-3	13-0
	19.2	7-8	7-6	7-1	5-4	12-0	11-9	10-2	7-11	15-10	14-9	13-2	10-1	20-2	17-6	15-9	11-11
	24	7-1	7-0	6-4	4-9	11-2	10-6	9-2	7-1	14-8	13-2	11-9	9-0	18-9	15-8	14-1	10-8

LIGHT ROOF COVERING, NO CEILING: 50 PSF LIVE, 10 PSF DEAD
Maximum Allowable Span, feet - inches

Species Group	Spacing inches oc	2×6 Sel Str	No.1	No.2	No.3	2×8 Sel Str	No.1	No.2	No.3	2×10 Sel Str	No.1	No.2	No.3	2×12 Sel Str	No.1	No.2	No.3
Douglas fir-larch	12	8-5	8-2	7-8	5-10	13-3	12-0	11-3	8-6	17-6	15-3	14-3	10-9	22-4	18-7	17-5	13-2
	16	7-8	7-1	6-8	5-0	12-1	10-5	9-9	7-4	15-10	13-2	12-4	9-4	19-5	16-1	15-1	11-5
	19.2	7-3	6-6	6-1	4-7	11-4	9-6	8-11	6-9	14-6	12-0	11-3	8-6	17-8	14-8	13-9	10-5
	24	6-8	5-10	5-5	4-1	10-3	8-6	7-11	6-0	13-0	10-9	10-1	7-7	15-10	13-2	12-4	9-4
Hem-fir	12	8-0	7-10	7-5	5-10	12-6	11-9	11-1	8-6	16-6	14-10	14-0	10-9	21-1	18-1	17-2	13-2
	16	7-3	6-11	6-7	5-0	11-5	10-2	9-7	7-4	15-0	12-10	12-2	9-4	19-1	15-8	14-10	11-5
	19.2	6-10	6-4	6-0	4-7	10-9	9-3	8-9	6-9	14-2	11-9	11-1	8-6	17-5	14-4	13-7	10-5
	24	6-4	5-8	5-4	4-1	9-11	8-3	7-10	6-0	12-9	10-6	9-11	7-7	15-7	12-10	12-1	9-4
Southern pine	12	8-4	8-2	8-0	6-2	13-1	12-10	11-9	9-2	17-2	16-10	15-3	11-8	21-11	20-3	18-2	13-9
	16	7-6	7-5	7-1	5-4	11-10	11-8	10-2	7-11	15-7	14-9	13-2	10-1	19-11	17-6	15-9	11-11
	19.2	7-1	7-0	6-6	4-11	11-2	10-8	9-4	7-3	14-8	13-5	12-0	9-2	18-9	16-0	14-4	10-10
	24	6-7	6-5	5-10	4-4	10-4	9-7	8-4	6-5	13-8	12-0	10-9	8-3	17-5	14-4	12-10	9-9

Rafters

MEDIUM ROOF COVERING, NO CEILING: 20 PSF LIVE, 15 PSF DEAD
Maximum Allowable Span, feet - inches

Species Group	Spacing inches oc	2 × 6				2 × 8				2 × 10				2 × 12			
		Sel Str	No.1	No.2	No.3	Sel Str	No.1	No.2	No.3	Sel Str	No.1	No.2	No.3	Sel Str	No.1	No.2	No.3
Douglas fir-larch	12	11-6	10-9	10-1	7-7	18-0	15-9	14-9	11-2	23-9	19-11	18-8	14-1	29-4	24-4	22-9	17-3
	16	10-5	9-4	8-9	6-7	16-4	13-8	12-9	9-8	20-9	17-3	16-2	12-2	25-5	21-1	19-9	14-11
	19.2	9-10	8-6	7-11	6-0	15-0	12-5	11-8	8-10	19-0	15-9	14-9	11-2	23-2	19-3	18-0	13-7
	24	9-1	7-7	7-1	5-5	13-5	11-2	10-5	7-10	17-0	14-1	13-2	10-0	20-9	17-3	16-1	12-2
Hem-fir	12	10-10	10-6	9-11	7-7	17-0	15-4	14-6	11-2	22-5	19-5	18-4	14-1	28-7	23-9	22-5	17-3
	16	9-10	9-1	8-7	6-7	15-6	13-3	12-7	9-8	20-5	16-10	15-11	12-2	24-11	20-7	19-5	14-11
	19.2	9-3	8-3	7-10	6-0	14-7	12-2	11-6	8-10	18-8	15-4	14-6	11-2	22-9	18-9	17-9	13-7
	24	8-7	7-5	7-0	5-5	13-2	10-10	10-3	7-10	16-8	13-9	13-0	10-0	20-4	16-9	15-10	12-2
Southern pine	12	11-3	11-1	10-9	8-1	17-8	17-4	15-5	11-11	23-4	22-3	19-11	15-3	29-9	26-6	23-9	18-0
	16	10-3	10-0	9-4	7-0	16-1	15-4	13-4	10-4	21-2	19-4	17-3	13-2	27-1	22-11	20-7	15-7
	19.2	9-8	9-5	8-6	6-5	15-2	14-0	12-2	9-5	19-11	17-7	15-9	12-0	25-5	20-11	18-10	14-3
	24	8-11	8-5	7-7	5-9	14-1	12-6	10-11	8-5	18-6	15-9	14-1	10-9	23-6	18-9	16-10	12-9

MEDIUM ROOF COVERING, NO CEILING: 30 PSF LIVE, 15 PSF DEAD
Maximum Allowable Span, feet - inches

Species Group	Spacing inches oc	2 × 6				2 × 8				2 × 10				2 × 12			
		Sel Str	No.1	No.2	No.3	Sel Str	No.1	No.2	No.3	Sel Str	No.1	No.2	No.3	Sel Str	No.1	No.2	No.3
Douglas fir-larch	12	10-0	9-6	8-10	6-8	15-9	13-11	13-0	9-10	20-9	17-7	16-5	12-5	25-10	21-6	20-1	15-2
	16	9-1	8-3	7-8	5-10	14-4	12-0	11-3	8-6	18-4	15-3	14-3	10-9	22-5	18-7	17-5	13-2
	19.2	8-7	7-6	7-0	5-4	13-3	11-0	10-3	7-9	16-9	13-11	13-0	9-10	20-5	17-0	15-11	12-0
	24	7-11	6-8	6-3	4-9	11-10	9-10	9-2	6-11	15-0	12-5	11-8	8-9	18-3	15-2	14-2	10-9
Hem-fir	12	9-6	9-3	8-9	6-8	14-10	13-6	12-10	9-10	19-7	17-2	16-2	12-5	25-0	20-11	19-10	15-2
	16	8-7	8-0	7-7	5-10	13-6	11-9	11-1	8-6	17-10	14-10	14-0	10-9	22-0	18-1	17-2	13-2
	19.2	8-1	7-4	6-11	5-4	12-9	10-8	10-1	7-9	16-5	13-7	12-10	9-10	20-1	16-7	15-8	12-0
	24	7-6	6-6	6-2	4-9	11-7	9-7	9-1	6-11	14-8	12-1	11-6	8-9	18-0	14-10	14-0	10-9
Southern pine	12	9-10	9-8	9-6	7-2	15-6	15-2	13-7	10-6	20-5	19-8	17-7	13-5	26-0	23-4	21-0	15-10
	16	8-11	8-9	8-3	6-2	14-1	13-6	11-9	9-2	18-6	17-0	15-3	11-8	23-8	20-3	18-2	13-9
	19.2	8-5	8-3	7-6	5-8	13-3	12-4	10-9	8-4	17-5	15-6	13-11	10-7	22-3	18-5	16-7	12-6
	24	7-10	7-5	6-8	5-1	12-3	11-1	9-7	7-5	16-2	13-11	12-5	9-6	20-8	16-6	14-10	11-3

Rafters

MEDIUM ROOF COVERING, NO CEILING: 40 PSF LIVE, 15 PSF DEAD
Maximum Allowable Span, feet - inches

Species Group	Spacing inches oc	2 × 6				2 × 8				2 × 10				2 × 12			
		Sel Str	No.1	No.2	No.3	Sel Str	No.1	No.2	No.3	Sel Str	No.1	No.2	No.3	Sel Str	No.1	No.2	No.3
Douglas fir-larch	12	9-1	8-7	8-0	6-1	14-4	12-7	11-9	8-11	18-10	15-11	14-10	11-3	23-5	19-5	18-2	13-9
	16	8-3	7-5	6-11	5-3	13-0	10-10	10-2	7-8	16-7	13-9	12-11	9-9	20-3	16-10	15-9	11-11
	19.2	7-9	6-9	6-4	4-10	11-11	9-11	9-3	7-0	15-2	12-7	11-9	8-11	18-6	15-4	14-4	10-10
	24	7-3	6-1	5-8	4-3	10-8	8-11	8-4	6-3	13-6	11-3	10-6	7-11	16-6	13-9	12-10	9-9
Hem-fir	12	8-7	8-4	7-11	6-1	13-6	12-3	11-7	8-11	17-10	15-6	14-8	11-3	22-9	18-11	17-11	13-9
	16	7-10	7-3	6-10	5-3	12-3	10-7	10-0	7-8	16-2	13-5	12-8	9-9	19-11	16-5	15-6	11-11
	19.2	7-4	6-7	6-3	4-10	11-7	9-8	9-2	7-0	14-10	12-3	11-7	8-11	18-2	15-0	14-2	10-10
	24	6-10	5-11	5-7	4-3	10-6	8-8	8-2	6-3	13-4	10-11	10-4	7-11	16-3	13-5	12-8	9-9
Southern pine	12	8-11	8-9	8-7	6-6	14-1	13-9	12-4	9-6	18-6	17-9	15-11	12-2	23-8	21-1	19-0	14-4
	16	8-1	8-0	7-5	5-7	12-9	12-3	10-8	8-3	16-10	15-5	13-9	10-6	21-6	18-3	16-5	12-5
	19.2	7-8	7-6	6-9	5-1	12-0	11-2	9-9	7-6	15-10	14-1	12-7	9-7	20-2	16-8	15-0	11-4
	24	7-1	6-9	6-1	4-7	11-2	10-0	8-8	6-9	14-8	12-7	11-3	8-7	18-9	14-11	13-5	10-2

MEDIUM ROOF COVERING, NO CEILING: 50 PSF LIVE, 15 PSF DEAD
Maximum Allowable Span, feet - inches

Species Group	Spacing inches oc	2 × 6				2 × 8				2 × 10				2 × 12			
		Sel Str	No.1	No.2	No.3	Sel Str	No.1	No.2	No.3	Sel Str	No.1	No.2	No.3	Sel Str	No.1	No.2	No.3
Douglas fir-larch	12	8-5	7-11	7-5	5-7	13-3	11-7	10-10	8-2	17-6	14-7	13-8	10-4	21-6	17-10	16-9	12-8
	16	7-8	6-10	6-5	4-10	12-1	10-0	9-4	7-1	15-3	12-8	11-10	8-11	18-8	15-6	14-6	10-11
	19.2	7-3	6-3	5-10	4-5	11-0	9-2	8-6	6-5	13-11	11-7	10-10	8-2	17-0	14-1	13-3	10-0
	24	6-8	5-7	5-3	3-11	9-10	8-2	7-8	5-9	12-5	10-4	9-8	7-4	15-3	12-8	11-10	8-11
Hem-fir	12	8-0	7-8	7-3	5-7	12-6	11-3	10-8	8-2	16-6	14-3	13-6	10-4	21-1	17-5	16-6	12-8
	16	7-3	6-8	6-4	4-10	11-5	9-9	9-3	7-1	15-0	12-4	11-8	8-11	18-4	15-1	14-3	10-11
	19.2	6-10	6-1	5-9	4-5	10-9	8-11	8-5	6-5	13-8	11-3	10-8	8-2	16-9	13-9	13-0	10-0
	24	6-4	5-5	5-2	3-11	9-8	8-0	7-6	5-9	12-3	10-1	9-6	7-4	14-11	12-4	11-8	8-11
Southern pine	12	8-4	8-2	7-11	5-11	13-1	12-10	11-4	8-9	17-2	16-4	14-7	11-2	21-11	19-5	17-5	13-2
	16	7-6	7-5	6-10	5-2	11-10	11-3	9-10	7-7	15-7	14-2	12-8	9-8	19-11	16-10	15-1	11-5
	19.2	7-1	6-11	6-3	4-8	11-2	10-3	8-11	6-11	14-8	12-11	11-7	8-10	18-9	15-4	13-10	10-5
	24	6-7	6-2	5-7	4-2	10-4	9-2	8-0	6-2	13-8	11-7	10-4	7-11	17-3	13-9	12-4	9-4

Rafters

HEAVY ROOF COVERING, NO CEILING: 20 PSF LIVE, 20 PSF DEAD
Maximum Allowable Span, feet - inches

Species Group	Spacing inches oc	2 × 6				2 × 8				2 × 10				2 × 12			
		Sel Str	No.1	No.2	No.3	Sel Str	No.1	No.2	No.3	Sel Str	No.1	No.2	No.3	Sel Str	No.1	No.2	No.3
Douglas fir- larch	12	11-6	10-1	9-5	7-1	17-9	14-9	13-9	10-5	22-5	18-8	17-5	13-2	27-5	22-9	21-4	16-1
	16	10-5	8-9	8-2	6-2	15-4	12-9	11-11	9-0	19-5	16-2	15-1	11-5	23-9	19-9	18-5	13-11
	19.2	9-7	7-11	7-5	5-8	14-0	11-8	10-11	8-3	17-9	14-9	13-9	10-5	21-8	18-0	16-10	12-9
	24	8-7	7-1	6-8	5-0	12-6	10-5	9-9	7-4	15-10	13-2	12-4	9-4	19-5	16-1	15-1	11-5
Hem-fir	12	10-10	9-10	9-3	7-1	17-0	14-4	13-7	10-5	22-1	18-2	17-2	13-2	26-11	22-2	21-0	16-1
	16	9-10	8-6	8-0	6-2	15-1	12-5	11-9	9-0	19-1	15-9	14-11	11-5	23-4	19-3	18-2	13-11
	19.2	9-3	7-9	7-4	5-8	13-9	11-4	10-9	8-3	17-5	14-4	13-7	10-5	21-4	17-7	16-7	12-9
	24	8-5	6-11	6-7	5-0	12-4	10-2	9-7	7-4	15-7	12-10	12-2	9-4	19-1	15-8	14-10	11-5
Southern pine	12	11-3	11-1	10-1	7-7	17-8	16-7	14-5	11-2	23-4	20-10	18-8	14-3	29-9	24-9	22-3	16-10
	16	10-3	9-8	8-9	6-7	16-1	14-4	12-6	9-8	21-2	18-1	16-2	12-4	26-11	21-5	19-3	14-7
	19.2	9-8	8-10	7-11	6-0	15-2	13-1	11-5	8-10	19-11	16-6	14-9	11-3	24-7	19-7	17-7	13-4
	24	8-11	7-11	7-1	5-4	14-1	11-9	10-2	7-11	18-3	14-9	13-2	10-1	22-0	17-6	15-9	11-11

HEAVY ROOF COVERING, NO CEILING: 30 PSF LIVE, 20 PSF DEAD
Maximum Allowable Span, feet - inches

Species Group	Spacing inches oc	2 × 6				2 × 8				2 × 10				2 × 12			
		Sel Str	No.1	No.2	No.3	Sel Str	No.1	No.2	No.3	Sel Str	No.1	No.2	No.3	Sel Str	No.1	No.2	No.3
Douglas fir- larch	12	10-0	9-0	8-5	6-4	15-9	13-2	12-4	9-4	20-1	16-8	15-7	11-9	24-6	20-4	19-1	14-5
	16	9-1	7-10	7-3	5-6	13-9	11-5	10-8	8-1	17-5	14-5	13-6	10-3	21-3	17-8	16-6	12-6
	19.2	8-7	7-1	6-8	5-0	12-6	10-5	9-9	7-4	15-10	13-2	12-4	9-4	19-5	16-1	15-1	11-5
	24	7-8	6-4	4-11	4-6	11-3	9-4	8-9	6-7	14-2	11-9	11-0	8-4	17-4	14-5	13-6	10-2
Hem-fir	12	9-6	8-9	8-4	6-4	14-10	12-10	12-2	9-4	19-7	16-3	15-4	11-9	24-1	19-10	18-9	14-5
	16	8-7	7-7	7-2	5-6	13-6	11-1	10-6	8-1	17-1	14-1	13-4	10-3	20-10	17-2	16-3	12-6
	19.2	8-1	6-11	6-7	5-0	12-4	10-2	9-7	7-4	15-7	12-10	12-2	9-4	19-1	15-8	14-10	11-5
	24	7-6	6-2	5-10	4-6	11-0	9-1	8-7	6-7	13-11	11-6	10-10	8-4	17-1	14-0	13-3	10-2
Southern pine	12	9-10	9-8	9-0	6-9	15-6	14-10	12-11	10-0	20-5	18-8	16-8	12-9	26-0	22-2	19-11	15-1
	16	8-11	8-8	7-10	5-10	14-1	12-10	11-2	8-8	18-6	16-2	14-5	11-0	23-8	19-2	17-3	13-0
	19.2	8-5	7-11	7-1	5-4	13-3	11-9	10-2	7-11	17-5	14-9	13-2	10-1	22-0	17-6	15-9	11-11
	24	7-10	7-1	6-4	4-9	12-3	10-6	9-2	7-1	16-2	13-2	11-9	9-0	19-8	15-8	14-1	10-8

Headers

EXTERIOR-WALL HEADER SPANS FOR DOUGLAS FIR-LARCH, HEM-FIR, AND SOUTHERN PINE

Roof Live Load (psf)		20 Building Width, ft.			30 Building Width, ft.			40 Building Width, ft.			50 Building Width, ft.		
	Size	20	28	36	20	28	36	20	28	36	20	28	36
Headers	2-2×4	3-6	3-2	2-10	3-3	2-10	2-7	3-0	2-7	2-4	2-10	2-5	2-2
supporting	2-2×6	5-5	4-8	4-2	4-10	4-2	3-9	4-5	3-10	3-5	4-1	3-7	3-2
roof and	2-2×8	6-10	5-11	5-4	6-2	5-4	4-9	5-7	4-10	4-4	5-2	4-6	4-0
ceiling	2-2×10	8-5	7-3	6-6	7-6	6-6	5-10	6-10	5-11	5-4	6-4	5-6	4-11
	2-2×12	9-9	8-5	7-6	8-8	7-6	6-9	7-11	6-10	6-2	7-4	6-4	5-8
	3-2×8	8-4	7-5	6-8	7-8	6-8	5-11	7-0	6-1	5-5	6-6	5-8	5-0
	3-2×10	10-6	9-1	8-2	9-5	8-2	7-3	8-7	7-5	6-8	7-11	6-10	6-2
	3-2×12	12-2	10-7	9-5	10-11	9-5	8-5	9-11	8-7	7-8	9-2	8-0	7-2
	4-2×8	9-2	8-4	7-8	8-6	7-8	6-11	8-0	7-0	6-3	7-6	6-6	5-10
	4-2×10	11-8	10-6	9-5	10-10	9-5	8-5	9-11	8-7	7-8	9-2	7-11	7-1
	4-2×12	14-1	12-2	10-11	12-7	10-11	9-9	11-6	9-11	8-11	10-8	9-2	8-3

EXTERIOR-WALL HEADER SPANS FOR DOUGLAS FIR-LARCH, HEM-FIR, AND SOUTHERN PINE

Roof Live Load (psf)		20 Building Width, ft.			30 Building Width, ft.			40 Building Width, ft.			50 Building Width, ft.		
	Size	20	28	36	20	28	36	20	28	36	20	28	36
Headers	2-2×4	3-1	2-9	2-5	2-10	2-6	2-3	2-8	2-4	2-1	2-6	2-2	2-0
supporting	2-2×6	4-6	4-0	3-7	4-2	3-8	3-3	3-11	3-5	3-1	3-8	3-2	2-11
roof, ceiling,	2-2×8	5-9	5-0	4-6	5-3	4-8	4-2	4-11	4-4	3-11	4-8	4-1	3-8
and one floor	2-2×10	7-0	6-2	5-6	6-5	5-8	5-1	6-0	5-3	4-9	5-8	4-11	4-5
(with center	2-2×12	8-1	7-1	6-5	7-6	6-7	5-11	7-0	6-1	5-6	6-7	5-9	5-2
bearing wall)	3-2×8	7-2	6-3	5-8	6-7	5-10	5-3	6-2	5-5	4-10	5-10	5-1	4-7
	3-2×10	8-9	7-8	6-11	8-1	7-1	6-5	7-6	6-7	5-11	7-1	6-2	5-7
	3-2×12	10-2	8-11	8-0	9-4	8-2	7-5	8-9	7-8	6-11	8-3	7-2	6-6
	4-2×8	8-1	7-3	6-7	7-8	6-8	6-0	7-1	6-3	5-7	6-8	5-10	5-3
	4-2×10	10-1	8-10	8-0	9-4	8-2	7-4	8-8	7-7	6-10	8-2	7-2	6-5
	4-2×12	11-9	10-3	9-3	10-10	9-6	8-6	10-1	8-10	7-11	9-6	8-4	7-6

Headers

EXTERIOR-WALL HEADER SPANS FOR DOUGLAS FIR-LARCH, HEM-FIR, AND SOUTHERN PINE

Roof Live Load (psf)		20 Building Width, ft.			30 Building Width, ft.			40 Building Width, ft.			50 Building Width, ft.		
	Size	20	28	36	20	28	36	20	28	36	20	28	36
Headers	2-2×4	2-8	2-4	2-1	2-8	2-3	2-0	2-6	2-2	2-0	2-5	2-1	1-10
supporting	2-2×6	3-11	3-5	3-0	3-10	3-4	3-0	3-8	3-2	2-11	3-6	3-0	2-9
roof, ceiling,	2-2×8	5-0	4-4	3-10	4-10	4-3	3-9	4-8	4-1	3-8	4-5	3-10	3-5
and 1	2-2×10	6-1	5-3	4-8	5-11	5-2	4-7	5-8	4-11	4-5	5-5	4-8	4-3
floor	2-2×12	7-1	6-1	5-5	6-11	6-0	5-4	6-7	5-9	5-2	6-3	5-5	4-11
	3-2×8	6-3	5-5	4-10	6-1	5-3	4-9	5-10	5-1	4-7	5-6	4-10	4-4
	3-2×10	7-7	6-7	5-11	7-5	6-5	5-9	7-1	6-2	5-7	6-9	5-10	5-3
	3-2×12	8-10	7-8	6-10	8-8	7-6	6-8	8-3	7-2	6-6	7-10	6-10	6-1
	4-2×8	7-2	6-3	5-7	7-1	6-1	5-6	6-9	5-10	5-3	6-4	5-7	5-0
	4-2×10	8-9	7-7	6-10	8-7	7-5	6-8	8-3	7-2	6-5	7-9	6-9	6-1
	4-2×12	10-2	8-10	7-11	10-0	8-8	7-9	9-6	8-4	7-6	9-0	7-10	7-1

EXTERIOR-WALL HEADER SPANS FOR DOUGLAS FIR-LARCH, HEM-FIR, AND SOUTHERN PINE

Roof Live Load (psf)		20 Building Width, ft.			30 Building Width, ft.			40 Building Width, ft.			50 Building Width, ft.		
	Size	20	28	36	20	28	36	20	28	36	20	28	36
Headers	2-2×4	2-7	2-3	2-0	2-6	2-2	2-0	2-5	2-1	1-11	2-3	2-0	1-10
supporting	2-2×6	3-9	3-3	2-11	3-8	3-3	2-11	3-6	3-1	2-10	3-4	2-11	2-8
roof, ceiling,	2-2×8	4-9	4-2	3-9	4-8	4-1	3-8	4-5	3-11	3-6	4-3	3-9	3-4
and 2	2-2×10	5-9	5-1	4-7	5-8	5-0	4-6	5-5	4-9	4-4	5-2	4-7	4-1
floors	2-2×12	6-8	5-10	5-3	6-7	5-9	5-2	6-4	5-7	5-0	6-0	5-3	4-9
(with center													
bearing wall)	3-2×8	5-11	5-2	4-8	5-10	5-1	4-7	5-7	4-11	4-5	5-4	4-8	4-2
	3-2×10	7-3	6-4	5-9	7-1	6-3	5-7	6-10	6-0	5-5	6-6	5-8	5-2
	3-2×12	8-5	7-4	6-7	8-3	7-3	6-6	7-11	6-11	6-3	7-6	6-7	6-0
	4-2×8	6-10	6-0	5-5	6-8	5-11	5-4	6-5	5-8	5-1	6-1	5-5	4-10
	4-2×10	8-4	7-4	6-7	8-2	7-2	6-6	7-10	6-11	6-3	7-6	6-7	5-11
	4-2×12	9-8	8-6	7-8	9-6	8-4	7-6	9-1	8-0	7-3	8-8	7-7	6-11

Headers

EXTERIOR-WALL HEADER SPANS FOR DOUGLAS FIR-LARCH, HEM-FIR, AND SOUTHERN PINE

Roof Live Load (psf)		20 Building Width, ft.			30 Building Width, ft.			40 Building Width, ft.			50 Building Width, ft.	
Size	20	28	36	20	28	36	20	28	36	20	28	36
Headers 2-2×4	2-1	1-10	1-7	2-1	1-9	1-7	2-0	1-9	1-7	2-0	1-9	1-7
supporting 2-2×6	3-1	2-8	2-4	3-0	2-7	2-4	3-0	2-7	2-4	2-11	2-7	2-3
roof, ceiling 2-2×8	3-10	3-4	3-0	3-10	3-4	2-11	3-9	3-3	2-11	3-9	3-3	2-11
and 2 2-2×10	4-9	4-1	3-8	4-8	4-0	3-7	4-7	4-0	3-7	4-7	3-11	3-6
floors 2-2×12	5-6	4-9	4-3	5-5	4-8	4-2	5-4	4-8	4-2	5-3	4-7	4-1
3-2×8	4-10	4-2	3-9	4-9	4-2	3-8	4-9	4-1	3-8	4-8	4-1	3-7
3-2×10	5-11	5-1	4-7	5-10	5-1	4-6	5-9	5-0	4-6	5-8	4-11	4-5
3-2×12	6-10	5-11	5-4	6-9	5-10	5-3	6-8	5-10	5-2	6-7	5-9	5-1
4-2×8	5-7	4-10	4-4	5-6	4-9	4-3	5-5	4-9	4-3	5-5	4-8	4-2
4-2×10	6-10	5-11	5-3	6-9	5-10	5-3	6-8	5-9	5-2	6-7	5-8	5-1
4-2×12	7-11	6-10	6-2	7-10	6-9	6-1	7-9	6-8	6-0	7-8	6-7	5-11

INTERIOR-WALL HEADER SPANS FOR DOUGLAS FIR-LARCH, HEM-FIR, AND SOUTHERN PINE

Size	Building Width, ft. 20	28	36	Size	Building Width, ft. 20	28	36
Headers 2-2×4	3-5	2-10	2-6	Headers 2-2×4	2-3	1-11	1-9
supporting 2-2×6	4-11	4-2	3-8	supporting 2-2×6	3-4	2-10	2-6
1 floor 2-2×8	6-3	5-4	4-8	2 floors 2-2×8	4-3	3-7	3-3
(with center 2-2×10	7-8	6-6	5-9	(with center 2-2×10	5-2	4-5	3-11
bearing wall) 2-2×12	8-11	7-6	6-7	bearing wall) 2-2×12	6-0	5-2	4-7
3-2×8	7-10	6-8	5-10	3-2×8	5-4	4-6	4-0
3-2×10	9-7	8-1	7-2	3-2×10	6-6	5-6	4-11
3-2×12	11-1	9-5	8-3	3-2×12	7-6	6-5	5-8
4-2×8	9-0	7-8	6-9	4-2×8	6-1	5-3	4-8
4-2×10	11-1	9-4	8-3	4-2×10	7-6	6-5	5-8
4-2×12	12-10	10-10	9-7	4-2×12	8-8	7-5	6-7

Floor Joists

SLEEPING ROOMS AND ATTICS: 30 PSF LIVE, 10 PSF DEAD
Maximum Allowable Span, feet - inches

Group Species	Spacing inches oc	2 × 6 Sel Str	No.1/ No.2	No.3	2 × 8 Sel Str	No.1/ No.2	No.3	2 × 10 Sel Str	No.1/ No.2	No.3	2 × 12 Sel Str	No.1/ No.2	No.3
Spruce-pine-fir	12	11-7	11-3	9-8	15-3	14-11	12-4	19-5	19-0	15-0	23-7	23-0	17-5
	16	10-6	10-3	8-5	13-10	13-6	10-8	17-8	17-2	13-0	21-6	19-11	15-1
	24	9-2	8-11	6-10	12-1	11-6	8-8	15-5	14-1	10-7	18-9	16-3	12-4
Douglas fir-larch (N)	12	12-6	11-10	9-6	16-6	15-7	12-0	21-0	19-3	14-8	25-7	22-4	17-0
	16	11-4	10-9	8-2	15-0	13-8	10-5	19-1	16-8	12-8	23-3	19-4	14-8
	24	9-11	8-10	6-8	13-1	11-2	8-6	16-8	13-8	10-4	19-10	15-10	12-0
Hem-fir (N)	12	12-0	11-10	10-5	15-10	15-7	13-2	20-3	19-10	16-1	24-8	24-2	18-8
	16	10-11	10-9	9-0	14-5	14-2	11-5	18-5	18-0	13-11	22-5	21-4	16-2
	24	9-7	9-4	7-4	12-7	12-4	9-4	16-1	15-0	11-5	19-7	17-5	13-2
Northern species	12	10-5	10-5	8-1	13-9	13-2	10-3	17-6	16-1	12-7	21-4	18-8	14-7
	16	9-6	9-0	7-0	12-6	11-5	8-11	15-11	13-11	10-11	19-4	16-2	12-7
	24	8-3	7-4	5-9	10-11	9-4	7-3	13-11	11-5	8-11	16-11	13-2	10-4

ALL ROOMS EXCEPT SLEEPING ROOMS AND ATTICS: 40 PSF LIVE, 10 PSF DEAD
Maximum Allowable Span, feet - inches

Group Species	Spacing inches oc	2 × 6 Sel Str	No.1/ No.2	No.3	2 × 8 Sel Str	No.1/ No.2	No.3	2 × 10 Sel Str	No.1/ No.2	No.3	2 × 12 Sel Str	No.1/ No.2	No.3
Spruce-pine-fir	12	10-6	10-3	8-8	13-10	13-6	11-0	17-8	17-3	13-5	21-6	20-7	15-7
	16	9-6	9-4	7-6	12-7	12-3	9-6	16-0	15-5	11-8	19-6	17-10	13-6
	24	8-4	8-1	6-2	11-0	10-3	7-9	14-0	12-7	9-6	17-0	14-7	11-0
Douglas fir-larch (N)	12	11-4	10-9	8-6	15-0	14-1	10-9	19-1	17-3	13-1	23-3	20-0	15-2
	16	10-4	9-8	7-4	13-7	12-3	9-3	17-4	14-11	11-4	21-1	17-4	13-2
	24	9-0	7-11	6-0	11-11	10-0	7-7	15-2	12-2	9-3	17-9	14-2	10-9
Hem-fir (N)	12	10-11	10-9	9-4	14-5	14-2	11-9	18-5	18-0	14-5	22-5	21-11	16-8
	16	9-11	9-9	8-1	13-1	12-10	10-3	16-9	16-5	12-6	20-4	19-1	14-6
	24	8-8	8-6	6-7	11-5	11-0	8-4	14-7	13-5	10-2	17-9	15-7	11-10
Northern species	12	9-6	9-4	7-3	12-6	11-9	9-2	15-11	14-5	11-3	19-4	16-8	13-0
	16	8-7	8-1	6-3	11-4	10-3	8-0	14-6	12-6	9-9	17-7	14-6	11-3
	24	7-6	6-7	5-2	9-11	8-4	6-6	12-8	10-2	7-11	15-2	11-10	9-3

Floor Joists

OFFICE SPACE: 50 PSF LIVE, 10 PSF DEAD

Maximum Allowable Span, feet - inches

Group Species	Spacing inches oc	2 × 6 Sel Str	No.1/ No.2	No.3	2 × 8 Sel Str	No.1/ No.2	No.3	2 × 10 Sel Str	No.1/ No.2	No.3	2 × 12 Sel Str	No.1/ No.2	No.3
Spruce-	12	9-9	9-6	7-11	12-10	12-7	10-0	16-5	16-0	12-3	19-11	18-10	14-3
pine-fir	16	8-10	8-8	6-10	11-8	11-5	8-8	14-11	14-1	10-7	18-1	16-3	12-4
	24	7-9	7-5	5-7	10-2	9-5	7-1	13-0	11-6	8-8	15-10	13-4	10-1
Douglas fir-	12	10-6	9-11	7-9	13-11	12-11	9-9	17-9	15-9	11-11	21-7	18-3	13-10
larch (N)	16	9-7	8-10	6-8	12-7	11-2	8-6	16-1	13-8	10-4	19-7	15-10	12-0
	24	8-4	7-2	5-6	11-0	9-1	6-11	14-0	11-2	8-5	16-3	12-11	9-10
Hem-fir (N)	12	10-2	9-11	8-6	13-5	13-1	10-9	17-1	16-9	13-2	20-9	20-1	15-3
	16	9-3	9-1	7-4	12-2	11-11	9-4	15-6	15-0	11-5	18-10	17-5	13-2
	24	8-1	7-11	6-0	10-8	10-0	7-7	13-7	12-3	9-4	16-3	14-3	10-9
Northern	12	8-9	8-6	6-8	11-7	10-9	8-5	14-9	13-2	10-3	18-0	15-3	11-11
species	16	8-0	7-4	5-9	10-6	9-4	7-3	13-5	11-5	8-11	16-4	13-2	10-4
	24	7-0	6-0	4-8	9-2	7-7	5-11	11-9	9-4	7-3	13-10	10-9	8-5

ALL ROOMS WITH 1.5" OR LESS LIGHTWEIGHT CONCRETE TOPPING: 40 PSF LIVE, 20 PSF DEAD

Maximum Allowable Span, feet - inches

Group Species	Spacing inches oc	2 × 6 Sel Str	No.1/ No.2	No.3	2 × 8 Sel Str	No.1/ No.2	No.3	2 × 10 Sel Str	No.1/ No.2	No.3	2 × 12 Sel Str	No.1/ No.2	No.3
Spruce-	12	10-6	10-3	7-11	13-10	13-3	10-0	17-8	16-3	12-3	21-6	18-10	14-3
pine-fir	16	9-6	9-1	6-10	12-7	11-6	8-8	16-0	14-1	10-7	19-6	16-3	12-4
	24	8-4	7-5	5-7	11-0	9-5	7-1	13-8	11-6	8-8	15-11	13-4	10-1
Douglas fir-	12	11-4	10-2	7-9	15-0	12-11	9-9	19-1	15-9	11-11	22-11	18-3	13-10
larch (N)	16	10-4	8-10	6-8	13-7	11-2	8-6	17-1	13-8	10-4	19-10	15-10	12-0
	24	9-0	7-2	5-6	11-5	9-1	6-11	14-0	11-2	8-5	16-3	12-11	9-10
Hem-fir (N)	12	10-11	10-9	8-6	14-5	14-2	10-9	18-5	17-4	13-2	22-5	20-1	15-3
	16	9-11	9-8	7-4	13-1	12-4	9-4	16-9	15-0	11-5	19-10	17-5	13-2
	24	8-8	7-11	6-0	11-5	10-0	7-7	14-0	12-3	9-4	16-3	14-3	10-9
Northern	12	9-6	8-6	6-8	12-6	10-9	8-5	15-11	13-2	10-3	19-4	15-3	11-11
species	16	8-7	7-4	5-9	11-4	9-4	7-3	14-6	11-5	8-11	17-0	13-2	10-4
	24	7-6	6-0	4-8	9-9	7-7	5-11	11-11	9-4	7-3	13-10	10-9	8-5

Ceiling Joists

DRYWALL–NO FUTURE ROOMS AND NO ATTIC STORAGE: 10 PSF LIVE, 5 PSF DEAD

Maximum Allowable Span, feet - inches

Group Species	Spacing inches oc	2 × 6 Sel Str	2 × 6 No.1/ No.2	2 × 6 No.3	2 × 8 Sel Str	2 × 8 No.1/ No.2	2 × 8 No.3	2 × 10 Sel Str	2 × 10 No.1/ No.2	2 × 10 No.3	2 × 12 Sel Str	2 × 12 No.1/ No.2	2 × 12 No.3
Spruce-	12	12-2	11-10	10-10	19-1	18-8	15-10	25-2	24-7	20-1	32-1	31-4	24-6
pine-fir	16	11-0	10-9	9-5	17-4	16-11	13-9	22-10	22-4	17-5	29-2	28-1	21-3
	24	9-8	9-5	7-8	15-2	14-9	11-2	19-11	18-9	14-2	25-5	22-11	17-4
Douglas fir-	12	13-2	12-5	10-7	20-8	19-6	15-5	27-2	25-8	19-7	34-8	31-6	23-11
larch (N)	16	11-11	11-3	9-2	18-9	17-8	13-5	24-8	22-4	16-11	31-6	27-3	20-8
	24	10-5	9-10	7-6	16-4	14-5	10-11	21-7	18-3	13-10	27-6	22-3	16-11
Hem-fir (N)	12	12-8	12-5	11-7	19-11	19-6	17-0	26-2	25-8	21-6	33-5	32-9	26-4
	16	11-6	11-3	10-1	18-1	17-8	14-9	23-10	23-4	18-8	30-5	29-9	22-9
	24	10-0	9-10	8-3	15-9	15-6	12-0	20-10	20-1	15-3	26-6	24-6	18-7
Northern	12	10-11	10-11	9-1	17-2	17-0	13-3	22-8	21-6	16-10	28-11	26-4	20-6
species	16	9-11	9-11	7-10	15-7	14-9	11-6	20-7	18-8	14-7	26-3	22-9	17-9
	24	8-8	8-3	6-5	13-8	12-0	9-5	18-0	15-3	11-11	22-11	18-7	14-6

DRYWALL–NO FUTURE ROOMS AND LIMITED ATTIC STORAGE: 20 PSF LIVE, 10 PSF DEAD

Maximum Allowable Span, feet - inches

Group Species	Spacing inches oc	2 × 6 Sel Str	2 × 6 No.1/ No.2	2 × 6 No.3	2 × 8 Sel Str	2 × 8 No.1/ No.2	2 × 8 No.3	2 × 10 Sel Str	2 × 10 No.1/ No.2	2 × 10 No.3	2 × 12 Sel Str	2 × 12 No.1/ No.2	2 × 12 No.3
Spruce-	12	9-8	9-5	7-8	15-2	14-9	11-2	19-11	18-9	14-2	25-5	22-11	17-4
pine-fir	16	8-9	8-7	6-8	13-9	12-10	9-8	18-2	16-3	12-4	23-2	19-10	15-0
	24	7-8	7-2	5-5	12-0	10-6	7-11	15-10	13-3	10-0	19-5	16-3	12-3
Douglas fir-	12	10-5	9-10	7-6	16-4	14-5	10-11	21-7	18-3	13-10	27-6	22-3	16-11
larch (N)	16	9-6	8-6	6-6	14-11	12-6	9-6	19-7	15-9	12-0	24-3	19-3	14-8
	24	8-3	7-0	5-3	12-9	10-2	7-9	16-2	12-11	9-9	19-9	15-9	11-11
Hem-fir (N)	12	10-0	9-10	8-3	15-9	15-6	12-0	20-10	20-1	15-3	26-6	24-6	18-7
	16	9-1	8-11	7-1	14-4	13-9	10-5	18-11	17-5	13-2	24-1	21-3	16-1
	24	8-0	7-8	5-10	12-6	11-2	8-6	16-2	14-2	10-9	19-9	17-4	13-2
Northern	12	8-8	8-3	6-5	13-8	12-0	9-5	18-0	15-3	11-11	22-11	18-7	14-6
species	16	7-11	7-1	5-7	12-5	10-5	8-1	16-4	13-2	10-3	20-8	16-1	12-7
	24	6-11	5-10	4-6	10-10	8-6	6-8	13-10	10-9	8-5	16-11	13-2	10-3

Rafters

LIGHT ROOF COVERING, DRYWALL, NO ATTIC SPACE: 20 PSF LIVE, 10 PSF DEAD
Maximum Allowable Span, feet - inches

Group Species	Spacing inches oc	2 × 6 Sel Str	2 × 6 No.1/ No.2	2 × 6 No.3	2 × 8 Sel Str	2 × 8 No.1/ No.2	2 × 8 No.3	2 × 10 Sel Str	2 × 10 No.1/ No.2	2 × 10 No.3	2 × 12 Sel Str	2 × 12 No.1/ No.2	2 × 12 No.3
Spruce-pine-fir	12	15-2	14-9	12-0	19-11	19-6	15-3	25-5	24-7	18-7	30-11	28-6	21-7
	16	13-9	13-5	10-5	18-2	17-5	13-2	23-2	21-4	16-1	28-2	24-8	18-8
	24	12-0	11-3	8-6	15-10	14-3	10-9	20-2	17-5	13-2	24-1	20-2	15-3
Douglas fir-larch (N)	12	16-4	15-5	11-9	21-7	19-7	14-10	27-6	23-11	18-1	33-6	27-8	21-0
	16	14-11	13-4	10-2	19-7	16-11	12-10	25-0	20-8	15-8	30-1	24-0	18-2
	24	13-0	10-11	8-3	17-2	13-10	10-6	21-2	16-11	12-10	24-7	19-7	14-10
Hem-fir (N)	12	15-9	15-6	12-11	20-10	20-5	16-4	26-6	26-0	19-11	32-3	30-6	23-1
	16	14-4	14-1	11-2	18-11	18-6	14-2	24-1	22-9	17-3	29-4	26-5	20-0
	24	12-6	12-0	9-1	16-6	15-3	11-7	21-1	18-7	14-1	24-7	21-7	16-4
Northern species	12	13-8	12-11	10-1	18-0	16-4	12-9	22-11	19-11	15-7	27-11	23-1	18-0
	16	12-5	11-2	8-9	16-4	14-2	11-0	20-10	17-3	13-6	25-4	20-0	15-7
	24	10-10	9-1	7-1	14-3	11-7	9-0	18-1	14-1	11-0	21-0	16-4	12-9

LIGHT ROOF COVERING, DRYWALL, NO ATTIC SPACE: 30 PSF LIVE, 10 PSF DEAD
Maximum Allowable Span, feet - inches

Group Species	Spacing inches oc	2 × 6 Sel Str	2 × 6 No.1/ No.2	2 × 6 No.3	2 × 8 Sel Str	2 × 8 No.1/ No.2	2 × 8 No.3	2 × 10 Sel Str	2 × 10 No.1/ No.2	2 × 10 No.3	2 × 12 Sel Str	2 × 12 No.1/ No.2	2 × 12 No.3
Spruce-pine-fir	12	13-3	12-11	10-5	17-5	17-0	13-2	22-3	21-4	16-1	27-1	24-8	18-8
	16	12-0	11-9	9-0	15-10	15-1	11-5	20-2	18-5	13-11	24-7	21-5	16-2
	24	10-6	9-9	7-4	13-10	12-4	9-4	17-8	15-1	11-5	20-11	17-6	13-2
Douglas fir-larch (N)	12	14-4	13-4	10-2	18-10	16-11	12-10	24-1	20-8	15-8	29-3	24-0	18-2
	16	13-0	11-7	8-9	17-2	14-8	11-2	21-10	17-11	13-7	26-1	20-9	15-9
	24	11-4	9-5	7-2	15-0	12-0	9-1	18-4	14-8	11-1	21-3	17-0	12-10
Hem-fir (N)	12	13-9	13-6	11-2	18-2	17-10	14-2	23-2	22-9	17-3	28-2	26-5	20-0
	16	12-6	12-3	9-8	16-6	16-2	12-3	21-1	19-9	14-11	25-7	22-10	17-4
	24	10-11	10-5	7-11	14-5	13-2	10-0	18-4	16-1	12-3	21-3	18-8	14-2
Northern species	12	11-11	11-2	8-9	15-9	14-2	11-0	20-1	17-3	13-6	24-5	20-0	15-7
	16	10-10	9-8	7-7	14-3	12-3	9-7	18-3	14-11	11-8	22-2	17-4	13-6
	24	9-6	7-11	6-2	12-6	10-0	7-10	15-8	12-3	9-6	18-2	14-2	11-1

Rafters

LIGHT ROOF COVERING, DRYWALL, NO ATTIC SPACE: 40 PSF LIVE, 10 PSF DEAD

Maximum Allowable Span, feet - inches

Group Species	Spacing inches oc	2 × 6 Sel Str	2 × 6 No.1/ No.2	2 × 6 No.3	2 × 8 Sel Str	2 × 8 No.1/ No.2	2 × 8 No.3	2 × 10 Sel Str	2 × 10 No.1/ No.2	2 × 10 No.3	2 × 12 Sel Str	2 × 12 No.1/ No.2	2 × 12 No.3
Spruce-pine-fir	12	12-0	11-9	9-4	15-10	15-6	11-9	20-2	19-1	14-5	24-7	22-1	16-8
	16	10-11	10-8	8-1	14-5	13-6	10-3	18-4	16-6	12-6	22-4	19-2	14-6
	24	9-6	8-9	6-7	12-7	11-0	8-4	16-0	13-6	10-2	18-8	15-7	11-10
Douglas fir-larch (N)	12	13-0	12-0	9-1	17-2	15-2	11-6	21-10	18-6	14-0	26-7	21-5	16-3
	16	11-10	10-4	7-10	15-7	13-1	9-11	19-10	16-0	12-2	23-4	18-7	14-1
	24	10-4	8-5	6-5	13-5	10-9	8-2	16-5	13-1	9-11	19-1	15-2	11-6
Hem-fir (N)	12	12-6	12-3	10-0	16-6	16-2	12-8	21-1	20-4	15-5	25-7	23-7	17-11
	16	11-5	11-2	8-8	15-0	14-5	10-11	19-2	17-8	13-5	23-3	20-5	15-6
	24	9-11	9-4	7-1	13-1	11-9	8-11	16-5	14-5	10-11	19-1	16-8	12-8
Northern species	12	10-10	10-0	7-9	14-3	12-8	9-10	18-3	15-5	12-1	22-2	17-11	14-0
	16	9-10	8-8	6-9	13-0	10-11	8-7	16-7	13-5	10-5	19-11	15-6	12-1
	24	8-7	7-1	5-6	11-4	8-11	7-0	14-0	10-11	8-6	16-3	12-8	9-11

LIGHT ROOF COVERING, DRYWALL, NO ATTIC SPACE: 50 PSF LIVE, 10 PSF DEAD

Maximum Allowable Span, feet - inches

Group Species	Spacing inches oc	2 × 6 Sel Str	2 × 6 No.1/ No.2	2 × 6 No.3	2 × 8 Sel Str	2 × 8 No.1/ No.2	2 × 8 No.3	2 × 10 Sel Str	2 × 10 No.1/ No.2	2 × 10 No.3	2 × 12 Sel Str	2 × 12 No.1/ No.2	2 × 12 No.3
Spruce-pine-fir	12	11-2	10-11	8-6	14-8	14-3	10-9	18-9	17-5	13-2	22-10	20-2	15-3
	16	10-2	9-9	7-4	13-4	12-4	9-4	17-0	15-1	11-5	20-9	17-6	13-2
	24	8-10	7-11	6-0	11-8	10-1	7-7	14-8	12-4	9-4	17-1	14-3	10-9
Douglas fir-larch (N)	12	12-1	10-11	8-3	15-11	13-10	10-6	20-3	16-11	12-10	24-7	19-7	14-10
	16	11-0	9-5	7-2	14-5	12-0	9-1	18-4	14-8	11-1	21-3	17-0	12-10
	24	9-7	7-9	5-10	12-3	9-9	7-5	15-0	11-11	9-1	17-5	13-10	10-6
Hem-fir (N)	12	11-8	11-5	9-1	15-4	15-0	11-7	19-7	18-7	14-1	23-9	21-7	16-4
	16	10-7	10-4	7-11	13-11	13-2	10-0	17-9	16-1	12-3	21-3	18-8	14-2
	24	9-3	8-6	6-5	12-2	10-9	8-2	15-0	13-2	10-0	17-5	15-3	11-7
Northern species	12	10-1	9-1	7-1	13-3	11-7	9-0	16-11	14-1	11-0	20-7	16-4	12-9
	16	9-2	7-11	6-2	12-1	10-0	7-10	15-4	12-3	9-6	18-2	14-2	11-1
	24	8-0	6-5	5-0	10-6	8-2	6-4	12-10	10-0	7-9	14-10	11-7	9-0

Rafters

MEDIUM ROOF COVERING, DRYWALL, NO ATTIC SPACE: 20 PSF LIVE, 15 PSF DEAD
Maximum Allowable Span, feet - inches

Group Species	Spacing inches oc	2×6 Sel Str	No.1/ No.2	No.3	2×8 Sel Str	No.1/ No.2	No.3	2×10 Sel Str	No.1/ No.2	No.3	2×12 Sel Str	No.1/ No.2	No.3
Spruce-pine-fir	12	15-2	14-9	11-2	19-11	18-8	14-1	25-5	22-9	17-3	30-11	26-5	20-0
	16	13-9	12-9	9-8	18-2	16-2	12-2	23-2	19-9	14-11	27-4	22-10	17-3
	24	12-0	10-5	7-10	15-9	13-2	10-0	19-3	16-1	12-2	22-4	18-8	14-1
Douglas fir-larch (N)	12	16-4	14-4	10-10	21-7	18-1	13-9	27-6	22-1	16-9	32-2	25-8	19-5
	16	14-11	12-5	9-5	19-7	15-8	11-11	24-0	19-2	14-6	27-11	22-2	16-10
	24	12-8	10-1	7-8	16-1	12-10	9-9	19-8	15-8	11-10	22-9	18-2	13-9
Hem-fir (N)	12	15-9	15-6	11-11	20-10	19-11	15-1	26-6	24-4	18-6	32-2	28-3	21-5
	16	14-4	13-8	10-4	18-11	17-3	13-1	24-0	21-1	16-0	27-11	24-5	18-6
	24	12-6	11-2	8-5	16-1	14-1	10-8	19-8	17-3	13-1	22-9	20-0	15-2
Northern species	12	13-8	11-11	9-4	18-0	15-1	11-9	22-11	18-6	14-5	27-6	21-5	16-8
	16	12-5	10-4	8-1	16-4	13-1	10-3	20-7	16-0	12-6	23-10	18-6	14-6
	24	10-10	8-5	6-7	13-9	10-8	8-4	16-9	13-1	10-2	19-5	15-2	11-10

MEDIUM ROOF COVERING, DRYWALL, NO ATTIC SPACE: 30 PSF LIVE, 15 PSF DEAD
Maximum Allowable Span, feet - inches

Group Species	Spacing inches oc	2×6 Sel Str	No.1/ No.2	No.3	2×8 Sel Str	No.1/ No.2	No.3	2×10 Sel Str	No.1/ No.2	No.3	2×12 Sel Str	No.1/ No.2	No.3
Spruce-pine-fir	12	13-3	12-11	9-10	17-5	16-5	12-5	22-3	20-1	15-2	27-1	23-3	17-7
	16	12-0	11-3	8-6	15-10	14-3	10-9	20-2	17-5	13-2	24-1	20-2	15-3
	24	10-6	9-2	6-11	13-10	11-8	8-9	17-0	14-2	10-9	19-8	16-6	12-5
Douglas fir-larch (N)	12	14-4	12-7	9-7	18-10	16-0	12-1	24-1	19-6	14-10	28-5	22-7	17-2
	16	13-0	10-11	8-3	17-2	13-10	10-6	21-2	16-11	12-10	24-7	19-7	14-10
	24	11-2	8-11	6-9	14-2	11-3	8-7	17-4	13-9	10-6	20-1	16-0	12-2
Hem-fir (N)	12	13-9	13-6	10-6	18-2	17-7	13-4	23-2	21-6	16-3	28-2	24-11	18-11
	16	12-6	12-0	9-1	16-6	15-3	11-7	21-1	18-7	14-1	24-7	21-7	16-4
	24	10-11	9-10	7-5	14-2	12-5	9-5	17-4	15-2	11-6	20-1	17-7	13-4
Northern species	12	11-11	10-6	8-3	15-9	13-4	10-5	20-1	16-3	12-8	24-3	18-11	14-9
	16	10-10	9-1	7-1	14-3	11-7	9-0	18-1	14-1	11-0	21-0	16-4	12-9
	24	9-6	7-5	5-10	12-1	9-5	7-4	14-10	11-6	9-0	17-2	13-4	10-5

Rafters

MEDIUM ROOF COVERING, DRYWALL, NO ATTIC SPACE: 40 PSF LIVE, 15 PSF DEAD
Maximum Allowable Span, feet - inches

Group Species	Spacing inches oc	2 × 6 Sel Str	No.1/ No.2	No.3	2 × 8 Sel Str	No.1/ No.2	No.3	2 × 10 Sel Str	No.1/ No.2	No.3	2 × 12 Sel Str	No.1/ No.2	No.3
Spruce-pine-fir	12	12-0	11-9	8-11	15-10	14-10	11-3	20-2	18-2	13-9	24-7	21-1	15-11
	16	10-11	10-2	7-8	14-5	12-11	9-9	18-4	15-9	11-11	21-10	18-3	13-9
	24	9-6	8-4	6-3	12-7	10-6	7-11	15-4	12-10	9-9	17-10	14-11	11-3
Douglas fir-larch (N)	12	13-0	11-5	8-8	17-2	14-5	10-11	21-10	17-8	13-5	25-8	20-5	15-6
	16	11-10	9-10	7-6	15-7	12-6	9-6	19-2	15-3	11-7	22-3	17-9	13-5
	24	10-1	8-1	6-1	12-10	10-3	7-9	15-8	12-6	9-6	18-2	14-6	11-0
Hem-fir (N)	12	12-6	12-3	9-6	16-6	15-11	12-1	21-1	19-5	14-9	25-7	22-6	17-1
	16	11-5	10-10	8-3	15-0	13-9	10-5	19-2	16-10	12-9	22-3	19-6	14-9
	24	9-11	8-11	6-9	12-10	11-3	8-6	15-8	13-9	10-5	18-2	15-11	12-1
Northern species	12	10-10	9-6	7-5	14-3	12-1	9-5	18-3	14-9	11-6	21-11	17-1	13-4
	16	9-10	8-3	6-5	13-0	10-5	8-2	16-5	12-9	9-11	19-0	14-9	11-6
	24	8-7	6-9	5-3	10-11	8-6	6-8	13-5	10-5	8-1	15-6	12-1	9-5

MEDIUM ROOF COVERING, DRYWALL, NO ATTIC SPACE: 50 PSF LIVE, 15 PSF DEAD
Maximum Allowable Span, feet - inches

Group Species	Spacing inches oc	2 × 6 Sel Str	No.1/ No.2	No.3	2 × 8 Sel Str	No.1/ No.2	No.3	2 × 10 Sel Str	No.1/ No.2	No.3	2 × 12 Sel Str	No.1/ No.2	No.3
Spruce-pine-fir	12	11-2	10-10	8-2	14-8	13-8	10-4	18-9	16-9	12-8	22-10	19-5	14-8
	16	10-2	9-4	7-1	13-4	11-10	8-11	17-0	14-6	10-11	20-1	16-9	12-8
	24	8-10	7-8	5-9	11-7	9-8	7-4	14-1	11-10	8-11	16-5	13-8	10-4
Douglas fir-larch (N)	12	12-1	10-6	8-0	15-11	13-3	10-1	20-3	16-3	12-4	23-7	18-10	14-3
	16	11-0	9-1	6-11	14-5	11-6	8-9	17-8	14-1	10-8	20-5	16-4	12-4
	24	9-4	7-5	5-8	11-9	9-5	7-2	14-5	11-6	8-8	16-8	13-4	10-1
Hem-fir (N)	12	11-8	11-5	8-9	15-4	14-7	11-1	19-7	17-10	13-7	23-7	20-9	15-9
	16	10-7	10-0	7-7	13-11	12-8	9-7	17-8	15-6	11-9	20-5	17-11	13-7
	24	9-3	8-2	6-2	11-9	10-4	7-10	14-5	12-8	9-7	16-8	14-8	11-1
Northern species	12	10-1	8-9	6-10	13-3	11-1	8-8	16-11	13-7	10-7	20-2	15-9	12-3
	16	9-2	7-7	5-11	12-1	9-7	7-6	15-1	11-9	9-2	17-6	13-7	10-7
	24	8-0	6-2	4-10	10-1	7-10	6-1	12-4	9-7	7-6	14-3	11-1	8-8

Rafters

HEAVY ROOF COVERING, DRYWALL, NO ATTIC SPACE: 20 PSF LIVE, 20 PSF DEAD
Maximum Allowable Span, feet - inches

Group Species	Spacing inches oc	2 × 6 Sel Str	No.1/ No.2	No.3	2 × 8 Sel Str	No.1/ No.2	No.3	2 × 10 Sel Str	No.1/ No.2	No.3	2 × 12 Sel Str	No.1/ No.2	No.3
Spruce-pine-fir	12	15-2	13-9	10-5	19-11	17-5	13-2	25-5	21-4	16-1	29-6	24-8	18-8
	16	13-9	11-11	9-0	18-1	15-1	11-5	22-1	18-5	13-11	25-7	21-5	16-2
	24	11-8	9-9	7-4	14-9	12-4	9-4	18-0	15-1	11-5	20-11	17-6	13-2
Douglas fir-larch (N)	12	16-4	13-4	10-2	21-3	16-11	12-10	26-0	20-8	15-8	30-1	24-0	18-2
	16	14-6	11-7	8-9	18-5	14-8	11-2	22-6	17-11	13-7	26-1	20-9	15-9
	24	11-10	9-5	7-2	15-0	12-0	9-1	18-4	14-8	11-1	21-3	17-0	12-10
Hem-fir (N)	12	15-9	14-9	11-2	20-10	18-8	14-2	26-0	22-9	17-3	30-1	26-5	20-0
	16	14-4	12-9	9-8	18-5	16-2	12-3	22-6	19-9	14-11	26-1	22-10	17-4
	24	11-10	10-5	7-11	15-0	13-2	10-0	18-4	16-1	12-3	21-3	18-8	14-2
Northern species	12	13-8	11-2	8-9	18-0	14-2	11-0	22-2	17-3	13-6	25-9	20-0	15-7
	16	12-5	9-8	7-7	15-9	12-3	9-7	19-3	14-11	11-8	22-3	17-4	13-6
	24	10-2	7-11	6-2	12-10	10-0	7-10	15-8	12-3	9-6	18-2	14-2	11-1

HEAVY ROOF COVERING, DRYWALL, NO ATTIC SPACE: 30 PSF LIVE, 20 PSF DEAD
Maximum Allowable Span, feet - inches

Group Species	Spacing inches oc	2 × 6 Sel Str	No.1/ No.2	No.3	2 × 8 Sel Str	No.1/ No.2	No.3	2 × 10 Sel Str	No.1/ No.2	No.3	2 × 12 Sel Str	No.1/ No.2	No.3
Spruce-pine-fir	12	13-3	12-4	9-4	17-5	15-7	11-9	22-3	19-1	14-5	26-5	22-1	16-8
	16	12-0	10-8	8-1	15-10	13-6	10-3	19-9	16-6	12-6	22-10	19-2	14-6
	24	10-5	8-9	6-7	13-2	11-0	8-4	16-1	13-6	10-2	18-8	15-7	11-10
Douglas fir-larch (N)	12	14-4	12-0	9-1	18-10	15-2	11-6	23-3	18-6	14-0	26-11	21-5	16-3
	16	13-0	10-4	7-10	16-6	13-1	9-11	20-1	16-0	12-2	23-4	18-7	14-1
	24	10-7	8-5	6-5	13-5	10-9	8-2	16-5	13-1	9-11	19-1	15-2	11-6
Hem-fir (N)	12	13-9	13-2	10-0	18-2	16-8	12-8	23-2	20-4	15-5	26-11	23-7	17-11
	16	12-6	11-5	8-8	16-6	14-5	10-11	20-1	17-8	13-5	23-4	20-5	15-6
	24	10-7	9-4	7-1	13-5	11-9	8-11	16-5	14-5	10-11	19-1	16-8	12-8
Northern species	12	11-11	10-0	7-9	15-9	12-8	9-10	19-10	15-5	12-1	23-0	17-11	14-0
	16	10-10	8-8	6-9	14-1	10-11	8-7	17-2	13-5	10-5	19-11	15-6	12-1
	24	9-1	7-1	5-6	11-6	8-11	7-0	14-0	10-11	8-6	16-3	12-8	9-11

Rafters

LIGHT ROOF COVERING, NO CEILING: 20 PSF LIVE, 10 PSF DEAD
Maximum Allowable Span, feet - inches

Group Species	Spacing inches oc	2 × 6 Sel Str	No.1/ No.2	No.3	2 × 8 Sel Str	No.1/ No.2	No.3	2 × 10 Sel Str	No.1/ No.2	No.3	2 × 12 Sel Str	No.1/ No.2	No.3
Spruce-	12	10-7	10-4	8-3	16-8	15-11	12-0	21-11	20-2	15-3	28-0	24-7	18-7
pine-fir	16	9-8	9-5	7-1	15-2	13-9	10-5	19-11	17-5	13-2	25-5	21-4	16-1
	24	8-5	7-8	5-10	13-3	11-3	8-6	17-0	14-3	10-9	20-9	17-5	13-2
Douglas fir-	12	11-6	10-7	8-0	18-0	15-5	11-9	23-9	19-7	14-10	30-0	23-11	18-1
larch (N)	16	10-5	9-2	6-11	16-4	13-4	10-2	21-3	16-11	12-10	26-0	20-8	15-8
	24	9-1	7-6	5-8	13-8	10-11	8-3	17-4	13-10	10-6	21-2	16-11	12-10
Hem-fir (N)	12	11-1	10-10	8-10	17-4	17-0	12-11	22-11	21-6	16-4	29-2	26-4	19-11
	16	10-10	9-10	7-8	15-9	14-9	11-2	20-10	18-8	14-2	26-0	22-9	17-3
	24	8-9	8-3	6-3	13-8	12-0	9-1	17-4	15-3	11-7	21-2	18-7	14-1
Northern	12	9-7	8-10	6-10	15-0	12-11	10-1	19-10	16-4	12-9	25-3	19-11	15-7
species	16	8-8	7-8	5-11	13-8	11-2	8-9	18-0	14-2	11-0	22-2	17-3	13-6
	24	7-7	6-3	4-10	11-9	9-1	7-1	14-10	11-7	9-0	18-1	14-1	11-0

LIGHT ROOF COVERING, NO CEILING: 30 PSF LIVE, 10 PSF DEAD
Maximum Allowable Span, feet - inches

Group Species	Spacing inches oc	2 × 6 Sel Str	No.1/ No.2	No.3	2 × 8 Sel Str	No.1/ No.2	No.3	2 × 10 Sel Str	No.1/ No.2	No.3	2 × 12 Sel Str	No.1/ No.2	No.3
Spruce-	12	9-3	9-1	7-1	14-7	13-9	10-5	19-2	17-5	13-2	24-6	21-4	16-1
pine-fir	16	8-5	8-2	6-2	13-3	11-11	9-0	17-5	15-1	11-5	22-1	18-5	13-11
	24	7-4	6-8	5-0	11-7	9-9	7-4	14-9	12-4	9-4	18-0	15-1	11-5
Douglas fir-	12	10-0	9-2	6-11	15-9	13-4	10-2	20-9	16-11	12-10	26-0	20-8	15-8
larch (N)	16	9-1	7-11	6-0	14-4	11-7	8-9	18-5	14-8	11-2	22-6	17-11	13-7
	24	7-11	6-6	4-11	11-10	9-5	7-2	15-0	12-0	9-1	18-4	14-8	11-1
Hem-fir (N)	12	9-8	9-6	7-8	15-2	14-9	11-2	20-0	18-8	14-2	25-6	22-9	17-3
	16	8-9	8-7	6-7	13-9	12-9	9-8	18-2	16-2	12-3	22-6	19-9	14-11
	24	7-8	7-1	5-5	11-10	10-5	7-11	15-0	13-2	10-0	18-4	16-1	12-3
Northern	12	8-4	7-8	5-11	13-1	11-2	8-9	17-4	14-2	11-0	22-1	17-3	13-6
species	16	7-7	6-7	5-2	11-11	9-8	7-7	15-9	12-3	9-7	19-3	14-11	11-8
	24	6-8	5-5	4-3	10-2	7-11	6-2	12-10	10-0	7-10	15-8	12-3	9-6

Rafters

LIGHT ROOF COVERING, NO CEILING: 40 PSF LIVE, 10 PSF DEAD
Maximum Allowable Span, feet - inches

Group Species	Spacing inches oc	2 × 6			2 × 8			2 × 10			2 × 12		
		Sel Str	No.1/ No.2	No.3	Sel Str	No.1/ No.2	No.3	Sel Str	No.1/ No.2	No.3	Sel Str	No.1/ No.2	No.3
Spruce-pine-fir	12	8-5	8-3	6-4	13-3	12-4	9-4	17-5	15-7	11-9	22-3	19-1	14-5
	16	7-8	7-3	5-6	12-0	10-8	8-1	15-10	13-6	10-3	19-9	16-6	12-6
	24	6-8	5-11	4-6	10-5	8-9	6-7	13-2	11-0	8-4	16-1	13-6	10-2
Douglas fir-larch (N)	12	9-1	8-2	6-2	14-4	12-0	9-1	18-10	15-2	11-6	23-3	18-6	14-0
	16	8-3	7-1	5-4	13-0	10-4	7-10	16-6	13-1	9-11	20-1	16-0	12-2
	24	7-3	5-9	4-5	10-7	8-5	6-5	13-5	10-9	8-2	16-5	13-1	9-11
Hem-fir (N)	12	8-9	8-7	6-10	13-9	13-2	10-0	18-2	16-8	12-8	23-2	20-4	15-5
	16	8-0	7-10	5-11	12-6	11-5	8-8	16-6	14-5	10-11	20-1	17-8	13-5
	24	7-0	6-4	4-10	10-7	9-4	7-1	13-5	11-9	8-11	16-5	14-5	10-11
Northern species	12	7-7	6-10	5-4	11-11	10-0	7-9	15-9	12-8	9-10	19-10	15-5	12-1
	16	6-11	5-11	4-7	10-10	8-8	6-9	14-1	10-11	8-7	17-2	13-5	10-5
	24	6-0	4-10	3-9	9-1	7-1	5-6	11-6	8-11	7-0	14-0	10-11	8-6

LIGHT ROOF COVERING, NO CEILING: 50 PSF LIVE, 10 PSF DEAD
Maximum Allowable Span, feet - inches

Group Species	Spacing inches oc	2 × 6			2 × 8			2 × 10			2 × 12		
		Sel Str	No.1/ No.2	No.3	Sel Str	No.1/ No.2	No.3	Sel Str	No.1/ No.2	No.3	Sel Str	No.1/ No.2	No.3
Spruce-pine-fir	12	7-10	7-8	5-10	12-3	11-3	8-6	16-2	14-3	10-9	20-8	17-5	13-2
	16	7-1	6-8	5-0	11-2	9-9	7-4	14-8	12-4	9-4	18-0	15-1	11-5
	24	6-2	5-5	4-1	9-6	7-11	6-0	12-0	10-1	7-7	14-8	12-4	9-4
Douglas fir-larch (N)	12	8-5	7-6	5-8	13-3	10-11	8-3	17-4	13-10	10-6	21-2	16-11	12-10
	16	7-8	6-6	4-11	11-10	9-5	7-2	15-0	12-0	9-1	18-4	14-8	11-1
	24	6-7	5-3	4-0	9-8	7-9	5-10	12-3	9-9	7-5	15-0	11-11	9-1
Hem-fir (N)	12	8-2	8-0	6-3	12-10	12-0	9-1	16-10	15-3	11-7	21-2	18-7	14-1
	16	7-5	7-1	5-5	11-8	10-5	7-11	15-0	13-2	10-0	18-4	16-1	12-3
	24	6-6	5-10	4-5	9-8	8-6	6-5	12-3	10-9	8-2	15-0	13-2	10-0
Northern species	12	7-1	6-3	4-10	11-1	9-1	7-1	14-7	11-7	9-0	18-1	14-1	11-0
	16	6-5	5-5	4-3	10-1	7-11	6-2	12-10	10-0	7-10	15-8	12-3	9-6
	24	5-7	4-5	3-5	8-3	6-5	5-0	10-6	8-2	6-4	12-10	10-0	7-9

Rafters

MEDIUM ROOF COVERING, NO CEILING: 20 PSF LIVE, 15 PSF DEAD
Maximum Allowable Span, feet - inches

Group Species	Spacing inches oc	2 × 6 Sel Str	2 × 6 No.1/ No.2	2 × 6 No.3	2 × 8 Sel Str	2 × 8 No.1/ No.2	2 × 8 No.3	2 × 10 Sel Str	2 × 10 No.1/ No.2	2 × 10 No.3	2 × 12 Sel Str	2 × 12 No.1/ No.2	2 × 12 No.3
Spruce-pine-fir	12	10-7	10-1	7-7	16-8	14-9	11-2	21-11	18-8	14-1	27-3	22-9	17-3
	16	9-8	8-9	6-7	15-2	12-9	9-8	19-4	16-2	12-2	23-7	19-9	14-11
	24	8-5	7-1	5-5	12-5	10-5	7-10	15-9	13-2	10-0	19-3	16-1	12-2
Douglas fir-larch (N)	12	11-6	9-9	7-5	17-11	14-4	10-10	22-9	18-1	13-9	27-9	22-1	16-9
	16	10-5	8-6	6-5	15-6	12-5	9-5	19-8	15-8	11-11	24-0	19-2	14-6
	24	8-8	6-11	5-3	12-8	10-1	7-8	16-1	12-10	9-9	19-8	15-8	11-10
Hem-fir (N)	12	11-1	10-9	8-2	17-4	15-9	11-11	22-9	19-11	15-1	27-9	24-4	18-6
	16	10-0	9-4	7-1	15-6	13-8	10-4	19-8	17-3	13-1	24-0	21-1	16-0
	24	8-8	7-7	5-9	12-8	11-2	8-5	16-1	14-1	10-8	19-8	17-3	13-1
Northern species	12	9-7	8-2	6-4	15-0	11-11	9-4	19-5	15-1	11-9	23-9	18-6	14-5
	16	8-8	7-1	5-6	13-3	10-4	8-1	16-10	13-1	10-3	20-7	16-0	12-6
	24	7-5	5-9	4-6	10-10	8-5	6-7	13-9	10-8	8-4	16-9	13-1	10-2

MEDIUM ROOF COVERING, NO CEILING: 30 PSF LIVE, 15 PSF DEAD
Maximum Allowable Span, feet - inches

Group Species	Spacing inches oc	2 × 6 Sel Str	2 × 6 No.1/ No.2	2 × 6 No.3	2 × 8 Sel Str	2 × 8 No.1/ No.2	2 × 8 No.3	2 × 10 Sel Str	2 × 10 No.1/ No.2	2 × 10 No.3	2 × 12 Sel Str	2 × 12 No.1/ No.2	2 × 12 No.3
Spruce-pine-fir	12	9-3	8-10	6-8	14-7	13-0	9-10	19-2	16-5	12-5	24-0	20-1	15-2
	16	8-5	7-8	5-10	13-3	11-3	8-6	17-0	14-3	10-9	20-9	17-5	13-2
	24	7-4	6-3	4-9	11-0	9-2	6-11	13-11	11-8	8-9	17-0	14-2	10-9
Douglas fir-larch (N)	12	10-0	8-7	6-6	15-9	12-7	9-7	20-0	16-0	12-1	24-6	19-6	14-10
	16	9-1	7-6	5-8	13-8	10-11	8-3	17-4	13-10	10-6	21-2	16-11	12-10
	24	7-8	6-1	4-7	11-2	8-11	6-9	14-2	11-3	8-7	17-4	13-9	10-6
Hem-fir (N)	12	9-8	9-6	7-2	15-2	13-11	10-6	20-0	17-7	13-4	24-6	21-6	16-3
	16	8-9	8-3	6-3	13-8	12-0	9-1	17-4	15-3	11-7	21-2	18-7	14-1
	24	7-8	6-8	5-1	11-2	9-10	7-5	14-2	12-5	9-5	17-4	15-2	11-6
Northern species	12	8-4	7-2	5-7	13-1	10-6	8-3	17-2	13-4	10-5	20-11	16-3	12-8
	16	7-7	6-3	4-10	11-9	9-1	7-1	14-10	11-7	9-0	18-1	14-1	11-0
	24	6-6	5-1	4-0	9-7	7-5	5-10	12-1	9-5	7-4	14-10	11-6	9-0

Rafters

MEDIUM ROOF COVERING, NO CEILING: 40 PSF LIVE, 15 PSF DEAD
Maximum Allowable Span, feet - inches

Group Species	Spacing inches oc	2 × 6 Sel Str	2 × 6 No.1/ No.2	2 × 6 No.3	2 × 8 Sel Str	2 × 8 No.1/ No.2	2 × 8 No.3	2 × 10 Sel Str	2 × 10 No.1/ No.2	2 × 10 No.3	2 × 12 Sel Str	2 × 12 No.1/ No.2	2 × 12 No.3
Spruce-pine-fir	12	8-5	8-0	6-1	13-3	11-9	8-11	17-5	14-10	11-3	21-9	18-2	13-9
	16	7-8	6-11	5-3	12-0	10-2	7-8	15-5	12-11	9-9	18-10	15-9	11-11
	24	6-8	5-8	4-3	9-11	8-4	6-3	12-7	10-6	7-11	15-4	12-10	9-9
Douglas fir-larch (N)	12	9-1	7-10	5-11	14-4	11-5	8-8	18-2	14-5	10-11	22-2	17-8	13-5
	16	8-3	6-9	5-1	12-5	9-10	7-6	15-8	12-6	9-6	19-2	15-3	11-7
	24	6-11	5-6	4-2	10-1	8-1	6-1	12-10	10-3	7-9	15-8	12-6	9-6
Hem-fir (N)	12	8-9	8-7	6-6	13-9	12-7	9-6	18-2	15-11	12-1	22-2	19-5	14-9
	16	8-0	7-5	5-8	12-5	10-10	8-3	15-8	13-9	10-5	19-2	16-10	12-9
	24	6-11	6-1	4-7	10-1	8-11	6-9	12-10	11-3	8-6	15-8	13-9	10-5
Northern species	12	7-7	6-6	5-1	11-11	9-6	7-5	15-6	12-1	9-5	18-11	14-9	11-6
	16	6-11	5-8	4-5	10-7	8-3	6-5	13-5	10-5	8-2	16-5	12-9	9-11
	24	5-11	4-7	3-7	8-8	6-9	5-3	10-11	8-6	6-8	13-5	10-5	8-1

MEDIUM ROOF COVERING, NO CEILING: 50 PSF LIVE, 15 PSF DEAD
Maximum Allowable Span, feet - inches

Group Species	Spacing inches oc	2 × 6 Sel Str	2 × 6 No.1/ No.2	2 × 6 No.3	2 × 8 Sel Str	2 × 8 No.1/ No.2	2 × 8 No.3	2 × 10 Sel Str	2 × 10 No.1/ No.2	2 × 10 No.3	2 × 12 Sel Str	2 × 12 No.1/ No.2	2 × 12 No.3
Spruce-pine-fir	12	7-10	7-5	5-7	12-3	10-10	8-2	16-2	13-8	10-4	20-0	16-9	12-8
	16	7-1	6-5	4-10	11-2	9-4	7-1	14-2	11-10	8-11	17-4	14-6	10-11
	24	6-2	5-3	3-11	9-2	7-8	5-9	11-7	9-8	7-4	14-1	11-10	8-11
Douglas fir-larch (N)	12	8-5	7-2	5-5	13-2	10-6	8-0	16-8	13-3	10-1	20-4	16-3	12-4
	16	7-8	6-3	4-9	11-5	9-1	6-11	14-5	11-6	8-9	17-8	14-1	10-8
	24	6-4	5-1	3-10	9-4	7-5	5-8	11-9	9-5	7-2	14-5	11-6	8-8
Hem-fir (N)	12	8-2	7-11	6-0	12-10	11-7	8-9	16-8	14-7	11-1	20-4	17-10	13-7
	16	7-5	6-10	5-2	11-5	10-0	7-7	14-5	12-8	9-7	17-8	15-6	11-9
	24	6-4	5-7	4-3	9-4	8-2	6-2	11-9	10-4	7-10	14-5	12-8	9-7
Northern species	12	7-1	6-0	4-8	11-1	8-9	6-10	14-3	11-1	8-8	17-5	13-7	10-7
	16	6-5	5-2	4-1	9-9	7-7	5-11	12-4	9-7	7-6	15-1	11-9	9-2
	24	5-5	4-3	3-4	8-0	6-2	4-10	10-1	7-10	6-1	12-4	9-7	7-6

Rafters

HEAVY ROOF COVERING, NO CEILING: 20 PSF LIVE, 20 PSF DEAD
Maximum Allowable Span, feet - inches

Group Species	Spacing inches oc	2 × 6 Sel Str	2 × 6 No.1/No.2	2 × 6 No.3	2 × 8 Sel Str	2 × 8 No.1/No.2	2 × 8 No.3	2 × 10 Sel Str	2 × 10 No.1/No.2	2 × 10 No.3	2 × 12 Sel Str	2 × 12 No.1/No.2	2 × 12 No.3
Spruce-pine-fir	12	10-7	9-5	7-1	16-5	13-9	10-5	20-10	17-5	13-2	25-6	21-4	16-1
	16	9-8	8-2	6-2	14-3	11-11	9-0	18-1	15-1	11-5	22-1	18-5	13-11
	24	7-11	6-8	5-0	11-8	9-9	7-4	14-9	12-4	9-4	18-0	15-1	11-5
Douglas fir-larch (N)	12	11-6	9-2	6-11	16-9	13-4	10-2	21-3	16-11	12-10	26-0	20-8	15-8
	16	9-11	7-11	6-0	14-6	11-7	8-9	18-5	14-8	11-2	22-6	17-11	13-7
	24	8-1	6-6	4-11	11-10	9-5	7-2	15-0	12-0	9-1	18-4	14-8	11-1
Hem-fir (N)	12	11-1	10-1	7-8	16-9	14-9	11-2	21-3	18-8	14-2	26-0	22-9	17-3
	16	10-0	8-9	6-7	14-6	12-9	9-8	18-5	16-2	12-3	22-6	19-9	14-11
	24	8-1	7-1	5-5	11-10	10-5	7-11	15-0	13-2	10-0	18-4	16-1	12-3
Northern species	12	9-7	7-8	5-11	14-4	11-2	8-9	18-2	14-2	11-0	22-2	17-3	13-6
	16	8-8	6-7	5-2	12-5	9-8	7-7	15-9	12-3	9-7	19-3	14-11	11-8
	24	6-11	5-5	4-3	10-2	7-11	6-2	12-10	10-0	7-10	15-8	12-3	9-6

HEAVY ROOF COVERING, NO CEILING: 30 PSF LIVE, 20 PSF DEAD
Maximum Allowable Span, feet - inches

Group Species	Spacing inches oc	2 × 6 Sel Str	2 × 6 No.1/No.2	2 × 6 No.3	2 × 8 Sel Str	2 × 8 No.1/No.2	2 × 8 No.3	2 × 10 Sel Str	2 × 10 No.1/No.2	2 × 10 No.3	2 × 12 Sel Str	2 × 12 No.1/No.2	2 × 12 No.3
Spruce-pine-fir	12	9-3	8-5	6-4	14-7	12-4	9-4	18-8	15-7	11-9	22-9	19-1	14-5
	16	8-5	7-3	5-6	12-9	10-8	8-1	16-2	13-6	10-3	19-9	16-6	12-6
	24	7-1	5-11	4-6	10-5	8-9	6-7	13-2	11-0	8-4	16-1	13-6	10-2
Douglas fir-larch (N)	12	10-0	8-2	6-2	15-0	12-0	9-1	19-0	15-2	11-6	23-3	18-6	14-0
	16	8-11	7-1	5-4	13-0	10-4	7-10	16-6	13-1	9-11	20-1	16-0	12-2
	24	7-3	5-9	4-5	10-7	8-5	6-5	13-5	10-9	8-2	16-5	13-1	9-11
Hem-fir (N)	12	9-8	9-0	6-10	15-0	13-2	10-0	19-0	16-8	12-8	23-3	20-4	15-5
	16	8-9	7-10	5-11	13-0	11-5	8-8	16-6	14-5	10-11	20-1	17-8	13-5
	24	7-3	6-4	4-10	10-7	9-4	7-1	13-5	11-9	8-11	16-5	14-5	10-11
Northern species	12	8-4	6-10	5-4	12-10	10-0	7-9	16-3	12-8	9-10	19-10	15-5	12-1
	16	7-7	5-11	4-7	11-1	8-8	6-9	14-1	10-11	8-7	17-2	13-5	10-5
	24	6-2	4-10	3-9	9-1	7-1	5-6	11-6	8-11	7-0	14-0	10-11	8-6

Headers

EXTERIOR-WALL HEADER SPANS FOR SPRUCE-PINE-FIR, DOUGLAS FIR-LARCH (N), AND HEM-FIR (N)

Roof Live Load (psf)		20 Building Width, ft.			30 Building Width, ft.			40 Building Width, ft.			50 Building Width, ft.		
	Size	20	28	36	20	28	36	20	28	36	20	28	36
Headers	2-2×4	3-7	3-2	2-10	3-3	2-10	2-6	3-0	2-7	2-4	2-9	2-5	2-2
supporting	2-2×6	5-4	4-7	4-2	4-9	4-2	3-8	4-4	3-9	3-5	4-0	3-6	3-2
roof and	2-2×8	6-9	5-10	5-3	6-1	5-3	4-8	5-6	4-9	4-3	5-1	4-5	4-0
ceiling	2-2×10	8-3	7-2	6-5	7-5	6-5	5-9	6-9	5-10	5-3	6-3	5-5	4-10
	2-2×12	9-7	8-4	7-5	8-7	7-5	6-8	7-10	6-9	6-1	7-3	6-3	5-7
	3-2×8	8-6	7-4	6-7	7-7	6-7	5-10	6-11	6-0	5-4	6-5	5-7	4-11
	3-2×10	10-4	8-11	8-0	9-3	8-0	7-2	8-5	7-4	6-6	7-10	6-9	6-1
	3-2×12	12-0	10-5	9-3	10-9	9-3	8-4	9-10	8-6	7-7	9-1	7-10	7-0
	4-2×8	9-4	8-6	7-7	8-8	7-7	6-9	8-0	6-11	6-2	7-5	6-5	5-9
	4-2×10	11-11	10-4	9-3	10-8	9-3	8-3	9-9	8-5	7-7	9-0	7-10	7-0
	4-2×12	13-10	12-0	10-9	12-5	10-9	9-7	11-4	9-10	8-9	10-6	9-1	8-1

EXTERIOR-WALL HEADER SPANS FOR SPRUCE-PINE-FIR, DOUGLAS FIR-LARCH (N), AND HEM-FIR (N)

Roof Live Load (psf)		20 Building Width, ft.			30 Building Width, ft.			40 Building Width, ft.			50 Building Width, ft.		
	Size	20	28	36	20	28	36	20	28	36	20	28	36
Headers	2-2×4	3-0	2-8	2-5	2-10	2-6	2-3	2-7	2-4	2-1	2-6	2-2	1-11
supporting	2-2×6	4-5	3-11	3-6	4-1	3-7	3-3	3-10	3-4	3-0	3-7	3-2	2-10
roof, ceiling,	2-2×8	5-8	4-11	4-6	5-2	4-7	4-1	4-10	4-3	3-10	4-7	4-0	3-7
and 1	2-2×10	6-11	6-1	5-5	6-4	5-7	5-0	5-11	5-2	4-8	5-7	4-11	4-5
floor (with	2-2×12	8-0	7-0	6-4	7-4	6-5	5-10	6-11	6-0	5-5	6-6	5-8	5-1
center													
bearing wall)	3-2×8	7-1	6-2	5-7	6-6	5-8	5-2	6-1	5-4	4-9	5-9	5-0	4-6
	3-2×10	8-7	7-7	6-10	7-11	7-0	6-3	7-5	6-6	5-10	7-0	6-1	5-6
	3-2×12	10-0	8-9	7-11	9-3	8-1	7-3	8-7	7-6	6-9	8-1	7-1	6-4
	4-2×8	8-2	7-2	6-5	7-6	6-7	5-11	7-0	6-2	5-6	6-7	5-9	5-2
	4-2×10	9-11	8-9	7-11	9-2	8-1	7-3	8-7	7-6	6-9	8-1	7-1	6-4
	4-2×12	11-6	10-2	9-2	10-8	9-4	8-5	9-11	8-8	7-10	9-4	8-2	7-4

Headers

EXTERIOR-WALL HEADER SPANS FOR SPRUCE-PINE-FIR, DOUGLAS FIR-LARCH (N), AND HEM-FIR (N)

Roof Live Load (psf)		20 Building Width, ft.			30 Building Width, ft.			40 Building Width, ft.			50 Building Width, ft.		
	Size	20	28	36	20	28	36	20	28	36	20	28	36
Headers	2-2×4	2-8	2-3	2-1	2-7	2-3	2-0	2-6	2-2	1-11	2-4	2-0	1-10
supporting	2-2×6	3-10	3-4	3-0	3-10	3-3	2-11	3-7	3-2	2-10	3-5	3-0	2-8
roof, ceiling,	2-2×8	4-11	4-3	3-10	4-10	4-2	3-9	4-7	4-0	3-7	4-4	3-9	3-5
and 1	2-2×10	6-0	5-2	4-8	5-10	5-1	4-7	5-7	4-11	4-5	5-4	4-7	4-2
floor	2-2×12	6-11	6-0	5-4	6-10	5-11	5-3	6-6	5-8	5-1	6-2	5-4	4-10
	3-2×8	6-2	5-4	4-9	6-0	5-2	4-8	5-9	5-0	4-6	5-5	4-9	4-3
	3-2×10	7-6	6-6	5-10	7-4	6-4	5-8	7-0	6-1	5-6	6-8	5-9	5-2
	3-2×12	8-8	7-6	6-9	8-6	7-5	6-7	8-2	7-1	6-4	7-8	6-9	6-0
	4-2×8	7-1	6-2	5-6	6-11	6-0	5-5	6-8	5-9	5-2	6-3	5-6	4-11
	4-2×10	8-8	7-6	6-8	8-6	7-4	6-7	8-1	7-1	6-4	7-8	6-8	6-0
	4-2×12	10-0	8-8	7-9	9-10	8-6	7-7	9-5	8-2	7-4	8-11	7-9	6-11

EXTERIOR-WALL HEADER SPANS FOR SPRUCE-PINE-FIR, DOUGLAS FIR-LARCH (N), AND HEM-FIR (N)

Roof Live Load (psf)		20 Building Width, ft.			30 Building Width, ft.			40 Building Width, ft.			50 Building Width, ft.		
	Size	20	28	36	20	28	36	20	28	36	20	28	36
Headers	2-2×4	2-6	2-2	2-0	2-6	2-2	1-11	2-4	2-1	1-11	2-3	2-0	1-9
supporting	2-2×6	3-8	3-3	2-11	3-7	3-2	2-10	3-6	3-1	2-9	3-3	2-11	2-7
roof, ceiling,	2-2×8	4-8	4-1	3-8	4-7	4-0	3-7	4-5	3-10	3-6	4-2	3-8	3-4
and 2	2-2×10	5-8	5-0	4-6	5-7	4-11	4-5	5-4	4-9	4-3	5-1	4-6	4-1
floors	2-2×12	6-7	5-9	5-3	6-6	5-8	5-1	6-2	5-6	4-11	5-11	5-2	4-8
(with center													
bearing wall)	3-2×8	5-10	5-1	4-7	5-9	5-0	4-6	5-6	4-10	4-4	5-3	4-7	4-2
	3-2×10	7-1	6-3	5-7	7-0	6-2	5-6	6-8	5-11	5-4	6-5	5-7	5-1
	3-2×12	8-3	7-3	6-6	8-1	7-1	6-5	7-9	6-10	6-2	7-5	6-6	5-10
	4-2×8	6-9	5-11	5-4	6-7	5-9	5-3	6-4	5-7	5-1	6-0	5-4	4-9
	4-2×10	8-3	7-2	6-6	8-1	7-1	6-4	7-9	6-10	6-2	7-4	6-6	5-10
	4-2×12	9-6	8-4	7-6	9-4	8-2	7-5	9-0	7-11	7-2	8-6	7-6	6-9

Headers

EXTERIOR-WALL HEADER SPANS FOR SPRUCE-PINE-FIR, DOUGLAS FIR-LARCH (N), AND HEM-FIR (N)

Roof Live Load (psf)		20 Building Width, ft.			30 Building Width, ft.			40 Building Width, ft.			50 Building Width, ft.		
	Size	20	28	36	20	28	36	20	28	36	20	28	36
Headers	2-2×4	2-1	1-9	1-7	2-0	1-9	1-7	2-0	1-9	1-7	2-0	1-9	1-6
supporting	2-2×6	3-0	2-7	2-4	3-0	2-7	2-4	2-11	2-6	2-3	2-11	2-6	2-3
roof, ceiling,	2-2×8	3-10	3-4	2-11	3-9	3-3	2-11	3-9	3-3	2-11	3-8	3-2	2-10
and 2	2-2×10	4-8	4-0	3-7	4-7	4-0	3-7	4-6	3-11	3-6	4-6	3-11	3-6
floors	2-2×12	5-5	4-8	4-2	5-4	4-7	4-2	5-3	4-7	4-1	5-2	4-6	4-0
	3-2×8	4-9	4-2	3-8	4-9	4-1	3-8	4-8	4-0	3-7	4-7	4-0	3-7
	3-2×10	5-10	5-1	4-6	5-9	5-0	4-5	5-8	4-11	4-5	5-7	4-10	4-4
	3-2×12	6-9	5-10	5-3	6-8	5-9	5-2	6-7	5-9	5-1	6-6	5-8	5-1
	4-2×8	5-6	4-9	4-3	5-5	4-9	4-3	5-5	4-8	4-2	5-4	4-7	4-1
	4-2×10	6-9	5-10	5-3	6-8	5-9	5-2	6-7	5-8	5-1	6-6	5-7	5-0
	4-2×12	7-10	6-9	6-0	7-8	6-8	6-0	7-7	6-7	5-11	7-6	6-6	5-10

INTERIOR-WALL HEADER SPANS FOR SPRUCE-PINE-FIR, DOUGLAS FIR-LARCH (N), AND HEM-FIR (N)

	Size	Building Width, ft. 20	28	36		Size	Building Width, ft. 20	28	36
Headers	2-2×4	3-4	2-10	2-6	Headers	2-2×4	2-3	1-11	1-9
supporting	2-2×8	6-2	5-3	4-7	supporting	2-2×8	4-2	3-7	3-2
1 floor	2-2×10	7-7	6-5	5-7	2 floors	2-2×10	5-1	4-4	3-10
(with center	2-2×12	8-9	7-5	6-6	(with center	2-2×12	5-11	5-1	4-6
bearing wall)					bearing wall)				
	3-2×8	7-9	6-6	5-9		3-2×8	5-3	4-6	4-0
	3-2×10	9-5	8-0	7-1		3-2×10	6-4	5-5	4-10
	3-2×12	10-11	9-3	8-2		3-2×12	7-5	6-4	5-7
	4-2×8	8-11	7-7	6-8		4-2×8	6-0	5-2	4-7
	4-2×10	10-11	9-3	8-2		4-2×10	7-4	6-4	5-7
	4-2×12	12-8	10-8	9-5		4-2×12	8-6	7-4	6-6

WOOD TRUSSES

With decreasing availability of large lumber, trusses offer cost savings in roofs and floors.

The illustration at top right shows how a simple triangle formed by two rafters, a ceiling joist, and a vertical tie (a king post truss) divides a building's total clear span into two rafter clear spans, each half as long. The vertical load of the roof is transformed into forces of tension (T) and compression (C) in the directions of the members.

The bottom illustration shows how, by the simple addition of diagonal members, the queen post truss further divides the rafter clear span.

Page 166 shows a small sample (30) of the variety of trusses available from truss manufacturers nationwide. Trusses offer an engineered and precise cost-effective solution to any span problem.

Pages 167–177 contain representative span tables showing clear spans as limited by the species/grades and sizes of the top and bottom chord members. All of the roof-truss tables (pages 167–175) assume moisture content of less than 19 percent, truss spacing of 24 inches on-center, and roof live and dead loads of 30 psf and 17 psf respectively. The floor-truss tables (pages 176–177) assume moisture content of less than 19 percent and the spacings, dead loads, and live loads shown. Remember, these tables are for preliminary design guidance only and must be confirmed by your truss manufacturer. For further information on the design and manufacture of trusses, consult the *Metal Plate Connected Wood Truss Handbook* of the Wood Truss Council of America.

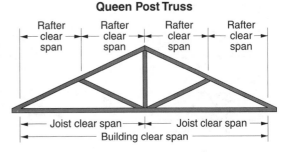

King Post Truss

Queen Post Truss

Example: What is the maximum allowable span of queen post trusses (page 167) of #2 southern pine with 24-inch on-center spacing, roof pitch of 4/12, top-chord (rafter) dead and live loads of 7 psf and 30 psf respectively, and bottom-chord (ceiling joist) dead load of 10 psf, if the top chord is a 2×6 and the bottom chord a 2×4?

On page 167, read across the row labeled Species Group *Southern pine* and Grade *#2* to the group of four spans under Pitch of Top Chord *4/12*. For a 2×6 top chord the maximum span is 28-2. For a 2×4 bottom chord the maximum span is 18-9. The truss span is thus limited to 18-9 by the 2×4 bottom chord. If we increase the bottom chord to a 2×6, the span will be limited by the bottom chord to 26-3.

Source: *Metal Plate Connected Wood Truss Handbook* (Madison, Wis: Wood Truss Council of America, 1990).

A Sample of Roof Trusses

King post

Double fan

Clear story (clerestory)

Queen post (fan)

Modified fan

Pitched Warren

Modified queen post

Howe

Mansard

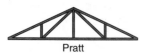
Pratt

Double Howe (KK)

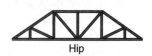
Hip

Simple Fink

Howe scissors

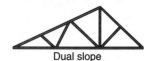
Dual slope

Fink (W)

Double Howe scissors

Gambrel

Double Fink (WW)

Vaulted

Polynesian (duo pitch)

Triple Fink (WWW)

Cambered

Room-in-attic

Umbrella

Scissored Warren

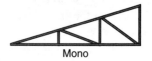
Mono

Gable end

Double inverted

Scissors mono

Queen Post Truss

Spacing 24 inches on-center
30 psf top chord (rafter) live
7 psf top chord (rafter) dead
10 psf bottom chord (joist) dead

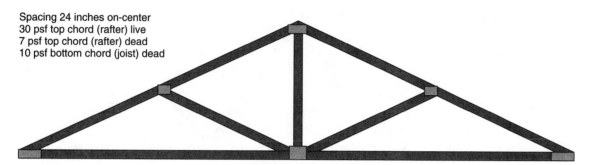

MAXIMUM ALLOWABLE SPAN, FEET - INCHES
(maximum span is smallest of four figures)

		Pitch of Top Chord											
		3/12				**4/12**				**5/12**			
Species Group	**Grade**	**Top Chord**		**Bottom Chord**		**Top Chord**		**Bottom Chord**		**Top Chord**		**Bottom Chord**	
		2×4	2×6	2×4	2×6	2×4	2×6	2×4	2×6	2×4	2×6	2×4	2×6
Southern pine	No.1 dense	21-6	28-2	21-6	28-2	21-6	28-2	21-6	28-2	21-6	28-2	21-6	28-2
	No.1	21-6	28-2	20-8	27-5	21-6	28-2	20-8	27-5	21-6	28-2	20-8	27-5
	No.2 dense	21-6	28-2	20-1	27-3	21-6	28-2	20-1	27-3	21-6	28-2	20-1	27-3
	No.2	21-6	28-2	18-9	26-3	21-6	28-2	18-9	26-3	21-6	28-2	18-9	26-3
Douglas fir-larch	Sel str	22-4	29-0	22-4	29-0	22-4	29-0	22-4	29-0	22-4	29-0	22-4	29-0
	No.1 and better	22-4	29-0	20-7	28-0	22-4	29-0	20-7	28-0	22-4	29-0	20-7	28-0
	No.1	22-4	29-0	19-0	27-0	22-4	29-0	19-0	27-0	22-4	29-0	19-0	27-0
	No.2	22-4	29-0	17-3	26-0	22-4	29-0	17-3	26-0	22-4	29-0	17-3	26-0
Spruce-pine-fir	Sel str	19-9	25-10	19-9	25-10	19-9	25-10	19-9	25-10	19-9	25-10	19-9	25-10
	No.1	19-9	25-10	16-10	24-6	19-9	25-10	16-10	24-6	19-9	25-10	16-10	24-6
	No.2	19-9	25-10	16-10	24-6	19-9	25-10	16-10	24-6	19-9	25-10	16-10	24-6
Hem-fir	Sel str	20-8	26-8	20-8	26-8	20-8	26-8	20-8	26-8	20-8	26-8	20-8	26-8
	No.1	20-8	26-8	18-1	25-5	20-8	26-8	18-1	25-5	20-8	26-8	18-1	25-5
	No.2	20-8	26-8	16-6	23-8	20-8	26-8	16-6	23-8	20-8	26-8	16-6	23-8

Fink Truss

Spacing 24 inches on-center
30 psf top chord (rafter) live
7 psf top chord (rafter) dead
10 psf bottom chord (joist) dead

MAXIMUM ALLOWABLE SPAN, FEET - INCHES
(maximum span is smallest of four figures)

Species Group	Grade	Top Chord 2×4	2×6	Bottom Chord 2×4	2×6	Top Chord 2×4	2×6	Bottom Chord 2×4	2×6	Top Chord 2×4	2×6	Bottom Chord 2×4	2×6
		3/12				*4/12*				*5/12*			
Southern pine	No.1 dense	27-7	40-11	30-10	40-11	30-7	42-0	31-11	42-0	31-11	42-0	31-11	42-0
	No.1	26-10	39-11	29-7	40-9	29-9	42-0	30-8	40-9	31-1	42-0	30-8	40-9
	No.2 dense	26-5	39-1	26-6	37-10	29-4	42-0	29-10	40-7	30-7	42-0	29-10	40-7
	No.2	25-5	37-5	25-0	35-4	28-2	41-5	27-10	39-1	29-5	42-0	27-10	39-1
Douglas fir-larch	Sel str	28-2	41-10	33-2	41-10	31-4	43-2	33-2	43-2	32-8	43-2	33-2	43-2
	No.1 and better	26-8	39-7	30-2	41-9	29-7	43-2	30-6	41-9	30-10	43-2	30-6	41-9
	No.1	25-8	38-1	27-5	39-1	28-6	42-2	28-3	40-3	29-8	43-2	28-3	40-3
	No.2	24-6	36-4	24-10	35-1	27-3	40-3	25-7	38-8	28-5	41-10	25-7	38-8
Spruce-pine-fir	Sel str	26-0	38-5	29-0	38-5	29-0	38-5	29-4	38-5	29-4	38-5	29-4	38-5
	No.1	23-8	34-11	21-4	29-8	26-5	38-5	24-7	34-5	27-7	38-5	25-0	36-5
	No.2	23-8	34-11	21-4	29-8	26-5	38-5	24-7	34-5	27-7	38-5	25-0	36-5
Hem-fir	Sel str	26-11	39-9	30-9	39-9	30-0	39-9	30-9	39-9	30-0	39-9	30-9	39-9
	No.1	24-9	36-7	25-10	36-5	27-6	39-9	26-10	37-11	28-9	39-9	26-10	37-11
	No.2	23-8	34-10	23-0	32-5	26-3	38-7	24-5	35-2	27-5	39-9	24-5	35-2

Pitch of Top Chord

Modified Queen Post Truss

Spacing 24 inches on-center
30 psf top chord (rafter) live
7 psf top chord (rafter) dead
10 psf bottom chord (joist) dead

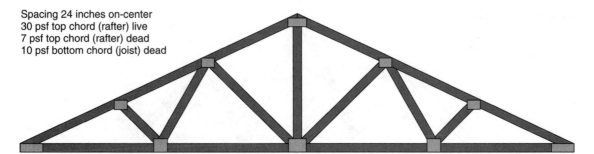

MAXIMUM ALLOWABLE SPAN, FEET - INCHES
(maximum span is smallest of four figures)

| | | | Pitch of Top Chord | | | | | | | | | | |
| | | 3/12 | | | | 4/12 | | | | 5/12 | | | |
Species Group	Grade	Top Chord 2×4	2×6	Bottom Chord 2×4	2×6	Top Chord 2×4	2×6	Bottom Chord 2×4	2×6	Top Chord 2×4	2×6	Bottom Chord 2×4	2×6
Southern pine	No.1 dense	33-2	46-5	33-2	46-5	38-4	54-6	38-8	54-6	41-10	55-10	42-5	55-10
	No.1	32-5	46-5	31-9	44-6	37-4	54-6	37-0	52-4	40-9	55-10	40-9	54-2
	No.2 dense	32-1	46-5	27-11	39-7	36-10	54-6	33-0	46-11	40-2	55-10	36-11	52-7
	No.2	30-9	45-7	26-4	37-0	35-5	52-5	31-1	43-10	38-8	55-10	34-9	49-1
Douglas fir-larch	Sel str	34-0	50-10	40-0	50-10	39-2	57-5	44-0	57-5	42-9	57-5	44-0	57-5
	No.1 and better	32-2	48-1	33-0	46-6	37-1	55-4	37-11	53-10	40-6	57-5	40-6	55-5
	No.1	31-1	46-4	29-7	41-10	35-10	53-4	34-4	48-9	39-1	57-5	37-5	53-6
	No.2	29-8	44-1	26-8	37-2	34-2	50-9	31-1	43-7	37-5	55-5	33-10	48-5
Spruce-pine-fir	Sel str	30-11	43-7	30-11	43-7	36-1	51-0	36-3	51-0	38-11	51-0	38-11	51-0
	No.1	28-3	41-10	21-11	30-1	32-9	48-6	26-4	36-5	36-0	51-0	29-10	41-5
	No.2	28-3	41-10	21-11	30-1	32-9	48-6	26-4	36-5	36-0	51-0	29-10	41-5
Hem-fir	Sel str	32-5	48-3	37-7	48-3	37-4	52-10	40-10	52-10	40-10	52-10	40-10	52-10
	No.1	29-10	44-3	27-7	38-6	34-5	51-0	32-3	45-3	37-8	52-10	35-6	50-4
	No.2	28-6	42-3	24-1	33-9	32-10	48-7	28-6	40-1	35-11	52-10	31-11	45-1

Double Fink Truss

Spacing 24 inches on-center
30 psf top chord (rafter) live
7 psf top chord (rafter) dead
10 psf bottom chord (joist) dead

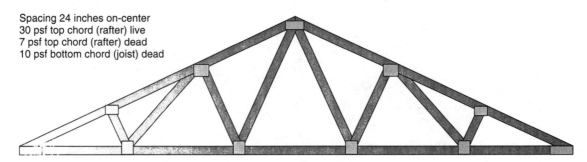

MAXIMUM ALLOWABLE SPAN, FEET - INCHES
(Maximum span is smallest of four figures)

Species Group	Grade	3/12 Top Chord 2×4	3/12 Top Chord 2×6	3/12 Bottom Chord 2×4	3/12 Bottom Chord 2×6	4/12 Top Chord 2×4	4/12 Top Chord 2×6	4/12 Bottom Chord 2×4	4/12 Bottom Chord 2×6	5/12 Top Chord 2×4	5/12 Top Chord 2×6	5/12 Bottom Chord 2×4	5/12 Bottom Chord 2×6
Southern pine	No.1 dense	33-3	49-3	35-5	49-3	38-3	57-0	42-4	57-0	41-8	60-0	47-9	60-0
	No.1	32-4	48-4	33-11	47-1	37-2	55-7	40-7	56-10	40-7	60-0	45-9	60-0
	No.2 dense	31-11	47-6	29-5	41-7	36-9	54-6	35-9	50-7	40-1	59-4	40-9	57-10
	No.2	30-7	45-5	27-9	38-11	35-3	52-3	33-8	47-3	38-6	56-11	38-5	54-0
Douglas fir-larch	Sel str	33-11	50-8	43-9	50-8	39-0	58-2	50-11	58-2	42-7	60-0	54-11	60-0
	No.1 and better	32-1	47-11	35-7	49-11	36-11	55-1	42-0	58-2	40-4	60-0	46-11	60-0
	No.1	31-0	46-2	31-9	44-7	35-8	53-1	37-9	53-4	39-0	57-11	42-5	60-0
	No.2	29-6	43-11	28-4	39-3	34-1	50-7	34-0	47-4	37-3	55-3	38-4	53-8
Spruce-pine-fir	Sel str	31-1	46-1	32-10	46-1	35-11	53-6	39-6	53-6	39-3	58-4	44-9	58-4
	No.1	28-1	41-8	22-10	31-1	32-7	48-4	28-1	38-6	35-10	53-0	32-5	44-8
	No.2	28-1	41-8	22-10	31-1	32-7	48-4	28-1	38-6	35-10	53-0	32-5	44-8
Hem-fir	Sel str	32-3	48-1	40-10	48-1	37-2	55-4	47-11	55-4	40-8	60-0	50-10	60-0
	No.1	29-8	44-1	29-4	40-9	34-3	50-10	35-3	49-1	37-6	55-6	39-9	55-9
	No.2	28-4	42-1	25-5	35-4	32-9	48-5	30-10	43-0	35-10	52-11	35-2	49-4

Mono Truss

Spacing 24 inches on-center
30 psf top chord (rafter) live
7 psf top chord (rafter) dead
10 psf bottom chord (joist) dead

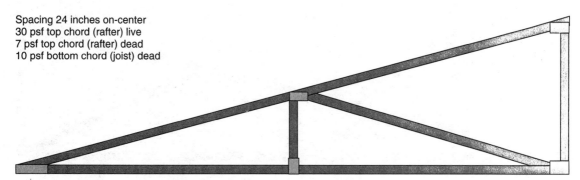

MAXIMUM ALLOWABLE SPAN, FEET - INCHES
(maximum span is smallest of four figures)

Species Group	Grade	Top Chord 2×4	Top Chord 2×6	Bottom Chord 2×4	Bottom Chord 2×6	Top Chord 2×4	Top Chord 2×6	Bottom Chord 2×4	Bottom Chord 2×6	Top Chord 2×4	Top Chord 2×6	Bottom Chord 2×4	Bottom Chord 2×6
		3/12				4/12				5/12			
Southern pine	No.1 dense	16-0	23-5	21-6	23-5	16-6	24-1	21-6	24-1	16-10	24-6	21-6	24-6
	No.1	15-6	22-11	20-8	23-5	16-0	23-7	20-8	24-1	16-3	23-11	20-8	24-6
	No.2 dense	15-2	22-1	20-1	23-5	15-7	22-8	20-1	24-1	15-10	23-0	20-1	24-6
	No.2	14-6	21-0	18-9	23-5	14-11	21-6	18-9	24-1	15-1	21-9	18-9	24-6
Douglas fir-larch	Sel str	16-5	24-1	22-4	24-1	17-0	24-9	22-4	24-9	17-4	25-3	22-4	25-3
	No.1 and better	15-3	22-5	20-7	24-1	15-9	23-0	20-7	24-9	16-0	23-4	20-7	25-3
	No.1	14-7	21-4	19-0	24-1	15-0	21-10	19-0	24-9	15-2	22-2	19-0	25-3
	No.2	13-10	20-3	17-3	24-1	14-2	20-8	17-3	24-9	14-5	20-11	17-3	25-3
Spruce-pine-fir	Sel str	15-4	22-4	19-9	22-4	15-10	23-1	19-9	23-1	16-1	23-6	19-9	23-6
	No.1	13-8	19-11	16-10	22-4	14-0	20-4	16-10	23-1	14-3	20-8	16-10	23-6
	No.2	13-8	19-11	16-10	22-4	14-0	20-4	16-10	23-1	14-3	20-8	16-10	23-6
Hem-fir	Sel str	15-11	23-3	20-8	23-3	16-6	24-0	20-8	24-0	16-10	24-6	20-8	24-6
	No.1	14-2	20-8	18-1	23-3	14-7	21-2	18-1	24-0	14-10	21-6	18-1	24-6
	No.2	13-6	19-8	16-6	23-3	13-10	20-1	16-6	23-8	14-1	20-5	16-6	23-8

Pitch of Top Chord

Mono Truss

Spacing 24 inches on-center
30 psf top chord (rafter) live
7 psf top chord (rafter) dead
10 psf bottom chord (joist) dead

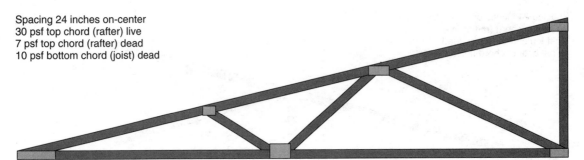

MAXIMUM ALLOWABLE SPAN, FEET - INCHES
(maximum span is smallest of four figures)

Species Group	Grade	3/12 Top Chord 2×4	2×6	3/12 Bottom Chord 2×4	2×6	4/12 Top Chord 2×4	2×6	4/12 Bottom Chord 2×4	2×6	5/12 Top Chord 2×4	2×6	5/12 Bottom Chord 2×4	2×6
Southern pine	No.1 dense	21-6	28-2	21-6	28-2	21-6	28-2	21-6	28-2	21-6	28-2	21-6	28-2
	No.1	21-6	28-2	20-8	27-5	21-6	28-2	20-8	27-5	21-6	28-2	20-8	27-5
	No.2 dense	21-6	28-2	20-1	27-3	21-6	28-2	20-1	27-3	21-6	28-2	20-1	27-3
	No.2	21-3	28-2	18-9	26-3	21-6	28-2	18-9	26-3	21-6	28-2	18-9	26-3
Douglas fir-larch	Sel str	22-4	29-0	22-4	29-0	22-4	29-0	22-4	29-0	22-4	29-0	22-4	29-0
	No.1 and better	22-4	29-0	20-7	28-0	22-4	29-0	20-7	28-0	22-4	29-0	20-7	28-0
	No.1	21-5	29-0	19-0	27-0	22-4	29-0	19-0	27-0	22-4	29-0	19-0	27-0
	No.2	20-6	29-0	17-3	26-0	22-4	29-0	17-3	26-0	22-4	29-0	17-3	26-0
Spruce-pine-fir	Sel str	19-9	25-10	19-9	25-10	19-9	25-10	19-9	25-10	19-9	25-10	19-9	25-10
	No.1	19-9	25-10	16-10	24-6	19-9	25-10	16-10	24-6	19-9	25-10	16-10	24-6
	No.2	19-9	25-10	16-10	24-6	19-9	25-10	16-10	24-6	19-9	25-10	16-10	24-6
Hem-fir	Sel str	20-8	26-8	20-8	26-8	20-8	26-8	20-8	26-8	20-8	26-8	20-8	26-8
	No.1	20-8	26-8	18-1	25-5	20-8	26-8	18-1	25-5	20-8	26-8	18-1	25-5
	No.2	19-10	22-1	16-6	23-8	20-8	26-8	16-6	23-8	20-8	26-8	16-6	23-8

Pitch of Top Chord

Howe Scissors Truss

Spacing 24 inches on-center
30 psf top chord (rafter) live
7 psf top chord (rafter) dead
10 psf bottom chord (joist) dead

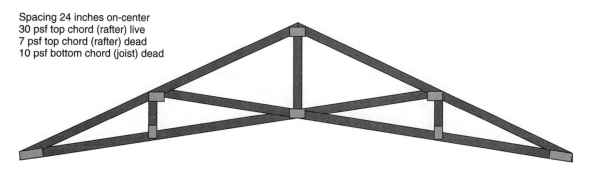

MAXIMUM ALLOWABLE SPAN, FEET - INCHES
(maximum span is smallest of four figures)

		Pitch of Top Chord											
		3/12				4/12				5/12			
Species Group	Grade	Top Chord		Bottom Chord		Top Chord		Bottom Chord		Top Chord		Bottom Chord	
		2×4	2×6	2×4	2×6	2×4	2×6	2×4	2×6	2×4	2×6	2×4	2×6
Southern pine	No.1 dense	20-2	30-0	21-8	30-0	22-5	33-5	26-7	33-5	24-1	35-11	30-8	35-11
	No.1	19-7	29-4	20-8	28-7	21-10	32-7	25-5	33-5	23-6	35-0	29-4	35-11
	No. 2 dense	19-4	28-10	17-9	25-1	21-6	32-0	22-0	31-2	23-2	34-4	25-7	35-11
	No.2	18-6	27-7	16-9	23-6	20-8	30-8	20-9	29-2	22-3	32-11	24-2	34-0
Douglas fir-larch	Sel str	20-6	30-8	27-8	30-8	22-10	34-1	33-2	34-1	24-7	36-8	36-6	36-8
	No.1 and better	19-5	29-0	22-1	30-8	21-8	32-4	26-10	34-1	23-4	34-9	30-8	36-8
	No.1	18-9	28-0	19-6	27-4	20-11	31-2	23-11	33-8	22-7	33-6	27-5	36-8
	No.2	17-10	26-8	17-4	23-11	20-0	29-9	21-4	29-6	21-7	32-0	24-7	34-3
Spruce-pine-fir	Sel str	18-9	27-11	19-11	27-11	21-0	31-4	24-7	31-4	22-8	33-9	28-6	33-9
	No.1	16-11	25-3	13-7	18-8	19-1	28-4	17-0	23-2	20-8	30-8	19-11	27-5
	No.2	16-11	25-3	13-7	18-8	19-1	28-4	17-0	23-2	20-8	30-8	19-11	27-5
Hem-fir	Sel str	19-6	29-1	25-6	29-1	21-9	32-5	30-10	32-5	23-6	34-11	34-9	34-11
	No.1	17-11	26-9	17-11	24-9	20-1	29-10	22-1	30-7	21-8	32-2	25-5	34-11
	No.2	17-2	25-6	15-4	21-4	19-2	28-5	19-0	26-5	20-9	30-7	22-1	30-11

Queen Scissors Truss

Spacing 24 inches on-center
30 psf top chord (rafter) live
7 psf top chord (rafter) dead
10 psf bottom chord (joist) dead

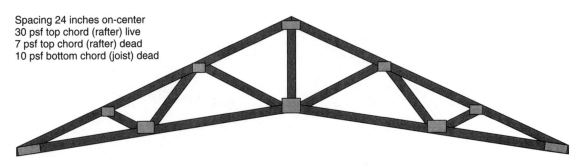

MAXIMUM ALLOWABLE SPAN, FEET - INCHES
(maximum span is smallest of four figures)

Species Group	Grade	Top Chord 3/12 2×4	2×6	Bottom Chord 2×4	2×6	Top Chord 4/12 2×4	2×6	Bottom Chord 2×4	2×6	Top Chord 5/12 2×4	2×6	Bottom Chord 2×4	2×6
Southern pine	No.1 dense	20-5	28-5	20-5	28-5	25-2	35-1	25-2	35-1	28-6	40-8	29-1	40-8
	No.1	20-5	28-5	19-7	27-1	25-1	35-1	24-1	33-6	27-8	40-8	27-10	38-11
	No.2 dense	20-5	28-5	16-9	23-10	24-10	35-1	20-9	29-5	27-5	40-8	24-3	34-5
	No.2	20-5	28-5	15-10	22-4	23-9	35-1	19-7	27-6	26-3	39-1	22-10	32-2
Douglas fir-larch	Sel str	22-11	34-5	26-2	34-5	26-4	39-6	31-7	39-6	29-0	43-5	35-10	43-5
	No.1 and better	21-7	32-5	20-10	29-2	24-10	37-3	25-5	35-9	27-5	41-1	29-2	41-1
	No.1	20-10	31-4	18-5	25-11	24-1	36-0	22-8	31-10	26-7	39-9	26-1	36-10
	No.2	19-9	29-7	16-5	22-9	22-10	34-2	20-2	27-11	25-4	37-9	23-4	32-6
Spruce-pine-fir	Sel str	18-10	26-5	18-10	26-5	23-3	32-8	23-3	32-8	26-7	38-1	27-0	38-1
	No.1	18-5	26-5	12-11	17-9	21-6	32-1	16-0	21-10	24-0	35-8	18-10	25-10
	No.2	18-5	26-5	12-11	17-9	21-6	32-1	16-0	21-10	24-0	35-8	18-10	25-10
Hem-fir	Sel str	21-7	32-6	24-1	32-6	24-11	37-4	29-3	37-4	27-6	41-2	33-6	41-2
	No.1	19-10	29-8	16-11	23-6	22-11	34-3	20-10	28-11	25-5	37-10	24-2	33-8
	No.2	19-0	28-5	14-6	20-3	22-0	32-9	17-11	24-11	24-4	36-1	20-11	29-3

Double Howe Scissors Truss

Spacing 24 inches on-center
30 psf top chord (rafter) live
7 psf top chord (rafter) dead
10 psf bottom chord (joist) dead

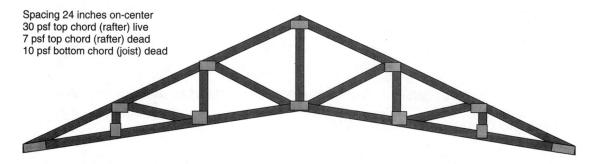

MAXIMUM ALLOWABLE SPAN, FEET - INCHES
(maximum span is smallest of four figures)

Species Group	Grade	Pitch of Top Chord 3/12 Top Chord 2×4	3/12 Top Chord 2×6	3/12 Bottom Chord 2×4	3/12 Bottom Chord 2×6	4/12 Top Chord 2×4	4/12 Top Chord 2×6	4/12 Bottom Chord 2×4	4/12 Bottom Chord 2×6	5/12 Top Chord 2×4	5/12 Top Chord 2×6	5/12 Bottom Chord 2×4	5/12 Bottom Chord 2×6
Southern pine	No.1 dense	20-11	29-0	20-11	29-0	25-9	36-6	26-6	36-6	28-4	42-4	31-6	42-4
	No.1	20-11	29-0	20-0	27-7	24-11	36-6	25-4	34-9	27-6	41-2	30-1	41-6
	No.2 dense	20-11	29-0	17-0	24-2	24-8	36-6	21-6	30-4	27-2	40-7	25-9	36-3
	No.2	20-5	29-0	16-1	22-8	23-7	35-3	20-3	28-5	26-1	38-10	24-3	33-11
Douglas fir-larch	Sel str	22-9	34-2	27-5	34-2	26-2	39-3	34-4	39-3	28-10	43-2	40-4	43-2
	No.1 and better	21-5	32-2	21-6	30-0	24-8	37-0	27-2	37-8	27-3	40-10	32-1	43-2
	No.1	20-9	31-1	18-11	26-7	23-11	35-9	23-11	33-4	26-5	39-5	28-4	39-9
	No.2	19-7	29-5	16-9	23-2	22-8	33-11	21-2	29-0	25-2	37-6	25-2	34-7
Spruce-pine-fir	Sel str	19-2	26-11	19-2	26-11	23-10	33-10	24-3	33-10	26-4	39-6	28-11	39-6
	No.1	18-3	26-11	13-0	17-11	21-4	31-10	16-4	22-3	23-10	35-5	19-7	26-7
	No.2	18-3	26-11	13-0	17-11	21-4	31-10	16-4	22-3	23-10	35-5	19-7	26-7
Hem-fir	Sel str	21-5	32-3	25-0	32-3	24-9	37-1	31-6	37-1	27-4	40-10	37-2	40-10
	No.1	19-8	29-6	17-3	23-11	22-9	34-0	21-10	30-0	25-3	37-7	26-0	35-10
	No.2	18-10	28-2	14-8	20-6	21-10	32-6	18-6	25-8	24-2	35-10	22-2	30-8

Floor Truss, 16" On-Center

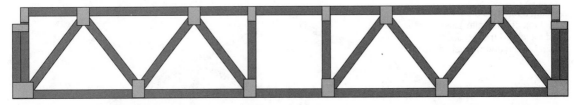

MAXIMUM ALLOWABLE SPAN, FEET - INCHES

Species Group	Grade	16" Deep Top Chord	Bottom Chord	20" Deep Top Chord	Bottom Chord	24" Deep Top Chord	Bottom Chord
Top chord: 40 psf live, 5 psf dead — Bottom chord: 0 psf live, 5 psf dead							
Southern pine	No.1 dense	29-0	26-2	34-0	29-0	38-0	33-0
	No.1	28-0	25-0	33-4	28-0	37-9	31-0
	No.2	24-4	24-0	29-0	27-0	33-0	30-0
Douglas fir-larch	Sel str	27-3	27-3	34-2	34-2	41-0	41-0
	No.1	27-3	26-8	34-2	30-2	41-0	33-2
	No.2	27-3	24-6	34-2	27-8	38-3	30-6
Spruce-pine-fir	Sel str	24-6	24-0	29-0	29-0	33-0	33-0
	No.1	24-0	21-0	28-0	23-6	32-2	26-0
	No.2	24-0	21-0	28-0	23-6	32-2	26-0
Hem-fir	Sel str	25-0	25-0	29-4	29-4	34-0	34-0
	No.1	24-6	24-6	29-0	28-4	33-0	31-0
	No.2	23-6	22-6	27-6	25-6	31-6	28-4
Top chord: 40 psf live, 10 psf dead — Bottom chord: 0 psf live, 10 psf dead							
Southern pine	No.1 dense	27-8	25-0	32-8	28-0	36-9	31-0
	No.1	27-0	24-0	31-6	26-9	35-8	30-0
	No.2	24-6	21-0	29-0	24-0	33-0	26-2
Douglas fir-larch	Sel str	26-6	26-6	31-5	31-5	36-0	36-0
	No.1	25-9	23-4	30-0	26-4	34-9	29-2
	No.2	25-0	21-1	29-7	24-1	34-0	26-7
Spruce-pine-fir	Sel str	24-2	23-2	29-0	26-9	33-0	29-2
	No.1	24-0	18-2	28-2	20-6	31-2	23-0
	No.2	24-0	18-2	28-2	20-6	31-2	23-0
Hem-fir	Sel str	25-0	25-0	29-8	29-8	33-9	33-9
	No.1	24-6	22-0	29-0	25-0	33-0	27-2
	No.2	23-5	20-0	27-7	22-2	31-4	24-9

Floor Truss, 24" On-Center

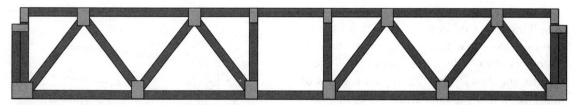

MAXIMUM ALLOWABLE SPAN, FEET - INCHES

Species Group	Grade	16" Deep Top Chord	16" Deep Bottom Chord	20" Deep Top Chord	20" Deep Bottom Chord	24" Deep Top Chord	24" Deep Bottom Chord
\multicolumn{8}{Top chord: 40 psf live, 5 psf dead — Bottom chord: 0 psf live, 5 psf dead}								
Southern pine	No.1 dense	25-0	22-4	29-2	25-3	34-0	28-0
	No.1	24-9	22-0	28-9	24-6	33-0	27-0
	No.2	24-0	19-1	28-3	21-3	31-2	24-0
Douglas fir-larch	Sel str	26-8	26-6	31-6	30-0	36-0	33-0
	No.1	25-6	21-4	28-8	24-3	31-8	26-6
	No.2	23-6	19-6	26-8	22-2	29-6	24-3
Spruce-pine-fir	Sel str	24-0	21-3	29-0	24-1	32-0	26-9
	No.1	21-0	16-9	24-1	18-9	27-0	20-8
	No.2	21-0	16-9	24-1	18-9	27-0	20-8
Hem-fir	Sel str	25-0	25-0	29-6	28-2	33-8	31-2
	No.1	24-3	20-0	27-4	22-7	30-3	25-0
	No.2	23-0	18-0	26-0	20-5	28-8	22-8
\multicolumn{8}{Top chord: 40 psf live, 10 psf dead — Bottom chord: 0 psf live, 10 psf dead}								
Southern pine	No.1 dense	24-0	19-6	28-0	22-0	30-9	24-4
	No.1	23-6	18-9	27-6	20-9	30-0	23-6
	No.2	22-4	16-4	25-4	18-4	28-0	20-4
Douglas fir-larch	Sel str	26-0	23-0	29-4	26-2	32-4	29-0
	No.1	23-0	18-4	25-8	20-6	28-4	23-0
	No.2	21-0	16-8	24-0	18-9	26-3	20-9
Spruce-pine-fir	Sel str	23-0	18-8	26-0	21-0	29-0	23-4
	No.1	19-2	14-3	22-0	16-1	24-1	17-9
	No.2	19-2	14-3	22-0	16-1	24-1	17-9
Hem-fir	Sel str	24-3	22-0	27-4	24-9	30-3	27-4
	No.1	21-9	17-0	24-6	19-4	27-0	21-4
	No.2	20-7	15-3	23-2	17-3	25-8	19-2

SPANS FOR PREFABRICATED-WOOD I-JOISTS

The Trus Joist Corporation's TJI/25 joist is representative of a new generation of framing members. Available in 9-1/2 and 11-7/8-inch depths, the Residential TJI/25 joist consists of a plywood web pressure-fitted between laminated veneer-lumber flanges.

Advantages include

• longer spans (up to 60 feet), which eliminate lapping over girders or wall lines
• greater strength per weight
• reduced warping, shrinkage, twisting, and splitting
• convenient prepunched web knock-outs for utilities

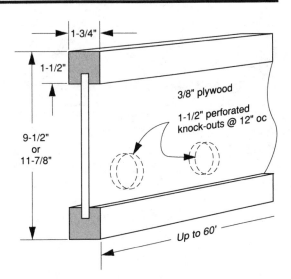

MANUFACTURER'S MAXIMUM RECOMMENDED SPANS

Depth inches	Spacing inches oc	at Dead Load at Live Load	Floor 10 40	Roof (slope 6/12 or less) 10 20	10 30	10 40	10 50
9-1/2	12		16 - 10	23 - 10	21 - 07	19 - 09	18 - 03
	16		15 - 04	21 - 06	19 - 06	17 - 10	16 - 06
	24		13 - 04	18 - 08	16 - 11	15 - 05	13 - 02
11-7/8	12		20 - 00	28 - 04	25 - 09	23 - 06	21 - 09
	16		18 - 02	25 - 08	23 - 03	21 - 03	19 - 07
	24		15 - 10	22 - 03	20 - 02	17 - 02	14 - 07

Source: *Residential Product Reference Guide* (Boise, Idaho: Trus Joist Corporation, 1988).
Notes: Spans are given as feet - inches. Dead loads and live loads are given in pounds per square foot. Check the literature of specific manufacturers of prefabricated-wood I-joists for additional application requirements.

SPAN TABLE FOR STRESSED-SKIN PANELS

One-sided plywood stressed-skin panels have plywood skins glued to the tops of longitudinal framing members. The skin and framing act as a single unit in carrying loads.

When designed with both top (plywood) and bottom (gypsum drywall) skins and filled with insulation, the stressed-skin panel offers insulation and a ready-to-finish ceiling.

Numerous companies offer variations on the theme: wall panels, panels filled with rigid foam insulation, and panels substituting oriented-strand board (OSB) for the structural skin. Consult manufacturers for specifications.

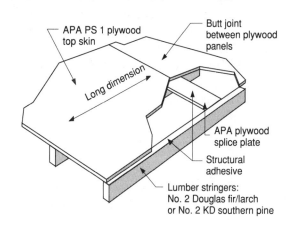

RECOMMENDED SPANS FOR PS 1 PLYWOOD STRESSED-SKIN ROOF PANELS

Top Skin	Stringer Size	Number of Stringers (spacing)	Allowable Single-Span Uniform Live Load, psf (10 psf dead load assumed)						
			20	25	30	35	40	45	50
	2x4		12' 9"	11' 10"	11' 1"	10' 6"			
	2x6		18' 10"	17' 5"	16' 5"	15' 7"			
3/8" APA Structural I	2x8	4 (12" oc)	24' 0"	22' 3"	20' 11"	19' 10"			
Rated Sheathing Exp 1	2x10		29' 11"	27' 9"	26' 1"	24' 9"			
	2x12		35' 8"	33' 1"	31' 0"	29' 3"			
	2x4		12' 2"	11' 3"	10' 7"	10' 0"	9' 7"	9' 2"	8' 10"
15/32" APA	2x6		17' 9"	16' 5"	15' 4"	14' 6"	13' 9"	13' 1"	12' 7"
Rated Sheathing Exp 1	2x8	3 (16" oc)	22' 7"	20' 11"	19' 6"	18' 5"	17' 6"	16' 8"	15' 11"
32/16 5-ply	2x10		27' 3"	25' 3"	23' 8"	22' 3"	21' 2"	20' 2"	19' 3"
	2x12		31' 11"	29' 4"	27' 8"	26' 1"	24' 9"	23' 7"	22' 7"
	2x4		10' 7"	9' 10"	9' 2"	8' 8"			
15/32" APA Structural I	2x6		14' 8"	13' 7"	12' 8"	12' 0"			
Rated Sheathing Exp 1	2x8	2 (24" oc)	17' 11"	16' 7"	15' 7"	14' 8"			
5-ply	2x10		21' 7"	20' 0"	18' 8"	17' 7"			
	2x12		25' 1"	23' 3"	21' 9"	20' 6"			
	2x4		13' 1"	12' 2"	11' 5"	10' 10"	10' 4"	9' 11"	9' 7"
19/32" APA	2x6		19' 2"	17' 9"	16' 8"	15' 10"	15' 2"	14' 7"	14' 0"
Rated Sheathing Exp 1	2x8	4 (12" oc)	24' 4"	22' 7"	21' 3"	20' 2"	19' 3"	18' 6"	17' 10"
40/20 4-ply or 5-ply	2x10		30' 2"	28' 0"	26' 4"	25' 0"	23' 10"	22' 10"	21' 10"
	2x12		36' 0"	33' 4"	31' 4"	29' 8"	28' 1"	26' 10"	25' 8"

SPAN TABLES FOR PLANK FLOORS AND ROOFS

Plank floors and roofs are used in plank-and-beam and post-and-beam constructions. The tables below are for tongue-and-groove planks applied in the alternating two-span style, as shown.

Note that two spans must be determined

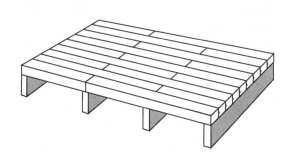

- limited by Fb
- limited by deflection

FLOOR: 40 PSF LIVE LOAD; DEFLECTION RATIO 1/360

Thickness, inches Nominal	(Actual)	Dead Load psf	Fb, psi 900	1,200	1,500	1,800	E, 10^6 psi 1.0	1.2	1.4
1	(3/4)	5	3 - 10	4 - 06	5 - 00	5 - 06	2 - 09	2 - 11	3 - 01
2	(1-1/2)	7	7 - 07	8 - 09	9 - 09	10 - 08	5 - 06	5 - 10	6 - 02
3	(2-1/2)	12	12 - 00	13 - 10	15 - 06	16 - 11	9 - 02	9 - 09	10 - 03
4	(3-1/2)	15	16 - 04	18 - 10	21 - 01	23 - 01	12 - 10	13 - 08	14 - 05
6	(5-1/2)	20	24 - 07	28 - 05	31 - 09	34 - 09	20 - 04	21 - 08	22 - 09

Source: *Wood Structural Design Data* (Washington, DC: National Forest Products Association, 1978).
Note: Spans are given as feet–inches.

ROOF: 20 PSF LIVE LOAD; DEFLECTION RATIO 1/180

Thickness, inches Nominal	(Actual)	Dead Load psf	Fb, psi 900	1,200	1,500	1,800	E, 10^6 psi 1.0	1.2	1.4
1	(3/4)	7	5 - 00	5 - 09	6 - 05	7 - 01	4 - 04	4 - 08	4 - 11
2	(1-1/2)	10	9 - 06	10 - 11	12 - 03	13 - 05	8 - 09	9 - 04	9 - 09
3	(2-1/2)	15	14 - 7	16 - 10	18 - 10	20 - 08	14 - 07	15 - 06	16 - 04
4	(3-1/2)	20	19 - 02	22 - 02	24 - 09	27 - 01	20 - 05	21 - 08	22 - 10
6	(5-1/2)	25	28 - 05	32 - 09	36 - 08	40 - 02	32 - 04	34 - 04	36 - 02

Source: *Wood Structural Design Data* (Washington, DC: National Forest Products Association, 1978).
Note: Spans are given as feet–inches.

ROOF: 30 PSF LIVE AND SNOW LOAD; DEFLECTION RATIO 1/180

Thickness, inches Nominal	(Actual)	Dead Load psf	Fb, psi				E, 10⁶ psi		
			900	1,200	1,500	1,800	1.0	1.2	1.4
1	(3/4)	7	4 - 03	4 - 10	5 - 06	6 - 00	3 - 10	4 - 01	4 - 03
2	(1-1/2)	10	8 - 02	9 - 05	10 - 07	11 - 07	7 - 08	8 - 01	8 - 06
3	(2-1/2)	15	12 - 10	14 - 10	16 - 08	18 - 03	12 - 09	13 - 06	14 - 03
4	(3-1/2)	20	17 - 02	19 - 09	22 - 02	24 - 03	17 - 10	18 - 11	19 - 11
6	(5-1/2)	25	25 - 08	29 - 07	33 - 02	36 - 04	28 - 03	30 - 00	31 - 07

Source: *Wood Structural Design Data* (Washington, DC: National Forest Products Association, 1978).
Note: Spans are given as feet–inches.

ROOF: 40 PSF LIVE AND SNOW LOAD; DEFLECTION RATIO 1/180

Thickness, inches Nominal	(Actual)	Dead Load psf	Fb, psi				E, 10⁶ psi		
			900	1,200	1,500	1,800	1.0	1.2	1.4
1	(3/4)	7	3 - 09	4 - 04	4 - 10	5 - 04	3 - 05	3 - 08	3 - 10
2	(1-1/2)	10	7 - 04	8 - 05	9 - 05	10 - 04	6 - 11	7 - 04	7 - 09
3	(2-1/2)	15	11 - 08	13 - 05	15 - 01	16 - 06	11 - 07	12 - 04	12 - 11
4	(3-1/2)	20	15 - 07	18 - 01	20 - 02	22 - 02	16 - 03	17 - 03	18 - 02
6	(5-1/2)	25	23 - 07	27 - 03	30 - 06	33 - 05	25 - 08	27 - 03	28 - 08

Source: *Wood Structural Design Data* (Washington, DC: National Forest Products Association, 1978).
Note: Spans are given as feet–inches.

SPAN TABLES FOR GLUED LAMINATED BEAMS

The term *structural glued laminated timber* refers to an engineered assembly of selected wood laminations bonded with adhesives. The advantages of glued laminated timbers are greater spans (up to 60 feet), greater strength (laminations allow elimination of defects such as knots), and virtual elimination of shrinkage, check, twist, and splitting.

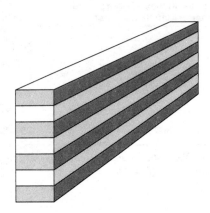

The floor and roof beam tables of the following pages are applicable for straight, simply supported beams that are laterally supported by decking, joists, or purlins and restrained against rotation at beam ends. The strength values are for southern pine, in pounds per square inch (psi):

- Fiber stress in bending, Fb = 2,400 psi.
- Shear stress, Fv = 200 psi.
- Modulus of elasticity, $E = 1.7 \times 10^6$ psi.
- Moisture content = 16 percent. For greater moisture, multiply Fb by 0.80.
- Deflections are limited to 1/180 of span for roof beams, based on total load. Deflection for floor beams is limited to 1/360 of span, based on total load.

The allowable loads shown in the tables are in pounds per linear foot and include dead (including the beam) and live loads.

The tables that follow do not include all of the sizes and wood species available. For species and sizes available in your area, consult your lumberyard.

GLUED LAMINATED FLOOR BEAMS
Total Load in Pounds per Linear Foot; Deflection 1/360 of Span Based on Live Load of 80% of Total Load

Actual Size b x d, inches	Clear Span, feet								
	8	12	16	20	24	28	32	36	40
3 x 5.5	256	76	32						
3 x 6.9	500	148	62	32					
3 x 8.3	851	256	108	55	32				
3 x 9.6	1,158	406	171	88	51	32			
3 x 11.0	1,427	606	256	131	76	48	32		
3 x 12.4	1,667	851	364	186	108	68	46	32	
3 x 13.8	1,927	1,050	500	256	148	93	62	44	32
3 x 15.1	2,208	1,271	665	340	197	124	83	58	43
5 x 9.6	1,930	677	286	146	85	53	36		
5 x 11.0	2,378	1,010	426	218	126	80	53	37	
5 x 12.4	2,779	1,418	607	311	180	113	76	53	39
5 x 13.8	3,212	1,751	833	426	247	155	104	73	53
5 x 15.1	3,681	2,118	1,108	567	328	207	139	97	71
5 x 16.5	4,190	2,378	1,415	737	426	268	180	126	92
5 x 17.9	4,747	2,642	1,648	936	542	341	229	161	117
5 x 19.3	5,357	2,919	1,897	1,170	677	426	286	201	146
6.75 x 12.4	3,752	1,914	819	419	243	153	102	72	52
6.75 x 13.8	4,336	2,363	1,124	575	333	210	140	99	72
6.75 x 15.1	4,969	2,860	1,496	766	443	279	187	131	96
6.75 x 16.5	5,657	3,211	1,848	994	575	362	243	170	124
6.75 x 17.9	6,408	3,567	2,151	1,264	732	461	309	217	158
6.75 x 19.3	7,231	3,941	2,477	1,550	914	575	385	271	197
6.75 x 20.6	8,137	4,336	2,824	1,767	1,124	708	474	333	243
6.75 x 22.0	9,138	4,752	3,192	1,998	1,362	859	575	404	295
6.75 x 23.4	10,252	5,192	3,476	2,242	1,529	1,030	690	485	353

Source: *Glued Laminated Timbers* (Vancouver, Wash: American Institute of Timber Construction, 1988).

GLUED LAMINATED ROOF BEAMS
Total Load in Pounds per Linear Foot; Deflection 1/180 of Span Based on Total Load

Actual Size b x d, inches	Clear Span, feet								
	8	12	16	20	24	28	32	36	48
3 x 5.5	409	121	51	26					
3 x 6.9	679	237	100	51	30				
3 x 8.3	978	409	173	88	51	32			
3 x 9.6	1,332	592	274	140	81	51	34		
3 x 11.0	1,641	773	409	210	121	76	51	36	26
3 x 12.4	1,917	978	550	298	173	109	73	51	37
3 x 13.8	2,216	1,208	679	409	237	149	100	70	51
3 x 15.1	2,540	1,462	822	526	315	198	133	93	68
5 x 9.6	2,220	986	457	234	135	85	57	40	29
5 x 11.0	2,735	1,288	682	349	202	127	85	60	44
5 x 12.4	3,196	1,631	917	497	288	181	121	85	62
5 x 13.8	3,693	2,013	1,132	682	395	249	167	117	85
5 x 15.1	4,233	2,436	1,370	863	525	331	222	156	113
5 x 16.5	4,819	2,735	1,628	1,019	682	429	288	202	147
5 x 17.9	5,459	3,038	1,895	1,186	809	546	366	257	187
5 x 19.3	6,160	3,357	2,181	1,365	931	674	457	321	234
6.75 x 12.4	4,314	2,201	1,230	671	388	245	164	115	84
6.75 x 13.8	4,986	2,718	1,503	921	533	336	225	158	115
6.75 x 15.1	5,714	3,289	1,801	1,127	709	447	299	210	153
6.75 x 16.5	6,506	3,692	2,125	1,330	907	580	388	273	199
6.75 x 17.9	7,370	4,102	2,474	1,549	1,056	737	494	347	253
6.75 x 19.3	8,316	4,532	2,848	1,783	1,216	879	617	433	316
6.75 x 20.6	9,358	4,986	3,247	2,032	1,386	1,003	757	533	388
6.75 x 22.0	10,509	5,465	3,671	2,298	1,567	1,133	856	647	471
6.75 x 23.4	11,790	5,971	3,997	2,578	1,758	1,272	961	750	565

Source: *Glued Laminated Timbers* (Vancouver, Wash: American Institute of Timber Construction, 1988).

SPAN TABLE FOR STEEL BEAMS

Steel beams often provide the most cost-effective solution, even in residential construction, when the loads to be carried are heavy. The table below shows the uniformly distributed loads that can be carried by the most common steel beam: shape W, type A36 steel (actually I-shaped and called an I-beam).

The table can also be used for concentrated loads by first converting the concentrated loads, C, to their equivalent uniform loads, U, as shown at right. If the application is unusu-al, or if the beam is being used near its limit, consult with a structural engineer to check for shear, deflection, and end constraints.

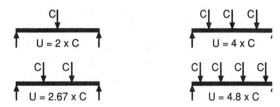

C = load concentrated at a point
U = equivalent uniformly distributed load

Beam Type	Actual Size b x d, inches		Clear Span, feet 8	12	16	20	24	28
W8x10	4	7-7/8	15,600	10,400	7,800			
W8x13	4	8	19,800	13,200	9,900			
W8x15	4	8-1/8	23,600	15,700	11,800			
W8x18	5-1/4	8-1/8	30,400	20,300	15,200			
W8x21	5-1/4	8-1/4	36,400	24,300	18,200			
W8x24	6-1/2	7-7/8	41,800	27,900	20,900			
W10x12	4	9-7/8	21,800	14,500	10,900	8,700		
W10x15	4	10	27,600	18,400	13,800	11,000		
W10x17	4	10-1/8	32,400	21,600	16,200	13,000		
W10x19	4	10-1/4	37,600	25,100	18,800	15,000		
W10x22	5-3/4	10-1/8	46,400	30,900	23,200	18,600		
W10x26	5-3/4	10-3/8	55,800	37,200	27,900	22,300		
W10x30	5-3/4	10-1/2	64,800	43,200	32,400	25,900		
W12x14	4	11-7/8	29,800	19,900	14,900	11,900	9,930	8,510
W12x16	4	12	32,400	22,800	17,100	13,700	11,400	9,770
W12x19	4	12-1/8	42,600	28,400	21,300	17,000	14,200	12,200
W12x22	4	12-1/4	50,800	33,900	25,400	20,300	16,900	14,500
W12x26	6-1/2	12-1/4	66,800	44,500	33,400	26,700	22,300	19,100

Source: *Manual of Steel Construction* (New York: American Institute of Steel Construction, 1969).
Note: Loads are given in pounds.

POST-AND-BEAM FRAMING

Evolution

Also known as *post-and-girt* and *post-and-lintel,* the *post-and-beam* framing system was already well developed in Europe before the discovery of America. The main method of fastening the heavy timbers, the mortise-and-tenon joint (see following pages), was developed sometime between 200 BC and 500 BC, and the self-supporting braced frame was developed around AD 900.

The system is characterized by the use of large, widely spaced load-carrying timbers. Vertical loads are carried by *posts* at the corners and intersections of load-bearing walls. *Plates* and *girts* collect distributed roof and floor loads from *rafters* and *joists* and transfer them to the posts. Wall *studs* carry no vertical loads, being used merely for attachment of wall sheathing.

Post-and-beam framing requires a higher degree of carpentry skill than the so-called stick systems. First, the integrity of the frame depends on the choice and meticulous execution of the wood-peg-fastened joints. Second, the entire frame is precut, each intersecting member being labeled. Finally – usually in a single day – the members are assembled into large wall sections called bents, and then raised with much human power and conviviality. Raising day is the moment of truth for the lead framer.

The post-and-beam system evolved more from economic necessity than from aesthetic sensibility. Large timbers were more plentiful, circular-saw and band-saw mills were not available to slice timbers into sticks, and nails were still made by hand, which made them much too costly to use in the quantities needed for stick framing.

Modern Developments

The post-and-beam frame enjoys a new popularity today, but not for economic reasons. Even proponents acknowledge that a well-executed timber frame is more expensive than a frame employing wood-saving and labor-saving trusses, metal fasteners, and power nailing. The appeal of the timber frame lies in its material and craftsmanship, as well as nostalgia.

The post-and-beam frame of today remains unchanged. In fact, most framers try simply to emulate the best of the past. What have changed, however, are the insulation and the sheathing systems. Few contractors still practice the classic in-fill system, whereby wall studs are placed between the load-carrying posts to support exterior sheathing and interior finish.

Foam-filled stressed-skin panels dominate the market now. Widely available from dozens of manufacturers, most consist of exterior panels of oriented-strand board (OSB) and interior sheets of gypsum drywall, separated by either urethane or expanded polystyrene foam.

Wiring and plumbing were developed well after the timber frame. To this day, artful concealment of pipes and wires requires detailed planning on the part of the designer.

Typical Post-and-Beam Frame

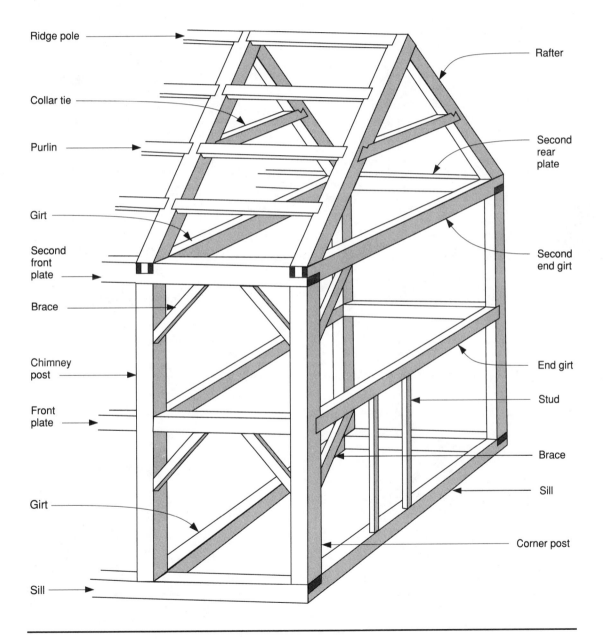

Ridge pole

Collar tie

Purlin

Girt

Second front plate

Brace

Chimney post

Front plate

Girt

Sill

Rafter

Second rear plate

Second end girt

End girt

Stud

Brace

Sill

Corner post

Common Post-and-Beam Trusses

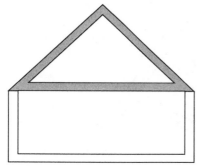

SIMPLE

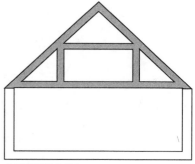

QUEENPOST

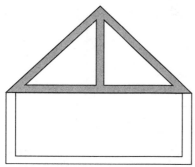

KINGPOST

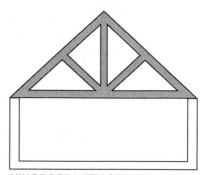

KINGPOST WITH STRUTS

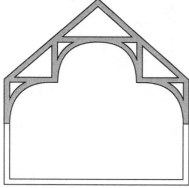

HAMMER BEAM

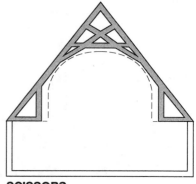

SCISSORS

Post-and-Beam Lap Joints

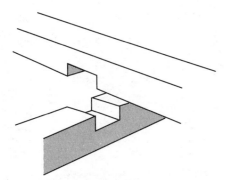

THROUGH HALF LAP

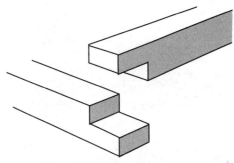

CORNER HALF LAP

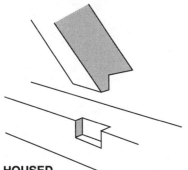

**HOUSED
BIRD'S MOUTH**

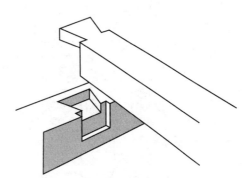

HOUSED LAPPED DOVETAIL

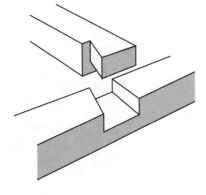

LAPPED HALF DOVETAIL

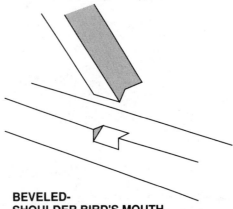

**BEVELED-
SHOULDER BIRD'S MOUTH**

Mortise-and-Tenon Joints

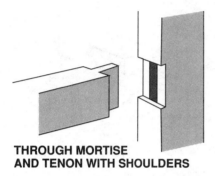

**THROUGH MORTISE
AND TENON WITH SHOULDERS**

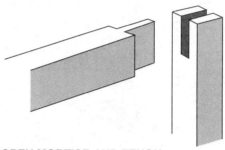

OPEN MORTISE AND TENON

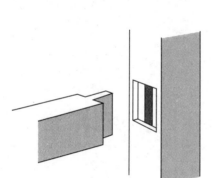

HOUSED MORTISE AND TENON

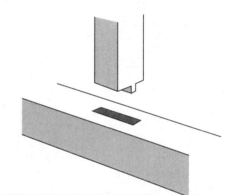

STUB MORTISE AND TENON

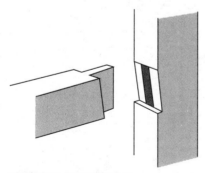

**MORTISE AND TENON
WITH DIMINISHED HAUNCH**

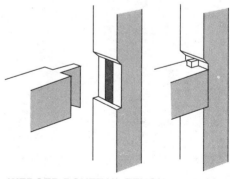

WEDGED DOVETAIL TENON

BALLOON FRAMING

The *balloon frame* developed in response to at least three innovations in the construction industry:

- invention of the machine-made nail, which made nailing far less expensive than hand-pegged joinery
- more advanced circular-saw mills, which lowered the cost of standardized sawn lumber
- development of a housing industry, which promoted standardization of materials and methods

In the balloon frame, there is no requirement for lumber more than 2 inches thick. Plates, sills, and posts are all built up from 2-inch stock. Front and rear girts are replaced by ribbons (or ribbands), 1 x 6-inch or 1 x 8-inch boards let into the studs as a support for the joists. Studs now carry most of the vertical load. Bracing is supplied by sheathing panels or by diagonally applied sheathing boards.

The balloon frame gets its name from the fact that the studs run unbroken from sill to top plate, regardless of the number of stories, making the frame like a membrane or balloon. Both the major advantage and disadvantage of the system derive from this fact. In drying, lumber may shrink up to 8 percent in width, but only 0.1 percent in length. Since the overall dimensions of a balloon frame are controlled by length, shrinkage cracks in stucco and plaster finishes are minimized.

The major disadvantage of the balloon frame is the necessity of fire-stopping the vertical wall cavities, which otherwise would act as flues in rapidly spreading fire from basement to attic. A related disadvantage is the difficulty of insulating the wall cavities, which typically open into the basement at the foundation sill.

Because of the fire danger, and because of the scarcity of long framing members, the balloon frame has been replaced almost entirely by the platform frame, discussed in the next section. Framing details for both systems are similar and are shown in later pages.

Typical Balloon Frame

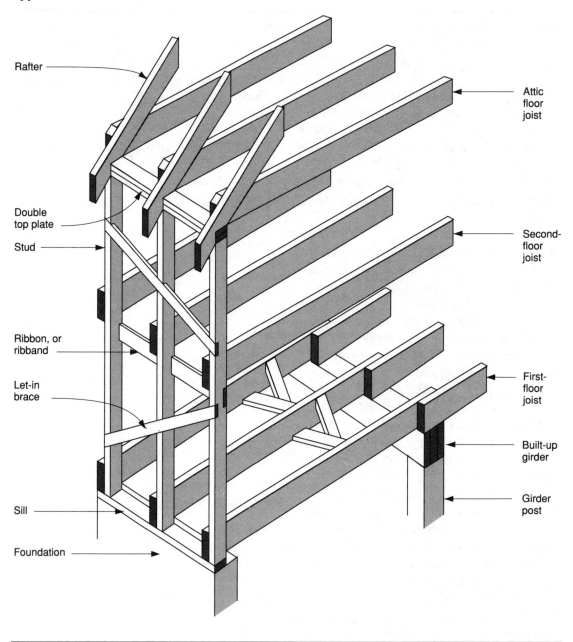

Rafter

Attic floor joist

Double top plate

Stud

Second-floor joist

Ribbon, or ribband

Let-in brace

First-floor joist

Built-up girder

Girder post

Sill

Foundation

PLATFORM FRAMING

The platform frame solved the fire-stop problem of the balloon frame. Each story of a platform frame is built upon a platform consisting
of joists, band joists, and subfloor. Story
builds upon story. After completion of the
first-story walls, the second platform is built
identical to the first as if the first-story walls
were the foundation. Short of modular panels,
the platform house is the ultimate in standardization, requiring the lowest levels of carpentry skill and the fewest standard sizes of
lumber.

The platform frame standardized the use of
band joists and sole plates. The sole plate and
top plate of each wall automatically provide
fire-stopping and nailing surfaces for both exterior sheathing and interior drywall. The ad-
vent of gypsum drywall and other forms of
paneling made frame shrinkage less important. Bracing is provided either by the structurally rated exterior sheathing panels, by diagonally applied sheet metal wind braces, or
more rarely, by let-in 1 x 4-inch wood braces.

Optimum Value Engineering

Many cost-saving refinements to the platform
frame have been promoted under the name of
"optimum value engineering":

Roof trusses, spanning greater distances but
using less material, save both material and labor. In addition, trusses eliminate the need for
interior walls to support long ceiling joists.

Uniform 24-inch on-center framing uses a
third less wood, reduces heat loss through the
frame, and simplifies the modular system.

One-inch sill and sole plates are structurally adequate and cut wood use by half.

Three-inch interior walls save a quarter of
the 5 percent of interior space lost to walls.

Off-center spliced joists, where butting
joists are spliced with plywood and the girder
overhangs alternate, span greater distances for
the same depth.

Elimination of bridging eliminates a very
time-consuming task. Tests have shown that
bridging does little to stiffen a floor and that
long joists can be prevented from twisting
with a single 1x3 nailed across the bottom
edges of the joists.

Single-layer floors, such as APA Sturd-I-
Floor nonplywood panels, are rigid enough to
span 24 inches and are smooth enough to
serve as a base for carpet and sheet flooring.

Simpler corner posts, essentially two single
end studs from the intersecting walls, eliminate waste and heat loss. Blocking for drywall
is provided by metal drywall backup clips.

Typical Platform Frame

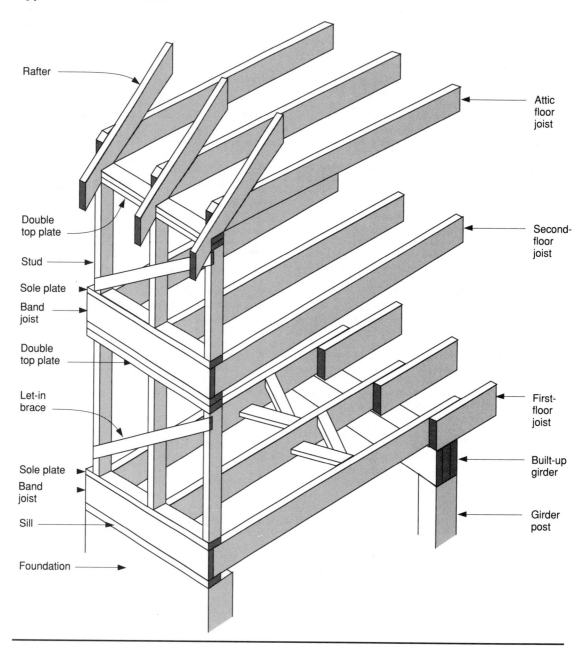

Rafter

Attic floor joist

Double top plate

Second-floor joist

Stud

Sole plate

Band joist

Double top plate

Let-in brace

First-floor joist

Sole plate

Built-up girder

Band joist

Sill

Girder post

Foundation

FRAMING DETAILS

Floor Joist Splices

BUTTED AND SPLICED

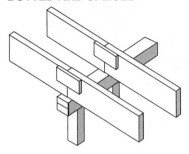

ON STEEL BEAM

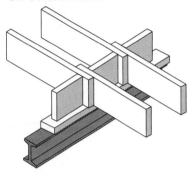

OVERLAPPED

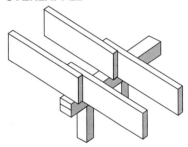

SET INTO STEEL BEAM

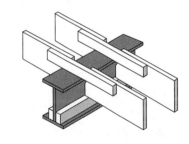

**OVERLAPPED AND BLOCKED
ON BUILT-UP BEAM**

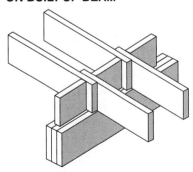

**OVERLAPPED AND BLOCKED
ON SOLID BEAM**

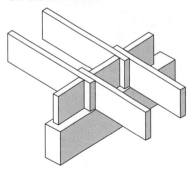

Joists Butted at Beam

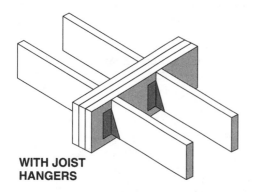

**WITH JOIST
HANGERS**

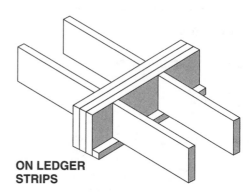

**ON LEDGER
STRIPS**

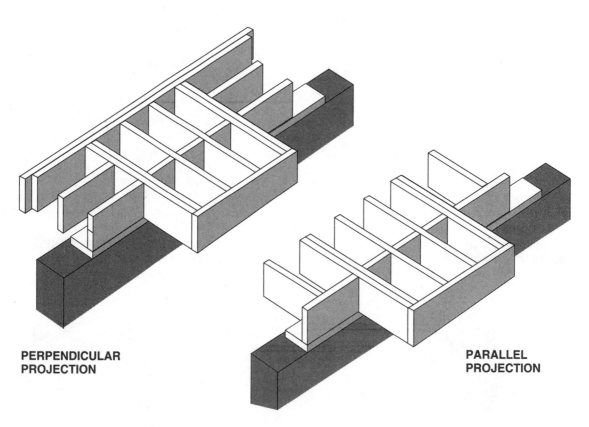

**PERPENDICULAR
PROJECTION**

**PARALLEL
PROJECTION**

Bridging and Blocking

SOLID BLOCKING

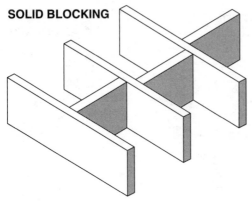

DIAGONAL BRIDGING

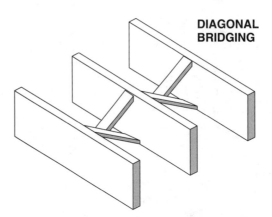

DIAGONAL BRIDGING

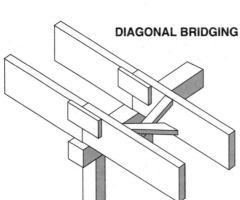

SOLID BLOCKING

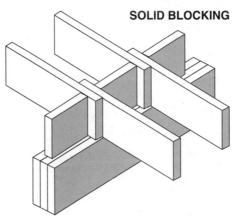

METAL BRIDGING

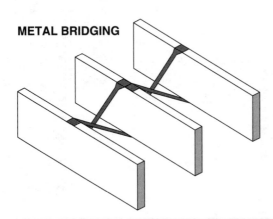

METAL BRIDGING

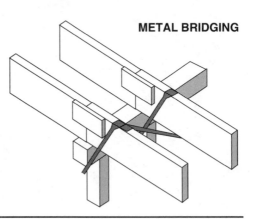

Walls

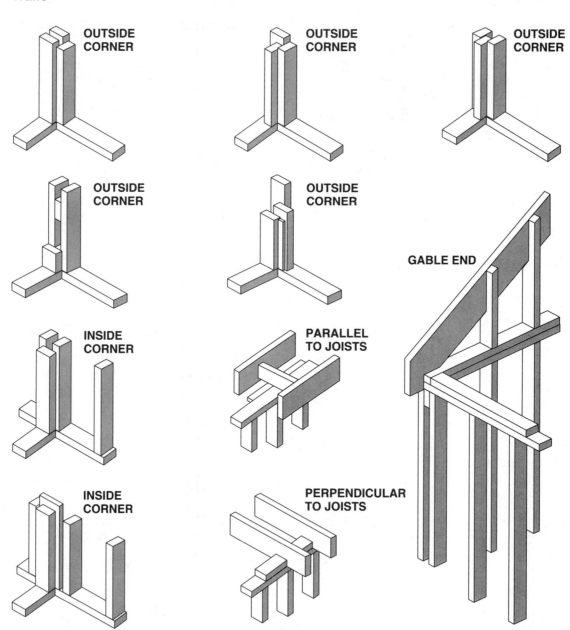

OUTSIDE CORNER

OUTSIDE CORNER

OUTSIDE CORNER

OUTSIDE CORNER

OUTSIDE CORNER

GABLE END

INSIDE CORNER

PARALLEL TO JOISTS

INSIDE CORNER

PERPENDICULAR TO JOISTS

Partitions

SECOND FLOOR BEARING WALL

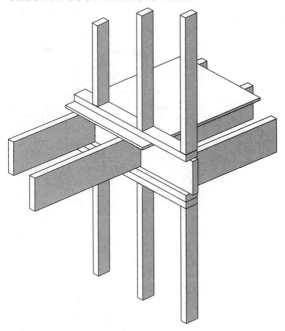

SECOND FLOOR BEARING WALL

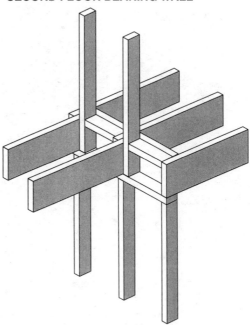

NONBEARING WALL

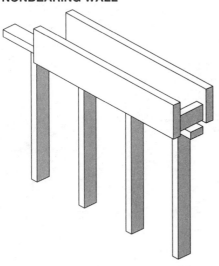

FIRST FLOOR BEARING WALL

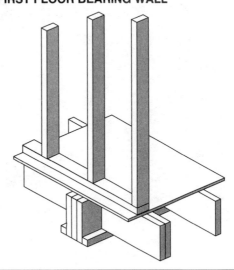

Headers

WALL OPENINGS

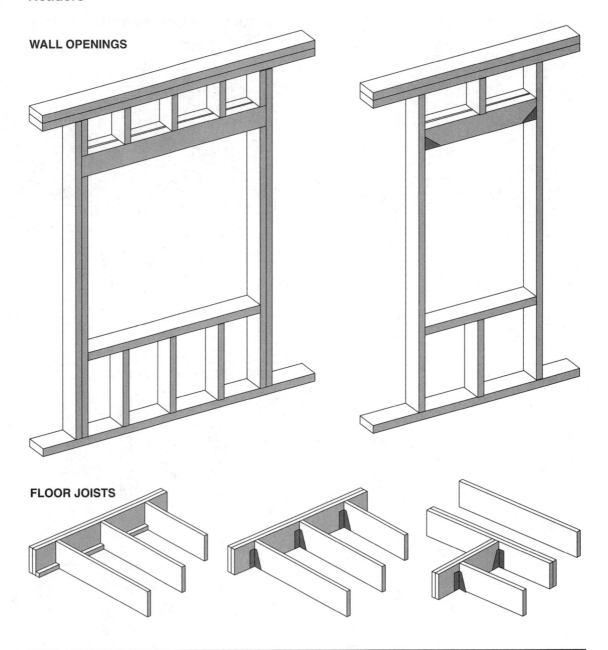

FLOOR JOISTS

Stairways

**PERPENDICULAR TO
FLOOR JOISTS**

**PARALLEL TO
FLOOR JOISTS**

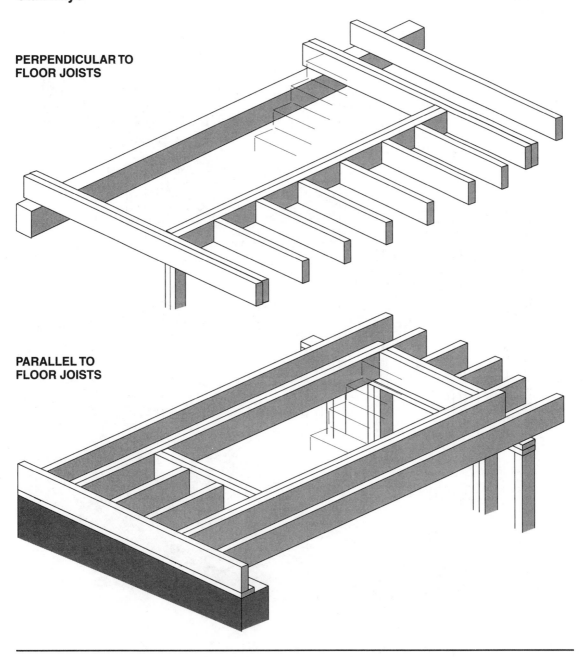

Bay Window

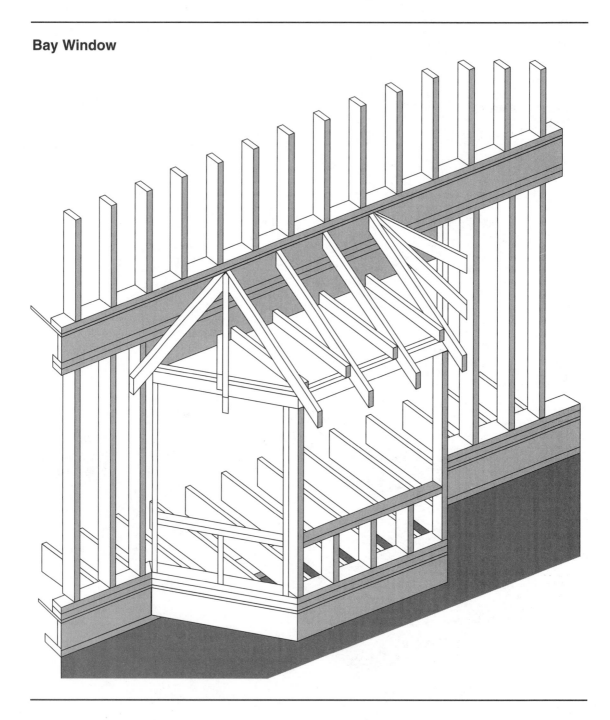

Roof Overhangs

**GABLE
END**

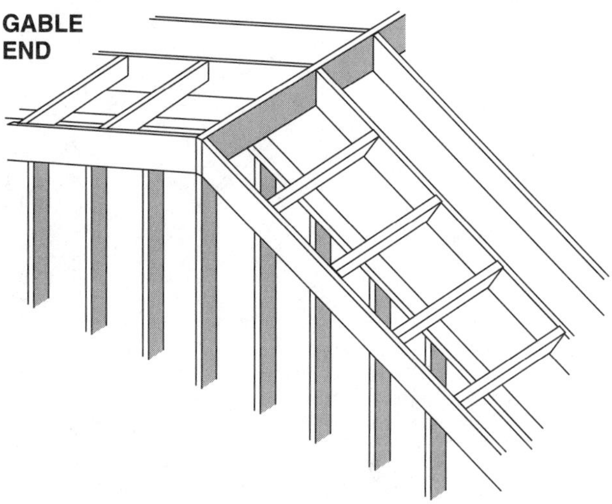

EAVE

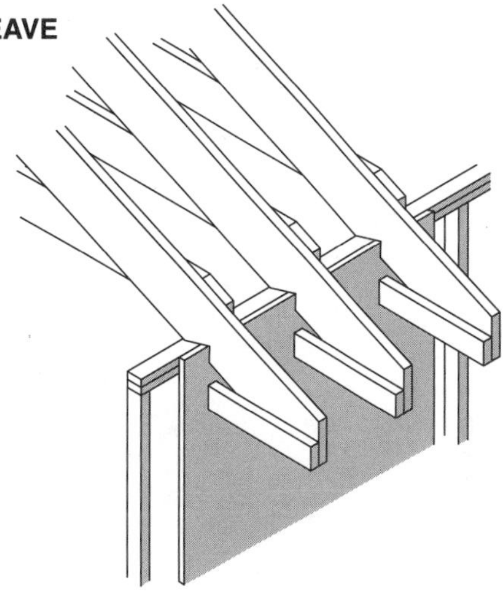

FLAT ROOF

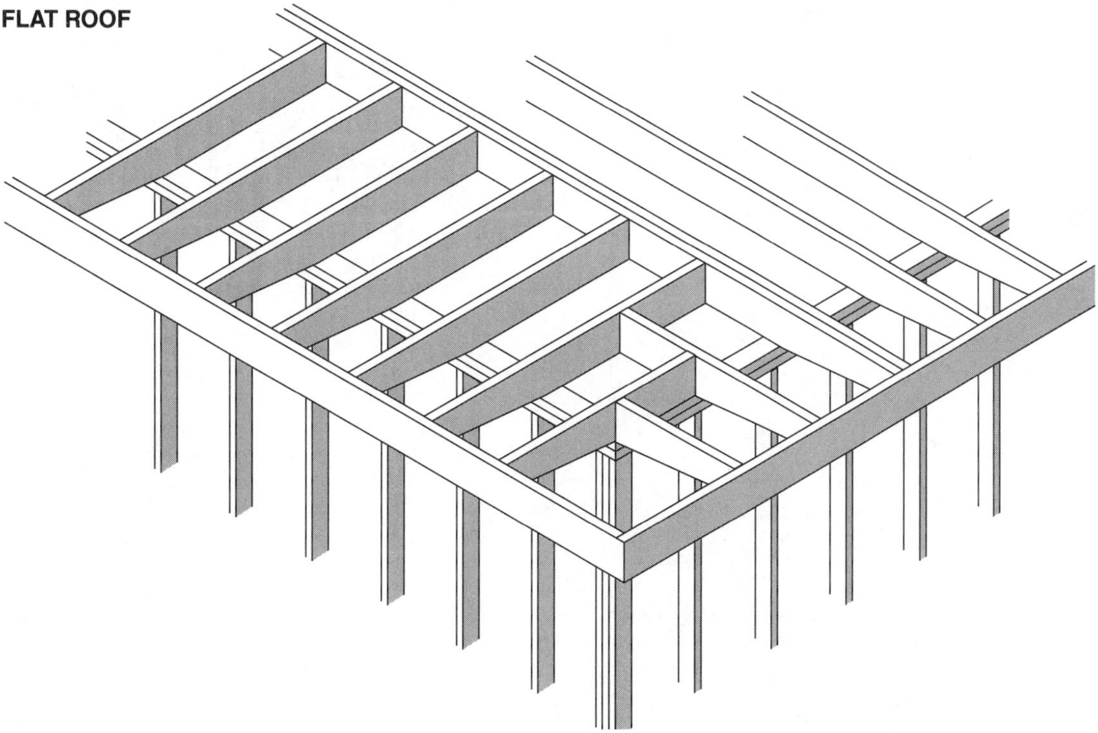

Dormers

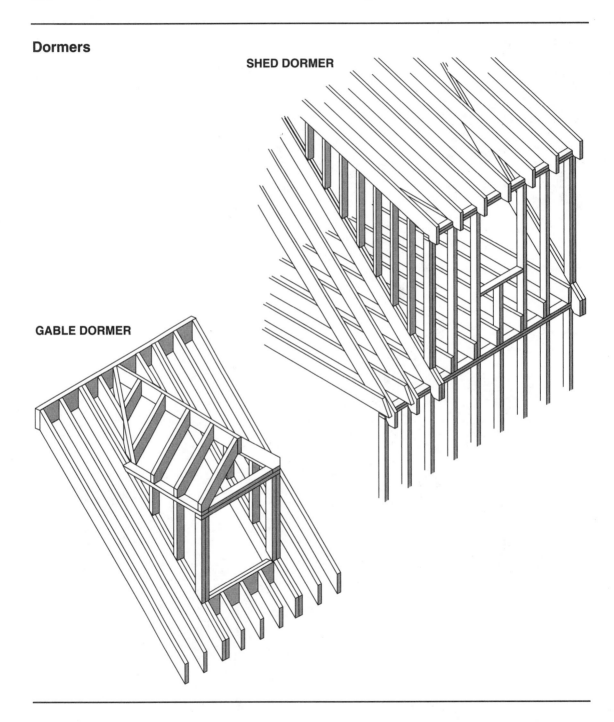

SHED DORMER

GABLE DORMER

Chimneys and Fireplaces

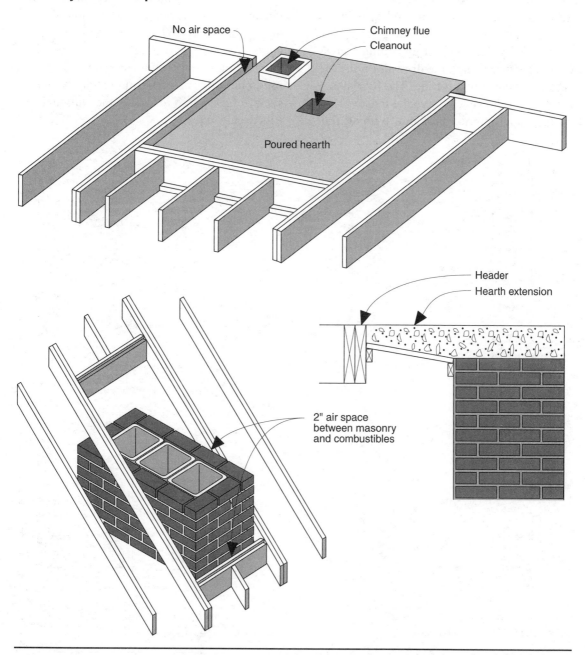

No air space

Chimney flue

Cleanout

Poured hearth

Header

Hearth extension

2" air space between masonry and combustibles

Framing for Superinsulation

Superinsulation can be achieved by using thicker studs, by double-studding (two walls with a cavity between), or by adding a layer of foam either behind the drywall on the inside or as sheathing outside. The four systems below, however, solve the thorny wiring/vapor retarder problem. With all four systems, the vapor barrier is placed over the inside of the studs before installation of the horizontal nailers. Wiring is run along the nailers without penetrating the vapor retarder, and the thickness of the nailers (1-1/2 inches minimum) places the wiring inside the wall surface far enough to satisfy the National Electrical Code.

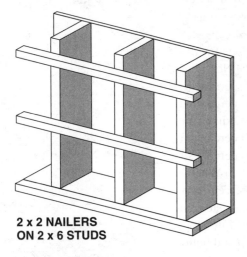

**2 x 2 NAILERS
ON 2 x 6 STUDS**

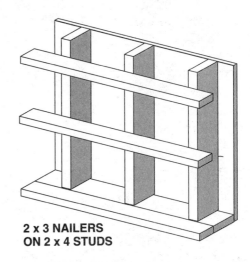

**2 x 3 NAILERS
ON 2 x 4 STUDS**

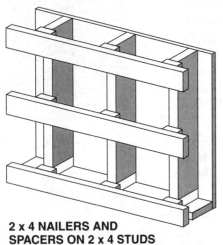

**2 x 4 NAILERS AND
SPACERS ON 2 x 4 STUDS**

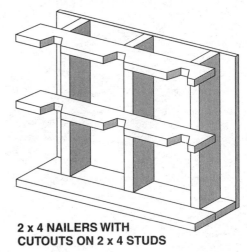

**2 x 4 NAILERS WITH
CUTOUTS ON 2 x 4 STUDS**

CHECKLIST OF CODE REQUIREMENTS

The following is a partial list of requirements from the 1995 Council of American Building Officials (CABO) *One and Two Family Dwelling Code*. Consult the publication for the full text and additional provisions.

Floor Framing

(502.3.2) Joists under bearing partitions:
- double joists or adequate beam
- block separated double joists ≤ 4' on-center

(502.4) Joist, beam, and girder end bearing:
- ≥ 1-1/2" on wood or metal
- ≥ 3" on masonry, except on ribbon or hanger

(502.5) Lateral joist restraint at supports:
- full-depth solid blocking, or
- attached to header, band, or rim joist, or
- nailed to adjacent stud

(502.5.1) Bridge > 2 × 12 joists at 10' on-center:
- full-depth solid blocking, or
- diagonal wood or metal, or
- 1 × 3 nominal strip at bottom

(502.5.1) Drilling/notching joists:
- not to exceed 1/6 depth top or bottom
- not in middle third of span
- not to exceed 1/4 depth for end ledger strip

(502.7) Holes in joists not within 2" of top or bottom and diameter ≤ 1/3 depth

Wall Framing

(602.2) Grade minimum of No. 3, standard, or stud, except nonbearing studs may be utility

(602.3.3) Top plate:
- doubled
- overlapped at corners and intersections
- joints overlapped 48" minimum

(602.3.4) Joists/rafters ≤ 5" of bearing stud if:

- joists/rafters spaced > 16" on-center, and
- bearing studs spaced 24" on-center except if:
- top plates are two 2 × 6s or 3 × 4s, or
- there is a third top plate, or
- solid blocking reinforces top plate

(602.4.1) Interior nonbearing partitions may be:
- 2 × 3 studs spaced 24" on-center, or
- 2 × 4 flat studs spaced 16" on-center
- capped with single top plate

(602.5) Notching studs not to exceed:
- 25 percent width of bearing studs
- 40 percent width of nonbearing studs

Drilling diameter not to exceed:
- 40 percent stud width, ≥ 5/8" from edge
- 60 percent stud width for doubled studs

(602.5.1) When drilling/notching top plate > 50 percent, top plate to be reinforced with 24-gauge steel angle

(602.7) Fire-stopping required at:
- top and bottom of walls
- intersections of walls and ceilings
- top and bottom of stair stringers
- openings in floor and ceiling

(602.7.1.1) Fire-stopping of unfaced fiberglass must fill entire cavity to depth of 16" minimum

MAXIMUM STUD SPACING (TABLE 602.3d)

Size	Studs Supporting		
	Roof + Ceiling	Roof + Floor + Ceiling	Roof + 2 Floors + Ceiling
2 x 4	24"	16"	–
3 x 4	24"	24"	16"
2 x 6	24"	24"	16"

WALL BRACING (TABLE 602.9)

Seismic Zone	Wall	Bracing	Amount
0, 1, 2	1 story Top of 2 or 3 stories 1st of 2 stories 2nd of 3 stories	1 x 4 let-in bracing or structural sheathing	Each end and at least every 25' of wall length
	1st of 3 stories	Structural sheathing	Min 48" panels located as required for let-in bracing
3, 4	1 story Top of 2 or 3 stories	1 x 4 let-in bracing or structural sheathing	Each end and at least every 25' of wall length
	1st of 2 stories 2nd of 3 stories	Structural sheathing	25% of length to be sheathed
	1st of 3 stories	Structural sheathing	40% of length to be sheathed

Roof Framing

(802.3) Details:
- rafters parallel to ceiling joists tied to joists
- rafters not parallel to joists have rafter ties
- rafter ties as close to plate as practical
- rafter ties spaced 4' maximum
- rafter tops tied to ridge board or gusseted
- ridge board 3/4" × cut depth of rafter min.
- valley/hip rafter 1-1/2" × cut depth of rafter

(802.3.1) Ceiling joists:
- lapped 3" minimum, or
- tied together to resist thrust of rafters

(802.5) Rafter/ceiling joist end bearing:
- 1-1/2" minimum on wood or metal
- 3" minimum on masonry

(802.6) Cutting/notching rafter/ceiling joist:
- 1/4 depth at ends
- 1/6 depth in top or bottom and not in middle third
- 1/3 depth in top, ≤ depth from support

(802.7) Bored holes in rafter/ceiling joist:
- not within 2" of top or bottom
- diameter ≤ 1/3 depth

(802.8) Lateral support at points of bearing for rafter/ceiling joists with depth-to-thickness ratio > 5/1

(802.8.1) Bridging rafters with > 6/1 depth-to-thickness ratio at 10' maximum intervals:
- full-depth solid blocking, or
- diagonal wood or metal, or
- 1 × 3 strip at bottom

(802.9) Openings in ceiling and roof framing to be framed with headers

(802.10) Trusses to be constructed and braced to specifications of Truss Plate Institute

(802.11) Roof assemblies subject to ≥ 20 psf of wind uplift must have truss or rafter ties effectively connected to foundation

7   Sheathing

If the frame is the building's skeleton, then the sheathing is its skin. Sheathing functions to enclose the building in an airtight barrier, to strengthen its studs, joists, and rafters by tying them together, to brace the building against racking (twisting) under wind and seismic forces, and to provide a base for flooring, siding, and roofing.

By far, most sheathing is done with panels manufactured by a member of, and under the inspection of, The Engineered Wood Association (APA). The APA publishes a wide range of builder-oriented literature on its products. All of the material in this chapter is adapted from its booklet *Design/Construction Guide: Residential and Commercial.*

The chapter begins by showing and explaining the grade stamps for all of the *APA sheathing panels*.

Next, illustrations and tables show you all you'll ever need to know about APA *floor sheathing, underlayment, Sturd-I-Floor, glued floor, wall sheathing*, and panels for *roof sheathing*.

THE ENGINEERED WOOD ASSOCIATION (APA) SHEATHING PANELS

Panels for sheathing can be manufactured in a variety of ways: as plywood (cross-laminated wood veneer), as composites (veneer faces bonded to reconstituted wood cores), or as nonveneered panels such as oriented-strand board (OSB).

Some plywood panels are manufactured under the provisions of Voluntary Product Standard PS-1-95 for Construction and Industrial Plywood, a detailed manufacturing specification developed cooperatively by the plywood industry and the US Department of Commerce. Other veneered panels, however, as well as an increasing number of performance-rated composite and nonveneered panels, are manufactured under the provisions of APA PRP-108, Performance Standards and Policies for Structural-Use Panels, or under Voluntary Product Standard PS 2-92, Performance Standard for Wood-Based Structural-Use Panels. These APA performance-rated panels are easy to use and specify because the recommended end use and maximum support spacings are indicated in the APA trademarks.

The list at right describes the veneer grading system. The two panel sides are often of different grade so that less expensive veneers can be used on the side of the panel that will not show.

The tables on the following pages constitute a field (or lumberyard) guide to APA sheathing, including veneered and nonveneered panels, and panels intended for exterior and interior conditions. Read the labels carefully. They contain a lot of information.

Plywood Veneer Grades

A – This smooth, paintable veneer allows not more than 18 neatly made repairs—boat, sled, or router type, and parallel to the grain. Wood or synthetic repairs are permitted. It may be used as natural finish in less demanding applications.

B – Solid surface veneer permits shims, sled or router repairs, tight knots to 1 inch across the grain, and minor splits. Wood or synthetic repairs are permitted.

C – This veneer has tight knots to 1-1/2 inches. It has knotholes to 1 inch across the grain with some to 1-1/2 inches if the total width of knots and knotholes is within specified limits. Repairs are synthetic or wood. Discoloration and sanding defects that do not impair strength are permitted. Limited splits and stitching are allowed.

C Plugged – Improved C veneer has splits limited to 1/8-inch width and knotholes and other open defects limited to 1/4 × 1/2 inch. It admits some broken grain. Wood or synthetic repairs are also permitted.

D – Knots and knotholes to 2-1/2-inch width across the grain, and 1/2 inch larger within specified limits, are allowed. Limited splits and stitching are permitted. This face grade is limited to exposure 1 or interior panels.

Source: *Design/Construction Guide: Residential and Commercial* (Tacoma, Wash: APA–The Engineered Wood Association, 1996).

PERFORMANCE-RATED PANELS

Typical APA Grade Stamp	Thicknesses, Inches	Grade Designation, Description, and Uses
APA THE ENGINEERED WOOD ASSOCIATION RATED SHEATHING 24/16 7/16 INCH SIZED FOR SPACING EXPOSURE 1 000 PRP-108 HUD-UM-40C	5/16 3/8 7/16 15/32 1/2 19/32 5/8 23/32 3/4	**APA RATED SHEATHING** Exposure Durability Classifications: Exterior, Exposure 1 Specially designed for subflooring and wall and roof sheathing. Also good for a broad range of other construction and industrial applications. Can be manufactured as plywood, as a composite, or as OSB.
APA THE ENGINEERED WOOD ASSOCIATION RATED SHEATHING STRUCTURAL 1 32/16 15/32 INCH SIZED FOR SPACING EXPOSURE 1 000 PS 1-95 CD PRP-108	5/16 3/8 7/16 15/32 1/2 19/32 5/8 23/32 3/4	**APA STRUCTURAL I RATED SHEATHING** Exposure Durability Classifications: Exterior, Exposure 1 Unsanded grade for use where shear and cross-panel strength properties are of maximum importance, such as panelized roofs and diaphragms. Can be manufactured as plywood, as a composite, or as OSB.
APA THE ENGINEERED WOOD ASSOCIATION RATED STURD-I-FLOOR 20 oc 19/32 INCH SIZED FOR SPACING T&G NET WIDTH 47-1/2 EXPOSURE 1 000 PRP-108 HUD-UM-40C	19/32 5/8 23/32 3/4 1 1-1/8	**APA RATED STURD-I-FLOOR** Exposure Durability Classifications: Exterior, Exposure 1 Specially designed as combination subfloor-underlayment. Provides smooth surface for application of carpet and pad and possesses high concentrated and impact load resistance. Can be manufactured as plywood, as a composite, or as OSB. Available square-edge or tongue-and-groove.
APA THE ENGINEERED WOOD ASSOCIATION RATED SIDING 24 oc 19/32 INCH SIZED FOR SPACING EXPOSURE 1 000 PRP-108 HUD-UM-40C	11/32 3/8 7/16 15/32 1/2 19/32 5/8	**APA RATED SIDING** Exposure Durability Classification: Exterior For exterior siding, fencing, etc. Can be manufactured as plywood, as a composite, or as an overlaid OSB. Both panel and lap siding available. Special surface treatment, such as V-groove, channel groove, deep groove (such as APA Texture 1-11), brushed, rough sawn, and overlaid (Medium-Density) with smooth- or texture-embossed face. Span Rating (stud spacing for siding qualified for APA Sturd-I-Wall applications) and face grade classification (for veneer-faced siding) indicated in trademark.

SANDED AND TOUCH-SANDED PANELS

Typical APA Grade Stamp	Thicknesses, Inches	Grade Designation, Description, and Uses
APA THE ENGINEERED WOOD ASSOCIATION A-C GROUP 1 EXTERIOR 000 PS 1-95	1/4, 11/32 3/8, 15/32 1/2, 19/32 5/8, 23/32 3/4	**APA A-C** Exposure Durability Classification: Exterior For use where appearance of only one side is important in exterior or interior applications, such as soffits, fences, farm buildings, etc.
APA THE ENGINEERED WOOD ASSOCIATION A-D GROUP 1 EXPOSURE 1 000 PS 1-95	1/4, 11/32 3/8, 15/32 1/2, 19/32 5/8, 23/32 3/4	**APA A-D** Exposure Durability Classification: Interior, Exposure 1 For use where appearance of only one side is important in interior applications, such as paneling, built-ins, shelving, partitions, flow racks, etc.
APA THE ENGINEERED WOOD ASSOCIATION B-C GROUP 1 EXTERIOR 000 PS 1-95	1/4, 11/32 3/8, 15/32 1/2, 19/32 5/8, 23/32 3/4	**APA B-C** Exposure Durability Classification: Exterior Utility panel for farm service and work buildings, boxcar and truck linings, containers, tanks, agricultural equipment, as a base for exterior coatings and other exterior uses or applications subject to high or continuous moisture.
APA THE ENGINEERED WOOD ASSOCIATION B-D GROUP 2 EXPOSURE 1 000 PS 1-95	1/4, 11/32 3/8, 15/32 1/2, 19/32 5/8, 23/32 3/4	**APA B-D** Exposure Durability Classifications: Interior, Exposure 1 Utility panel for backing, sides of built-ins, industry shelving, slip sheets, separator boards, bins, and other interior or protected applications.
APA THE ENGINEERED WOOD ASSOCIATION UNDERLAYMENT GROUP 1 EXPOSURE 1 000 PS 1-95	1/4, 11/32 3/8, 15/32 1/2, 19/32 5/8, 23/32 3/4	**APA UNDERLAYMENT** Exposure Durability Classifications: Interior, Exposure 1 For application over structural subfloor. Provides smooth surface for application of carpet and pad and possesses high concentrated and impact load resistance. For areas to be covered with resilient flooring, specify panels with sanded face.

Typical APA Grade Stamp	Thicknesses, Inches	Grade Designation, Description, and Uses
APA THE ENGINEERED WOOD ASSOCIATION C-C PLUGGED GROUP 2 EXTERIOR 000 PS 1-95	11/32, 3/8 15/32, 1/2 19/32, 5/8 23/32, 3/4	APA C-C PLUGGED Exposure Durability Classification: Exterior For use as an underlayment over structural subfloor, refrigerated or controlled atmosphere storage rooms, open soffits, and other similar applications where continuous or severe moisture may be present. Provides smooth surface for application of carpet and pad and possesses high concentrated and impact load resistance. For areas to be covered with resilient flooring, specify panels with sanded face.
APA THE ENGINEERED WOOD ASSOCIATION C-D PLUGGED GROUP 2 EXPOSURE 1 000 PS 1-95	3/8, 15/32 1/2, 19/32 5/8, 23/32 3/4	APA C-D PLUGGED Exposure Durability Classifications: Interior, Exposure 1 For open soffits, built-ins, and other interior or protected applications. Not a substitute for APA Rated Sturd-I-Floor or Underlayment, as it lacks their puncture resistance.
A-A • G-1 • EXPOSURE 1-APA • 000 • PS1-95	1/4, 11/32 3/8, 15/32 1/2, 19/32 5/8, 23/32 3/4	APA A-A Exposure Durability Classifications: Interior, Exposure 1, Exterior For use where appearance of both sides is important for interior applications, such as built-ins, cabinets, furniture, partitions; and for exterior applications, such as fences, signs, boats, shipping containers, tanks, ducts, etc. Smooth surfaces suitable for painting.
A-B • G-1 • EXPOSURE 1-APA • 000 • PS1-95	1/4, 11/32 3/8, 15/32 1/2, 19/32 5/8, 23/32 3/4	APA A-B Exposure Durability Classifications: Interior, Exposure 1, Exterior For use where appearance of one side is less important but where two solid surfaces are necessary.
B-B • G-2 • EXPOSURE 1-APA • 000 • PS1-95	1/4, 11/32 3/8, 15/32 1/2, 19/32 5/8, 23/32 3/4	APA B-B Exposure Durability Classifications: Interior, Exposure 1, Exterior Utility panels with two solid sides.

SPECIALTY PANELS

Typical APA Grade Stamp	Thicknesses, Inches	Grade Designation, Description, and Uses
APA THE ENGINEERED WOOD ASSOCIATION DECORATIVE GROUP 2 EXPOSURE 1 000 PS 1-95	15/16, 3/8 1/2, 5/8	**APA DECORATIVE** Exposure Durability Classifications: Interior, Exposure I, Exterior Rough-sawn, brushed, grooved, or striated faces. For paneling, interior accent walls, built-ins, counter facing, and exhibit displays. Can also be made by some manufacturers in Exterior for exterior siding, gable ends, fences, and other exterior applications. Use recommendations for Exterior panels vary with the particular product. Check with the manufacturer.
APA THE ENGINEERED WOOD ASSOCIATION M. D. OVERLAY GROUP 1 EXTERIOR 000 PS 1-95	11/32, 3/8 15/32, 1/2 19/32, 5/8 23/32, 3/4	**APA MEDIUM-DENSITY OVERLAY (MDO)** Exposure Durability Classification: Exterior Smooth, opaque, resin-fiber overlay on one or both faces. Ideal base for paint, both indoors and outdoors. For exterior siding, paneling, shelving, exhibit displays, cabinets, and signs.
APA THE ENGINEERED WOOD ASSOCIATION PLYFORM B-B CLASS 1 EXTERIOR 000 PS 1-95	19/32, 5/8 23/32, 3/4	**APA B-B PLYFORM CLASS I** Exposure Durability Classification: Exterior Concrete form grades with high reuse factor. Sanded both faces and mill-oiled unless otherwise specified. Special restrictions on species. Also available in HDO for very smooth concrete finish, and with special overlays.
HDO • A-A • G-1 • EXT-APA • 000 • PS1-95	3/8, 1/2 5/8, 3/4	**APA HIGH-DENSITY OVERLAY (HDO)** Exposure Durability Classification: Exterior Has a hard, semi-opaque resin-fiber overlay on both faces. Abrasion-resistant. For concrete forms, cabinets, countertops, signs, tanks. Also available with skid-resistant screen-grid surface.
MARINE • A-A • EXT-APA • 000 • PS1-95	1/4, 3/8 1/2, 5/8 3/4	**APA MARINE** Exposure Durability Classification: Exterior Ideal for boat hulls. Made only with Douglas fir or western larch. Subject to special limitations on core gaps and face repairs. Also available with MDO or HDO faces.
PLYRON • EXPOSURE 1-APA • 000	1/2, 5/8 3/4	**APA PLYRON** Exposure Durability Classifications: Interior, Exposure I, Exterior Hardboard face on both sides. Faces tempered, untempered, smooth, or screened. For countertops, shelving, cabinet doors, and flooring.

FLOOR SHEATHING

APA Panel Subflooring

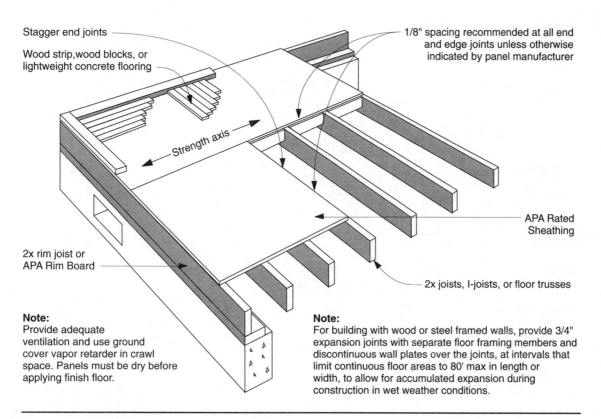

Stagger end joints

Wood strip, wood blocks, or lightweight concrete flooring

1/8" spacing recommended at all end and edge joints unless otherwise indicated by panel manufacturer

Strength axis

APA Rated Sheathing

2x rim joist or APA Rim Board

2x joists, I-joists, or floor trusses

Note:
Provide adequate ventilation and use ground cover vapor retarder in crawl space. Panels must be dry before applying finish floor.

Note:
For building with wood or steel framed walls, provide 3/4" expansion joints with separate floor framing members and discontinuous wall plates over the joints, at intervals that limit continuous floor areas to 80' max in length or width, to allow for accumulated expansion during construction in wet weather conditions.

APA PANEL SUBFLOORING

Panel Span Rating	Panel Thickness inches	Maximum Span inches	Nail Size and Type	Maximum Nail Spacing, inches	
				Supported Panel Edges	Intermediate Supports
24/16	7/16	16	6d common	6	12
32/16	15/32, 1/2	16	8d common	6	12
40/20	19/32, 5/8	20	8d common	6	12
48/24	23/32, 3/4	24	8d common	6	12
60/32	7/8	32	8d common	6	12

Source: *Design/Construction Guide: Residential and Commercial* (Tacoma, Wash: APA–The Engineered Wood Association, 1996).

UNDERLAYMENT

APA Plywood Underlayment

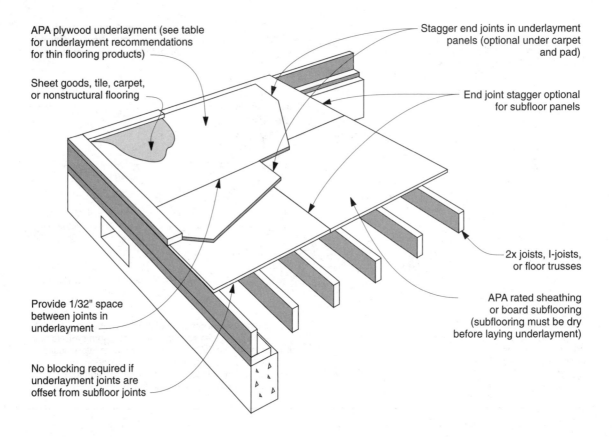

APA plywood underlayment (see table for underlayment recommendations for thin flooring products)

Sheet goods, tile, carpet, or nonstructural flooring

Stagger end joints in underlayment panels (optional under carpet and pad)

End joint stagger optional for subfloor panels

2x joists, I-joists, or floor trusses

APA rated sheathing or board subflooring (subflooring must be dry before laying underlayment)

Provide 1/32" space between joints in underlayment

No blocking required if underlayment joints are offset from subfloor joints

APA PLYWOOD UNDERLAYMENT

Plywood Grades	Application	Minimum Plywood Thickness, inches	Fastener Size and Type	Maximum Fastener Spacing, inches	
				Panel Edges	Intermediate
APA Underlayment APA C-C Plugged Ext	Over smooth subfloor	1/4	3d x 1-1/4" ring-shank nails	3	6 each way
APA Rated Sturd-I-Floor (19/32" or thicker)	Over lumber subfloor or uneven surfaces	11/32	min 12-1/2-ga (0.099") shank dia	6	8 each way

Source: *Design/Construction Guide: Residential and Commercial* (Tacoma, Wash: APA–The Engineered Wood Association, 1996)

STURD-I-FLOOR

APA Sturd-I-Floor 16, 20, and 24 oc

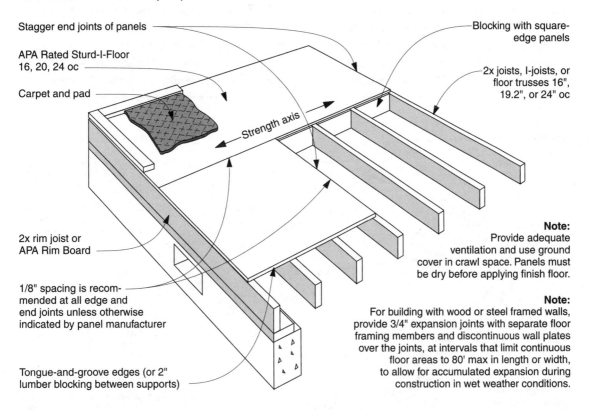

Stagger end joints of panels

APA Rated Sturd-I-Floor
16, 20, 24 oc

Carpet and pad

Strength axis

Blocking with square-
edge panels

2x joists, I-joists, or
floor trusses 16",
19.2", or 24" oc

Note:
Provide adequate
ventilation and use ground
cover in crawl space. Panels must
be dry before applying finish floor.

Note:
For building with wood or steel framed walls,
provide 3/4" expansion joints with separate floor
framing members and discontinuous wall plates
over the joints, at intervals that limit continuous
floor areas to 80' max in length or width,
to allow for accumulated expansion during
construction in wet weather conditions.

2x rim joist or
APA Rim Board

1/8" spacing is recom-
mended at all edge and
end joints unless otherwise
indicated by panel manufacturer

Tongue-and-groove edges (or 2"
lumber blocking between supports)

APA RATED STURD-I-FLOOR

Maximum Joist Spacing, inches	Panel Thickness, inches	Fastening: Glue-Nailed			Fastening: Nailed Only		
		Nail Size and Type	Maximum Spacing, inches		Nail Size and Type	Maximum Spacing, inches	
			Supported Panel Edges	Intermediate Supports		Supported Panel Edges	Intermediate Supports
16	19/32, 5/8	6d ring/screw shank	12	12	6d ring/screw shank	6	12
20	19/32, 5/8	6d ring/screw shank	12	12	6d ring/screw shank	6	12
24	23/32, 3/4	6d ring/screw shank	12	12	6d ring/screw shank	6	12
24	7/8	6d ring/screw shank	6	12	6d ring/screw shank	6	12
32	7/8	6d ring/screw shank	6	12	6d ring/screw shank	6	12
48	1-3/32, 1-1/8	6d ring/screw shank	6	6	6d ring/screw shank	6	6

Source: *Design/Construction Guide: Residential and Commercial* (Tacoma, Wash: APA–The Engineered Wood Association, 1996)

GLUED FLOOR

APA Glued Floor System

The APA glued floor system is based on thoroughly tested gluing techniques and field-applied construction adhesives that firmly and permanently secure a layer of wood structural panels to wood joists. The glue bond is so strong that floor and joists behave like integral T-beam units. Floor stiffness is increased appreciably over conventional construction, particularly when tongue-and-groove joints are glued. Gluing also helps eliminate squeaks, floor vibration, bounce, and nail-popping.

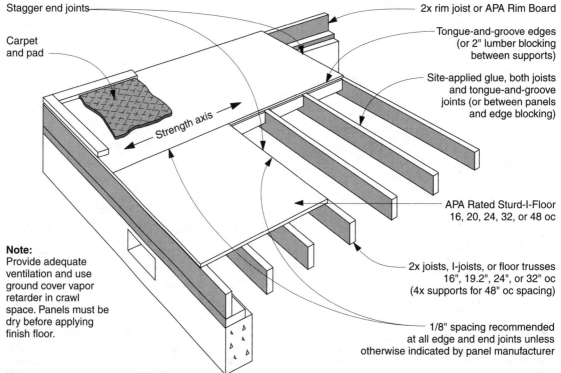

Stagger end joints

Carpet and pad

Strength axis

2x rim joist or APA Rim Board

Tongue-and-groove edges (or 2" lumber blocking between supports)

Site-applied glue, both joists and tongue-and-groove joints (or between panels and edge blocking)

APA Rated Sturd-I-Floor 16, 20, 24, 32, or 48 oc

2x joists, I-joists, or floor trusses 16", 19.2", 24", or 32" oc (4x supports for 48" oc spacing)

1/8" spacing recommended at all edge and end joints unless otherwise indicated by panel manufacturer

Note:
Provide adequate ventilation and use ground cover vapor retarder in crawl space. Panels must be dry before applying finish floor.

Note:
For building with wood or steel framed walls, provide 3/4" expansion joints with separate floor framing members and discontinuous wall plates over the joints, at intervals that limit continuous floor areas to 80' max in length or width, to allow for accumulated expansion during construction in wet weather conditions.

Source: *Design/Construction Guide: Residential and Commercial* (Tacoma, Wash: APA–The Engineered Wood Association, 1996).

WALL SHEATHING

APA Panel Wall Sheathing

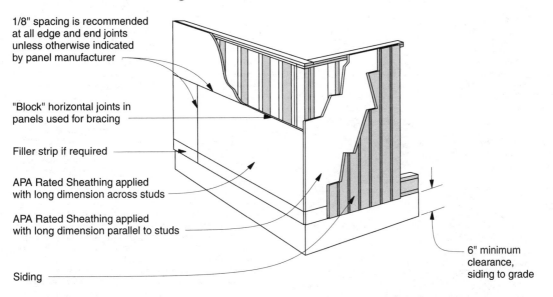

1/8" spacing is recommended at all edge and end joints unless otherwise indicated by panel manufacturer

"Block" horizontal joints in panels used for bracing

Filler strip if required

APA Rated Sheathing applied with long dimension across studs

APA Rated Sheathing applied with long dimension parallel to studs

Siding

6" minimum clearance, siding to grade

RECOMMENDED NAILING SCHEDULE FOR APA PANEL WALL SHEATHING

Span Rating	Stud Spacing, inches	Nail Size	Spacing, inches	
			Edge	Intermediate
12/0, 16/0, 20/0 or wall—16 oc	16	6d up to 1/2" 8d if thicker	6	12
24/0, 24/16, 32/16, or wall—24 oc	24	6d up to 1/2" 8d if thicker	6	12

Source: *Design/Construction Guide: Residential and Commercial* (Tacoma, Wash.: APA–The Engineered Wood Association, 1996).

RECOMMENDED STAPLING SCHEDULE FOR APA PANEL WALL SHEATHING

Panel Thickness, inches	Staple Length, inches	Spacing around Perimeter of Panel, inches	Spacing at Middle of Panel, inches
5/16	1-1/4	4	6
3/8	1-3/8	4	8
1/2	1-1/2	4	8

Source: *Design/Construction Guide: Residential and Commercial* (Tacoma, Wash: APA–The Engineered Wood Association, 1996).

ROOF SHEATHING

APA Panel Roof Sheathing

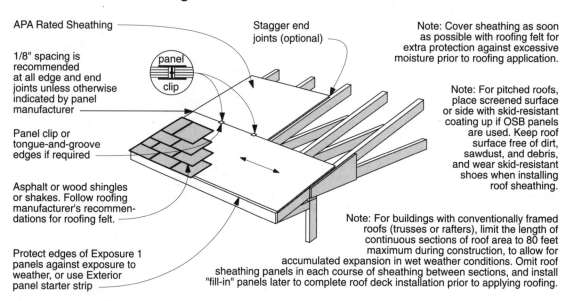

APA Rated Sheathing

1/8" spacing is recommended at all edge and end joints unless otherwise indicated by panel manufacturer

Panel clip or tongue-and-groove edges if required

Asphalt or wood shingles or shakes. Follow roofing manufacturer's recommendations for roofing felt.

Protect edges of Exposure 1 panels against exposure to weather, or use Exterior panel starter strip

Stagger end joints (optional)

Note: Cover sheathing as soon as possible with roofing felt for extra protection against excessive moisture prior to roofing application.

Note: For pitched roofs, place screened surface or side with skid-resistant coating up if OSB panels are used. Keep roof surface free of dirt, sawdust, and debris, and wear skid-resistant shoes when installing roof sheathing.

Note: For buildings with conventionally framed roofs (trusses or rafters), limit the length of continuous sections of roof area to 80 feet maximum during construction, to allow for accumulated expansion in wet weather conditions. Omit roof sheathing panels in each course of sheathing between sections, and install "fill-in" panels later to complete roof deck installation prior to applying roofing.

RECOMMENDED UNIFORM ROOF LIVE LOADS FOR APA RATED SHEATHING[a] WITH LONG DIMENSION PERPENDICULAR TO SUPPORTS

Panel Span Rating	Minimum Panel Thickness, inches	Maximum Span, inches		Allowable Live Loads, psf[c] Spacing of Supports Center-to-Center, inches							
		With Edge Support[b]	Without Edge Support	12	16	20	24	32	40	48	60
12/0	5/16	12	12	30							
16/0	5/16	16	16	70	30						
20/0	5/16	20	20	120	50	30					
24/0	3/8	24	20[d]	190	100	60	30				
24/16	7/16	24	24	190	100	65	40				
32/16	15/32, 1/2	32	28	325	180	120	70	30			
40/20	19/32, 5/8	40	32	—	305	205	130	60	30		
48/24	23/32, 3/4	48	36	—	—	280	175	95	45	35	
60/32	7/8	60	48	—	—	—	305	165	100	70	35

[a] Includes APA rated sheathing/ceiling deck.
[b] Tongue-and-groove edges, panel edge clips (one midway between each support, except two equally spaced between supports 48 inches oc), lumber blocking, or other.
[c] 10 psf dead load assumed.
[d] 24 inches for 15/32-inch and 1/2-inch panels.

Source: *Design/Construction Guide: Residential and Commercial* (Tacoma, Wash: APA–The Engineered Wood Association, 1996).

8   Siding

The first section identifies the *function of siding* (protection of the walls from moisture) and illustrates the three principles that all sidings must follow.

Next, the advantages and disadvantages of all of the common *siding options* are compared.

The rest of this chapter is filled with illustrations and tables designed to help you successfully install *vinyl siding, hardboard lap siding, cedar shingles*, the various *horizontal wood sidings, vertical wood sidings*, combined *plywood siding/*sheathing, and *stucco*.

Finally, we provide you with a *checklist of code requirements* relating to siding.

THE FUNCTION OF SIDING

Aside from decoration and a possible dual role as structural sheathing, the function of siding is to keep the structure and interior of a building dry. If the siding is painted, it must be allowed to dry from both sides.

Water penetrates siding in one or more of three ways:

• as bulk water flowing downward under the force of *gravity*
• as rain water driven horizontally by the pressure of *wind*
• as rain water drawn upward by *capillarity* (surface tension acting in small spaces)

The three problems are prevented in three very different ways:

• gravity - by flashing at horizontal junctures of building surfaces and materials
• wind - by venting the back side of siding to equalize air pressures
• capillarity - by eliminating capillary-sized gaps between siding courses, using round head nails or wedges

The illustrations below show examples of each technique. Details specific to each type of siding are presented in the following pages.

Siding Moisture Control

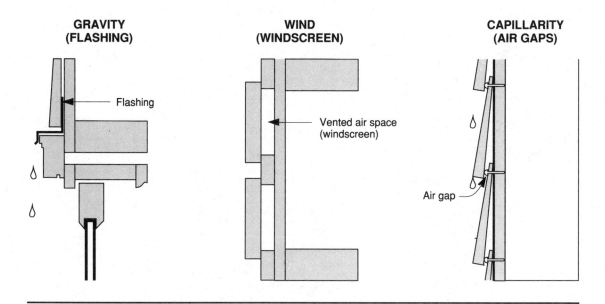

GRAVITY
(FLASHING)

Flashing

WIND
(WINDSCREEN)

Vented air space
(windscreen)

CAPILLARITY
(AIR GAPS)

Air gap

SIDING OPTIONS

In the chart below, *care* includes the cost of maintaining appearance as well as integrity. Vinyl and aluminum are not subject to maintenance, but bright colors may fade over time.

Life assumes proper maintenance and may vary widely under differing conditions.

Cost is for materials only and does not include the cost of professional installation. In general, labor costs are lower for vinyl, aluminum, and plywood. Labor also varies with region: Stucco, for example, would cost least in the Southwest (SW), where it predominates.

SIDINGS COMPARED

Material	Care	Life, yr	Cost	Advantages	Disadvantages
Aluminum	None	30	Medium	Ease of installation over existing sidings Fire resistance	Susceptibility to denting, rattling in wind
Hardboard	Paint Stain	30	Low	Low cost Fast installation	Susceptibility to moisture in some
Horizontal wood	Paint Stain None	50+	Medium to high	Good looks if of high quality	Slow installation Moisture/paint problems
Plywood	Paint Stain	20	Low	Low cost Fast installation	Short life Susceptibility to moisture in some
Shingles	Stain None	50+	High	Good looks Long life Low maintenance	Slow installation
Stucco	None	50+	Low to medium	Long life Good looks in SW Low maintenance	Susceptibility to moisture
Vertical wood	Paint Stain None	50+	Medium	Fast installation	Barn look if not of highest quality Moisture/paint problems
Vinyl	None	30	Low	Low cost Ease of installation over existing sidings	Fading of bright colors No fire resistance

VINYL SIDING

Below are the most common vinyl and aluminum sidings and installation accessories. Some variations exist between manufacturers. The basic rules for installation are universal, however:

- Nail in the center of slots.
- Do not nail too tightly.
- Leave at least 1/4-inch clearance at all stops.
- Do not pull horizontal sidings up tight.
- Strap and shim all uneven walls.

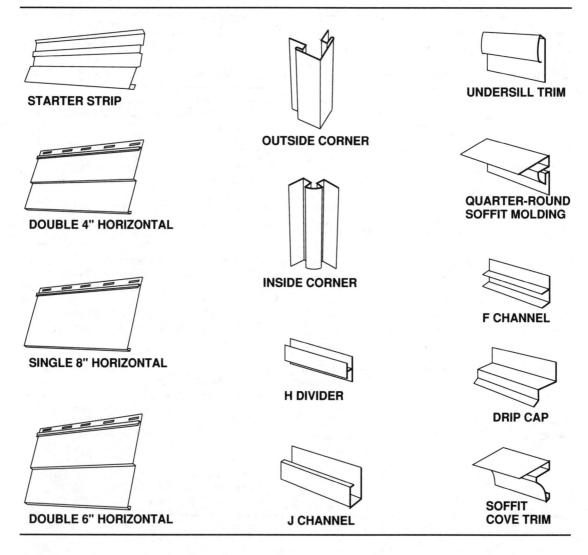

STARTER STRIP

DOUBLE 4" HORIZONTAL

SINGLE 8" HORIZONTAL

DOUBLE 6" HORIZONTAL

OUTSIDE CORNER

INSIDE CORNER

H DIVIDER

J CHANNEL

UNDERSILL TRIM

QUARTER-ROUND SOFFIT MOLDING

F CHANNEL

DRIP CAP

SOFFIT COVE TRIM

Installation of Vinyl Siding

STEP 1. STARTER STRIP

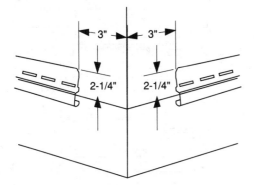

STEP 2. INSIDE CORNER

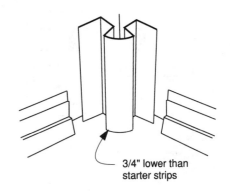

3/4" lower than
starter strips

STEP 3. OUTSIDE CORNER

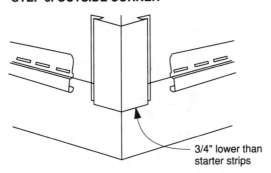

3/4" lower than
starter strips

STEP 4. FIRST PANEL

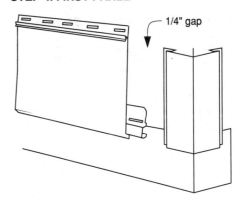

1/4" gap

STEP 5. ADDITIONAL PANELS

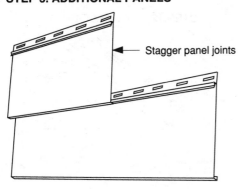

Stagger panel joints

STEP 6. OVERLAPPING JOINTS

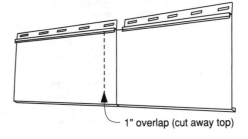

1" overlap (cut away top)

STEP 7. WINDOW TOPS AND SIDES

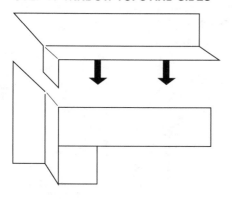

STEP 8. FIT PANELS TO WINDOWS

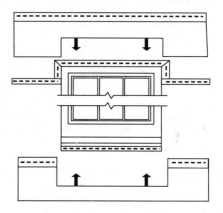

STEP 9. UNDEREAVES TOP COURSE

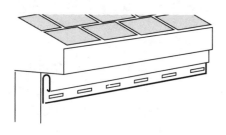

STEP 10. CUT AND PUNCH TOP COURSE

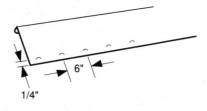

STEP 11. SNAP TOP COURSE IN PLACE

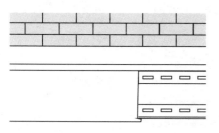

STEP 12. FIT GABLE ENDS

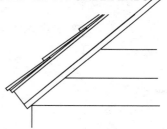

HARDBOARD LAP SIDING

Hardboard lap siding may be cut with either fine-tooth hand or power saws. The cutting action should be into the appearance face of the siding, i.e., face up with hand saws and table saws and face down with portable circular saws.

Lap siding may be applied directly to studs or over sheathing. Studs should be spaced 16 inches on-center in both cases. An air-barrier membrane should be used when the siding is applied directly to studs or over board sheathings.

Use only corrosion-resistant nails. Nails should penetrate framing members by 1-1/2 inches minimum (8d nails for studs only; 10d nails over sheathing).

Allow at least 6 inches between the siding and the ground or any area where water may collect. A starter strip 1-1/2 inches wide and the same thickness as the siding should be installed level with the bottom edge of the sill plate. Nail the starter strip with the recommended siding nails. Cut, fit, and install the first course of siding to extend at least 1/4 inch, but no more than 1 inch, below the starter strip. Nail the bottom edge of the first course of siding 16 inches on-center and through the starter strip at each stud location. Maintain contact at the joints without forcing. Leave a space of 1/8 inch between the siding and window or door frames and corner boards. Caulk this space after the siding is installed.

All joints must fall over studs and be nailed on both top and bottom on each side of the joint. Stagger succeeding joints for best appearance.

The second and all succeeding courses of siding must overlap the previous course a minimum of 1 inch. Locate nails 1/2 inch from the bottom edge and not more than 16 inches on-center. Nail through both courses and into the framing members.

Install shim strips for continuous horizontal support behind the siding wherever it is notched out above or below openings.

Use wooden corner boards at least 1-1/8 inches thick or formed metal corners (available from distributors) at all inside and outside corners.

Factory-primed siding should be painted within 60 days after installation. If it is exposed for a longer period, lightly sand the primer, or reprime the siding with a good-quality exterior primer that is compatible with the final finish coat.

Unprimed siding should be finished within 30 days after installation. If the finish will be paint, prime the siding with a good-quality compatible exterior primer.

More detailed application instructions are available form hardboard siding manufacturers.

Installation of Hardboard Lap Siding

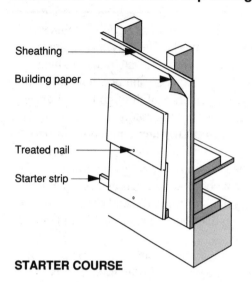

Sheathing

Building paper

Treated nail

Starter strip

STARTER COURSE

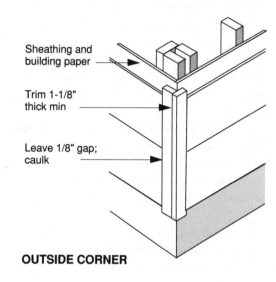

Sheathing and
building paper

Trim 1-1/8"
thick min

Leave 1/8" gap;
caulk

OUTSIDE CORNER

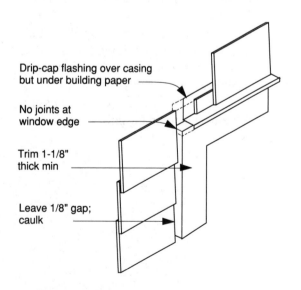

Drip-cap flashing over casing
but under building paper

No joints at
window edge

Trim 1-1/8"
thick min

Leave 1/8" gap;
caulk

DOORS AND WINDOWS

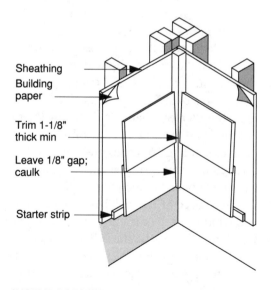

Sheathing

Building
paper

Trim 1-1/8"
thick min

Leave 1/8" gap;
caulk

Starter strip

INSIDE CORNER

CEDAR SHINGLES

SPECIFICATIONS

Grade	Length, inches	Butt, inches	Bundles/ Square	Maximum Exposure and Nails			
				Single Course		Double Course	
Red Cedar							
No. 1 blue label	16	0.40	4	7-1/2	3d	12	5d
(premium grade, 100% heartwood,	18	0.45	4	8-1/2	3d	14	5d
100% clear, 100% edge grain)	24	0.50	4	11-1/2	4d	16	6d
No. 2 red label	16	0.40	4	7-1/2	3d	12	5d
(good grade, 10" clear on 16" shingle,	18	0.45	4	8-1/2	3d	14	5d
16" clear on 24" shingle)	24	0.50	4	11-1/2	4d	16	6d
No. 3 black label	16	0.40	4	7-1/2	3d	12	5d
(utility grade, 6" clear on 16" shingle,	18	0.45	4	8-1/2	3d	14	5d
10" clear on 24" shingle)	24	0.50	4	11-1/2	4d	16	6d
No. 4 undercoursing	16	0.40	2 or 4	7-1/2	5d	–	–
(for bottom course in	18	0.45	2 or 4	8-1/2	5d	–	–
double-coursed walls)							
No. 1 or 2 rebutted-rejoined	16	0.40	1	7-1/2	3d	12	5d
(machine trimmed, square	18	0.45	1	8-1/2	3d	14	5d
edged, top grade)							
White Cedar							
Extra (perfectly clear)	16	0.40	4	7-1/2	3d	12	5d
1st clear (7" clear, no sapwood)	16	0.40	4	7-1/2	3d	12	5d
2nd clear (sound knots, no sapwood)	16	0.40	4	7-1/2	3d	12	5d
Clear wall (sapwood, curls)	16	0.40	4	7-1/2	3d	12	5d
Utility (undercoursing only)	16	0.40	4		3d		5d

Note: Exposure is given in inches.

COVERAGE

Length, inches	Coverage of One Square at Exposure, inches								
	4	5	6	7	8	9	10	11	12
16	80	100	120	140	160				
18	72	90	109	127	145	163			
24			80	93	106	120	133	146	160

Source: *Exterior and Interior Product Glossary* (Bellevue, Wash: Red Cedar Shingle & Handsplit Shake Bureau, 1980).

Installation of Cedar Shingles

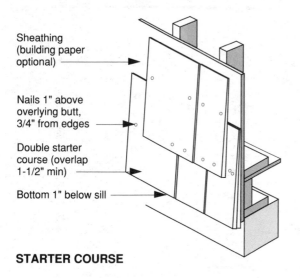

Sheathing (building paper optional)

Nails 1" above overlying butt, 3/4" from edges

Double starter course (overlap 1-1/2" min)

Bottom 1" below sill

STARTER COURSE

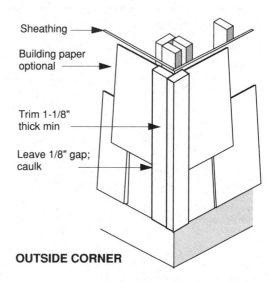

Sheathing

Building paper optional

Trim 1-1/8" thick min

Leave 1/8" gap; caulk

OUTSIDE CORNER

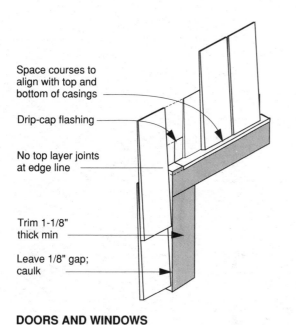

Space courses to align with top and bottom of casings

Drip-cap flashing

No top layer joints at edge line

Trim 1-1/8" thick min

Leave 1/8" gap; caulk

DOORS AND WINDOWS

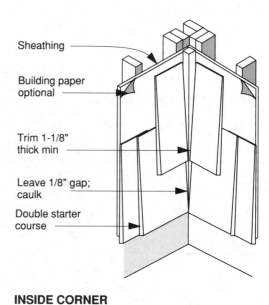

Sheathing

Building paper optional

Trim 1-1/8" thick min

Leave 1/8" gap; caulk

Double starter course

INSIDE CORNER

HORIZONTAL WOOD SIDING

Materials and Patterns

Most horizontal siding is of either redwood or western red cedar. The heartwoods of both have superior decay resistance and dimensional stability. Local species are also used, however, in regions where the price difference is great enough. Eastern white pine is popular in the Northeast. Not as decay resistant, it is easily worked and holds paint nearly as well as redwood and cedar. Consult your lumberyard for local variations in both species and pattern.

Moisture

Wood changes in width and thickness with changes in moisture content. To minimize dimensional change after installation, siding should be installed after its moisture content has come to equilibrium with the air.

Minimize moisture problems:

- Specify kiln-dried vertical-grain siding.
- Use a narrow siding.
- Precondition the siding for moisture.
- Select patterns that allow for shrinkage and expansion.
- Permit the siding to stabilize before applying any finish.

After siding has been installed, it can still pick up moisture before it is painted or stained. Later, when the siding dries, joints may open up, or buckling may occur. It is good practice to prime or prefinish all sides, edges, and ends after it has reached equilibrium with the air and before it is installed.

Nailing

Nails for applying wood siding should be rust resistant: stainless steel, hot-dipped galvanized, or high-tensile-strength aluminum. Don't use electroplated galvanized or unfinished (bright) nails.

Nails should be strong enough for nailing without predrilling. They should not make an unsightly pattern, cause splitting when driven near the end or edge of the siding, or pop after being driven flush.

Recommended penetration into a solid wood base (either studs or wood sheathing) is 1-1/2 inch minimum, or 1-1/4 inch with ring shank nails. Longer nails are required for installation over other than solid wood sheathing and may require predrilling to avoid splits.

Horizontal wood siding should be applied to studs 24 inches on-center maximum when applied over solid sheathing and 16 inches on-center when applied without sheathing.

Estimating Coverage

The area factors in the table on the following page make it easy to determine the approximate board footage of siding needed for the various patterns and sizes shown. Simply multiply the length and width of the area to be covered, times the appropriate area factor. Add a 10 percent allowance for trim and waste to the resulting figure.

HORIZONTAL WOOD SIDING

Pattern	Nominal Size, inches	Dressed Size, inches		Area Factor
		Total Width	Face Width	
PLAIN BEVEL (CLAPBOARD)	1/2 x 4	3-1/2	3-1/2	1.60
	1/2 x 6	5-1/2	5-1/2	1.33
	3/4 x 8	7-1/4	7-1/4	1.28
	3/4 x 10	9-1/4	9-1/4	1.21
RABBETED BEVEL (DOLLY VARDEN)	3/4 x 6	5-1/2	5	1.20
	1 x 8	7-1/4	6-3/4	1.19
	1 x 10	9-1/4	8-3/4	1.18
	1 x 12	11-1/4	10-3/4	1.12
TONGUE & GROOVE	1 x 4	3-3/8	3-1/8	1.28
	1 x 6	5-3/8	5-1/8	1.17
	1 x 8	7-1/8	6-7/8	1.16
	1 x 10	9-1/8	8-7/8	1.13
DROP (T&G OR SHIPLAP)	1 x 6	5-3/8	5-1/8	1.17
	1 x 8	7-1/8	6-3/4	1.16
	1 x 10	9-1/8	8-3/4	1.13
	1 x 12	11-1/8	10-3/4	1.10
SHIPLAP	1 x 6	5-3/8	5	1.17
	1 x 8	7-1/8	6-3/4	1.16
	1 x 10	9-1/8	8-3/4	1.13
	1 x 12	11-1/8	10-3/4	1.10
CHANNEL SHIPLAP	1 x 6	5-3/8	5	1.17
	1 x 8	7-1/8	6-3/4	1.16
	1 x 10	9-1/8	8-3/4	1.13
	1 x 12	11-1/8	10-3/4	1.10
V - SHIPLAP	1 x 6	5-3/8	5	1.17
	1 x 8	7-1/8	6-3/4	1.16
	1 x 10	9-1/8	8-3/4	1.13
	1 x 12	11-1/8	10-3/4	1.10
LOG CABIN	1 x 6	5-7/16	4-15/16	1.22
	1 x 8	7-1/8	6-5/8	1.21
	1 x 10	9-1/8	8-5/8	1.16

Note: These sizes are typical, but sizes vary with manufacturer.

Installation of Horizontal Wood Siding

PLAIN BEVEL

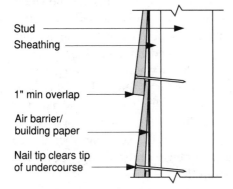

Stud
Sheathing

1" min overlap

Air barrier/
building paper

Nail tip clears tip
of undercourse

RABBETED BEVEL

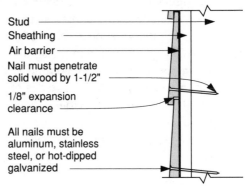

Stud
Sheathing
Air barrier
Nail must penetrate
solid wood by 1-1/2"

1/8" expansion
clearance

All nails must be
aluminum, stainless
steel, or hot-dipped
galvanized

TONGUE & GROOVE

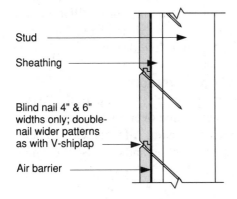

Stud

Sheathing

Blind nail 4" & 6"
widths only; double-
nail wider patterns
as with V-shiplap

Air barrier

V - SHIPLAP

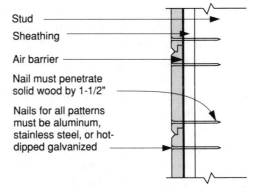

Stud

Sheathing

Air barrier

Nail must penetrate
solid wood by 1-1/2"

Nails for all patterns
must be aluminum,
stainless steel, or hot-
dipped galvanized

OUTSIDE CORNER

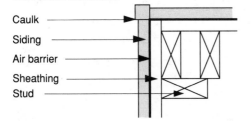

Caulk
Siding
Air barrier
Sheathing
Stud

INSIDE CORNER

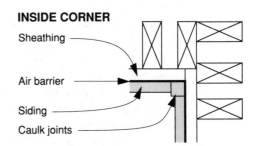

Sheathing

Air barrier

Siding

Caulk joints

Source: *Redwood Siding Patterns and Application* (Mill Valley, Calif: California Redwood Association).

VERTICAL WOOD SIDING

Materials and Patterns

Vertical siding, due to its more rustic character, is less often of redwood or western red cedar and more often of a local, rough-sawn species. With the exception of the shiplap and channel shiplap patterns, all recommended patterns are of simple square-edge design. For utility buildings, unfinished (rough-sawn) boards are most often used. For a less rustic appearance use S1S (surfaced one side) boards.

Moisture

Wood changes in width and thickness with changes in moisture content. To minimize dimensional change after installation, siding should be installed after its moisture content has come to equilibrium with the air.

The recommended procedure is this:

• Use as narrow a siding as practical (a rule of thumb is width < 8 x thickness).
• Precondition the siding for moisture.
• Select patterns that allow for movement.
• Treat both sides of siding with water repellent before installation.

Vertical wood siding is rarely painted. Unpainted wood naturally changes color with exposure to sunlight and water. One method for preserving the appearance of either stained or unfinished siding is to immediately bleach the wood to approximate its ultimate aged appearance. Most paint and stain manufacturers offer some sort of bleach for this purpose.

Nailing

Recommended nailing patterns are shown on the following page. Nails for applying vertical wood siding should be rust resistant: stainless steel, hot-dipped galvanized, or high-tensile-strength aluminum. Do not use electroplated galvanized or unfinished (bright) nails.

Nails should be strong enough for nailing without predrilling. They should not make an unsightly pattern, cause splitting when driven near the end or edge of the siding, or pop after being driven flush.

Recommended penetration into a solid wood base (either studs or wood sheathing) is 1-1/4 inches minimum with ring shank nails. Longer nails are required for installation over other than solid wood sheathing and may require predrilling to avoid splits.

Estimating Coverage

Coverage depends on the width of the boards, battens (if any), and overlap. Overlaps should be 1/2 inch minimum. Allow at least 10 percent waste for cutting and defects; allow more for poorer grades.

Installation of Vertical Wood Siding

BOARD & BATTEN

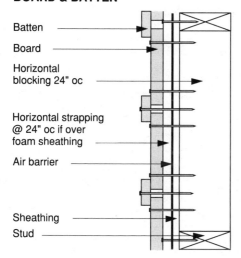

Batten

Board

Horizontal
blocking 24" oc

Horizontal strapping
@ 24" oc if over
foam sheathing

Air barrier

Sheathing

Stud

REVERSE BOARD & BATTEN

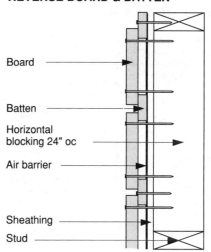

Board

Batten

Horizontal
blocking 24" oc

Air barrier

Sheathing

Stud

BOARD & BOARD

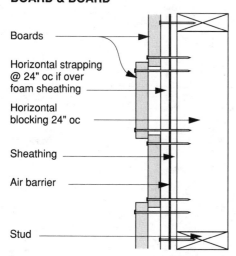

Boards

Horizontal strapping
@ 24" oc if over
foam sheathing

Horizontal
blocking 24" oc

Sheathing

Air barrier

Stud

CHANNEL SHIPLAP

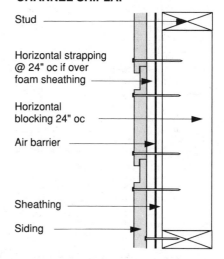

Stud

Horizontal strapping
@ 24" oc if over
foam sheathing

Horizontal
blocking 24" oc

Air barrier

Sheathing

Siding

Source: *Redwood Siding Patterns and Application* (Mill Valley, Calif: California Redwood Association).

Vertical Wood Siding Joints

OUTSIDE CORNER

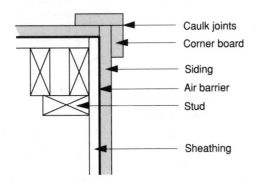

- Caulk joints
- Corner board
- Siding
- Air barrier
- Stud
- Sheathing

INSIDE CORNER

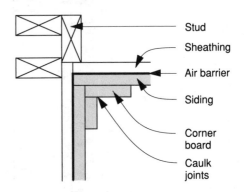

- Stud
- Sheathing
- Air barrier
- Siding
- Corner board
- Caulk joints

BELTLINE JOINT

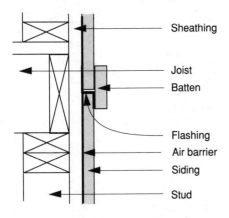

- Sheathing
- Joist
- Batten
- Flashing
- Air barrier
- Siding
- Stud

BEVELED BUTT JOINT

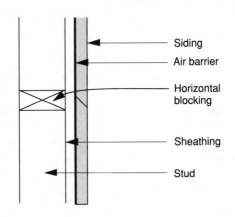

- Siding
- Air barrier
- Horizontal blocking
- Sheathing
- Stud

Source: *Redwood Siding Patterns and Application* (Mill Valley, Calif: California Redwood Association).

PLYWOOD SIDING

APA Sturd-I-Wall

The APA Sturd-I-Wall system consists of APA rated siding (panel or lap) applied directly to studs or over nonstructural fiberboard, gypsum, or rigid foam insulation sheathing. Nonstructural sheathing is sheathing not recognized by building codes as meeting both bending and racking strength requirements.

Since the single layer of plywood panel siding is strong and rack-resistant, it eliminates the cost of installing separate structural sheathing or diagonal wall bracing. Panel sidings are normally installed vertically but may also be placed horizontally (face grain across supports) if horizontal joints are blocked. Maximum stud spacings are 16 inches on-center for vertical panels and 24 inches on-center for horizontal panels, unless otherwise indicated in the grade stamp.

Over Foam Sheathing

When installing panel siding over rigid foam insulation sheathing, drive the nails flush with the siding surface, but avoid over-driving, which can result in dimpling of the siding due to the compressible nature of the foam sheathing.

Plywood sidings are occasionally treated with water repellents or wood preservatives to improve finishing characteristics or durability. If the siding has been treated, be sure the surface treatment is dry, to avoid chemical reaction with the foam sheathing.

Because of the high resistance of foam sheathing to vapor transmission, an effective vapor barrier must be installed on the warm side of the wall to avoid condensation problems in the wall cavity. When rigid foam insulation sheathing is used, building codes also generally require l/2-inch gypsum drywall, or other materials of equivalent fire rating, on the inside surface of the wall for fire protection.

Nailing

All panel siding edges should be spaced 1/8 inch minimum and backed with solid framing or blocking. Use nonstaining, noncorrosive nails to prevent staining of the siding. Nail panel edges 6 inches on-center with 6d box or siding nails for panels up to 1/2 inch thick; use 8d nails for thicker panels.

In addition to 1/8-inch minimum edge spacing and the use of straight studs, nailing sequence can also be a factor in achieving a flat wall surface. The recommended nailing procedure is to first position the siding panel, maintaining the recommended edge spacing, and lightly tack at each corner. Next, install the first row of nails along the edge adjacent to the preceding panel, from top to bottom. Remove the remaining tacking nails, then nail the row at the first intermediate stud. Continue with the other studs, finishing at the edge opposite the preceding panel. Complete the installation by nailing along the top and bottom plates.

APA 303 Siding Patterns

APA 303 sidings include a wide variety of surface textures and patterns, most developed for best performance with stains. The most common patterns are shown below. Actual dimensions of groove spacing, width, and depth may vary with the manufacturer.

Plywood sidings also come in a wide variety of face veneer grades. Consult your local siding distributor or retailer for the most appropriate grade, considering the pattern, species, and finish desired.

The use of plywood siding is necessarily a trade-off. Against the convenience and savings of a combined sheathing and siding, one must weigh the more pleasing bold relief and longer life of board sidings.

ROUGH SAWN

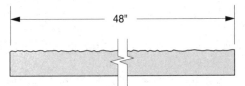

Thicknesses: 11/32", 3/8", 15/32", 1/2", 19/32", 5/8"

KERFED ROUGH SAWN

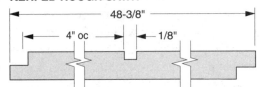

Thicknesses: 11/32", 3/8", 15/32", 1/2", 19/32", 5/8"

TEXTURE 1-11 (T1-11)

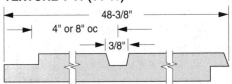

Thicknesses: 19/32", 5/8"

REVERSE BOARD AND BATTEN

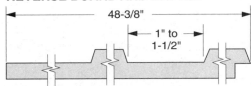

Thicknesses: 19/32", 5/8"

CHANNEL GROOVE

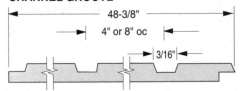

Thicknesses: 11/32", 3/8", 15/32", 1/2"

BRUSHED

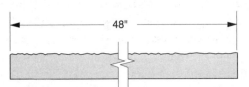

Thicknesses: 11/32", 3/8", 15/32", 1/2", 19/32", 5/8"

Plywood Siding Joint Details

(VIEWED FROM TOP)

**VERTICAL
BUTT & CAULK**

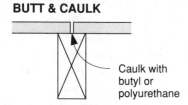

Caulk with
butyl or
polyurethane

**T1-11 AND
CHANNEL GROOVE**

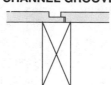

**VERTICAL
BATTEN**

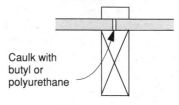

Caulk with
butyl or
polyurethane

**INSIDE
CORNER JOINT**

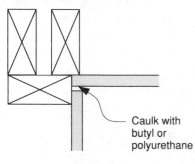

Caulk with
butyl or
polyurethane

**OUTSIDE
CORNER JOINT**

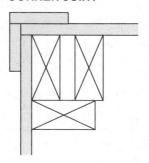

(VIEWED FROM SIDE)

**HORIZONTAL
SHIPLAP**

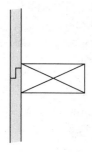

**HORIZONTAL
BUTT & FLASH**

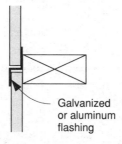

Galvanized
or aluminum
flashing

**HORIZONTAL
BANDBOARD**

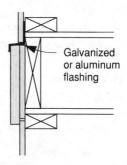

Galvanized
or aluminum
flashing

STUCCO

Stucco is a mixture of sand, cement, lime, and water. The most common formula is four parts sand to one part portland cement, with a smaller amount of lime. The amount of water is adjusted for workability. A good starting point for the mix is shown in the table on the following page.

Bases
Stucco can be applied over any suitably rigid base. Cast-in-place concrete and concrete masonry block walls are ideal. Wood frame walls can be used, provided they are rigidly braced and covered with metal reinforcement.

Metal reinforcement comes in several styles: welded wire, woven wire, and expanded metal lath. The latter consists of sheet metal, slit and deformed to provide an open grid that is usually self-furring (held out from the base wall by a constant distance). If the metal reinforcement is not self-furring, it should be attached and held 1/4 inch from the base with special furring nails. The metal reinforcement should be galvanized or otherwise treated to be noncorrosive.

The metal reinforcement must be firmly attached and rigid. Joints should overlap a minimum of 1 inch and be made only over a solid backing. For open framing without sheathing, this means at the studs.

Mixes
Achieving the proper mix is the aspect requiring the most experience. (Stucco is not a good candidate for do-it-yourself application.) As in most masonry work, the key is workability.

The mix must flow well enough to form a smooth and level coat, but not well enough to sag after application. Also, the amount of sand in the mix influences both strength and susceptibility to later cracking. For convenience, the same mix is used for both the scratch (first) coat and the brown (second) coat, with more sand being added to the brown coat.

A factory mix is usually used for the finish coat. The manufacturer's recommendations should be strictly followed.

Control Joints
Control joints allow movement without cracking of the stucco due to thermal expansion and contraction, wetting and drying, and slight movements of the underlying structure. Over concrete masonry, control joints in the stucco are only required over the control joints in the masonry. Over wood walls, control joints should be spaced no more than 18 feet apart, but in no case so as to create unjointed panels of over 150 square feet.

Application
The scratch coat should completely fill the metal reinforcement and be scored or scratched horizontally for good bonding. It should be kept moist for a minimum of 12 hours and allowed to set 48 hours before the next coat. The brown coat (if there is one) should be kept moist for 12 hours and allowed to set for 7 days. The finish coat requires wetting for 12 hours.

Painting stucco is not recommended, since complete paint removal would be required before repair or recoating of the stucco.

OPTIMUM CURING TIMES

Coat	Keep Moist, hours	Total Set, days
Scratch	12	2
Brown	12	7
Finish	12	2

TYPICAL STUCCO MIX

Component	Cubic Feet	Gallons	Pounds
Sharp sand	2	15	200
Portland cement	1/2	3-3/4	47
Lime	1/3	2-1/2	12
Water	3/4	6	48

Stucco Accessories

**SOFFIT DRIP
SCREEN**

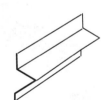

**FOUNDATION OR
WEEP SCREED**

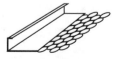

**SQUARE CASING
BEAD**

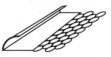

**BULLNOSE CASING
BEAD**

INSIDE CORNER AND SOFFIT

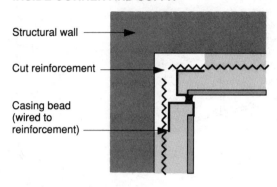

Structural wall

Cut reinforcement

Casing bead
(wired to
reinforcement)

Control Joints

VERTICAL (EVERY 18' MAX)

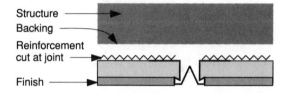

Structure

Backing

Reinforcement
cut at joint

Finish

HORIZONTAL

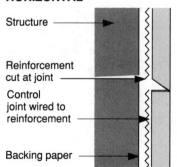

Structure

Reinforcement
cut at joint

Control
joint wired to
reinforcement

Backing paper

Application of Stucco

OVER CONCRETE MASONRY

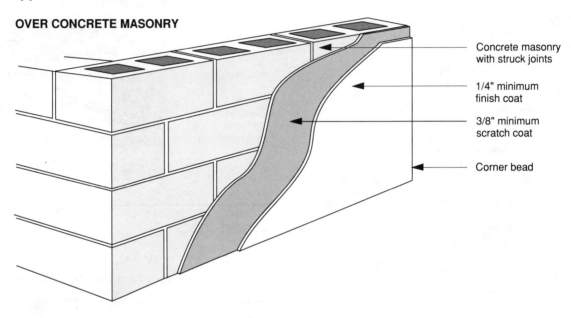

Concrete masonry
with struck joints

1/4" minimum
finish coat

3/8" minimum
scratch coat

Corner bead

OVER WOOD-SHEATHED WALL

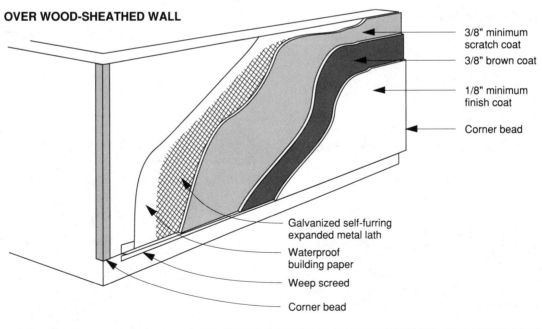

3/8" minimum
scratch coat

3/8" brown coat

1/8" minimum
finish coat

Corner bead

Galvanized self-furring
expanded metal lath

Waterproof
building paper

Weep screed

Corner bead

CHECKLIST OF CODE REQUIREMENTS

The following is a partial list of requirements from the 1995 Council of American Building Officials (CABO) *One and Two Family Dwelling Code*. Consult the publication for the full text and additional provisions.

(PARTIAL) WEATHER-RESISTANT SIDING ATTACHMENT AND MINIMUM THICKNESS

Siding Material	Nominal Thickness	Sheathing Paper Required	Type of Support and Fasteners [1,2,3]				
			Wood, Wood Structural Panel	Fiberboard Sheathing into Stud	Gypsum Sheathing into Stud	Direct to Stud	Number or Spacing of Fasteners
Horizontal aluminum	0.019"–0.024"	No	.120 nail $1^1/_2$" long	.120 nail 2" long	.120 nail 2" long	Not allowed	Stud spacing
Hardboard vertical board/batten	$7/_{16}$"	(4)	.099 nail 2" long	.099 nail $2^1/_2$" long	.099 nail 2" long	.099 nail $1^3/_4$" long	6" edges 8" interior
Hardboard horizontal lap siding	$7/_{16}$"	(4)	.099 nail 2" long	.099 nail $2^1/_2$" long	.099 nail $2^1/_4$" long	.099 nail 2" long	2 per stud
Steel	29 gauge	No	.113 nail–$1^3/_4$" $1^3/_4$" staple	.113 nail–$2^3/_4$" $2^1/_2$" staple	.113 nail–$2^1/_2$" $2^1/_4$" staple	Not allowed	Stud spacing
Particleboard panel	$3/_8$"–$1/_2$"	(4)	6d box nail	6d box nail	6d box nail	6d box nail (not $3/_8$")	6" edges 12" interior
	$5/_8$"	(4)	6d box nail	8d box nail	8d box nail	6d box nail	
Plywood panel (EXT grade)[5]	$3/_8$"	(4)	.099 nail–2" $1^3/_8$" staple	.113 nail–$2^1/_2$" $2^1/_4$" staple	.099 nail–2" 2" staple	.099 nail–2" $1^3/_8$" staple	6" edges 12" interior
Vinyl siding	0.035"	No	.120 nail–$1^1/_2$" $1^3/_4$" staple	.120 nail–2" $2^1/_2$" staple	.120 nail–2" 2-$1/_2$" staple	Not allowed	Stud spacing
Wood—rustic, drop, or shiplap[6]	$3/_8$" minimum $19/_{32}$" average	No	1" into stud	1" into stud	1" into stud	.113 nail–$2^1/_2$" 2" staple	1/stud ≤ 6" 2/stud ≥ 8"

[1] Nails may be T-head, modified round head, or round head with smooth or deformed shank.

[2] Staples shall be 16-gauge wire minimum and have crown width of 7/16" OD.

[3] Nails or staples shall be aluminum, galvanized, or rust-preventive coated and driven into studs for fiberboard or gypsum backing.

[4] Joints over sheathing or weather-resistant membrane are allowable. Otherwise, joints must occur at studs and be lapped or covered with battens.

[5] Three-eighths-inch plywood shall be nailed directly to studs 16" on-center. One-half-inch plywood shall be nailed directly to studs 24" on-center.

[6] Vertical woodboard sidings shall be nailed to horizontal nailing strips or blocking 24" on-center. Nails shall penetrate 1-1/2" into studs, strips, or blocking.

9　Roofing

Terms such as *eaves, soffit, fascia*, and *ridge* are referred to throughout this chapter, so before you look up the installation details of your favorite roof, read the first section, *"Roof Terminology."*

If you are trying to decide what kind of roofing to install, read the second section, *"Roofing Materials."*

The following sections, the real meat of the chapter, describe in words and illustrations how to install 11 different types of roofing. They range from the rather industrial *built-up roof,* through three varieties of *roll roofing,* ubiquitous *asphalt shingles,* classic *cedar shingles* and *cedar shakes,* and regional materials such as *slate* and *Spanish tile,* to *preformed metal panel* and *standing-seam* roofing.

Although *ventilation* is not roofing, proper ventilation of the space beneath the roofing is imperative for the proper operation and maximum lifetime of the roof. Ventilation removes moisture in winter and heat in summer. It is also absolutely the best way to prevent the destructive buildup of ice dams.

The best time to install, replace, or repair *gutters* is when you are roofing, so a description of the typical gutter system and all of its parts is included, too.

Finally, we provide you with a *checklist of code requirements* relating to roofing.

ROOF TERMINOLOGY

Function of the Roof
The primary function of a roof is to shield the building beneath from moisture damage, whether from rain, snow, or ice. The primary design characteristics controlling the success of a roof are pitch (angle) and coverage (overlap) of the roofing material.

Pitch and Slope
The pitch of a roof is the vertical rise divided by the total span. The slope of a roof is vertical rise divided by horizontal run.

Example: A roof peak is 8 feet above the top plate. The total span (building width) is 24 feet. The pitch is 8/24, or 1/3; the slope is 8/12, usually expressed as inches of rise per 12 inches of run, or 8/12. Exposure is the down-slope width of roofing material exposed after installation. Coverage is the number of layers of roofing that cover from surface to underlayment.

Example: A roof is covered with asphalt shingles measuring 12 inches x 36 inches. The bottom 5 inches of each shingle are exposed. Thus, the exposure is 5 inches and the coverage is double (the coverage varies from double to triple, but the average is double).

Parts of a Roof

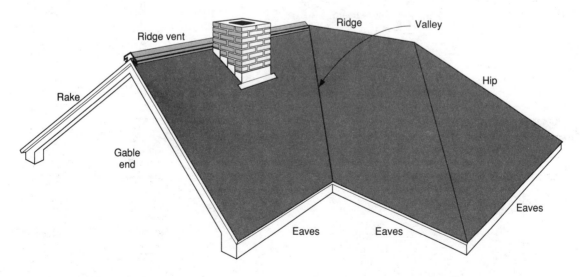

Cornice Terminology

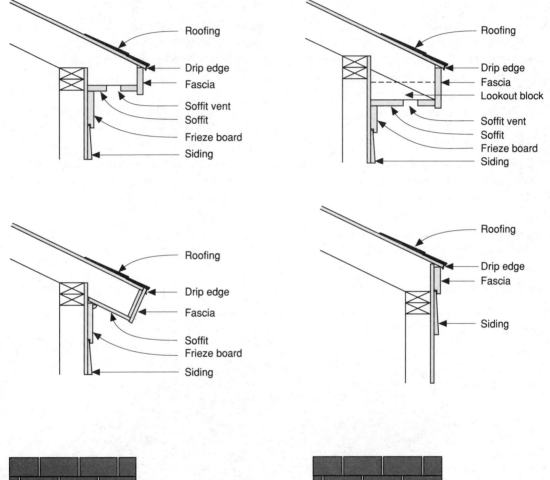

Roofing
Drip edge
Fascia
Soffit vent
Soffit
Frieze board
Siding

Roofing
Drip edge
Fascia
Lookout block
Soffit vent
Soffit
Frieze board
Siding

Roofing
Drip edge
Fascia
Soffit
Frieze board
Siding

Roofing
Drip edge
Fascia
Siding

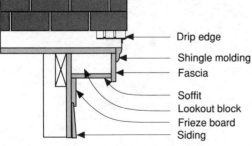

Drip edge
Shingle molding
Fascia
Soffit
Lookout block
Frieze board
Siding

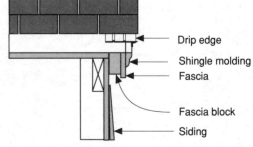

Drip edge
Shingle molding
Fascia
Fascia block
Siding

ROOFING MATERIALS

The type of roofing material selected is a function of regional architectural style as well as performance. The most common material by far is the ubiquitous asphalt shingle. However, the mark of a quality home in Vermont may be slate; in much of ski country, standing-seam; in the Southwest, Spanish tile.

Be aware also of rapid changes in the industry. Preformed metal panels are rapidly replacing standing-seam, and single-ply membrane is replacing the old-fashioned built-up roof.

ROOFING MATERIALS

Roofing Type	Minimum Slope	Life years	Relative Cost	Weight, pounds/ 100 square feet
ASPHALT SHINGLE 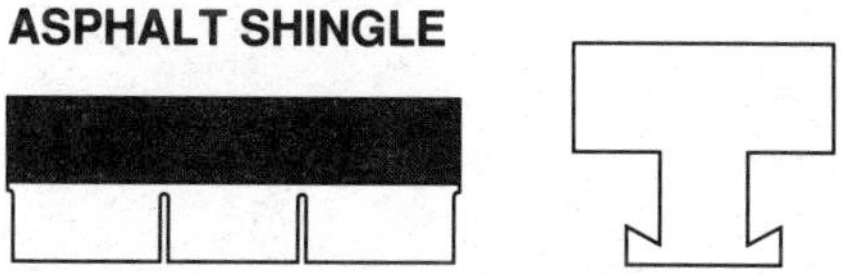	4	15 - 20	Low	200 - 300
BUILT-UP 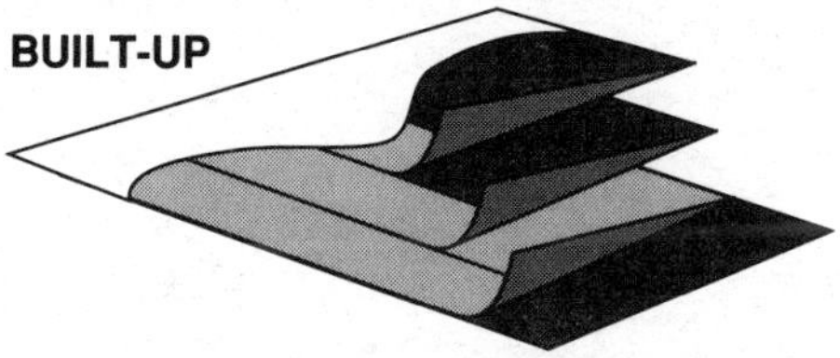	0	20	Medium	200
PREFORMED METAL 	3	30 - 50	Low	50
ROLL 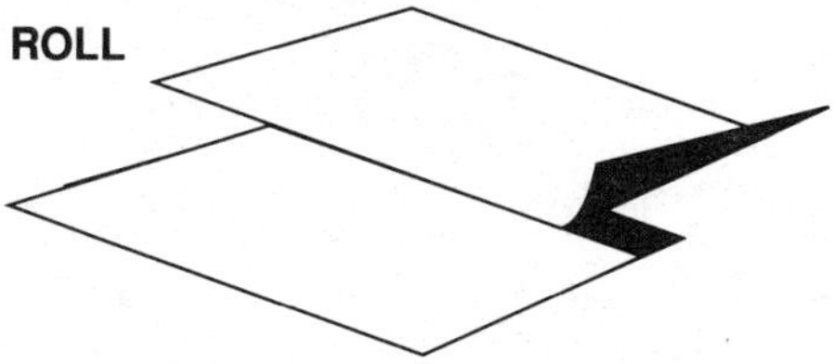	2	10	Low	90

(continued)

ROOFING MATERIALS—*Continued*

Roofing Type	Minimum Slope	Life years	Relative Cost	Weight, pounds/ 100 square feet
SELVAGE	1	15	Low	130
SLATE	5	100	High	750 - 4,000
SPANISH TILE	5	100	High	1,000 - 4,000
STANDING-SEAM	3	30 - 50	Medium	75
WOOD SHAKE	3	50	High	300
WOOD SHINGLE	3	25	Medium	150

BUILT-UP ROOF (BUR)

Built-up roofing is commonly used on flat or very-low-slope roofs where a completely impervious membrane is required. Two-ply, three-ply, or four-ply coverage is achieved by overlapping the 36-inch-wide felts by 19, 25, or 27 inches.

Proper application is very labor intensive and requires experience to properly seal roof edges and penetrations. Because leaks are difficult to diagnose and repair, professional installation is recommended.

The most common BUR (illustrated on the following page) offers no way to ventilate the insulation between the roof deck and the waterproof membrane. An effective vapor barrier is mandatory beneath the insulation wherever average January temperatures are below 40°F or indoor relative humidities are above 50 percent. Even with a vapor barrier, the roofing manufacturer's instructions may call for a vent mechanism in order to avoid blistering of the roof membrane.

A variation on the roof illustrated is the Insulated Roof Membrane Assembly (IRMA). In the IRMA roof, the insulation consists of tongue-and-groove sheets of waterproof extruded polystyrene placed *over,* rather than *under,* the built-up membrane. Stone ballast anchors the foam insulation against wind and protects the foam from ultraviolet degradation.

The IRMA roof has several significant advantages:

• No vapor barrier is required, since the waterproof membrane (itself a vapor barrier) is on the warm side of the insulation.

• Leaks are easily found and repaired after removal of the unfastened foam panels.

• The BUR membrane lasts longer, being protected against sunlight and the thermal stresses of wide temperature swings.

Built-Up Roof (BUR)

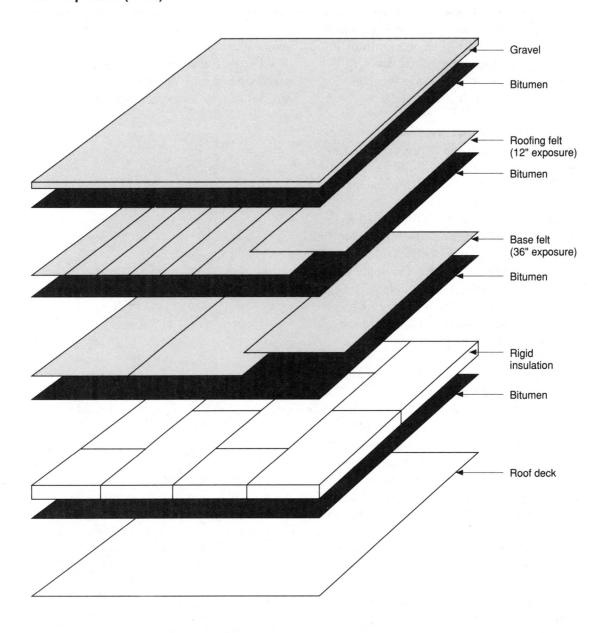

Gravel

Bitumen

Roofing felt
(12" exposure)

Bitumen

Base felt
(36" exposure)

Bitumen

Rigid
insulation

Bitumen

Roof deck

EXPOSED-NAIL ROLL ROOFING

Mineral-surfaced roll roofing is best applied in warm weather. If the temperature is below 45°F, the roofing should first be unrolled and laid flat in a warm space. Extreme care must be used in handling the roofing in cold weather to avoid cracking. In warm weather, unroll the roofing, cut it into 12 to 18-foot lengths, and stack it on the roof for several hours, or until the top sheet lies flat. This precaution will reduce the tendency of the roofing to buckle during application as it expands and relaxes from its tightly rolled condition.

Proper sealing of lap joints is critical to avoid leaks on low-slope roofs, so use only the plastic cement recommended by the manufacturer. Warm it if necessary to facilitate even spreading. Also, use only the amount recommended, as excess cement tends to bubble.

The illustration on the following page shows application parallel to the eaves. Application perpendicular to the eaves is also possible.

Begin by flashing all roof edges as shown. Valleys are flashed with mineral-surfaced roll roofing of the same color. First apply a half strip (18 inches wide), mineral surface down, the full length of the valley, using a minimum number of nails 1 inch from the edges. Next cover that with a full-width sheet, mineral surface up, with minimum nailing 1 inch from the edges.

Apply the first course as a full-width sheet overhanging the deck by 3/8 inch at the eaves and rakes. Nail the sheet 1 inch from the top edge, 3 inches on-center. If more than one length is required, cement and overlap the next length by 6 inches, then fasten it with a double row of nails 4 inches on-center.

The second and succeeding courses are applied in exactly the same way as the first, except each course overlaps the preceding by 2 inches minimum. The sheet is nailed first at the top, allowing for easier correction of ripples, before cementing and nailing of the bottom edge and end laps.

At valleys and ridges, the sheets are trimmed to butt at the ridge or valley intersections. Chalk lines are snapped 5-1/2 inches to both sides of the intersection, and 2-inch-wide strips of plastic cement are applied. Cut sheets of roofing into 12-inch-wide strips, apply them lengthwise over the intersection, and nail them 3 inches on-center, 3/4 inch from the edges.

Exposed-Nail Roll Roofing

FLASHING AND COVERING

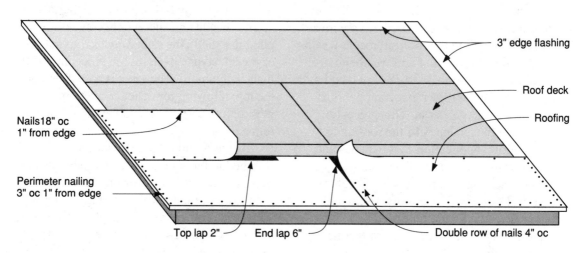

3" edge flashing

Roof deck

Roofing

Nails 18" oc 1" from edge

Perimeter nailing 3" oc 1" from edge

Top lap 2"

End lap 6"

Double row of nails 4" oc

VALLEY FLASHING

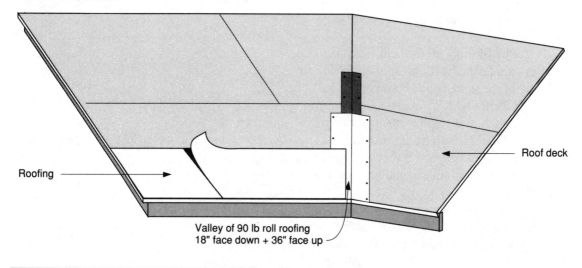

Roof deck

Roofing

Valley of 90 lb roll roofing
18" face down + 36" face up

CONCEALED-NAIL ROLL ROOFING

Mineral-surfaced roll roofing is best applied in warm weather. If the temperature is below 45°F, the roofing should first be unrolled and laid flat in a warm space. Extreme care must be used in handling the roofing in cold weather to avoid cracking. In warm weather, unroll the roofing, cut it into 12 to 18-foot lengths, and stack it on the roof for several hours, or until the top sheet lies flat. This precaution will reduce the tendency of the roofing to ripple during application as it expands and relaxes from its tightly rolled condition.

Proper sealing of lap joints is critical to avoid leaks on low-slope roofs, so use only the plastic cement recommended by the manufacturer. Warm it if necessary to facilitate even spreading. Also, use only the amount recommended, as excess cement tends to bubble.

The illustration on the following page shows application parallel to the eaves. Application perpendicular to the eaves is also possible.

Begin by flashing all roof edges as shown. Valleys are flashed with mineral-surfaced roll roofing of the same color. First apply a half strip (18 inches wide), mineral surface down, the full length of the valley, using a minimum number of nails 1 inch from the edges. Next cover that with a full-width sheet, mineral surface up, with minimum nailing 1 inch from the edges.

Next apply 9-inch-wide (one-quarter roll width) strips of roofing to eaves and rakes, nailing them 4 inches on-center 1 inch in from each edge. The strip edges should overhang eaves and rakes by approximately 3/8 inch to form a drip edge.

The first full course of roofing is applied with edges even with those of the underlying edge strips. Nail the top edge 4 inches on-center, taking care that the bottom of the succeeding course will completely cover the nails. If more than one length is required, cement and overlap the next length by 6 inches, then fasten it with a double row of nails 4 inches on-center. After the top edge is fully nailed, one person lifts the bottom edge while a second fully coats the eaves and rake edge strips with plastic cement. Thoroughly press the bottom and ends of the sheet into the coated edge strips.

The second and succeeding courses are applied in exactly the same way as the first, except that each course overlaps the preceding by at least 2 inches; 3 inches is better. The sheet is nailed first at the top, allowing for easier correction of ripples, before cementing and nailing of the bottom edge and end laps.

At valleys and ridges, the sheets are trimmed to butt at the ridge or valley intersections. Chalk lines are snapped 5-1/2 inches to both sides of the intersection, and 2-inch-wide strips of plastic cement are applied. Cut sheets of roofing into 12-inch-wide strips, apply them lengthwise over the intersection, and nail them 3 inches on-center, 3/4 inch from the edges.

Concealed-Nail Roll Roofing

FLASHING AND COVERING

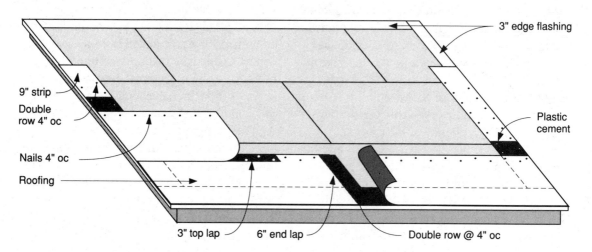

3" edge flashing

9" strip

Double
row 4" oc

Plastic
cement

Nails 4" oc

Roofing

3" top lap   6" end lap   Double row @ 4" oc

VALLEY FLASHING

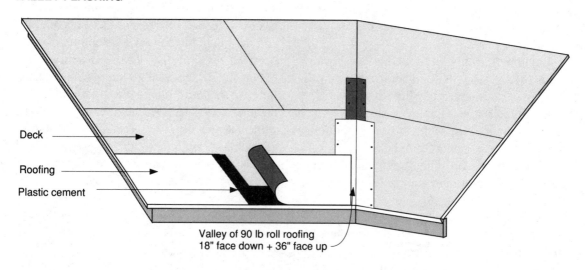

Deck

Roofing

Plastic cement

Valley of 90 lb roll roofing
18" face down + 36" face up

DOUBLE-COVERAGE ROLL ROOFING

Double-coverage roll roofing consists of 17 inches of mineral surfacing, intended for exposure, and 19 inches of selvage (nonmineral-surfaced) to be cemented and lapped. Like the other forms of roll roofing, it is best applied in warm weather in order to avoid cracking and buckling.

In warm weather (over 45°F), unroll the material, cut it into lengths of 12 to 18 feet, and stack it on a flat area of the roof. Allow it to stand until the top sheet lies flat. In cold weather do the same indoors, but be careful not to tear or crack the sheets during transport.

Cementing of the laps is the most critical part of the operation, so use only a plastic cement recommended by the roofing manufacturer, and prewarm the buckets by immersion in warm water if the weather is cold.

The roofing is customarily applied parallel to the eaves, as shown in the illustration. Begin by flashing all roof edges and valleys. Valleys are flashed with mineral-surfaced roll roofing of the same color. First apply a half strip (18 inches wide), mineral surface down, the full length of the valley, using a minimum number of nails 1 inch from the edges. Next cover it with a full-width sheet, mineral surface up, with minimum nailing 1 inch from the edges. The regular courses are then trimmed 3 inches back from the valley intersection, fully cemented to the flashing, and nailed through the selvage portion only.

The starter course is formed by splitting a sheet into a 17-inch-wide mineral-surfaced strip and a 19-inch-wide selvage strip. Put the mineral-surfaced strip aside for the final course, and apply the selvage as a starter strip, overhanging both eaves and rakes by 3/8 inch. Nail the strip 12 inches on-center in two rows: 1 inch above the eaves, and 5 inches from the top. Do *not* cement the starter strip directly to the deck. Coat the entire exposed surface of the starter strip with plastic cement.

The first regular course is a full-width sheet placed with the bottom edge and ends flush with the edges of the starter strip. Press the mineral-surfaced portion of the sheet into the cement (a roller is handy), and nail the selvage portion to the deck in two rows, 5 inches and 13 inches from the top edge. Succeeding courses are applied in the same fashion, except in this order: Nail top sheet, lift bottom of top sheet and apply cement, and press top sheet into cement.

End laps are formed by nailing the surfaced portion of the bottom strip 4 inches on-center and 1 inch from the end, applying a 6-inch-wide band of plastic cement to the entire width, and nailing the top strip through the selvage only.

Hips and ridges are finished by covering them with 12-inch-wide strips of roofing. Trim and butt the underlying courses at the intersections. Apply the 12 x 36-inch pieces in exactly the same way as the regular courses, starting with a 12 x 17-inch piece of selvage.

Double-Coverage Roll Roofing

FLASHING AND COVERING

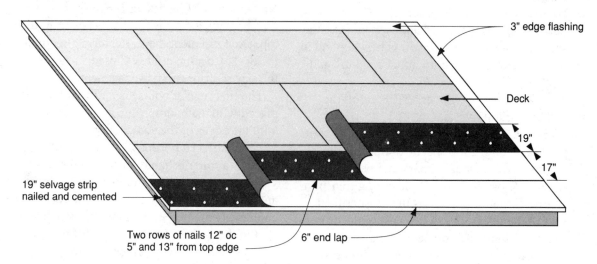

3" edge flashing

Deck

19"

17"

19" selvage strip
nailed and cemented

Two rows of nails 12" oc
5" and 13" from top edge

6" end lap

VALLEY FLASHING

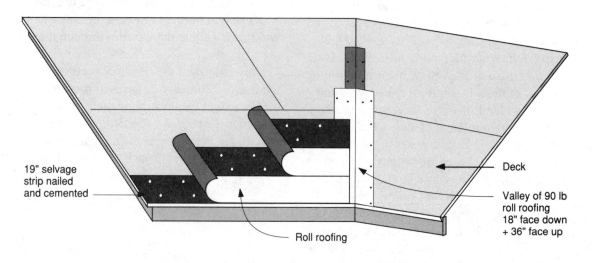

19" selvage
strip nailed
and cemented

Deck

Valley of 90 lb
roll roofing
18" face down
+ 36" face up

Roll roofing

ASPHALT SHINGLES

Underlayment and Drip Edges
Begin by installing a drip edge at the eaves. Next apply 15-pound asphalt-saturated felt underlayment, using a minimum number of nails. Take care to align the felt with the eaves so that the printed lines can serve as guidelines in installing the shingles. Overlap underlayment courses at least 2 inches and end laps 4 inches. Finally, install a drip edge over the underlayment along the rakes.

Eaves Flashing
Where there is any possibility of ice dams, install an eaves flashing of mineral-surfaced roll roofing or special plastic or rubber ice-shield membrane from the eaves to a point at least 12 inches inside the wall line.

Valley Flashing
Valleys are flashed with mineral-surfaced roll roofing of the same color. First apply a half strip (18 inches wide), mineral surface down, the full length of the valley, using a minimum number of nails 1 inch from the edges. Next cover with a full-width sheet, mineral surface up, with minimum nailing 1 inch from the edges. Shingle courses will be bevelled and trimmed along a line 3 inches from the center of the valley.

Shingle Application
Begin with a starter strip of shingles from which the mineral-surfaced portion has been removed. Remove 3 inches from the end of the first strip so that the cutouts of the first regular course will not fall over a starter joint.

Nail the strip 12 inches on-center 3 inches above the eaves, placing the nails to miss the cutouts of the course to follow.

The first course begins with a full-length shingle strip. Install it with the butts of starter and first course aligned. If there is a dormer, snap vertical chalk lines to both sides of the dormer so that vertical alignment of the interrupted courses can be maintained.

For the most common three-tab shingle, nail it in a line 5/8 inch above the cutouts and 1 inch from each end. With each succeeding course, remove an additional 6-inch width from the first shingle in the row, and line up the butts over the tops of the underlying cutouts. For other types of shingle, follow the manufacturer's directions for nailing.

A less regular effect can be created by removing 4 inches from each succeeding course. The tabs then line up every third course rather than every other course. The disadvantage is the need to measure more often.

For flashing details at pipes, chimneys, and abutting walls, see the following pages.

For further information about all types of asphalt roofing, see the Asphalt Roofing Manufacturer's Association's *Residential Roofing Manual.*

Asphalt Shingles

FLASHING AND COVERING

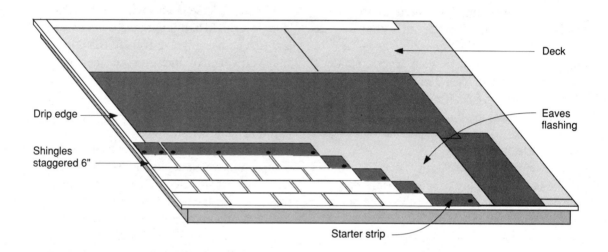

VALLEY FLASHING

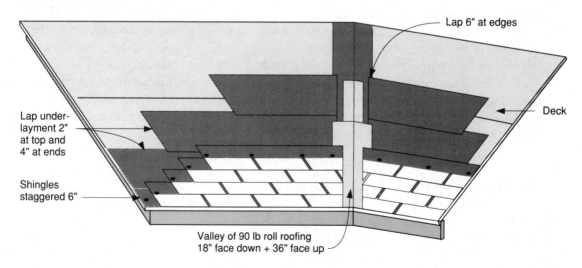

Flashing Walls and Penetrations

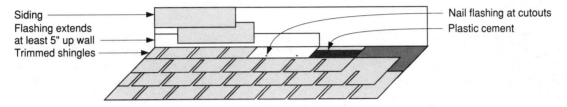

Siding
Flashing extends at least 5" up wall
Trimmed shingles
Nail flashing at cutouts
Plastic cement

ROOF BUTTING WALL

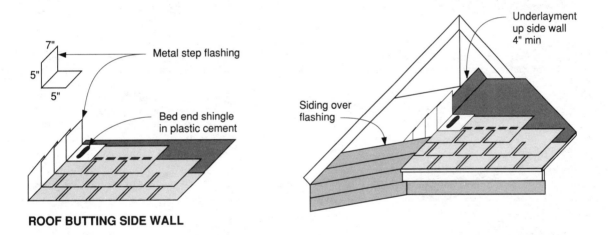

7"
5"
5"
Metal step flashing
Bed end shingle in plastic cement

Underlayment up side wall 4" min
Siding over flashing

ROOF BUTTING SIDE WALL

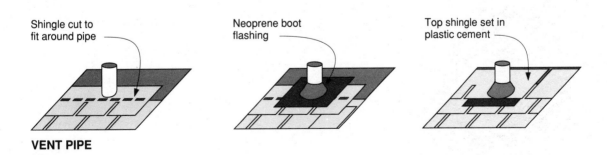

Shingle cut to fit around pipe
Neoprene boot flashing
Top shingle set in plastic cement

VENT PIPE

Flashing a Chimney

STEP 1
Apply asphalt primer to bricks; apply metal base flashing to front, overlapping shingles 4"

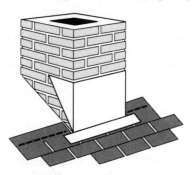

STEP 2
Nail metal step flashing over plastic cement; bed overlapping shingles in plastic cement

STEP 3
Install wood cricket at rear, and shingle to edge; embed rear corner flashings in cement

STEP 4
Embed rear base flashing in plastic cement; nail flashing to deck only

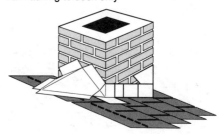

STEP 5
Set front and side cap flashings 1-1/2" into raked joints; refill with mortar

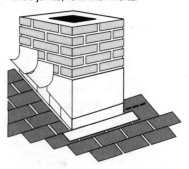

STEP 6
Install rear corner cap flashings; install rear cap flashing suitable to situation

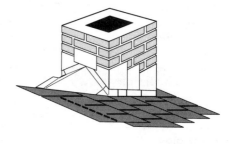

CEDAR SHINGLES

Being wood, cedar shingles periodically become soaked. For the longest life, wood shingles should be applied over sheathing strips spaced the same as the exposure of the shingles. To allow for swelling, shingles should be spaced at least 1/4 inch, regardless of estimated moisture content.

In areas where ice dams occur, the eaves should be flashed with 30-pound asphalt-saturated felt or special ice-shield membrane to 24 inches inside the wall line.

The first course should be doubled, with an eaves projection (drip edge) of 1 inch and joints spaced 1-1/2 inches minimum. Succeeding courses should be laid with exposure and nailing as shown in the table below.

RED CEDAR SHINGLES

Shingle Grade	Length inches	Maximum Exposure @ Roof Slope		Nails for		Description
		<4 in 12	>4 in 12	New Roof	Re-roof	
No. 1 blue label	16	3-3/4	5	3d	5d	Premium grade of shingle for roofs; 100% heartwood, 100% clear, 100% edge grain
	18	4-1/4	5-1/2	3d	5d	
	24	5-1/2	7-1/2	4d	6d	
No. 2 red label	16	3-1/2	4	3d	5d	Good grade; flat grain permitted; 10" clear on 16" shingle; 11" clear on 18" shingle; 16" clear on 24" shingle
	18	4	4-1/2	3d	5d	
	24	5-1/2	6-1/2	4d	6d	
No. 3 black label	16	3	3-1/2	3d	5d	Utility grade; flat grain permitted; 6" clear on 16" and 18" shingles; 10" clear on 24" shingle
	18	3-1/2	4	3d	5d	
	24	5	5-1/2	4d	6d	

Note: Exposure is given in inches.

ESTIMATING COVERAGE

Shingle Length	Square-Foot Coverage of 4 Bundles (nominal square) at Weather Exposure of									
	3-1/2"	4"	4-1/2"	5"	5-1/2"	6"	7"	8"	9"	10"
16"	70	80	90	100	110	120	140	160		
18"		72	81	90	100	109	127	145	163	
24"						80	93	106	120	133

Source for tables: *Exterior and Interior Product Glossary* (Bellevue, Wash: Red Cedar Shingle & Handsplit Shake Bureau, 1980).

Cedar Shingles

APPLICATION

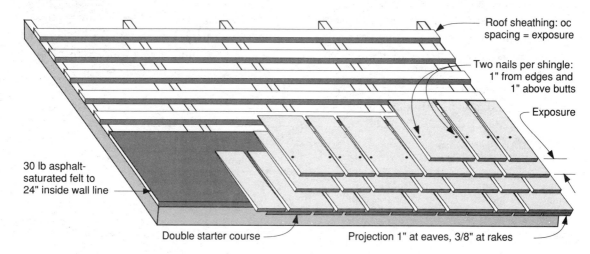

Roof sheathing: oc spacing = exposure

Two nails per shingle: 1" from edges and 1" above butts

Exposure

30 lb asphalt-saturated felt to 24" inside wall line

Double starter course

Projection 1" at eaves, 3/8" at rakes

OPEN-VALLEY FLASHING

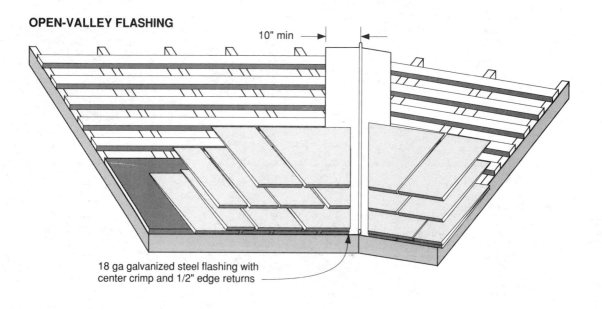

10" min

18 ga galvanized steel flashing with center crimp and 1/2" edge returns

CEDAR SHAKES

For the longest life, shakes should be applied over sheathing strips spaced the same as the exposure of the shakes. To allow for swelling, shakes should be spaced at least 1/2 inch.

In areas where ice dams occur, eaves should be flashed with 30-pound asphalt-saturated felt to 24 inches inside the wall line.

The first course should be doubled, with an eaves projection of 1 inch and joints spaced 1-1/2 inches minimum. An 18-inch-wide strip of 30-pound asphalt-saturated felt should be applied over the top of each course, twice the exposure above the butt. Use the exposures and nailing listed in the tables below.

CEDAR SHAKE SPECIFICATIONS

Shake Grade	Length inches	Butt inches	Max Exposure @ Slope 4 in 12	Nails for New Roof	Re-roof	Description
No. 1 hand split & resawn	18	1/2-3/4	7-1/2	6d	7d	First split to uniform thickness with steel froe, then sawn to produce
	18	3/4-1-1/4	7-1/2	7d	8d	two tapered shakes
	24	3/8	10	6d	7d	
	24	1/2-3/4	10	6d	7d	
	24	3/4-1-1/4	10	7d	8d	
No. 1 taper split	24	1/2-5/8	10	6d	7d	Split with steel froe, then reversed and resplit with taper
No. 1 straight split	18	3/8	7-1/2	6d	7d	Same thickness throughout; split with steel froe and mallet
	24	3/8	10	6d	7d	

Note: Exposure is given in inches.

ESTIMATING COVERAGE

Shake Type	Length inches	Square-Foot Coverage of 5 Bundles (nominal square) at Exposure of				
		5-1/2"	7-1/2"	8-1/2"	10"	11-1/2"
Hand split & resawn	18	55	75	85	100	
	24		75	85	100	115
Taper split	24		75	85	100	115
Straight split	18	65	90	100		
	24		75	85	100	115

Source for tables: *Exterior and Interior Product Glossary* (Bellevue, Wash: Red Cedar Shingle & Handsplit Shake Bureau, 1980).

Cedar Shakes

APPLICATION

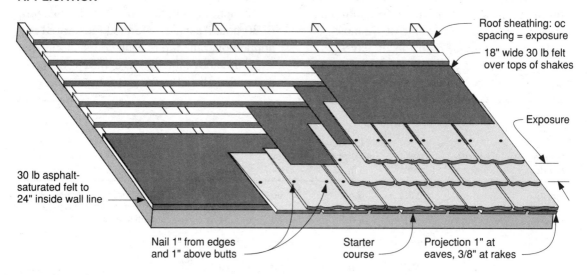

Roof sheathing: oc
spacing = exposure

18" wide 30 lb felt
over tops of shakes

Exposure

30 lb asphalt-
saturated felt to
24" inside wall line

Nail 1" from edges
and 1" above butts

Starter
course

Projection 1" at
eaves, 3/8" at rakes

OPEN-VALLEY FLASHING

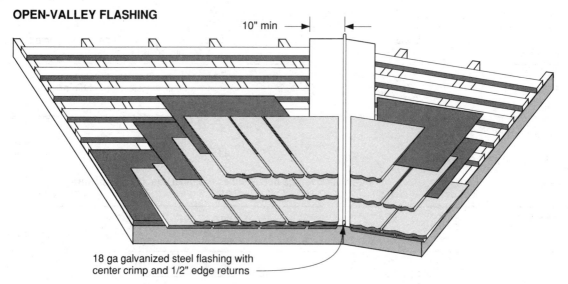

10" min

18 ga galvanized steel flashing with
center crimp and 1/2" edge returns

SLATE

"Standard" roofing slate is 3/16 inch thick. With the customary 3-inch lap, a square (100 square feet of roof covered) weighs up to 750 pounds. Thicker slates weigh proportionally more. Thus, the decision to roof with slate should be made before framing the roof.

Slate color is usually designated as falling into one of eight groups: black, blue-black, gray, blue-gray, purple, mottled purple and green, green, and red.

Application and flashing of slate is similar to that of asphalt shingles (see the sections on asphalt shingles). The differences are described below.

Size

Slates are cut to uniform size, varying in length or depth (in 2-inch increments) from 10 to 26 inches, and width (in 1 and 2-inch increments) from 6 to 14 inches.

Overlap and Exposure

Standard overlap is 3 inches, with 2 inches for slopes over 12/12 and 4 inches for slopes less than 8/12. Exposure is determined by the slate dimension and overlap using this formula:

Exposure = (slate length - overlap) / 2

Example: A 16-inch slate with standard 3-inch overlap would be applied with exposure (16 - 3) / 2, or 6-1/2 inches.

Joint overlap should be 3 inches minimum and one-half of a slate ideally.

Nailing

Slates come prepunched with nail holes 1/4 to 1/2 inch below the top edge and 1-1/4 to 2 inches from the edges. Special slate punches or a masonry drill must be used to create additional holes in the field. Most slate failures are due to the use of inferior shingle nails. Use *only* copper slating nails, available from slate distributors.

Use 3d nails for slates to 18 inches long, 4d for slates over 18 inches, and 6d nails at hips and ridges. Nail length should be twice the slate thickness plus 1 inch for sheathing penetration. Thus, a standard 3/16-inch slate calls for nails 1-3/4 inches long.

Slate

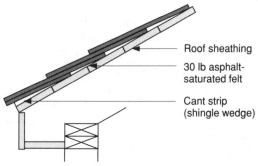

Roof sheathing

30 lb asphalt-
saturated felt

Cant strip
(shingle wedge)

STARTER COURSE

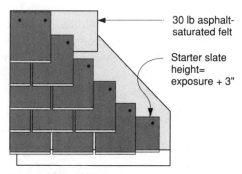

30 lb asphalt-
saturated felt

Starter slate
height=
exposure + 3"

STARTER COURSE

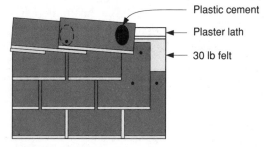

Plastic cement

Plaster lath

30 lb felt

SADDLE RIDGE

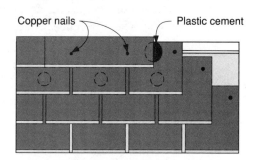

Copper nails

Plastic cement

STRIP SADDLE RIDGE

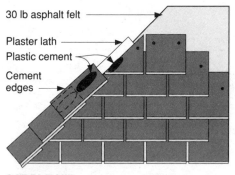

30 lb asphalt felt

Plaster lath

Plastic cement

Cement
edges

SADDLE HIP

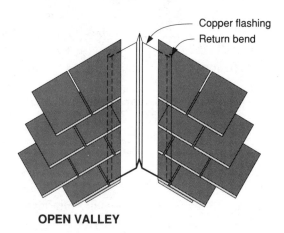

Copper flashing

Return bend

OPEN VALLEY

SPANISH TILE

The System and Its Parts

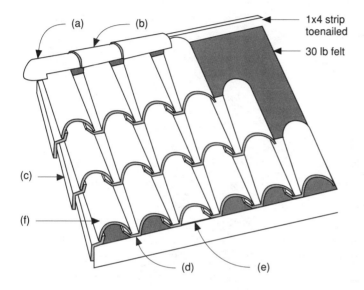

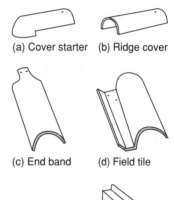

(a) Cover starter   (b) Ridge cover

(c) End band   (d) Field tile

(e) Eaves closure   (f) Gable rake

1x4 strip toenailed

30 lb felt

Installation

STEP 1
Install right gable-rake tiles, with first one overhanging eaves 2"; install eaves closures flush with eaves

STEP 2
Lay first field tile in mastic, covering gable-rake tile nails; space remaining field tiles for roof width

STEP 3
Install left gable-rake tiles; nail 1x3 on edge as nailer for end-closure tiles; set end tiles in mastic and nail to 1x3

STEP 4
Install 1x4 ridge nailer; starting and ending with cover-starter tiles, lay ridge-cover tiles in mastic tinted to match roof

PREFORMED METAL PANEL

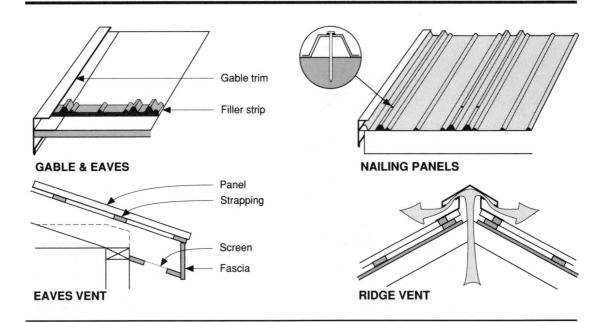

GABLE & EAVES   **NAILING PANELS**

EAVES VENT   **RIDGE VENT**

TYPICAL PATTERNS

Pattern Name	Panel Width inches	Net Width inches	Panel Length feet	Finishes	Pattern Cross Section
Strongbarn	26	24	6 - 30	Galvanized	
Paneldrain	32	30	6 - 36	Galvanized painted	
Channeldrain	38	30	6 - 36	Galvanized painted	
Channeldrain 2000	38	36	6 - 40	Galvanized painted	
Alutwin	–	24	–	Aluminum painted	
5 - V Crimp	26	24	5 - 36	Galvanized	

Note: Strongbarn is a registered trademark of Granite City; Paneldrain, Channeldrain, and 5-V Crimp are registered trademarks of Wheeling Corrugated; and Alutwin is a registered trademark of Alumax.

STANDING-SEAM

Type of Metal	Relative Cost	Thickness inches	Pounds per 100 Square Feet	Expansion inches[1]
Aluminum	Medium	0.032	60	0.46
Copper	Very high	0.022	125	0.34
Galvanized steel	Low	0.022	110	0.23
Stainless	High	0.018	80	0.35
Terne	Medium	0.015	90	0.25
Terne-coated stainless	Very high	0.015	90	0.32

[1] Indicates expansion per 30' of length as a result of 100°F change in temperature.

Seams

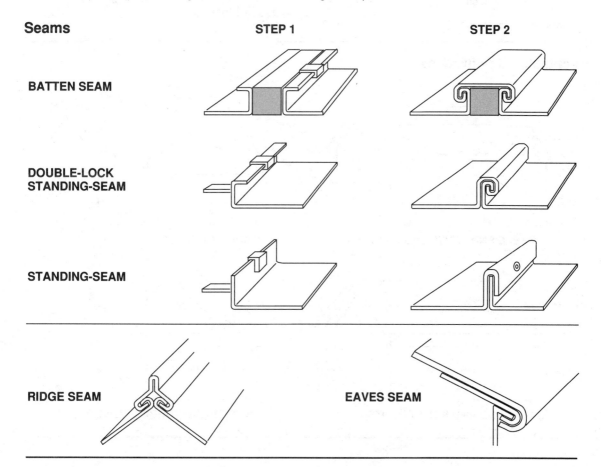

STEP 1   STEP 2

BATTEN SEAM

DOUBLE-LOCK
STANDING-SEAM

STANDING-SEAM

RIDGE SEAM   EAVES SEAM

VENTILATION

Building codes generally require both open attics and cathedral ceilings to have net free vent openings of 1/300 of the ceiling area. Ceilings without vapor barriers require a ventilation ratio of 1/150. Louvers and screens reduce net free areas by the factors below.

Vent Covering	Factor
1/8" screen	1.25
1/16" screen	2.00
Louvers and 1/8" screen	2.25
Louvers and 1/16" screen	3.00

Example: What is the required total gross vent opening for attic vents with louvers and 1/8-inch screen in a 25 x 40-foot house?
Ceiling area = 25' x 40' = 1,000 sq ft
Vent factor = 2.25
Gross area = 2.25 x 1/300 x 1,000 sq ft
          = 7.5 sq ft

The gross vent area should be split between inlets and outlets. The illustration below demonstrates the superiority of the soffit/ridge vent combination.

Ventilator Effectiveness

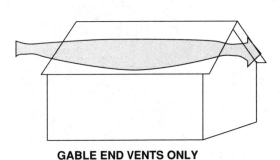

GABLE END VENTS ONLY

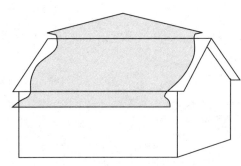

SOFFIT AND RIDGE VENTS

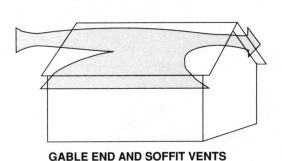

GABLE END AND SOFFIT VENTS

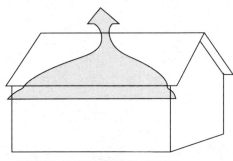

SOFFIT AND TURBINE VENTS

The Cause and Cure of Ice Dams

Ice dams are so common in snow country that many homeowners mistakenly believe them to be inevitable. As with most problems around the home, however, understanding the cause points to the cure.

The illustration at bottom left shows an ice dam in action. The attic floor is well insulated, but so is the roof, due to an accumulation of insulating snow. With insufficient attic ventilation, the attic temperature falls somewhere between that of the living space below and the air outside. Water from melting snow flows toward the eaves. The roofing at the eaves is closer to ambient air temperature, however, so the meltwater refreezes. As the ice builds up, it creates a dam for further melt-

water. If the backed-up meltwater extends beyond the coverage of the roofing material, the water penetrates the roofing.

The illustration at bottom right shows the cure: continuous soffit and ridge vents with insulation baffles providing a minimum 2-inch air channel. Attic air and roofing now follow the ambient air temperature, eliminating the simultaneous melting and refreezing of the snow cover.

An alternate, or perhaps backup, solution is installation of a continuous 36-inch-wide waterproof membrane under the starter course at the eaves. Several products are available for this special application.

Ice Dam Prevention

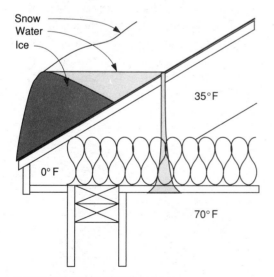

INSUFFICIENT VENTILATION

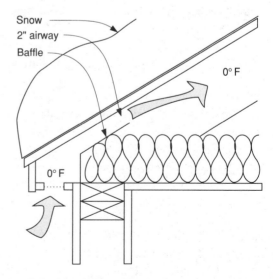

PROPER VENTILATION

GUTTERS

Gutters are widely available in plain or paint-ed aluminum or plastic. Less widely distributed are galvanized steel or copper. Gutters and downspouts usually come in 10-foot lengths, although some are available in lengths of 26 and 33 feet as well.

Several different gutter profiles and hanging systems are available. Your lumberyard or hardware store may offer one style in aluminum and another in plastic. Whatever type you choose, your source will be able to provide all of the accessories needed for a complete installation.

Use the illustration of a typical system on the following page to determine the number of each accessory needed for your installation. You would be wise to sketch the dimensions of your house, particularly the profile of the critical roof/eaves/fascia area. An experienced clerk can then verify the pieces needed.

Before installation, as a final check, lay out all pieces on the ground in their final relative positions.

Use a line level (level on a string) to mark the fascia with the proper slope. The slope should be 1 inch per 16 feet minimum, from one corner to another on a short building, or from center to both ends on a longer building. If the building is extremely long, intermediate low points and downspouts may be required.

Begin installation at a building corner. Attach end cap, downspout, or corner miter as required before attaching the first length of gutter. If a support molding is planned, install it before the gutter, to facilitate handling of the gutter.

Attach the gutter to the fascia with hangers every 2-1/2 feet on-center. Fascia hangers such as the style K fascia hanger illustrated are the most popular. Connect sections of gutter with slip joints as you go. The slip joints are later caulked with gutter mastic or silicone sealant.

If the downspout discharges into an underground drainage system, install strainers over the inlet of each gutter outlet to prevent clogging of the underground system. If the downspout discharges aboveground, install an elbow and a leader (horizontal section of downspout) leading away and down-slope from the building foundation.

Finally, if there are trees near your house, install a strainer cap over the gutter outlet to prevent clogging by leaves. Most gutter problems can be avoided by periodic removal of debris. Many gutter failures are due to the freezing of backed-up water, which destroys the mastic seal. Also, the weight of ice can prove more than the gutter hangers can bear.

Typical Gutter System Parts

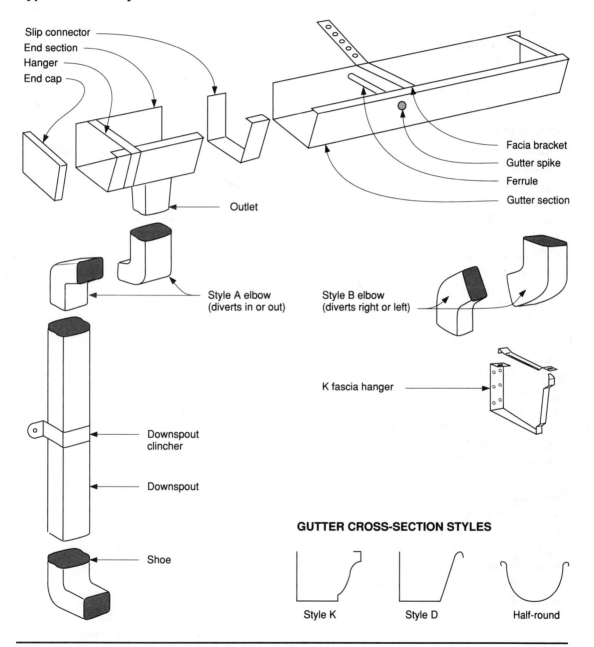

Slip connector
End section
Hanger
End cap

Facia bracket
Gutter spike
Ferrule
Gutter section

Outlet

Style A elbow
(diverts in or out)

Style B elbow
(diverts right or left)

K fascia hanger

Downspout
clincher

Downspout

Shoe

GUTTER CROSS-SECTION STYLES

Style K   Style D   Half-round

CHECKLIST OF CODE REQUIREMENTS

The following is a partial list of requirements from the 1995 Council of American Building Officials (CABO) *One and Two Family Dwelling Code*. Consult the publication for the full text and additional provisions.

Underlayments
(902.2) Single layer:
- laid parallel to eaves
- overlap 2" at tops and 4" at ends
- nailed only to hold in place

(902.3) Multiple layer:
- laid parallel to eaves
- 19" top overlap
- 12" end overlap
- end overlaps staggered 6' minimum
- blind-nailed only to hold in place

Asphalt Shingles
(903.2) Slopes ≥ 4/12 applied over at least one ply of No.15 felt

(903.3) Slopes ≥ 2/12 but < 4/12:
- self-sealing type or hand-sealed
- applied over two layers of No.15 felt
- in areas of ice dams, felt cemented together from eave to 24" inside interior wall line

Wood Shingles
(908.3) Installation:
- side lap 1-1/2" min.
- joints offset 1-1/2" min.
- shingle spacing 1/4" to 3/8"
- two corrosion-resistant fasteners per shingle

(908.3.1) Roof slope:
- 3/12 minimum

- slope < 4/12—reduce exposure, or apply over at least one ply of No. 15 felt

(908.3.2) Valley flashing:
- 28-gauge corrosion-resistant sheet metal
- end lap 4" minimum
- extend 10" under shingles for slope < 12/12
- extend 7" from center for slope ≥ 12/12

Wood Shakes
(909.3) Installation:
- side lap 1-1/2" min.
- joints offset 1-1/2" min.
- shingle spacing 3/8" to 5/8"
- two corrosion-resistant fasteners per shingle

(908.3.1) Shake placement:
- starter course doubled
- shakes interlaid with No. 30 minimum felt with no felt exposed to weather

(909.3.2) Roof slope:
- slope < 4/12—reduce exposure, or apply over at least one ply of No. 15 felt

(909.3.3) Valley flashing:
- 28-gauge corrosion-resistant sheet metal
- end lap 4" minimum
- extend 11" minimum from center for slope

Roof Ventilation
(806.1) Required:
- enclosed attic and rafter spaces
- corrosion-resistant 1/8" wire mesh

(806.2) Minimum net free vent area:
- ≥ 1/150
- ≥ 1/300 if VB on ceiling below
- ≥ 1/300 if 50–80% vent area ≥ 3' above eave and remainder is at eave or cornice

10   Windows and Doors

Can you imagine your house without windows? Windows perform more functions than any other component of a house. The more you know about windows, the more they can do for you.

We all know that windows are made of glass, but nowadays glass isn't always glass. Now there are space-age glazings. How do they compare with plain glass; how to each other? *"About Glass"* shows you the differences in amount of light passed and winter and summer insulation values. It also shows you how to avoid condensation on your windows in winter and what thickness of glass you need to withstand hurricane winds.

"About Windows" spells out all of the things windows can do and the types of windows you can buy at a lumberyard. Whether you are building a new house or replacing an existing window, you need to know how windows are measured. Roof windows and skylights are not as familiar as windows, so a section on a typical *skylight* line is included.

If you are installing a whole wall of windows (for a spectacular view or for a solar greenhouse, for example), you should be interested in *site-built windows*—patio door glazings that can cut your window costs by half.

The second half of this chapter is *about doors:* their functions, how they are constructed, and how to install a prehung door. *"Interior Doors"* illustrates a typical line (including folding doors) and the sizes in which they are available.

There is even a section on the door to the basement. Whether you are building new or converting the basement to a more accessible space, you can find the size of steel *bulkhead door* that will fit your house.

Finally, we provide you with a *checklist of code requirements* relating to windows and doors.

ABOUT GLASS

Types of Glass

Sheet glass is produced by drawing a sheet of glass from the molten liquid. It is the lowest in cost but has noticeable waviness due to variations in thickness.

Float glass is produced by floating glass on a tank of molten tin. It is much flatter than sheet glass and nearly as flat as plate glass. Float glass has largely replaced both sheet and plate glass.

Plate glass is produced by grinding and polishing to a high degree of flatness. It is used primarily for large display windows.

Tempered glass is produced by rapid cooling of the glass to achieve unrelieved tension in its surfaces. The process results in five times the impact strength of untempered glass and relatively harmless small fragments when it is shattered. Building codes generally require tempered glass for glazed doors (patio and French doors) and overhead windows (skylights and sloped sun space glazings).

Heat-absorbing glass is produced in various degrees and shades by adding chemicals to the glass. It is designed to absorb solar energy and reduce building cooling loads.

Reflective glass is produced by depositing metallic film on one or more of the glass surfaces. It is designed to reflect solar energy and reduce building cooling loads. Low-E glass utilizes a selective coating that reflects infrared waves, or heat energy, but not visible light. Heat Mirror is a selectively coated plastic film suspended between ordinary glazings for the same purpose.

Thermal Characteristics of Glazings

Aside from the choice of tempered glass for safety, the key decision in selection of glazing type involves heat and light transmission characteristics. The table on the following page lists these characteristics for a number of glazing types. The terms used in the table are explained below:

Visible light transmission is the percentage of light in the visible spectrum that passes through rather than being absorbed or reflected by the glazing.

Shading coefficient is the ratio of the total solar heat gain of the glazing to the gain of a single 1/8-inch-thick clear glazing.

Relative gain is the calculated total heat gain, in British thermal units (Btu) per hour, of 1 square foot of glazing when exposed to solar radiation of 200 Btu/square foot-hour, no exterior shading, and outside air 14 F° warmer than inside air.

Winter nighttime R is the R-value at the center of the glazing (ignoring the effects of the window frame) with no sunlight and a glass temperature of 70°F.

Summer daytime R is the R-value at the center of the glazing (ignoring the effects of the window frame) with solar radiation of 248.3 Btu/square foot-hour and a temperature difference of 14 F°.

THERMAL CHARACTERISTICS OF GLAZINGS

Glazing Type	Visible Light Transmission, %	Shading Coefficient	Relative Gain Btu/sq ft-hr	Winter Nighttime R	Summer Daytime R
Single Glazing					
Clear 1/8"	90	1.00	215	0.86	0.96
Clear 1/4"	88	0.95	204	0.88	0.96
Heat-absorbing 1/4"[1]	41	0.69	154	0.88	0.91
Heat-reflecting 1/4"[2]	52	0.71	157	0.88	0.91
Acrylic plastic 1/8"	92	1.02		0.94	1.02
Acrylic plastic 1/4"	90	1.00		1.04	1.12
Glass block 6"x6"x4"	60 - 80		167	1.76	
Double Glazing					
Clear 1/4", 1/2" space	80	0.82	172	2.00	1.70
Heat-absorbing, 1/2" space[1]	36	0.54	116	2.04	1.79
Heat-reflecting, 1/2" space[2]	46	0.56	119	2.04	1.79
Heat Mirror, 1/4" space					
HM44 Clear	37	0.34	73	3.2	2.8
HM55 Clear	46	0.40	85	3.2	2.8
HM66 Clear	54	0.48	101	3.2	2.8
HM88 Clear	69	0.66	137	3.1	2.7
HM44 Gray	17	0.23	51	3.2	2.8
HM44 Bronze	22	0.25	55	3.2	2.8
Low E, 1/2" space[3]					
Clear	74	0.79	163	3.2	2.8
Bronze	57	0.64	132	3.2	2.8

[1] Ford Glass Sunglas Grey.
[2] Ford Glass Sunglas Bronze.
[3] Peachtree Twinsul Low E.

Condensation and Glazings

Condensation is destructive to windows. If the window sash are untreated, condensate draws wood volatiles and pigments to the surface, leaving stains. An untreated sash will eventually deteriorate from dry rot.

Condensate appears on cold windows for the same reason it appears on the outside of a cold drink. The ability of air to hold water vapor drops with decreasing temperature. As room air comes into contact with the cold glass, it drops in temperature, depositing excess water vapor on the surface as liquid condensate.

The graph below shows the outdoor air temperature at which condensation first appears on windows, as a function of indoor relative humidity. An inside temperature of 70°F is assumed.

Example: What window R-value is required to prevent condensation at an indoor temperature and relative humidity of 70°F and 40 percent and an indoor temperature of 0°F? The intersection of 0°F outdoor temperature and 40 percent relative humidity occurs just below the R = 2 curve. A double-glazed window of R-2 will therefore prevent condensation.

Glazing R and Condensation

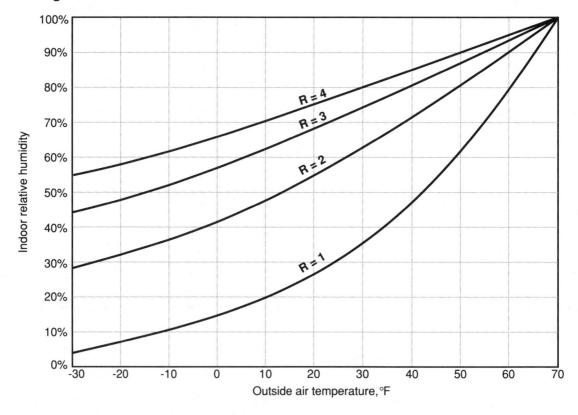

Outside air temperature, °F

Wind Loads on Glazing

For the purpose of load analysis, vertical windows in walls can be considered as walls. The maximum uniform load a glazing can safely support is a function of the glazed area, glazing thickness, and type of glass. The graph below assumes regular annealed float glass, supported on all four sides. For other glazings, multiply the loads in the chart by

- storm windows – 1.0
- factory-sealed double glazings – 1.5
- tempered single glazing – 4.0

Example: What uniform wind load will a 36 x 80-inch sheet of double-strength (DS) float glass withstand? What if we switch to tempered glass of the same thickness? The area is 3 x 6-2/3 feet = 20 square feet. From the chart, annealed DS will support 20 pounds per square foot (psf). From the factors above, tempered glass of the same size will support 4.0 x 20 psf = 80 psf.

If you live in a high-wind area and plan to install unusually large windows, consult a local commercial glazing distributor for advice on wind loads.

Maximum Glazing Area vs Pressure

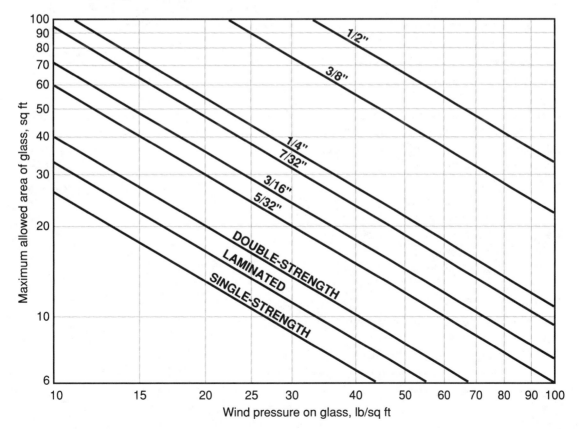

ABOUT WINDOWS

Window Functions
In common with walls, windows are expected
to keep out

- winter wind
- winter snow
- rain in all seasons
- bugs and other flying objects
- highway noise
- winter cold
- summer heat

They are expected, at the same time, to let in

- outside views
- natural light
- ventilating air
- winter solar gain

Window Types
The illustration on the next page shows eight
basic types of residential windows available
from dozens of manufacturers.

Double-hung windows contain two sash,
both of which slide up and down. A variation
is the single-hung window, in which the top
sash is generally fixed.

Casement windows hinge to one side,
which is specified when a unit is ordered.
They are very effective at capturing breezes,
provided they open toward the prevailing
breeze.

Fixed windows are often used in conjunction
with operable windows of other styles. Inexpensive "window walls" can be constructed of patio

doors and site-built fixed windows utilizing
patio door glazing units of the same size.

Awning windows are used for ventilation at
low levels, such as in a sun space, or as high
windows in bathrooms and kitchens.

Sliding windows are an inexpensive alternative for high windows in bathrooms and
kitchens.

Skylights, also known as *roof windows,* are
extremely effective summer exhaust ventila-
tors. They are also more effective in admitting
natural daylight than are vertical windows of
the same size.

Bay windows add space to rooms (often
with window seats) in addition to adding an
architectural design feature to an elevation.
They most often are assembled from a center
fixed unit and two double-hung or casement
flankers.

Bow windows are more elegant and expensive expressions of the bay. They are most often assembled from fixed and casement units
of the same unit dimensions.

Residential Window Unit Types

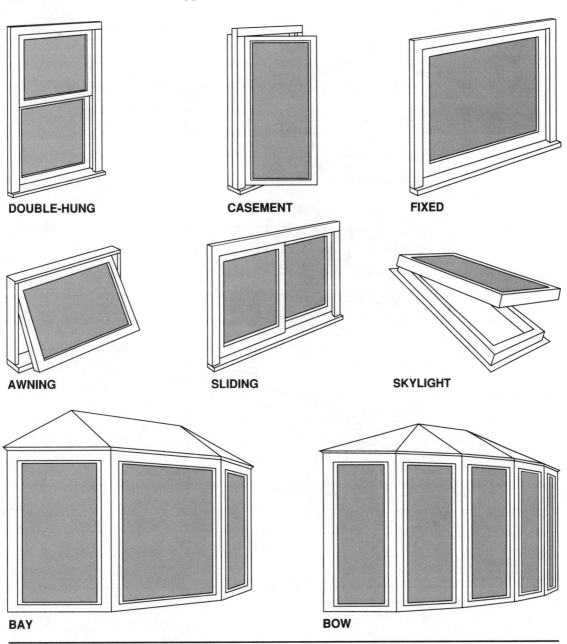

DOUBLE-HUNG

CASEMENT

FIXED

AWNING

SLIDING

SKYLIGHT

BAY

BOW

Window Anatomy and Terminology

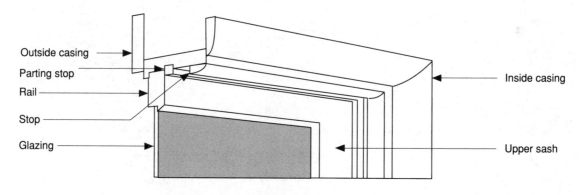

Outside casing
Parting stop
Rail
Stop
Glazing

Inside casing

Upper sash

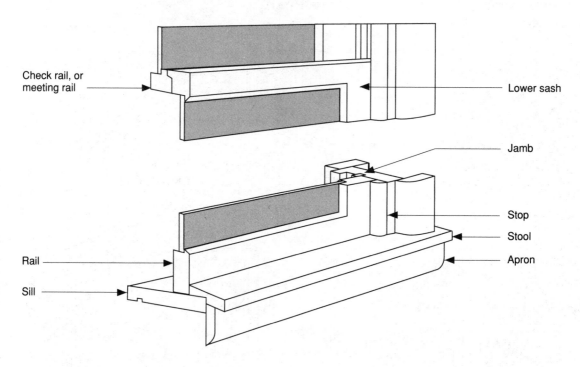

Check rail, or
meeting rail

Lower sash

Jamb

Stop
Stool
Apron

Rail

Sill

Specifying Window Measurements

Window manufacturers list four sets of dimensions (width x height) for each window. They are explained here, from the largest to the smallest:

Unit dimension is the overall size of the window unit, including casing if provided. With a casing, unit dimension will be larger than the rough, or framing, opening. With a nailing flange instead of casing, the unit dimension will be the dimensions of the jambs, or less than the rough opening.

Rough opening is the width and height of the framing opening. To allow for framing tolerance and leveling of the unit, manufacturers add from 1/2 to 1 inch to jamb width and height.

Sash size is the actual width and height of the sash. There may be more than one size of sash in a double-hung window or a window assembled from multiple units.

Glass size is the actual size of the window glazing.

Architects require unit size (dimension) to produce building elevations, and glass size for heat gain/loss calculations. Builders require rough opening, framing sill height, and location on the plan to the center of the unit.

Window Measurements

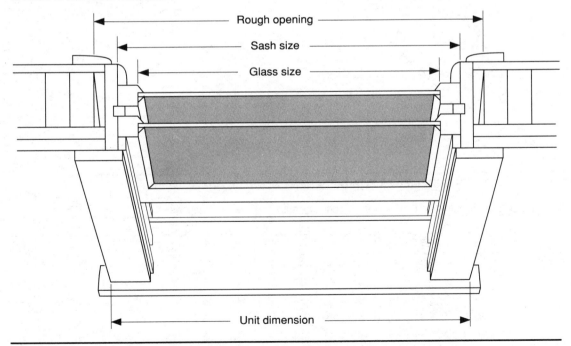

SKYLIGHTS

The illustrations and table below describe Velux-America, Inc., roof windows and skylights. Roof windows pivot so that both sides may be cleaned. Skylights may be fixed or pivot from the top.

The following models are available:

FS – fixed skylight
FSF – fixed skylight with ventilating flap
VS – ventilating skylight
GPL – egress roof window
GDL – balcony roof window

The illustration below left shows the definitions of unit dimension and rough opening. Rough opening in the roof sheathing is 1/2 inch larger than the unit width and height, regardless of whether the opening is relieved (angled) as shown.

The illustration below right and the table show critical installation measurements for units used for sitting and standing views.

Several other manufacturers offer skylights and roof windows. Ask for more information at your local building supply center.

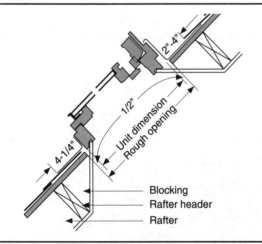
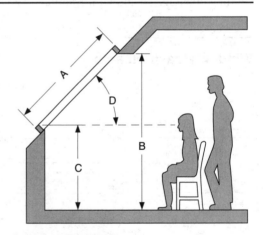

INSTALLATION MEASUREMENTS FOR DIFFERENT ROOF PITCHES, Inches

Size Number		104, 304				106, 306, 606				108, 308				112		
A		38-1/2				46-3/8				55				70-7/8		
B		74	76	78	80	74	76	78	80	74	76	78	80	74	76	78	80
C at D = 30°		60	62	64	66	56	58	60	62	52	54	56	58	44	46	48	50
35°		58	59	62	64	53	55	57	59	48	50	52	54	38	40	42	44
40°		55	57	59	61	50	52	54	56	44	46	48	50	33	35	37	39
45°		52	54	56	58	47	49	51	53	41	43	45	47	30	32	34	36
50°		50	52	54	56	44	46	48	50	37	39	41	43	24	26	28	30

Velux Roof Window and Skylight Sizes

Unit Height	Unit Width			
	21-1/2"	30-5/8"	37"	44-3/4"
27-1/2"	**101** FS FSF VS			**601** FS FSF VS
38-1/2"	**104** FS FSF VS	**304** FS FSF VS		
46-3/8"	**106** FS FSF VS	**306** FS FSF VS		**606** FS FSF VS GPL
55"	**108** FS FSF VS	**308** FS FSF VS GPL		
70-7/8"	**112** FS FSF			
100"			**419** GDL	

NOTES:
Bold numbers (101, etc.) are "sizes."
Lettered codes (FS, etc.) are "models."

For rough openings in:
New construction - add 1/2" to height and width.
Remodelling - add 1" to height and width.

SITE-BUILT WINDOWS

When there is no need for a window to open, particularly if there are several such units, installing bare factory-sealed glazing units in the field will save money.

Insulated Glazing Units

Many glass companies manufacture both custom and standard-size patio door replacement units. Due to volume, standard tempered-glass patio door units offer the greatest value. Standard patio door glazing sizes include: 28 x 76, 33 x 75, 34 x 76, 46 x 75, and 46 x 76 inches.

Insulated glass units are assembled by bonding two panes of glass to an aluminum spacer. The spacer is filled with a desiccant material, designed to absorb moisture and keep the cavity condensation free for the life of the unit (generally guaranteed for 10 years). The sealant may be one or more of the following: silicone, urethane, polysulfide, or polyisobutylene. If the unit is of patio door size, the glazing will most often be tempered, since building codes require tempered or safety glass in doors, within 10 inches of doors, within 18 inches of the floor, and for overhead or sloped glazings.

The greatest drawback to site-built windows is the high rate of failure of the glass seals. Failure is usually due to one of three causes:

- improper installation, where the unit is subjected to stress
- sloped installation, where the span is too great for the glass thickness, resulting in shear
- the use of site sealants that are incompatible with the glazing unit sealant

As a rule, installation at slopes of greater than 20° from vertical voids manufacturers' warrantees. Units can be double-sealed for approximately 20 percent higher cost.

The next two pages show details for installation of glazing direct to framing and within separate frames (jambs). In both cases the keys to success are these:

- *Float the glazing unit within the frame.* All of the weight of the unit should rest on two 4-inch-long strips of neoprene rubber (ask for setting blocks at a glass store) placed one-quarter of the unit width from each bottom corner.
- *Don't let the glass surfaces touch wood anywhere.* Glazing tape compressed between the unit and the wood stops distributes stresses and seals the unit against infiltration.
- *Seal against moisture incursion with a compatible caulk.* Ask the unit manufacturer for a specific recommendation.
- *Provide for moisture drainage.* Outside sill and stop must slope away from the unit. Inside sills should also slope away if in a high-humidity environment. Angle 1/8-inch weep holes from both sides of the setting blocks to outside the siding.

Installation Direct to Framing

MULLION

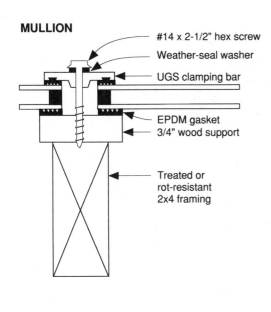

- #14 x 2-1/2" hex screw
- Weather-seal washer
- UGS clamping bar
- EPDM gasket
- 3/4" wood support
- Treated or rot-resistant 2x4 framing

JAMB

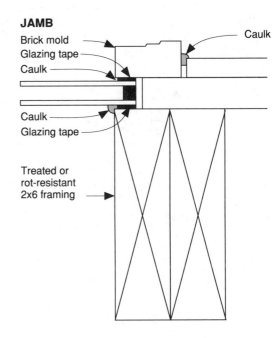

- Brick mold
- Glazing tape
- Caulk
- Caulk
- Caulk
- Glazing tape
- Treated or rot-resistant 2x6 framing

SILL

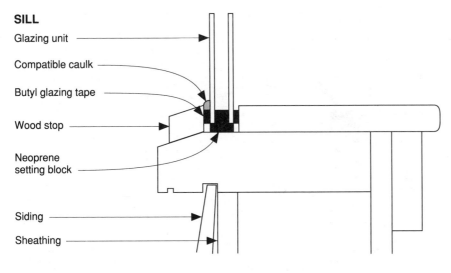

- Glazing unit
- Compatible caulk
- Butyl glazing tape
- Wood stop
- Neoprene setting block
- Siding
- Sheathing

Installation in Jambs

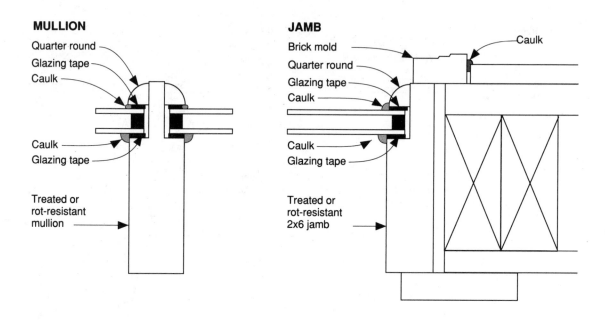

MULLION

Quarter round
Glazing tape
Caulk

Caulk
Glazing tape

Treated or
rot-resistant
mullion

JAMB

Brick mold
Quarter round
Glazing tape
Caulk

Caulk

Caulk
Glazing tape

Treated or
rot-resistant
2x6 jamb

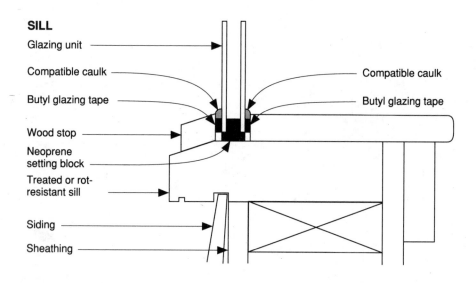

SILL

Glazing unit

Compatible caulk

Butyl glazing tape

Wood stop

Neoprene
setting block

Treated or rot-
resistant sill

Siding

Sheathing

Compatible caulk

Butyl glazing tape

ABOUT DOORS

Functions

Exterior doors function in numerous ways:

They let people in and out. This is not as trivial as it may seem but relates to the design of a welcoming entryway - an architectural subject by itself.

They let large objects in and out. The minimum width for an entry door (and some interior doors as well) should be 3 feet to facilitate moving furniture and appliances.

They keep intruders out. All entrance doors should have quality dead-bolt locks as well as the common latch set. In urban areas an additional lock, operated only from the inside, would be worthwhile.

They keep out winter wind and cold. Except for custom doors intended for historic preservation, the great majority of exterior doors sold today are steel with foam-insulated cores. These represent a giant advance over the classic wood door, in thermal performance if not appearance. Compared with an R-value of 1.5 for the classic wood-paneled door, the foam core door has an R-value of 6 to 12, reducing conductive heat loss by 75 to 85 percent. The best metal doors also incorporate magnetic weather strips, virtually eliminating infiltration.

They let in summer breezes, winter solar gain, and natural daylight. The original function of the storm door was the same as the storm window: to reduce winter heat loss by conduction and infiltration. These losses have largely been eliminated by the steel door. However, a combination "storm" door may still be desirable for summer ventilation.

Right-Handed or Left-Handed?

When ordering a prehung door, you must specify its "handedness." The illustration below shows how handedness is defined.

Door Handedness

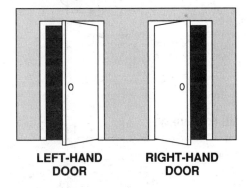

LEFT-HAND DOOR **RIGHT-HAND DOOR**

If a door opens toward you and the door knob is on the left, the door is *left-handed*.

If a door opens toward you and the knob is to the right, the door is *right-handed*.

The following pages will show the construction and installation of both interior and exterior classic wood and modern prehung doors. In addition, step-by-step instructions will be given for installing prehung doors.

Classic Wood Door Installation

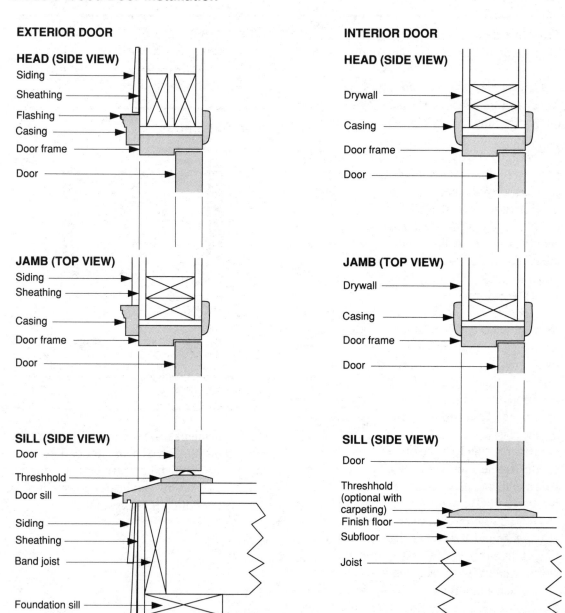

EXTERIOR DOOR

HEAD (SIDE VIEW)
- Siding
- Sheathing
- Flashing
- Casing
- Door frame
- Door

JAMB (TOP VIEW)
- Siding
- Sheathing
- Casing
- Door frame
- Door

SILL (SIDE VIEW)
- Door
- Threshhold
- Door sill
- Siding
- Sheathing
- Band joist
- Foundation sill

INTERIOR DOOR

HEAD (SIDE VIEW)
- Drywall
- Casing
- Door frame
- Door

JAMB (TOP VIEW)
- Drywall
- Casing
- Door frame
- Door

SILL (SIDE VIEW)
- Door
- Threshhold (optional with carpeting)
- Finish floor
- Subfloor
- Joist

Modern Prehung Door Installation

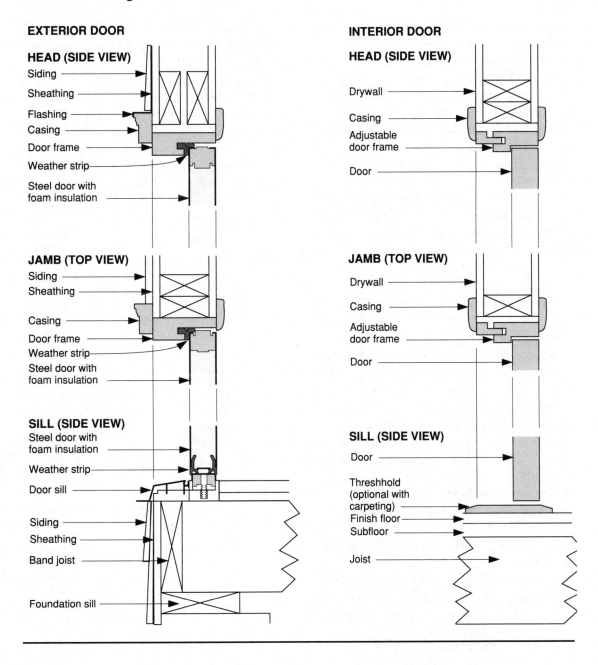

EXTERIOR DOOR

HEAD (SIDE VIEW)
- Siding
- Sheathing
- Flashing
- Casing
- Door frame
- Weather strip
- Steel door with foam insulation

JAMB (TOP VIEW)
- Siding
- Sheathing
- Casing
- Door frame
- Weather strip
- Steel door with foam insulation

SILL (SIDE VIEW)
- Steel door with foam insulation
- Weather strip
- Door sill
- Siding
- Sheathing
- Band joist
- Foundation sill

INTERIOR DOOR

HEAD (SIDE VIEW)
- Drywall
- Casing
- Adjustable door frame
- Door

JAMB (TOP VIEW)
- Drywall
- Casing
- Adjustable door frame
- Door

SILL (SIDE VIEW)
- Door
- Threshhold (optional with carpeting)
- Finish floor
- Subfloor
- Joist

Installing Prehung Doors

1. Make sure the rough opening in the wall framing is about 1 inch wider and 3/4 inch higher than the prehung door frame. The frame wall should be plumb (vertical), and the opening square. Squareness of the opening can be checked either by measuring the opposite diagonals (they should be equal) or with a large piece of plywood with two factory edges.

2. Make sure the bottom of the door will clear the finish floor. For interior doors, cut the bottom of the casing so that the door will just clear the finish floor. For exterior doors, make sure there is a spacer board under the sill to bring the sill bottom level with the finish floor.

3. Place the prehung door unit in the framing opening. Do not nail anything yet. Place solid shims between the hinge-side jamb and stud at all hinges and at the bottom.

4. Square the prehung unit within the rough opening. The unit is square when the space between the top of the door and the head jamb is uniform all the way across. Shim at the top or bottom of one of the side jambs to make the unit square. For example, to reduce the space at the lock side of the head jamb, shim the jamb at either the top of the lock side or the bottom of the hinge side.

5. Fasten the hinge jamb. Nail the hinge jamb through the shims to the stud, using two 16d galvanized finish nails at each hinge and the bottom.

6. Align the lock jamb. Do not use a level for this step: It's too late! Close the door gently. Adjust the lock-side jamb until it meets the door evenly.

7. Fasten the lock-side jamb. Shim the lock-side jamb at three points, and nail through the shims into the stud with two 16d galvanized finish nails at each point.

8. Fasten the exterior casing. If the casing does not sit flush on the exterior sheathing, shim between the casing and the sheathing, and nail through the casing and shims with 16d galvanized finish nails. After siding, caulk between the casing and the siding ends.

INTERIOR DOORS

Interior doors are available in three forms: door only, prehung door in a 4-9/16-inch-wide jamb, and prehung door with a split jamb, adjustable in width from 4-3/8 to 5-1/4 inches.

All doors are 1-3/8 inches thick. Rough-opening width and height are both the size of the door plus 2-1/2 inches. Most doors are available in the widths at right.

Height of Door	6' 6"	6' 8"
Width of Door	2' 0"	2' 0"
	2' 4"	2' 4"
	2' 6"	2' 6"
	2' 8"	2' 8"
	3' 0"	3' 0"

Typical Styles (Brockway-Smith Company)

Lauan
Hollow or solid-core

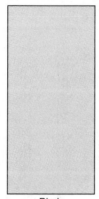

Birch
Hollow or solid-core

Fir
Two-panel

Fir
Five-cross-panel

Molded
Six-panel hollow

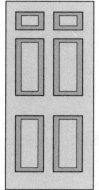

Pine
M-1051

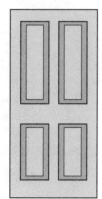

Pine
M-1053

Pine
M-1073

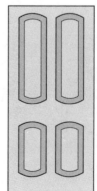

Pine
M-1074

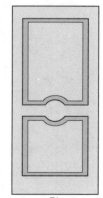

Pine
M-1075

Folding Doors

Folding doors are available without jambs in heights of both 6 feet 6 inches and 6 feet 8 inches. For jamb-opening heights, add 1/2 inch.

All units include all necessary hardware: guide tracks, top and bottom pivots, hinges, and knobs.

The illustration below and the table at right show the typical units available from the Brockway-Smith Company.

Width & Height of Door Unit feet & inches	Width of:	
	Door Only	Jamb Opening
Two-Door Units		
2-0 x 6-6 or 6-8	11-3/4"	2' 0"
2-4 x 6-6 or 6-8	1' 1-3/4"	2' 4"
2-6 x 6-6 or 6-8	1' 2-3/4"	2' 6"
2-8 x 6-6 or 6-8	1' 3-3/4"	2' 8"
3-0 x 6-6 or 6-8	1' 5-3/4"	3' 0"
Four-Door Units		
4-0 x 6-6 or 6-8	11-3/4"	4' 0"
5-0 x 6-6 or 6-8	1' 2-3/4"	5' 0"
6-0 x 6-6 or 6-8	1' 5-3/4"	6' 0"

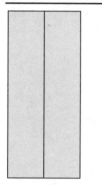
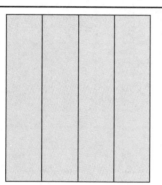

Lauan or birch flush-panel

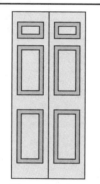

Pine colonial-panel

Pine louver-panel

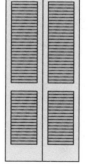
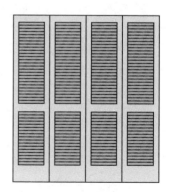

Pine full-louver

BULKHEAD DOORS

The table below and illustrations on the following page describe steel bulkhead doors and stair stringers available from the Bilco Company.

To design a basement bulkhead installation, take the following steps:

1. Determine the height of grade above the finished basement floor.

2. Find the appropriate height range in the table below.

3. Read across the table to find the dimensions of the concrete areaway (illustration at right) and the recommended sizes (illustrations on the following page).

Areaway Dimensions

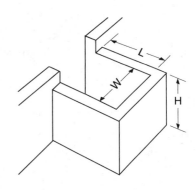

BILCO BASEMENT BULKHEAD INSTALLATION DIMENSIONS

Height of Grade above Floor	H [1]	L	W	Door Style	Extension Required	Stringer Style	Number of Treads
2' 8" to 3' 3"	3' 5-1/4"	3' 4"	3' 8"	SL	none	SL	4
3' 4" to 3' 11" [3]	4' 1-1/2"	3' 4"	3' 8"	SL	none	SL	4
4' 0" to 4' 7" [2]	4' 9-3/4"	4' 6"	3' 4"	O	none	O	6
4' 8" to 5' 4" [2]	5' 6"	5' 0"	3' 8"	B	none	B	7
5' 5" to 6' 0"	6' 2-1/4"	5' 8"	4' 0"	C	none	C	8
6' 1" to 6' 8" [3]	6' 10-1/2"	5' 8"	4' 0"	C	none	C	8
6' 1" to 6' 8"	6' 10-1/2"	6' 8"	4' 0"	C	12"	O + E	9
6' 9" to 7' 4" [3]	7' 6-3/4"	6' 8'	4' 0"	C	12"	O + E	9
6' 9" to 7' 4"	7' 6-3/4"	7' 2"	4' 0"	C	18"	B + E	10
7' 5" to 8' 1" [3]	8' 3"	7' 2"	4' 0"	C	18"	B + E	10
7' 5" to 8' 1"	8' 3"	7' 9"	4' 0"	C	24"	C + E	11

[1] Height above finished basement floor.
[2] Maximum height of house wall = 7' 4".
[3] Requires that one concrete step be added at finished basement floor.

Bulkhead Styles (Bilco Company)

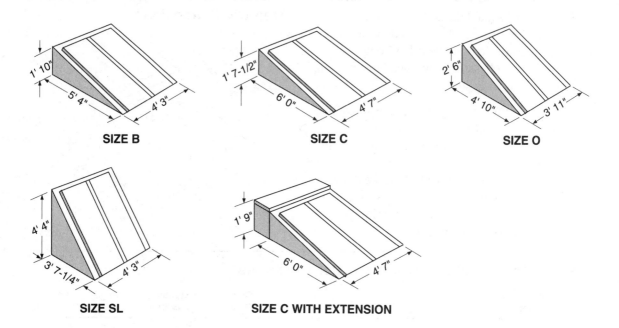

SIZE B

SIZE C

SIZE O

SIZE SL

SIZE C WITH EXTENSION

TYPICAL STANDARD INSTALLATION

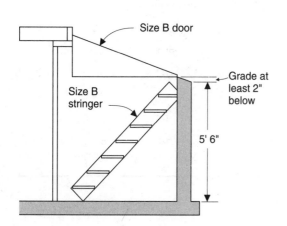

TYPICAL INSTALLATION WITH EXTENSION

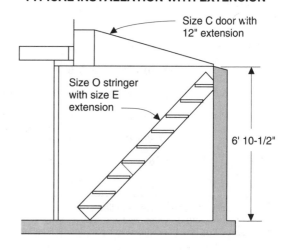

CHECKLIST OF CODE REQUIREMENTS

The following is a partial list of requirements from the 1995 Council of American Building Officials (CABO) *One and Two Family Dwelling Code*. Consult the publication for the full text and additional provisions.

Windows
(608.1) Testing and certification:
- aluminum AAMA (ANSI) 101
- wood ANSI/NWWDA I.S.2
- polyvinyl chloride ASTM D 4099

(608.2) Air infiltration 0.50 cfm/lin ft of crack maximum at 1.56 psf pressure differential

Sliding Glass Doors
(609.1) Testing and certification:
- aluminum AAMA (ANSI) 101
- wood ANSI/NWWDA I.S.3

(609.2) Air infiltration 0.50 cfm/lin ft of crack maximum at 1.56 psf pressure differential

Glazing
(308.4) Glazed areas, including glass mirrors, in hazardous locations must pass test requirements of CPSC 16-CFR, Part 1201.
Hazardous areas include:
- ingress and egress doors, except jalousies
- patio and swinging doors
- storm doors
- unframed swinging doors
- doors and enclosures for tubs, showers, hot tubs, saunas, steam rooms, and whirlpools where glazing bottom < 60" above drain
- pane of > 9 sq ft area with bottom edge < 18" above floor, top edge > 36" above floor, and within 36" horizontally of walking surface
- all glazings in railings
- walls and fences enclosing swimming pools where bottom edge of pool side < 60" above and < 36" horizontally from walking surface

Exceptions:
- door openings < 3" in diameter
- leaded glass panels
- faceted and decorative glass
- glazings in panels where there is permanent barrier between door and panel
- glazings in panels where there is protective bar 36" ± 2" above floor that can withstand pressure of 50 lb /lin ft
- Outside panes of multipane units when bottom of pane ≥ 25' above grade, roof, or walking surface
- louvered windows and jalousies ≥3/16" thick and ≤ 48" long with smooth edges
- mirrors mounted on flush or panel doors
- mirrors mounted or hung on walls

11 Plumbing

The piping in your basement may seem like a maze. Viewing it as three separate systems (cold supply, hot supply, and waste) makes it a lot clearer. That's the approach of this chapter.

The first section looks at the *supply distribution system:* the materials you are allowed to use, the appropriate size of pipe for each fixture, and how the whole system fits together.

The next section looks at the *supply layout and fittings.* Nothing will gain you more respect at the hardware store than being able to ask for a "drop ell with threaded outlet," instead of a "bent gizmo about 6 inches long with threads at each end." To that end, you'll find a complete field guide to polyethylene, copper, and polyvinyl chloride (PVC) supply fittings.

The remaining part of plumbing is the *drain, waste, and vent (DWV)* system. Rather than on water pressure, it depends on gravity to make its mixture of liquids and solids flow. For this reason it is subject to stringent rules regarding pipe sizes, slopes, and venting (introduction of air). You'll find both the rules and the physics behind them.

A second field guide illustrates 80 species of plastic *DWV fittings.* Whether you are installing a new system or adapting to a cast-iron system, you'll find just the fittings you'll need.

How do you know where the pipes go before you have the fixtures in hand? The *roughing-in* guide shows you.

Are you uncertain about how to connect all those pipes and fittings? *"Pipe Installation"* leads you, step-by-step, through all of the modern types: polyethylene, polybutylene, PVC, chlorinated PVC (CPVC), acrylonitrile-butadiene-styrene (ABS), and copper.

A guide to *water treatment* shows you what equipment you need to cure 11 common water-quality problems.

Finally, we provide you with a *checklist of code requirements* relating to plumbing.

SUPPLY DISTRIBUTION SYSTEM

Purpose

The supply system is the network of piping or tubing that distributes cold and hot water under pressure from the city water main or home well to each fixture in the building. Both the water heater and any water treatment equipment are part of the distribution system.

Materials

Supply water is under pressure of up to 160 pounds per square inch (psi) and may be consumed by humans, so the materials allowed are strictly specified by plumbing codes. Materials allowed under the Uniform Plumbing Code (the most widely used, although not universal) are listed in the table below.

ALLOWED SUPPLY PIPING MATERIALS

Material	Form	Advantages	Disadvantages
Chlorinated PVC (CPVC)	10', 20' lengths	Hot-water approved Jointed by cement or threads Lightweight	More expensive than PVC Must be supported 3' oc
Copper Pipe Type K (thick) green Type L (medium) blue Type M (thin) red	10', 20' lengths	Fast and easy to assemble	Water hammer may damage Susceptible to freeze damage
Copper Tubing Type K (thick) green Type L (medium) blue	30', 60', 100' coils	Requires fewer fittings Flare and compression fittings may be used Withstands a few freezings	More expensive than pipe
Galvanized steel	21' lengths	Strong	High cost (due to threaded joints) Corroded by soft (acidic) water Susceptible to scale from hard-water deposits
Polybutylene (PB)	Coils	Installed with fewer fittings No freeze damage Not susceptible to water hammer	Difficult to join Not allowed by all codes
Polyethylene (PE)	100, 120, 160 psi rated coils	Low cost Long lengths Good from well to house	Cold-water approved only
Polyvinyl chloride (PVC)	10', 20' lengths	Low cost Lightweight Jointed by cement or threads	Cold-water approved only

Source: *Uniform Plumbing Code Illustrated Training Manual* (Walnut, Calif: International Association of Plumbing and Mechanical Officials, 1985).

Sizing Supply Pipes

A water supply system is similar in function to an electrical wiring system, in distributing water to fixtures throughout the building. If the building is serviced by utility water, the water arrives through a 3/4 or 1-inch copper pipe from the water main. If the water comes from an individual well, it probably comes from the well via a flexible polyethylene pipe.

The illustration below shows how water is distributed to the individual fixtures. Starting from the point where water first enters the building, a pipe of 3/4-inch internal diameter (all pipes are specified by internal diameter, or ID) runs to the vicinity of every fixture requiring water. Because most fixtures consume hot water as well, a 3/4-inch branch pipe runs through a water heater before pairing with the cold-water pipe for the run through the building. From these two main supply pipes run smaller branch pipes to individual fixtures.

The primary reason for sizing pipes is to make the flow to each fixture relatively independent of the others. Who has not experienced a sudden temperature change while showering? The perfect temperature is achieved by balancing the hot and cold flows. Unless you have a temperature control valve, fluid friction causes a change in pressure at the shower when water is drawn elsewhere along the same pipe, resulting in a change in the proportion of hot and cold. Making the main 3/4 inch and the branches 1/2 inch or smaller minimizes this effect. The illustration shows the recommended supply pipe IDs for each fixture.

Typical Supply Pipe Sizes

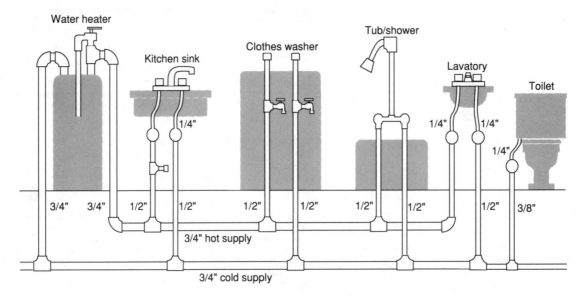

SUPPLY LAYOUT AND FITTINGS

The hot-water and cold-water supply piping for the simple bathroom below illustrates the names of the most common residential supply fittings.

Most supply fittings for polyethylene, copper pipe and tube, and polyvinyl chloride appear on the pages that follow.

1. Reducing tee, 3/4" x 3/4" x 1/2"
2. Reducing elbow, 3/4" x 1/2"
3. 90° elbow, 1/2"
4. Valve body
5. Drop ell with threaded outlet
6. Shower arm
7. Threaded nipple, 1/2"
8. Shut-off valve
9. Supply tube, 3/8"
10. Type L pipe, 3/4"
11. Type L pipe, 1/2"
12. Coupling, 3/4"

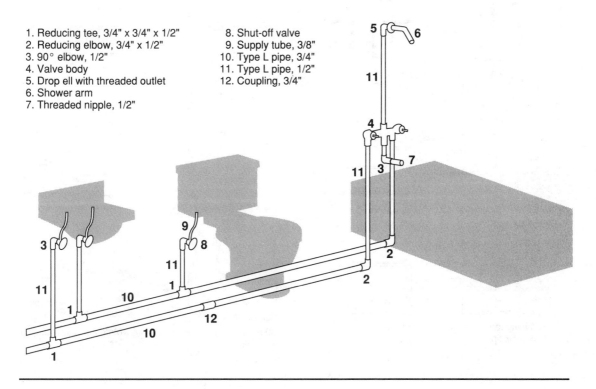

Polyethylene (PE) Supply Fittings

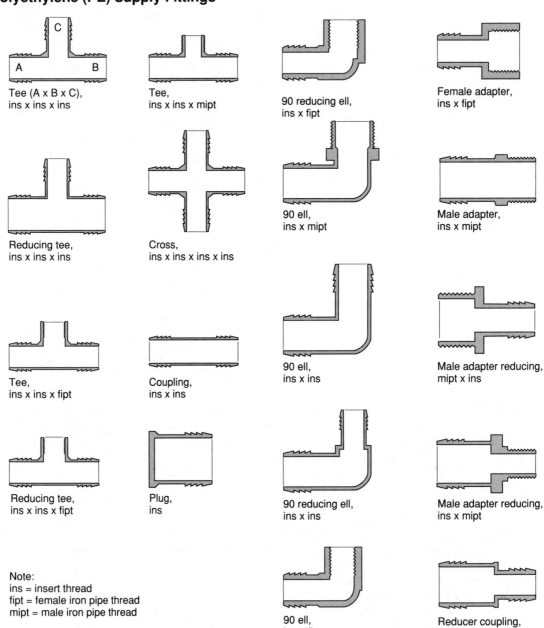

Tee (A x B x C),
ins x ins x ins

Tee,
ins x ins x mipt

90 reducing ell,
ins x fipt

Female adapter,
ins x fipt

Reducing tee,
ins x ins x ins

Cross,
ins x ins x ins x ins

90 ell,
ins x mipt

Male adapter,
ins x mipt

Tee,
ins x ins x fipt

Coupling,
ins x ins

90 ell,
ins x ins

Male adapter reducing,
mipt x ins

Reducing tee,
ins x ins x fipt

Plug,
ins

90 reducing ell,
ins x ins

Male adapter reducing,
ins x mipt

Note:
ins = insert thread
fipt = female iron pipe thread
mipt = male iron pipe thread

90 ell,
ins x fipt

Reducer coupling,
ins x ins

Copper Supply Fittings

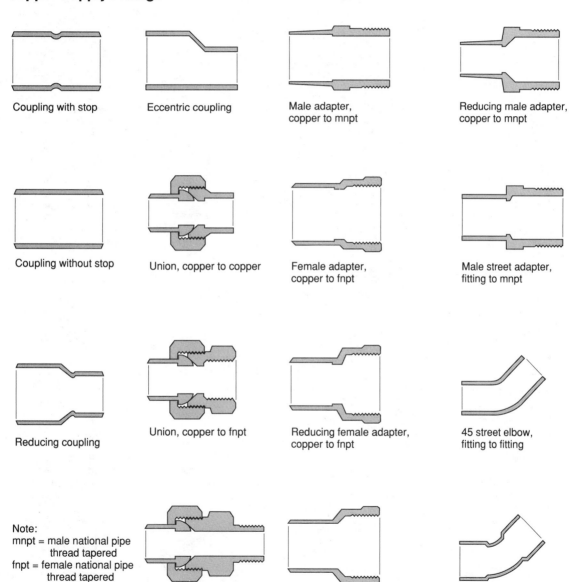

Coupling with stop

Eccentric coupling

Male adapter, copper to mnpt

Reducing male adapter, copper to mnpt

Coupling without stop

Union, copper to copper

Female adapter, copper to fnpt

Male street adapter, fitting to mnpt

Reducing coupling

Union, copper to fnpt

Reducing female adapter, copper to fnpt

45 street elbow, fitting to fitting

Note:
mnpt = male national pipe
　　　thread tapered
fnpt = female national pipe
　　　thread tapered

Union, copper to mnpt

Female street adapter, fitting to fnpt

45 street elbow, fitting to copper

(continued)

Copper Supply Fittings—*Continued*

45 street elbow,
fitting to copper

90 street elbow,
fitting to fitting

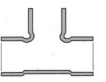

Tee,
all copper

Cross-over coupling,
copper to copper

90 street elbow,
fitting to fitting

90 elbow long-turn,
copper to copper

Street tee,
copper to fitting to copper

Return bend,
copper to copper

90 elbow,
copper to copper

90 reducing elbow,
copper to copper

Reducing tee,
by OD sizes

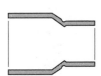

Reducer,
fitting to copper

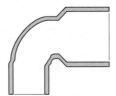

90 reducing elbow,
copper to copper

90 street elbow,
fitting to copper

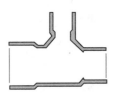

Reducing tee,
copper to copper to copper

End plug,
fitting end

90 street elbow,
copper to copper

90 elbow,
copper to fnpt

Air chamber,
fitting end

Tube cap,
tube end

Polyvinyl Chloride (PVC) Supply Fittings

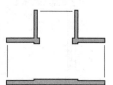

Tee,
slip x slip x slip

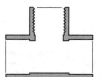

Tee,
slip x slip x fipt

90 ell reducing,
slip x fipt

90 street ell,
mipt x slip

Reducing tee,
slip x slip x slip

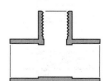

Reducing tee,
slip x slip x fipt

90 ell,
fipt x fipt

90 street ell,
mipt x fipt

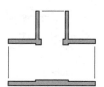

Reducing tee,
slip x slip x slip

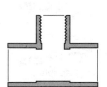

Reducing tee,
slip x slip x fipt

90 ell reducing,
slip x fipt

Side outlet ell,
slip x slip x fipt

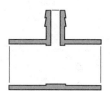

Transition tee,
slip x slip x insert

90 ell,
slip x slip

90 ell,
fipt x fipt

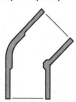

45 ell,
slip x slip

Note:
fipt = female iron pipe thread
IPS = iron pipe size
IPT = iron pipe thread
mipt = male iron pipe thread
slip = slip-fitting end
spig = spigot end
TT = thread-to-thread

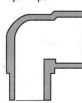

90 street ell,
spig x slip

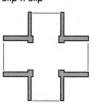

Cross,
slip x slip x slip x slip

(continued)

Polyvinyl Chloride (PVC) Supply Fittings—*Continued*

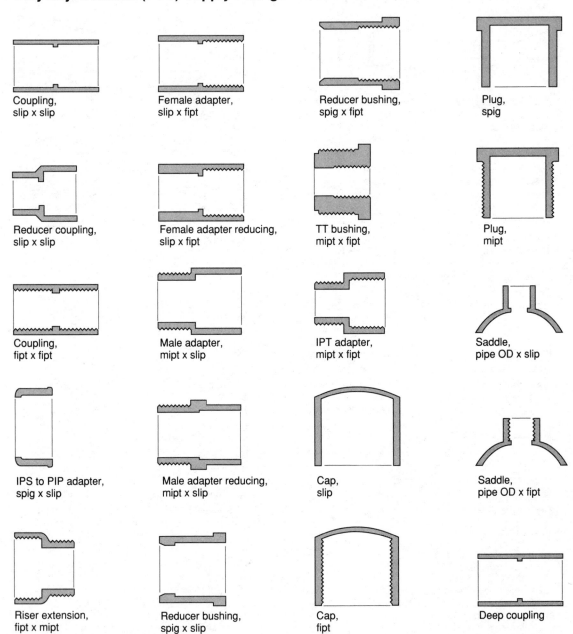

Coupling,
slip x slip

Female adapter,
slip x fipt

Reducer bushing,
spig x fipt

Plug,
spig

Reducer coupling,
slip x slip

Female adapter reducing,
slip x fipt

TT bushing,
mipt x fipt

Plug,
mipt

Coupling,
fipt x fipt

Male adapter,
mipt x slip

IPT adapter,
mipt x fipt

Saddle,
pipe OD x slip

IPS to PIP adapter,
spig x slip

Male adapter reducing,
mipt x slip

Cap,
slip

Saddle,
pipe OD x fipt

Riser extension,
fipt x mipt

Reducer bushing,
spig x slip

Cap,
fipt

Deep coupling

DRAIN, WASTE, AND VENT (DWV)

After water has been delivered to a fixture, it must be safely disposed of, regardless of whether it has been used or contaminated with waste. The drain, waste, and vent (DWV) system differs from the supply system in three important ways:

• Wastewater may include solids as well as liquids (so the pipes must be larger).
• The flow is by gravity alone (so the pipe must slope).
• Waste often generates sewer gas (so provision must be made to prevent sewer gas from flowing back into the building).

The DWV system always contains three things:

• *Traps,* retaining plugs of water, seal against the back flow of sewer gas.
• *Drain lines* carry liquid waste from fixtures. Vertical drains are called *stacks;* horizontal drains, *branches.* Pipes carrying discharge from water closets are also called *soil lines;* pipes not carrying toilet effluent are called *waste lines.*
• *Vent lines* introduce air from above the roof to drain lines wherever necessary to prevent loss of trap water seals. *Wet vents* are vents which, under very specific rules, can serve as drain lines as well.

The table below describes the DWV materials allowed by code.

DWV MATERIALS

Material	Advantages	Disadvantages
Cast iron	Durable Not attacked by chemicals Not affected by boiling water	Heavy (requires structural support) Difficult to cut Very time-consuming, expensive
Galvanized steel	(same as for cast iron)	Expensive Jointed by threads only Susceptible to rust on the inside
PVC and ABS	Low cost Solvent-welded or threaded Lightweight	Limited to 140°F Prone to physical damage Damaged by some chemicals

A Simple Drain, Waste, and Vent System

The illustration below shows the function and terminology of each element in a simple DWV system. Pipe sizes and slopes, types of traps, and placement of vent pipes are all strictly dictated by code.

Tables for sizing traps, drains, and vents are contained on the following page.

DWV Piping

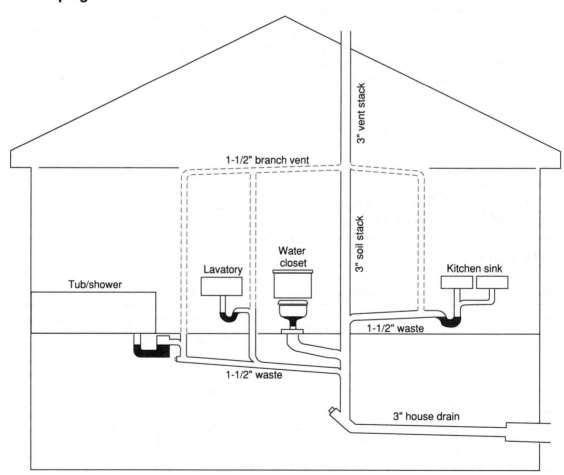

Fixture Units

The *fixture unit* was created in order to describe the peak rate of liquid discharge through a drainpipe. At the time of its definition, the smallest discharge was typically from a lavatory (bathroom sink), so the lavatory was assigned one fixture unit. For estimating purposes, a fixture unit is also equivalent to 7.5 gallons per minute. The table below assigns fixture units and minimum required trap sizes for common residential fixtures.

Fixture	Fixture Units	Minimum Trap Size, inches	Fixture	Fixture Units	Minimum Trap Size, inches
Bathtub	2	1-1/2	Sink, private bar	1	1-1/2
Bidet	2	1-1/2	Sink, kitchen	2	1-1/2
Clothes washer	2	2	Sink and dishwasher	3	1-1/2
Drinking fountain	1	1-1/4	Urinal, wall	2	1-1/2
Floor drain	2	2	Wash basin, single	1	1-1/4
Laundry tub	2	1-1/2	Wash basin pair	2	1-1/2
Shower stall	2	2	Water closet, private	4	3
Showers, per head	1	2	Water closet, public	6	3

Source: *Uniform Plumbing Code Illustrated Training Manual* (Walnut, Calif: International Association of Plumbing and Mechanical Officials, 1985).

Drain and Vent Pipe Sizing

The table below shows the maximum number of connected fixture units and maximum length for drain and vent pipes.

Pipe Size, inches		1-1/4	1-1/2	2	2-1/2	3	4
Drain Pipe							
Vertical	Max fixture units	1	2^1	16^2	32^2	48	256
	Max length, feet	45	65	85	148	212	300
Horizontal	Max fixture units	1	1	8^2	14^2	35	216
	Max length, feet	no limit					
Vent Pipe							
Vertical	Max fixture units	1	8	24	48	84	256
	Max length, feet	45	60	120	180	212	300
Horizontal	Max fixture units	1	8	24	48	84	256
	Max length, feet	45	60	120	180	212	300

Source: *Uniform Plumbing Code Illustrated Training Manual* (Walnut, Calif: International Association of Plumbing and Mechanical Officials, 1985).
[1] Except sinks and urinals.
[2] Except 6-unit traps or water closets.

Traps

A trap is a fitting designed to trap a volume of water in order to block the back passage of sewer gas. The illustration below shows a P-trap.

P-Trap

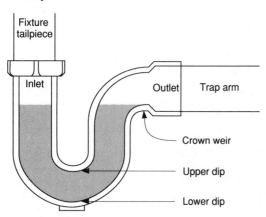

Fixture tailpiece

Inlet

Outlet   Trap arm

Crown weir

Upper dip

Lower dip

Traps can fail in three ways (see numbered illustrations of "Trap Failures" below):

1. If the fixture tailpiece is too tall, the falling water may develop enough momentum to carry it past the outlet weir. For this reason the code requires tailpieces to be as short as possible, but in no case longer than 24 inches (except clothes-washer standpipes, which should be between 18 and 24 inches).

2. If the wastewater completely fills the trap and outlet arm to a point below the outlet, the weight of water in the outlet may siphon the water behind it out of the trap.

3. If the trap arm is too long for its diameter, fluid friction may cause the waste to back up until it completely fills the outlet, resulting in siphoning, as in case 2.

Trap Failures

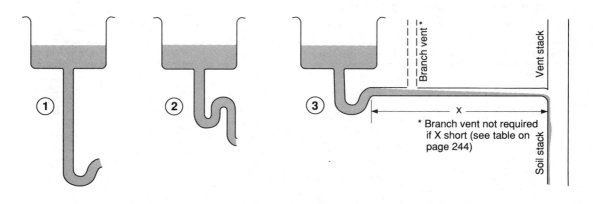

Branch vent*

Vent stack

①

②

③   x

* Branch vent not required
if X short (see table on
page 244)

Soil stack

Approved Trap Types

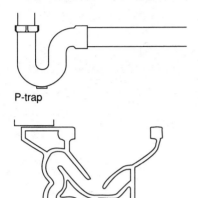

P-trap

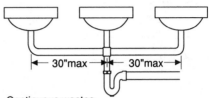

Continuous wastes
3 sinks max, in same room

Integral traps;
example: water closet

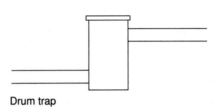

Drum trap
(may be prohibited locally)

Disapproved Trap Types

Full S-trap

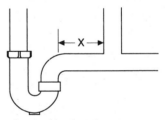

Crown-vented S

P-trap with x less than
twice trap arm diameter

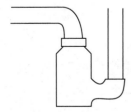

Trap larger than trap arm

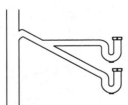

Two traps on same arm

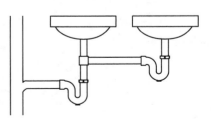

Serial traps

VENTING RULES

Every fixture must have a trap, and every trap must be vented. The table at right shows the maximum allowed length of unvented trap arm. The illustration below shows the three principal vent types:

Individual vents are lengths of vent pipe serving single-fixture traps. They may vent directly to the outside or connect to the rest of the vent system.

Common vents serve two fixture traps. They may vent directly to the outside or be connected to the rest of the vent system.

Wet vents are vertical lengths of pipe which serve both as drains for upper fixtures and as vents for lower fixtures. A wet vent must be at least one pipe size larger than the required drain for the upper fixtures, but not less than 2 inches. All fixtures served must be on the same floor of the building.

HORIZONTAL LENGTH OF TRAP ARMS
(slope of arm 1/4 inch per foot)

Trap Arm Pipe Size, inches	Maximum Arm Length (trap weir to vent)
1-1/4	2' 6"
1-1/2	3' 6"
2	5' 0"
3	6' 0"
4 and larger	10' 0"

Vent Types

INDIVIDUAL VENT   **COMMON VENT**   **WET VENT**

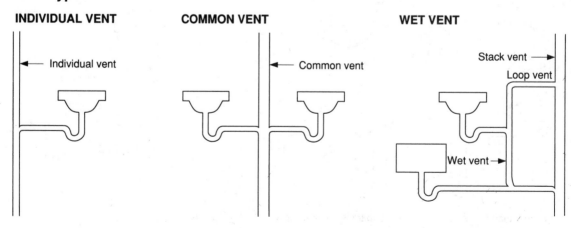

Two-Story Wet Venting

The illustration below shows wet venting on two floors. Note that the vent(s) for the first floor must all connect to the stack vent (vertical vent, which continues through the roof to the outside) at a point at least 6 inches higher than the rim of the highest fixture. This rule protects the vents in case of plugged drains.

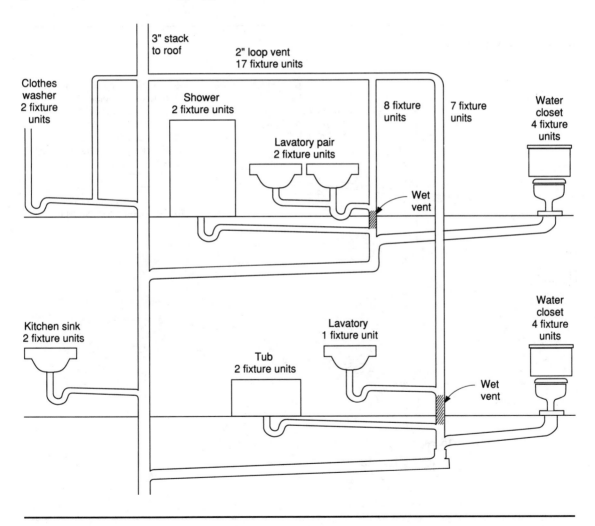

DRAIN, WASTE, AND VENT FITTINGS

The drain, waste, and vent piping for the simple bathroom below illustrates the use and terminology of the most common DWV fittings.

The pages which follow contain illustrations of most DWV fittings available through plumbing distributors.

Typical DWV Fittings

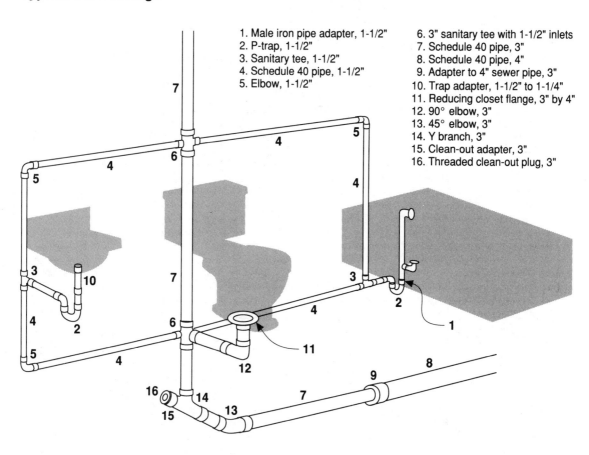

1. Male iron pipe adapter, 1-1/2"
2. P-trap, 1-1/2"
3. Sanitary tee, 1-1/2"
4. Schedule 40 pipe, 1-1/2"
5. Elbow, 1-1/2"
6. 3" sanitary tee with 1-1/2" inlets
7. Schedule 40 pipe, 3"
8. Schedule 40 pipe, 4"
9. Adapter to 4" sewer pipe, 3"
10. Trap adapter, 1-1/2" to 1-1/4"
11. Reducing closet flange, 3" by 4"
12. 90° elbow, 3"
13. 45° elbow, 3"
14. Y branch, 3"
15. Clean-out adapter, 3"
16. Threaded clean-out plug, 3"

Plastic DWV Fittings

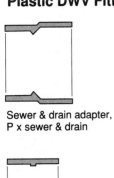

Sewer & drain adapter,
P x sewer & drain

Fitting flush bushing,
fitting x P

Fitting adapter,
fitting x F

Adapter,
P x M

Coupling,
P x P

Fitting trap adapter,
fitting x SJ

Fitting flush adapter,
fitting x F

Fitting adapter,
fitting x M

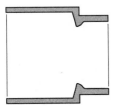

Repair coupling,
P x P

Trap adapter,
P x SJ

Fitting swivel adapter,
fitting x F

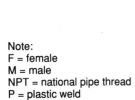

Soil pipe adapter,
P x hub

Adapter,
F x P

Fitting swivel adapter,
fitting x F

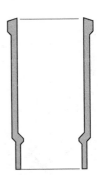

Soil pipe adapter,
P x spigot

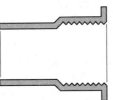

Fitting tray plug adapter,
fitting x NPT straight

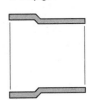

No hub soil pipe adapter,
P x no hub

Note:
F = female
M = male
NPT = national pipe thread
P = plastic weld
SJ = slip joint

(continued)

Plastic DWV Fittings—*Continued*

Vent tee,
P x P x P

Fitting tee,
fitting x P x P

Fixture tee,
P x P x F

Test tee,
P x P x F

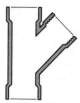

45 Y,
P x P x F

Two-way clean-out tee,
P x P x P

45 Y,
P x P x P

45 Y,
fitting x P x P

Tee,
P x P x P

22-1/2 ell,
P x P

22-1/2 fitting ell,
fitting x P

90 fitting ell,
fitting x P

90 fitting closet ell,
fitting x P

90 fitting long-turn ell,
fitting x P

90 fitting vent ell,
fitting x P

90 ell,
P x F

90 ell with side inlet,
P x P x P

90 vent ell,
P x P

90 ell,
P x P

90 closet ell,
P x P

90 long-turn ell,
P x P

45 ell, P X P

45 fitting ell,
fitting x P

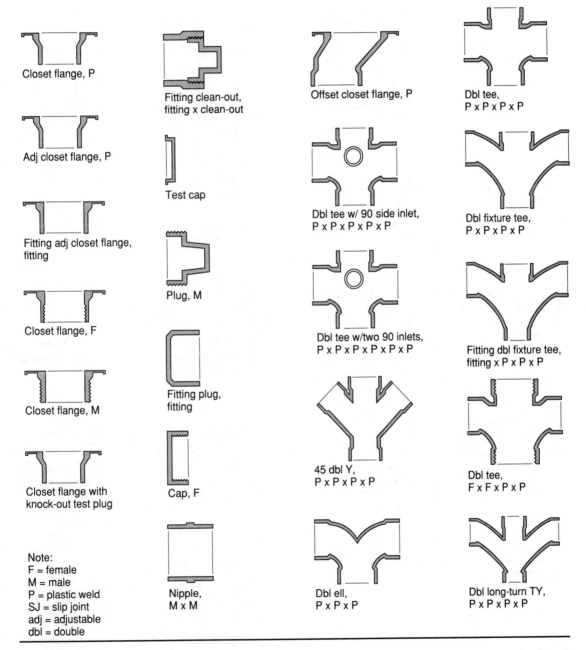

Closet flange, P

Adj closet flange, P

Fitting adj closet flange, fitting

Closet flange, F

Closet flange, M

Closet flange with knock-out test plug

Note:
F = female
M = male
P = plastic weld
SJ = slip joint
adj = adjustable
dbl = double

Fitting clean-out, fitting x clean-out

Test cap

Plug, M

Fitting plug, fitting

Cap, F

Nipple, M x M

Offset closet flange, P

Dbl tee w/ 90 side inlet, P x P x P x P

Dbl tee w/two 90 inlets, P x P x P x P x P

45 dbl Y, P x P x P x P

Dbl ell, P x P x P

Dbl tee, P x P x P x P

Dbl fixture tee, P x P x P x P

Fitting dbl fixture tee, fitting x P x P x P

Dbl tee, F x F x P x P

Dbl long-turn TY, P x P x P x P

(continued)

Plastic DWV Fittings—*Continued*

P-trap with clean-out,
P x SJ x clean-out

P-trap with union,
P x SJ

Tee with 90 R and L inlets,
P x P x P x P x P

Long radius TY,
P x P x P

P-trap,
P x SJ

P-trap with union,
P x P

Tee with 90 L inlet,
P x P x P x P

60 ell,
P x P

P-trap with clean-out,
P x P x clean-out

P-trap with union,
P x F

Tee with 90 R inlet,
P x P x P x P

Ell with high heel inlet,
P x P x P

P-trap,
P x P

Swivel drum trap,
P x P x clean-out

Return bend w/ clean-out,
P x P x clean-out

Ell with low heel inlet,
P x P x P

Note:
F = female
M = male
P = plastic weld
SJ = slip joint
adj = adjustable
dbl = double

Return bend,
P x P

Test tee,
P x P x clean-out

ROUGHING-IN

Both supply and DWV piping require cutting holes in floors, walls, and sometimes framing. The cutting of holes and running of pipes prior to installation of fixtures is called *roughing-in*.

The illustration below shows normal rough-in dimensions for bathroom fixtures. Use the lavatory dimensions shown for kitchen sinks as well, except for counter rim height, which is normally 36 inches for kitchen counters.

The most variable dimension is distance from the center line of the toilet discharge to the wall. For most water closets this dimension is 12 inches, but check with the supplier.

For bidets the equivalent distance averages 15 inches but is not as critical, since there is no tank to consider.

Normal Bathroom Rough-In Dimensions

Note:
a - varies 10" to 14"; most 12"
b - varies 3" to 11"

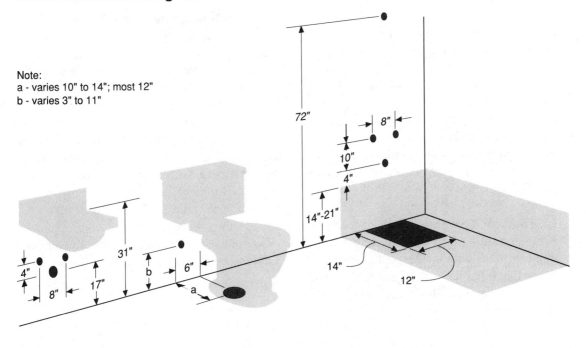

PIPE INSTALLATION

Flexible Polyethylene (PE)

Tools: backsaw or hacksaw, utility knife, regular screwdriver, bucket of hot water.

1. Cut the pipe squarely using a fine-tooth saw or knife.

2. Remove the burrs from inside the pipe with a utility knife.

3. Slip on a stainless steel hose clamp (two clamps if the joint will be inaccessible).

4. If the pipe is cold, soak the end in hot water until it becomes pliable. Wrapping it in a towel and pouring on hot water works, as well. Do not use a torch, as the plastic may permanently stretch and require starting over.

5. Insert the fitting entirely to the shoulder.

6. Position the hose clamp(s) 1/4 to 1/2 inch back from the shoulder of the fitting, and using a screwdriver, tighten them well but not enough to extrude plastic into the clamp slots.

Polybutylene (PB)

Tools: polybutylene-tubing cutting tool, utility knife, felt-tip pen, petroleum jelly, two adjustable wrenches.

1. Make sure the tubing and all of the compression fittings are marked. Reject any tubing or fittings with gouges, cuts, cracks, abrasions, or other defects. Reject tubing that has been kinked or that has faded from exposure to sunlight.

2. Cut the tubing squarely using a poly-butylene-tubing cutting tool.

3. Remove inside and outside tubing burrs with a utility knife.

4. Mark the end of the tubing with the felt-tip pen to show the depth of the compression fitting.

5. Lightly lubricate the end of the tubing with petroleum jelly.

6. Insert the tubing into the fitting up to the pen mark.

7. Tighten the compression fitting by hand.

8. Using a pair of wrenches, tighten the fitting further by a quarter turn (90°).

All bends in tubing must have radii of at least 10 times the outside diameter of the tubing. Support horizontal tubing runs every 32 inches and vertical runs every 48 inches, but loosely so that the tubing can move. Provide 6 inches of slack for every 50 feet of run, to allow for thermal expansion.

Brass or copper threaded fittings should be used to connect polybutylene tubing to other threaded metal fittings. Always solder sweat-adapter fittings before joining them with polybutylene tubing, as excess heat may damage the tubing.

Do not use fittings under slabs. Keep poly-butylene away from excess heat.

Polyvinyl Chloride (PVC) and Plastic

Tools: backsaw or hacksaw, miter jig, utility knife, pencil, cleaner/primer, and the appropriate solvent cement.

1. Make sure you have adequate ventilation. The solvent is extremely volatile.

2. Cut the pipe squarely using a fine-tooth saw and miter jig. A backsaw works best.

3. Remove burrs from both inside and outside of the cut pipe, using a utility knife. Do not clean the outside of the pipe with sandpaper, because the pipe tolerance is critical.

4. Preassemble, without cementing, as much piping as possible; line up and check slopes; and mark both sides of all joints with pencil.

5. (PVC only; not plastic) Paint the inside of the fitting and outside of the pipe with cleaner/primer to remove grease from surfaces.

6. Brush on solvent cement (solvent type CPV or all-purpose for PVC and CPVC plastic pipe; any type for ABS plastic). Apply a heavy coat first to the pipe, then a light coat to the fitting socket.

7. *Immediately* push the pipe into the socket with a twisting motion, and line up the pencil marks.

8. Look for a bead of excess solvent. If it's not there, you didn't use enough. Pull apart the joint immediately, and reapply solvent to the pipe.

9. If the bead is present, hold the pipe firmly in the joint for about a minute. The socket is slightly tapered, and the pipe may try to back out.

10. If you make a mistake, forget trying to disassemble the joint. It is welded. Cut the entire fitting out and start over.

Copper Pipe with Sweat Fittings

Tools: tube cutter, half-round file, steel wool or emery cloth, solder flux, solder, clean cloth, propane torch.

1. Cut the pipe to length with the tube cutter.

2. Remove inside burrs (a tube cutter leaves no outside burr) with the half-round file.

3. Clean the ends of the pipe and fitting socket with fine emery cloth or steel wool. To clean fitting sockets, plumbers often use special round wire brushes.

4. Blow out the fitting and wipe the pipe with a clean rag to remove any copper particles or bits of steel wool.

5. Apply solder flux to both surfaces. You can use a special flux brush or your finger. The flux prevents the clean copper surfaces from oxidizing when heated.

6. Slide the fitting all the way onto the pipe, and rotate it to distribute the flux.

7. Heat both pipe and fitting with the torch. Both should be heated to the same temperature, so expose each proportional to its mass. Do not overheat them. The joint is hot enough when the solder melts upon contact.

8. Apply solder until it appears all around the joint. A length of solder equal to the diameter of the pipe is enough. If the fitting is a valve, remove the stem to protect the washer.

9. Don't move the joint for at least 30 seconds.

10. If the joint leaks, drain all of the water from the line (solder melts at over 400°F and will never get beyond 212°F until all water is removed). Heat the joint, pull it apart, and remove excess solder by striking the pipe against a hard object. Repeat steps 5 through 9.

WATER TREATMENT

Water from both municipal systems and home wells is increasingly polluted with herbicides, pesticides, fertilizers, and organic solvents. This is in addition to the "natural" pollutants: iron, manganese, calcium, magnesium, hydrogen sulfide, radon, asbestos, suspended solids, and decayed organic matter.

The table below shows the available treatments for each contaminant. Many treatments reduce more than one contaminant, although with varying effectiveness.

When more than one treatment is employed, the order of treatment becomes important. Sediment should always be removed first, since it may clog or otherwise impair the effectiveness of other systems. Also, chlorination (to kill bacteria) should always be followed by a "taste and odor" activated-carbon filter if the water is for consumption.

Several companies specializing in water treatment equipment offer free testing and consultation (Sears is one). Given the complexity of treating multiple contaminants, it is advisable to use these services.

WATER TREATMENTS

Problem and/or Symptoms	Chlorine	Sediment Filter	Limestone Neutralizer	Water Softener	Carbon Filter	Reverse Osmosis	Distiller	Chemical Filter
Bacteria: disease	•					•	•	
Hard water: deposits, little lather				•		•		
Acid water: corroded pipes, green stain			•					
Cloudy water		•						
Red water: red stains, metallic taste				•	•		•	
Red slime: slime in toilet tank	•							
Brown water: stains				•	•		•	
Water with rotten egg smell	•				•			
Metals: copper, lead, mercury, sodium, zinc						•	•	
Pesticides						•	•	
Organics: TCE (trichloroethylene), THM (trihalomethane)					•	•	•	•

Sources: *Planning for an Individual Water System*, 3d ed (Athens, Ga: American Association for Vocational Instructional Materials, 1973) and *Sears Catalog* (Chicago: Sears, 1987).

CHECKLIST OF CODE REQUIREMENTS

The following is a partial list of requirements from the 1995 Council of American Building Officials (CABO) *One and Two Family Dwelling Code*. Consult the publication for the full text and additional provisions.

Water-Service Piping

(3403.2.2) When main pressure exceeds 80 psi, approved pressure-reducing valve to be installed on main at connection to service
(3403.4) Size:
 • sufficient size for demand
 • 3/4" nominal minimum
(2903.5.1) DWV system tested for 15 minutes at completion of rough installation by either:
 • water to 10' above highest fitting
 • 5 psi air

PIPING SUPPORT

Pipe Material	Maximum Horizontal Spacing
Cast-iron soil pipe	5' except may be 10' where 10' lengths of pipe are installed
Threaded steel pipe	3/4" diameter and under—10' 1" diameter and over—12'
Copper tube and pipe	1-1/4" diameter and under—6' 1-1/2" diameter and over—10'
Plastic DWV	4'
Plastic pipe & tube, hot & cold, rigid	3'
Plastic pipe & tube, hot & cold, flexible	32"

Traps

(3701.6) Separate trap for each fixture.
Exceptions:
 • fixtures with integral traps
 • multibowl sinks and lavatories (up to three) with outlets ≤ 30" apart
 • clothes washer/laundry tub not to be discharged into trap serving kitchen sink
 • laundry tray waste line may be discharged into clothes washer standpipe 30"– 48" high above weir and within 30" of tray discharge

Cleanouts

(3505.2.4) Cleanouts required:
 • at change of direction > 45 degrees
 • one per 40' regardless of turns
(3505.2.5) Accessibility of cleanouts:
 • 12" minimum in front of pipes ≥ 3" in diameter
 • 18" minimum in front of pipes < 3" in diameter

Vent Terminals

(3601.5.1) Location:
 • not under windows or doors
 • not ≤ 5' horizontally < 2' below opening
(3601.5.2) Extension above roof:
 • ≥ 6" above high side of penetration
 • ≥ 7' above roof if used for other purposes
(3601.55) Frost closure:
 • vent terminal ≥ 2" diameter if design temperature ≤ 0°F
 • size increase ≥ 12" below roof surface

Showers

(3210.1) General:
- floor area ≥ 900 sq in
- floor must contain circle of diameter ≥ 30"

(3210.3) Shower flow ≤ 2.5 gpm at 80 psi

(3210.4) Shower control valves:
- thermostatic and/or pressure balance type
- maximum water temperature 120°F

Water Heaters

(3301.3) Prohibited locations, except for direct-vent water heaters:
- bedrooms, bathrooms, and closets
- spaces opening into bedrooms or bathrooms

(1307.3) Appliances in garage:
- protected from impact by automobiles
- igniters and switches ≥ 18" above floor

Water Closets

(3206.1) General:
- floor-mounted fixtures secured to drainage connection and floor with corrosion-resistant fasteners
- wall-hung fixtures not to strain pipes
- joint with wall or floor to be watertight
- not to interfere with windows and doors

(307.2) Space required:
- 15" minimum center to wall
- 12" minimum center to tub
- 21" minimum to front

(3214.2) To be water-conserving type (1.6 gpm maximum)

Gas Valves

(2606.3) Approved shutoff valve location:
- outside appliance and ahead of union
- within 6' of appliance
- in same room as appliance

(2606.3.1) Shutoff valve may be inside or under appliance if appliance can be removed without removal of shutoff

(2606.4) Gas fireplace outlet valve:
- in same room as fireplace
- outside hearth
- ≤ 4' from outlet

Piping from valve penetrating hearth or wall:
- embedded 1-1/2" in masonry, or
- encased in metal sleeve
- sealed with high-temperature compound

(2607.6) Appliance connection piping:
- rigid metallic pipe, or
- semirigid metallic tubing, or
- listed gas appliance connector
- length ≤ 3', except range or dryer ≤ 6'
- not concealed in wall or floor
- outdoor appliance hose connector ≤ 15'

12   Wiring

As with most things in life, wiring is scary only if you don't understand it. That is why this chapter begins with simple *circuit theory,* comparing the electricity in a wire to water in a hose. The theory is simple, but it explains the functions of all of the elements in a typical residential system: hot wires, grounded wires, grounding wires (which are different), switches, receptacles, circuit breakers, and the service entrance.

"Residential Loads" shows you what circuits the National Electrical Code requires, the voltage and amperage of each, and how to determine the size of the service entrance your home needs.

"Materials for Wiring" explains the color code for wire insulation, the current-carrying capacity of each wire and cable size, and the number of wires electrical boxes are allowed to contain.

The code is very particular about how electricity gets from the utility pole to your service entrance box. *"Service Drops and Entrances"* illustrates the six possible methods, from temporary service during construction to underground service.

"Running Cable" and *"Running Conduit"* illustrate the code requirements for the two common ways of getting electricity around the building.

Whether you are adding circuits or troubleshooting a defective circuit, you should find the illustrations in *"Wiring Switches, Receptacles, and Lights"* a clear guide to which wire goes where.

Finally, we provide you with a *checklist of code requirements* relating to wiring.

CIRCUIT THEORY

Volts, Amps, and Ohms

A useful analogy can be made between the flow of water and the flow of electric current. In the illustration below, the pump creates water pressure in the pipe; the faucet turns the flow on and off; the energy in the falling water is converted to work in the paddle wheel; and the spent (with zero energy) water flows back to the pump intake. Similarly, in an electric circuit, the generator creates an electrical pressure (measured in volts); the switch turns the flow of electrons or current (measured in amps) on and off; the current flowing through the motor is converted to work; and the spent (zero energy) current flows back to the generator.

All of the useful theory in residential wiring can be summed up in two relationships:

Watts = volts × amps   (power formula)

$$\text{Amps} = \frac{\text{volts}}{\text{ohms}}$$ (Ohm's law)

The first relationship, the *power formula,* allows us to convert wattage ratings, found on electric devices such as lights and appliances, to amps. We need amps because this is the way wires and circuits are rated.

The second relationship, *Ohm's law,* allows us to understand why lights and appliances may draw different amperage even though connected to the same voltage.

A Water Analogy

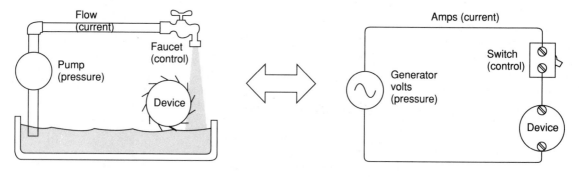

Circuits

With only a slight extension in complexity, we can increase the utility of a circuit (illustration below). The light bulb shows that we can switch on and off any sort of an electrical device. The plug and receptacle show that we can extend the circuit with power cords.

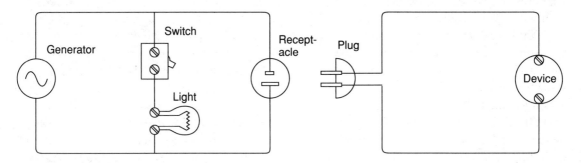

Circuit Breakers

The illustration below shows how we can connect a number of circuits to the set of wires entering a building. The service entrance box (also known as the circuit breaker box) contains a main circuit breaker for disconnecting all the power, and smaller individual breakers, each connected to a separate household circuit. A grounding bus bar (heavy copper bar with screws) and two breaker bus bars make connections simple.

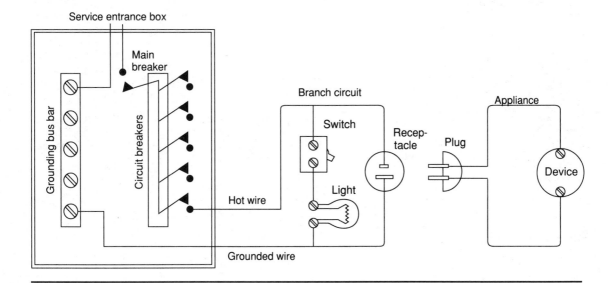

Grounding

Notice that the return wire at the bottom is la-
belled *grounded wire*. There must be a difference in voltage between the upper and lower
wires for current to flow. Rather than have
both wires float at an arbitrary voltage (such
as 1,000 volts!), the power company connects
the grounded wire to the earth, or ground.
That way, in a 110-volt circuit, the worst
shock you can receive is 110 volts, even if
you are standing barefoot on the ground.
Don't try it: It can still kill you. But at least
there is one wire we can safely touch (the
grounded wire) in every circuit.

The illustration below shows a further safety refinement. All receptacles, and most plugs,
have three prongs. The third prong is connected to a third wire, the *grounding wire*. The
grounding wire increases safety by connecting
the outside of any metal electrical box, tool,
or appliance directly to ground. Provided the
grounding wire is intact, there is no way we
can receive a shock by touching the electrical
box, tool housing, or appliance cabinet.

A final safety refinement is the recent code
requirement for a special circuit breaker,
called a ground fault interrupter (GFI), in any
circuit serving the bathroom, kitchen, garage,
swimming pool, or other outdoor location.
This circuit breaker trips, not only when the
hot wire carries too much current (is overload-
ed), but when the current in the hot wire differs from the return current in the grounded
wire. A current difference means that some
current is leaking, possibly returning to
ground through your body.

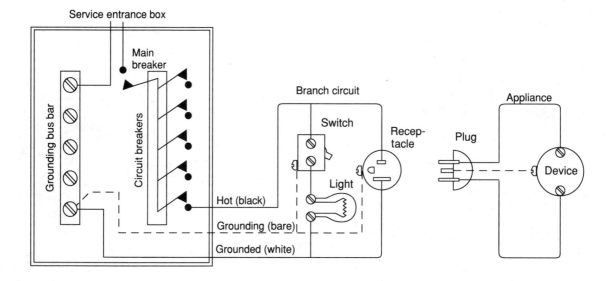

RESIDENTIAL LOADS

Required Circuits

The National Electrical Code requires four types of circuits in a residence: general lighting, small appliance, individual appliance, and ground fault circuit interrupt.

General Lighting—The name *lighting circuit* is misleading. The code means lighting plus wall receptacles into which lights, radios, vacuum cleaners, and so forth are plugged. The code requires 3 watts per square foot of living space (equivalent to one 15-amp circuit per 600 square feet or one 20-amp circuit per 800 square feet) but does not specify the number of receptacles which may be connected to each circuit. The rule of thumb is 12 outlets (receptacle pairs) for 15-amp circuits, 16 outlets for 20-amp circuits. While appliances are allowed, the sum of the appliance ratings must not exceed 50 percent of the circuit rating.

The code goes on to require a light controlled by a permanent wall switch in all habitable rooms, hallways, stairways, and attached garages. Such lights may be plugged into switched wall receptacles, but lights in kitchens and baths and at entrance doors must be permanently wired.

Small Appliance—Two separate 20-amp, 120-volt circuits are required for small appliances such as toasters, blenders, and coffeepots. Both circuits must be present in the kitchen, and one or more in the pantry and dining and family rooms. These circuits may not be used for lighting but may serve the refrigerator and freezer.

Individual Appliance—Some appliances and devices draw enough current to warrant individual circuits. The usual voltages and currents of these circuits follow:

Appliance	Volts	Amps
Clothes dryer	120/240	30
Clothes washer	120	20
Dishwasher	120	20
Garbage disposal	240	20
Kitchen range	120/240	50
Oil burner	120	20
Water heater	240	30
Water pump	240	20

Check with your appliance dealer for electrical ratings. For example, a dishwasher may contain a booster water heater, requiring a higher rating.

Computing Building Loads

The code specifies a method for computing the *building load,* which then determines the required capacity of the service entrance:

1. Compute the *heating, ventilating, and air-conditioning (HVAC) load,* the larger of 65 percent of central electric heating (not including baseboard) or 100 percent of air-conditioning (AC).

2. List and total *other loads:* baseboard or radiant heating, general lighting at 3 watts per square foot, small appliance circuits at 1,500 watts, the laundry circuit at 1,500 watts, and other major appliances.

3. Compute the *total derated load:* 10,000 watts, plus 40 percent of (*other loads -* 10,000), plus the *HVAC load.*

4. Service entrance amperage =
total derated load ÷ 240 volts.

EXAMPLE BUILDING-LOAD CALCULATION

Load		Calculation	Watts
HVAC Load	65% of central electric heat (none)	0.65 x 0	0
	100% of 35,000 Btu/hr AC (5,000 watts)	1.00 x 5,000	5,000
Other Loads	Electric baseboard, 70 lin ft	70 ft x 250 watts/ft	17,500
	General lighting for 2,000 sq ft	2,000 sq ft x 3 watts/sq ft	6,000
	Small appliances, 2 circuits	2 x 1,500 watts	3,000
	Laundry circuit	1 x 1,500 watts	1,500
	Clothes dryer		5,800
	Water heater		4,500
	Range/oven		12,000
			50,300
Derated Load	10,000 watts		10,000
	0.40 x (total other loads - 10,000 watts)	0.40 x 40,300	16,120
	HVAC load		5,000
			31,120

Service amps = total derated load ÷ 240 volts
= 31,120 ÷ 240
= 130 amps (150-amp service entrance)

TYPICAL APPLIANCE WATTAGES

Appliance		Watts	Appliance		Watts
Air conditioner	Room, 7,000 Btu/hr	800	Humidifier	Portable	80
	Central, 35,000 Btu/hr	5,000	Iron		1,100
Blanket		200	Light bulb		rated
Blender		375	Microwave	Small	1,200
Clothes dryer	Electric	5,800		Large	1,800
	Gas	500	Mixer		200
Clothes washer		600	Motor, running	1/4 hp	600
Coffee maker		1,000		1/3 hp	660
Computer	Personal	60		1/2 hp	840
Copy machine		1,500		3/4 hp	1,140
Dehumidifier	25 pints/day	575		1 hp	1,320
Dishwasher		1,000		1-1/2 hp	1,820
Disposal		400		2 hp	2,400
Electric heat	Baseboard/foot	250		3 hp	3,360
Fan	Attic	500	Projector		360
	Bath	100	Range		12,000
	Kitchen	250	Refrigerator	Frost-free	350
	Ceiling	50		Regular	300
	Window, 20-inch	275	Sewing machine		90
Freezer		500	Stereo	40 watts/channel	225
Fryer, deep fat		1,500	Sunlamp		275
Frying pan		1,200	TV, color	9-inch	30
Furnace	Blower	1,000		21-inch	120
	Oil burner	300	Toaster/oven		1,200
Hair curler		1,200	Vacuum cleaner		650
Hair dryer		1,200	Waffle iron	Single	600
Heat lamp		250		Double	1,200
Heater	Portable radiant	1,500	Water heater		4,500
Heating pad		75	Water pump	Shallow	660
Hot plate	Per burner	800		Deep	1,320

Source: *Home Wind Power* (Washington, DC: United States Department of Energy, 1981).

MATERIALS FOR WIRING

Color Code

The National Electrical Code specifies the colors of wire insulation so that it is obvious which wires serve which functions (hot, grounded, or grounding):

black – hot; connects to the darkest screw terminal on a receptacle or switch

red – second hot wire when there are two, as in a 240-volt circuit

white – grounded wire; connects to silver screw on receptacle

bare or *green* – grounding wire; connects to green screw

For convenience, the code allows white wires to be used as hot leads, provided the visible portion of the wire is painted or taped black.

Wire and Cable Types

All wire insulation must be labeled, identifying the type of insulation:

H	for heat-resistant
R	for rubber
T	for thermoplastic
W	for water-resistant

Cables (two or more wires sheathed in a single protective jacket) are similarly labeled:

AC	for armored (metal) cable
C	for corrosion-resistant
F	for feeder
NM	for nonmetallic
U	for underground
SE	for service entrance cable

WIRE (SINGLE-LEAD) LABELS

Wire Type	Label	Specifications
Thermoplastic vinyl	T	General purpose to 140°F (60°C)
	TW	General purpose to 140°F (60°C) and water resistant
	THW	General purpose to 167°F (75°C) and water resistant
Rubber	R	General purpose indoor to 140°F (60°C)
	RW	General purpose indoor to 140°F (60°C) and water resistant
Rubber and cotton braid	RH	General purpose to 167°F (75°C)
	RHH	General purpose to 194°F (90°C)
	RHW	General purpose to 167°F (75°C) and water resistant
	RH/RW	General purpose to 167°F (75°C) dry and 140°F (60°C) wet
Cotton braid	WP	Weatherproof for overhead outdoors

CABLE (MULTIPLE-LEAD) LABELS

Cable Type	Label	Specifications
Armored sheathing over plastic-insulated wire	ACT	General purpose
Armored sheathing over rubber-insulated wire	ACU	General purpose
Plastic or cotton-braid sheathing	NM	General purpose, dry-only to 140°F (60°C)
Plastic or neoprene sheathing	NMC	Wet or dry to 140°F (60°C); OK in barns
Plastic sheathing over neoprene-insulated wire	SE	Aboveground service entrance
Rubber and neoprene sheathing	USE	Underground service entrance
Thermoplastic sheathing	UF	Underground feeder and branch circuits

Ampacity of Wire

Because all wire has electrical resistance, the flow of electrons (current) generates heat and raises the temperature of the conductor. The code specifies the current-carrying capacity (ampacity) of each wire depending on its size, insulation type, and temperature rating, as shown in the table below.

AMPACITY OF COPPER WIRE IN CABLE AND CONDUIT

	Wire Size	Diameter inches	Resistance ohms/100 feet	Maximum Current / Voltage Drop per 100 Feet Types T and TW	Type THW	Type SE
●	14	0.064	0.248	15 A / 3.7 V	15 A / 3.7 V	
●	12	0.081	0.156	20 A / 3.1 V	20 A / 3.1 V	
●	10	0.102	0.098	30 A / 2.9 V	30 A / 2.9 V	
●	8	0.128	0.062	40 A / 2.5 V	45 A / 2.8 V	
●	6	0.184	0.039	55 A / 2.1 V	65 A / 2.5 V	
●	4	0.232	0.024	70 A / 1.7 V	85 A / 2.0 V	100 A / 2.4 V
●	2	0.292	0.015	95 A / 1.4 V	115 A / 1.7 V	125 A / 1.9 V
●	1	0.332	0.012		130 A / 1.6 V	150 A / 1.8 V
●	2/0	0.419	0.008		175 A / 1.4 V	200 A / 1.6 V

Note: A = ampacity; V = voltage drop.

Box Capacity

All wire connections, whether to switches, receptacles, and light fixtures or as junctions between branch circuits, must be made within a code-approved box having an approved cover plate. The code specifies the number of conductors allowed on the basis of box volume.

By *conductor,* the code means anything that occupies an equivalent volume. To arrive at the number of conductors, add

- each wire ending in the box
- each wire running unbroken through the box
  - each device, such as a switch or receptacle
  - each cable clamp or fixture stud
  - all grounding wires lumped as one

Box volume is usually stamped on the box. If not, compute the volume from the inside dimensions using the formulae in chapter 21. To arrive at the number of conductors allowed, divide the volume by the factors below:

#14 wire	2.00 cubic inches
#12 wire	2.25 cubic inches
#10 wire	2.50 cubic inches
#8 wire	3.00 cubic inches
#6 wire	5.00 cubic inches

The table below lists the results for the most common boxes. Extender rings and raised cover plates that increase volume are available for some box styles.

CONDUCTORS ALLOWED PER BOX

Box Shape	Outside Dimension inches	#6	#8	Wire Size #10	#12	#14
Square	4 x 1-1/4		6	7	8	9
	4 x 1-1/2	4	7	8	9	10
	4 x 2-1/8	6	10	12	13	15
	4-11/16 x 1-1/4	5	8	10	11	12
	4-11/16 x 1-1/2	5	9	11	13	14
Round & octagonal	4 x 1-1/4		4	5	5	6
	4 x 1-1/2		5	6	6	7
	4 x 2-1/8	4	7	8	9	10
Rectangular	2 x 3 x 2-1/4		3	4	4	5
	2 x 3 x 2-1/2		4	5	5	6
	2 x 3 x 2-3/4		4	5	6	7
	2 x 3 x 3-1/2	3	6	7	8	9

Common Electrical Boxes

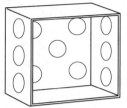

Square

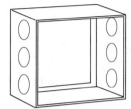

Square extender

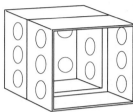

Square box with extender

Plastic handy box

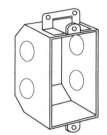

Beveled-switch

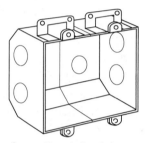

Ganged beveled-switch

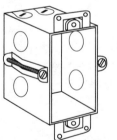

Box with ears and clamps

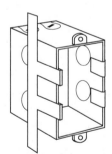

Wall box with brackets

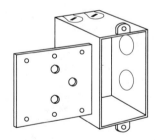

Box with side bracket

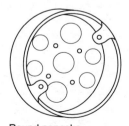

Round pancake

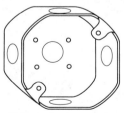

Octagon box

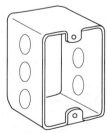

Metal handy box

SERVICE DROPS AND ENTRANCES

The *service drop* is the portion of wiring from the electric utility company's secondary distribution system (generally overhead wires on poles) to the first point of attachment on the building.

The *service entrance conductors* are the wires extending from the point of attachment of the service drop to the building's *service entrance equipment* (service entrance box).

The information presented here is based on the National Electrical Code. All local utilities adhere to the code, but some have even stricter requirements. So check with your local utility before installing a service drop or entrance.

The type of service installed depends on capacity (maximum amps), height of the building, type of exterior building surface (masonry, metal siding, wood, and so forth), and whether the owner or zoning calls for underground service.

The illustrations on the following pages show the specifications for various types of service:

Temporary service is used while a building is under construction. It generally ceases as soon as the building's service entrance equipment has been installed and inspected.

Cable service is commonly used on wood-sided buildings when the building height is sufficient to provide the required clearance above the ground.

Conduit service is used for capacities over 200 amps and for buildings with metal, stucco, or masonry siding. The conduit may be made of steel, aluminum, or PVC plastic.

Rigid steel masts are used when the building is too low to provide adequate clearance for the service drop. Neither aluminum nor PVC conduit may be substituted for rigid steel in masts.

Mobile-home service is used when the building (in most cases a mobile home) is temporary, subject to excessive movement, or structurally inadequate to support a rigid steel mast.

Underground service is provided at the owner's discretion (and added cost sometimes) or to comply with local zoning.

Minimum required clearances for all types of service drops are

- 10 feet above ground
- 10 feet above sidewalks
- 15 feet above residential driveways
- 18 feet above public ways (streets)

Temporary Service, 100 Amps Maximum

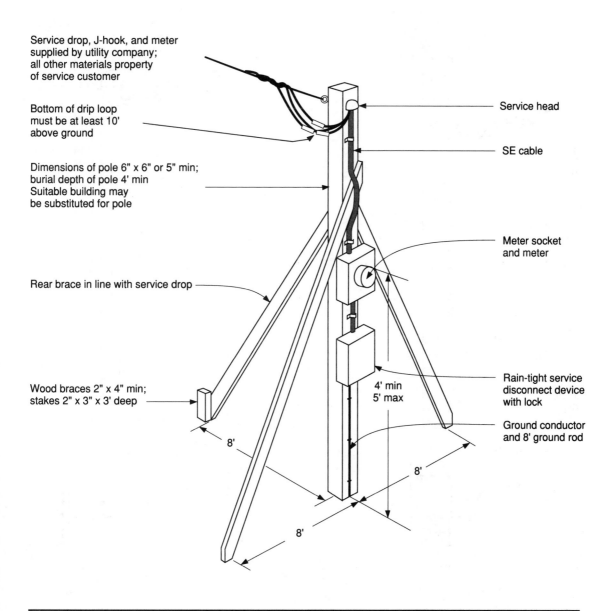

Service drop, J-hook, and meter supplied by utility company; all other materials property of service customer

Bottom of drip loop must be at least 10' above ground

Dimensions of pole 6" x 6" or 5" min; burial depth of pole 4' min
Suitable building may be substituted for pole

Rear brace in line with service drop

Wood braces 2" x 4" min; stakes 2" x 3" x 3' deep

Service head

SE cable

Meter socket and meter

Rain-tight service disconnect device with lock

Ground conductor and 8' ground rod

4' min
5' max

8'

8'

8'

Cable Service, 200 Amps Maximum

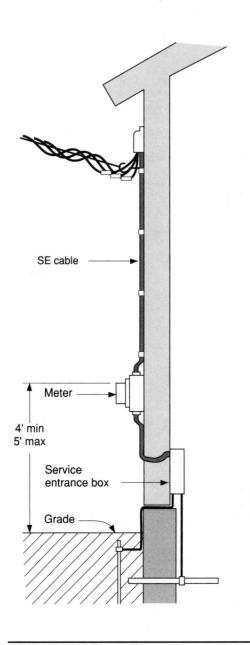

SE cable

Meter

4' min
5' max

Service
entrance box

Grade

Conduit Service, 400 Amps Maximum

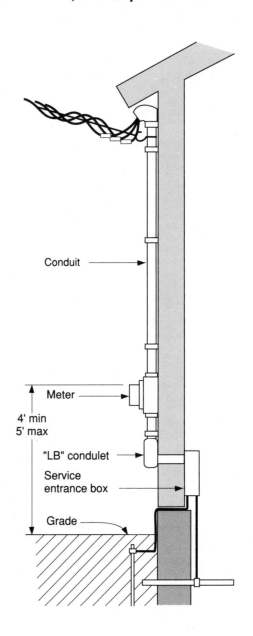

Conduit

Meter

4' min
5' max

"LB" condulet

Service
entrance box

Grade

Rigid Steel Mast for Low Building

Mobile-Home Service on Pole

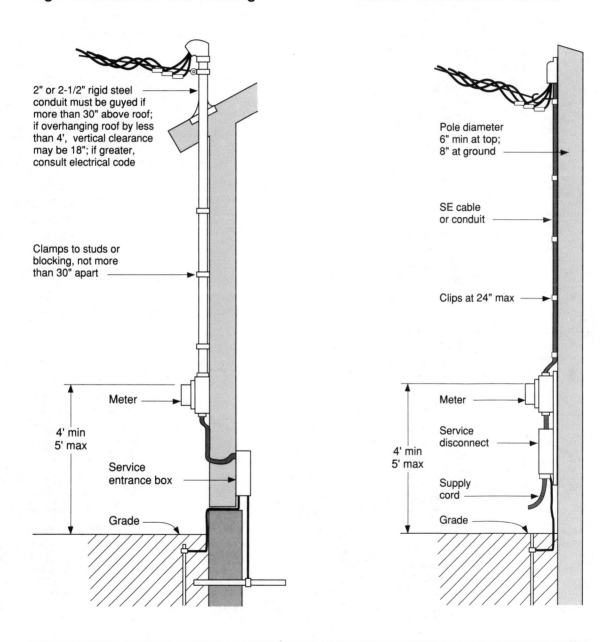

2" or 2-1/2" rigid steel conduit must be guyed if more than 30" above roof; if overhanging roof by less than 4', vertical clearance may be 18"; if greater, consult electrical code

Clamps to studs or blocking, not more than 30" apart

Meter

4' min
5' max

Service entrance box

Grade

Pole diameter 6" min at top; 8" at ground

SE cable or conduit

Clips at 24" max

Meter

Service disconnect

4' min
5' max

Supply cord

Grade

Underground Service, 400 Amps Maximum

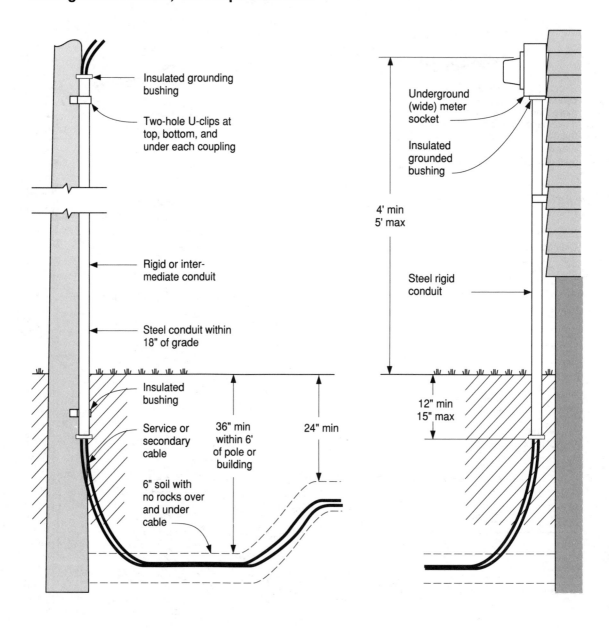

Insulated grounding bushing

Two-hole U-clips at top, bottom, and under each coupling

Rigid or inter-mediate conduit

Steel conduit within 18" of grade

Insulated bushing

Service or secondary cable

36" min within 6' of pole or building

24" min

6" soil with no rocks over and under cable

Underground (wide) meter socket

Insulated grounded bushing

4' min
5' max

Steel rigid conduit

12" min
15" max

Underground Cable

PRIMARY CABLE (OVER 240 VOLTS)

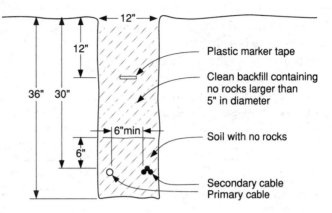

Plastic marker tape

Clean backfill containing no rocks larger than 5" in diameter

Soil with no rocks

Secondary cable
Primary cable

SECONDARY CABLE (UP TO 240 VOLTS)

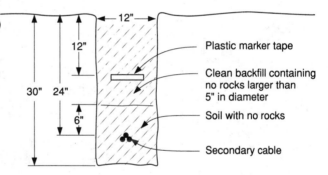

Plastic marker tape

Clean backfill containing no rocks larger than 5" in diameter

Soil with no rocks

Secondary cable

JOINT ELECTRIC AND TELEPHONE

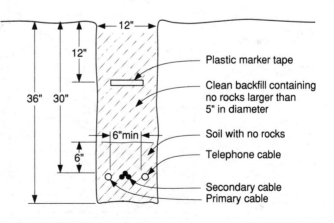

Plastic marker tape

Clean backfill containing no rocks larger than 5" in diameter

Soil with no rocks

Telephone cable

Secondary cable
Primary cable

RUNNING CABLE

With modern NM cable and plastic boxes,
running cable is a simple process. The illustration below shows the code guidelines. Note
the specification that cable must be at least
1-1/2 inches back from framing faces or be
protected by 18-gauge steel plates.

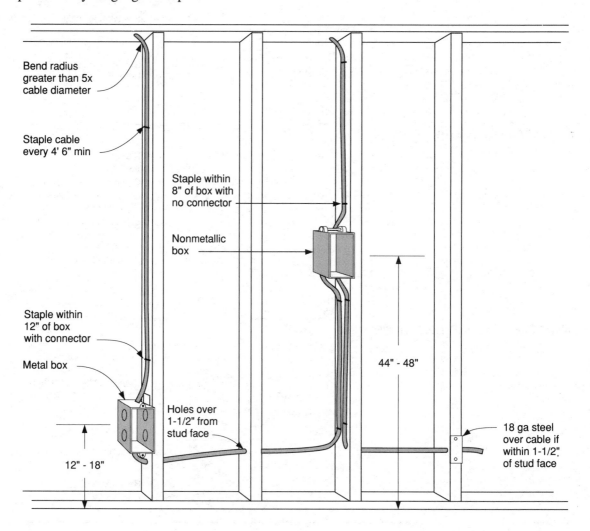

Bend radius
greater than 5x
cable diameter

Staple cable
every 4' 6" min

Staple within
8" of box with
no connector

Nonmetallic
box

Staple within
12" of box
with connector

Metal box

Holes over
1-1/2" from
stud face

44" - 48"

18 ga steel
over cable if
within 1-1/2"
of stud face

12" - 18"

RUNNING CONDUIT

Some codes require the use of rigid conduit throughout a building; others require conduit only for certain applications, such as wiring on masonry walls.

The three grades (thicknesses) of rigid metal conduit are

EMT – thin-wall metal conduit
IMC – intermediate metal conduit
RMC – rigid metal conduit

EMT is the easiest to use and the most common but may not be acceptable by local code. All conduit comes in straight 10-foot lengths and must be connected using only fittings appropriate to the application. Fittings (see illustration) are designed for either dry inside locations or wet outside locations. Outside fittings may be used either inside or outside.

Rules for Running Conduit

• Conduit is always installed, or run, before the wire it will contain. Wires are pulled through the conduit a section at a time, using a special, stiff pulling wire.

• A box or special pulling fitting must be provided for every 360° of accumulated bend in the conduit.

• Metal conduit is grounded, so no separate grounding wire is required; a white grounded-neutral wire *is* required, however.

• The size of the conduit is dictated by the number and type of wires it contains (see the tables at right).

Bending Conduit

Always use a conduit bender specifically designed for the size conduit being installed.

The illustration at the bottom of the following page shows the operation of bending conduit:

1. Measure the desired finished distance of the bend. This is usually the distance to a wall.

2. For 1/2-inch conduit, add 2-1/4 inches to the finished distance and mark the conduit. For other sizes see the marking on the conduit bender.

3. Line up the mark with the arrow on the bender, and pull back until the conduit makes a 90° angle with no pull on the bender handle.

CONDUIT WIRE CAPACITY

Type THW Wire							
Conduit Size, in	Wire Size						
	14	12	10	8	6	4	2
1/2	6	6	5	2			
3/4		6	6	4	2		
1			6	5	4	3	2
1-1/4				6	6	5	4
1-1/2					6	6	5
2						6	6

Type TW Wire							
Conduit Size, in	Wire Size						
	14	12	10	8	6	4	2
1/2	6	4	4				
3/4		6	6	3	2		
1			6	5	4	3	2
1-1/4				6	6	5	4
1-1/2					6	6	5
2						6	6

Bending Conduit

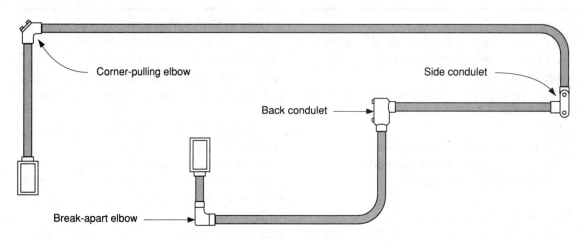

Corner-pulling elbow

Side condulet

Back condulet

Break-apart elbow

BENDING CONDUIT

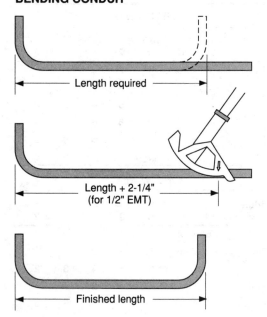

Length required

Length + 2-1/4"
(for 1/2" EMT)

Finished length

MAKING A KICK-BEND (OFFSET)

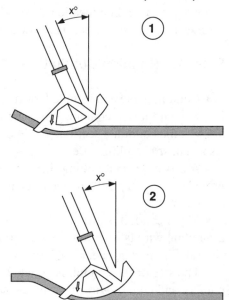

x°

①

x°

②

Conduit Fittings

INSIDE (DRY) USE ONLY

Coupling

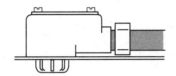

Set-screw connector

Offset connector

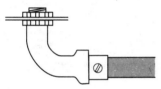

90° bushed elbow

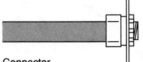

LB condulet

Pulling elbow

OUTSIDE (WET) USE

Coupling

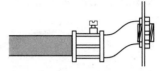

Connector

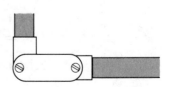

Offset connector

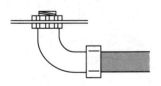

90° connector

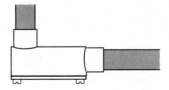

LB condulet

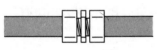

LL or LR condulet

WIRING SWITCHES, RECEPTACLES, AND LIGHTS

What to Do with the Grounding Wire

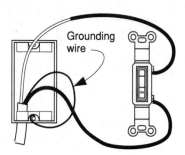

PLASTIC BOX, END OF CIRCUIT

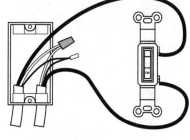

PLASTIC BOX, MIDCIRCUIT

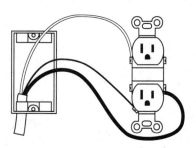

PLASTIC BOX, END OF CIRCUIT

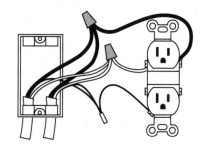

PLASTIC BOX, MIDCIRCUIT

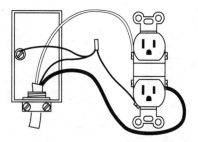

METAL BOX, END OF CIRCUIT

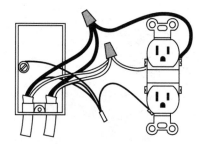

METAL BOX, MIDCIRCUIT

Receptacles

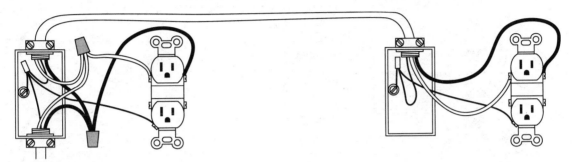

RECEPTACLES AT END OF CIRCUIT

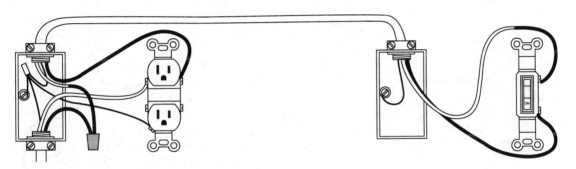

SWITCH-CONTROLLED RECEPTACLE

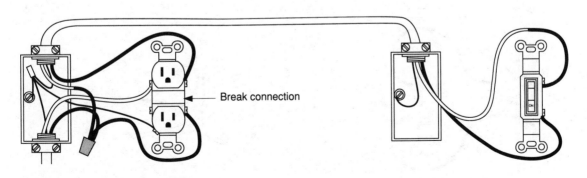

Break connection

SPLIT-CIRCUIT RECEPTACLE

Lights

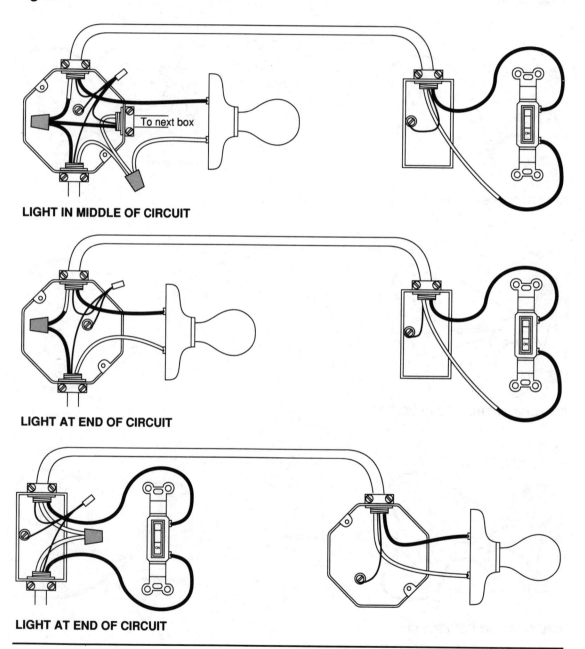

LIGHT IN MIDDLE OF CIRCUIT

To next box

LIGHT AT END OF CIRCUIT

LIGHT AT END OF CIRCUIT

Lights on Three-Way and Four-Way Switches

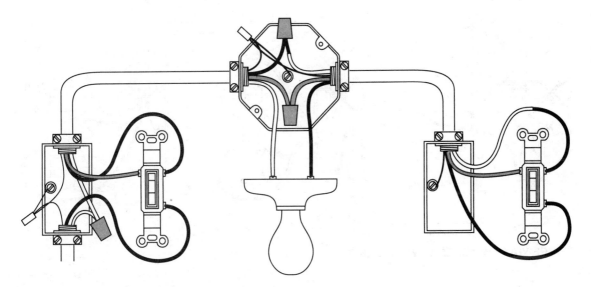

LIGHT BETWEEN THREE-WAY SWITCHES

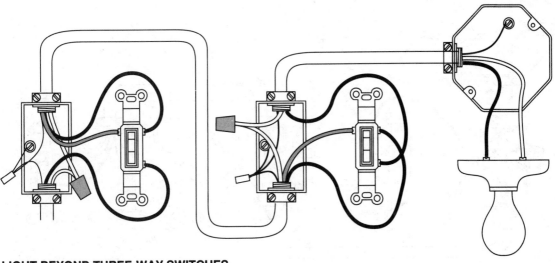

LIGHT BEYOND THREE-WAY SWITCHES

(continued)

Lights on Three-Way and Four-Way Switches—*Continued*

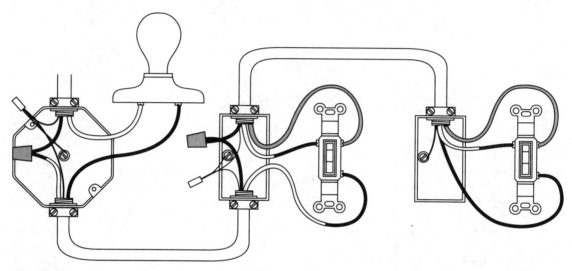

THREE-WAY SWITCHES BEYOND LIGHT

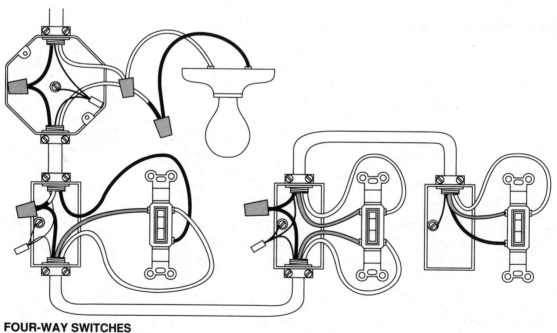

FOUR-WAY SWITCHES

CHECKLIST OF CODE REQUIREMENTS

The following is a partial list of requirements from the 1995 Council of American Building Officials (CABO) *One and Two Family Dwelling Code*. Consult the publication for the full text and additional provisions.

Service Drops
(4104.1) Clearance from building openings:
  • ≥ 3' from doors, porches, decks, balconies
  • ≥ 3' from sides/bottoms of opening windows
(4104.2.1) Clearance ≥ 8' above roof surface, including areas within 3' of edges
Exceptions:
  • ≥ 10' above pedestrian traffic
  • ≥ 3' above roofs sloped ≥ 4/12
  • ≥ 18" above roof overhang for ≤ 4'
  • final span to service drop at building
(4104.2.2) Clearance above grade:
  • ≥ 10' above pedestrian traffic areas
  • ≥ 12' above driveways and grounds
  • ≥ 18' above roads subject to truck traffic

Service Disconnects
(4101.6.2) To be installed in accessible location either outside or inside at nearest point of entrance of service conductors

Equipment
(3904.11) Each disconnecting means and each branch circuit to be labeled at its point of origin to indicate its purpose, except if obvious
(3905.1) Working space and clearances:
  • ≥ 30" wide, ≥ 36" deep, and ≥ 78" high
  • space not to be used for storage
  • space to be lighted if inside
  • space not to be in bathroom or clothes closet

Receptacle Outlets
(4401.2) Convenience outlets:
  • in every habitable room
  • no point along wall > 6' from outlet
  • includes every wall > 2' wide
  • includes freestanding counters and dividers
(4401.4) Small appliance outlets:
  • two 20-amp small appliance circuits
  • in kitchen, pantry, dining room, etc.
  • include refrigerator and freezer outlets
(4401.5) Countertop receptacles:
  • in kitchen and dining areas
  • for counters ≥ 12" wide
  • no point on counter > 24" from outlet
  • face-up outlets prohibited
(4401.5.1) Islands and peninsulas:
  • outlet needed if ≥ 12" wide and ≥ 24" long
  • no point on counter > 24" from outlet
  • above or within 12" below counter surface
(4401.6) Appliance outlets for specific appliances within 6' of appliance
(4401.7) Bathrooms and toilet rooms:
  • at least one outlet adjacent to lavatory
  • one outlet between two lavatories permissible
(4401.8) One outlet at grade level both front and back of dwelling
(4401.9) One outlet in laundry
(4401.10) One outlet in addition to laundry in:
  • basement
  • attached garage (detached with power)
(4401.11) One outlet in any hallway ≥ 10' long
(4401.12) One outlet for servicing HVAC equipment in attic or crawl space:
  • not on same circuit as HVAC equipment
  • ≤ 25' from equipment and on same level

Ground Fault Circuit Interrupter (GFCI)

(4402) GFCI protection to be provided for receptacles in:
- bathrooms
- garages, except if dedicated to appliances
- outdoors with direct grade access
- crawl spaces
- unfinished basements, except those dedicated to appliances and sump pumps
- kitchens within 6' of sink

Lighting Outlets

(4403.2) At least one wall switch-controlled lighting outlet in every habitable room, bathroom, hallway, stairway, attached garage, and detached garage with power, and at egress doors

Fixtures

(4504.8) Recessed fixture clearance:
- spaced ≥1/2" from combustibles, except fixtures listed for contact with insulation

(4504.9) Recessed fixture installation:
- spaced ≥ 3" from thermal insulation, except fixtures listed for contact with   insulation

(4503.11) Fixtures in clothes closets:
- limited to surface-mounted or recessed fixtures with completely enclosed lamps or fluorescent fixtures
- clearance between fixtures and storage space shall be ≥ 6" for fluorescent and recessed incandescent fixtures and ≥ 12" for surface-mounted incandescent fixtures

Nonmetallic (NM) Cable

(4302.1) Installation and support requirements:
- 1-1/4" from edge when parallel to framing
- holes in vertical framing recessed 1-1/4"

or protected by steel plate
- holes in horizontal framing recessed 2"
- metal framing holes bushed
- box openings bushed or have equivalent protection
- maximum on-center cable support 4'6"
- cable support ≤ 12" from box with clamp
- cable support ≤ 8" from box with no clamp

Thin-Wall Metal Conduit (EMT)

(4302.1) Installation and support requirements:
- holes in horizontal framing recessed 2"
- ends of tubing reamed to remove burrs
- box openings bushed or have equivalent protection
- maximum on-center tubing support 10'
- tubing support ≤ 36" from box
- four 90-degree turns maximum between junction boxes

Armored Cable (BX)

(4302.1) Installation and support requirements:
- 1-1/4" from edge when parallel to framing
- holes in vertical framing recessed 1-1/4" or protected by steel plate
- holes in horizontal framing recessed 2"
- box openings bushed or have equivalent protection
- maximum on-center cable support 4'6" (not required between ceiling lights spaced ≤ 6')
- cable support ≤ 12" from box (≤ 24" from box where flexibility is required)

13 Insulation

Sealing your house against both winter heat loss and summer heat gain requires the creation of what building scientists term *the thermal envelope.* Just as a knowledge of wood or fiberglass is required in order to construct a watertight boat, so an *understanding of R-values* is required to construct an energy-efficient home.

Walls, floors, and ceilings consist of building materials, air spaces, and insulations. So you'll find tables of *R-values of building materials, R-values of surfaces and trapped air spaces,* and *R-values of insulation materials.*

R-value, however, is not the only important quality of an insulation. Some insulations can be used only in walls and attics; some can be used under roofing shingles; and some can be buried between the foundation wall and the earth. A table of the *characteristics of insulation materials* will help you select the right one for every application. Another section will quantify the advantages of *compressing fiberglass* insulation.

You'll learn how to calculate the actual R-value of a construction. But to save you some calculation, this chapter includes an extensive illustrated table of the *R-values of walls, roofs, and floors.*

Of course, insulation is not the only component of a tight house. You will learn about the recent discoveries regarding *air/vapor barriers:* why and where they are needed and how they work.

A heat leak hit list guides you through your home, looking for 39 air leaks you probably never knew existed. But telling you where to look is not enough. You will find 42 clear illustrations of how to apply and install *caulks and weather strips,* plus tables detailing which type to use in every situation.

THE THERMAL ENVELOPE

The *thermal envelope* is the set of building surfaces separating the conditioned (heated and cooled) volume of a building from the outdoors and ground. It includes floors, above grade walls, foundation walls, ceilings (or roofs), windows, and doors.

To be effective, the thermal envelope must meet these conditions:

• All surfaces of the envelope must be insulated to an appropriate R-value (measure of thermal resistance) for the climate.

• The entire envelope must be sealed on the warm side by an effective air/vapor barrier.

• Doors and windows must be weather-stripped against infiltration.

This chapter describes constructing thermal envelopes in a heating climate. Cooling-climate techniques are the same, with the exception that air/vapor barriers should be on the outside rather than the inside of building surfaces.

Thermal Envelope

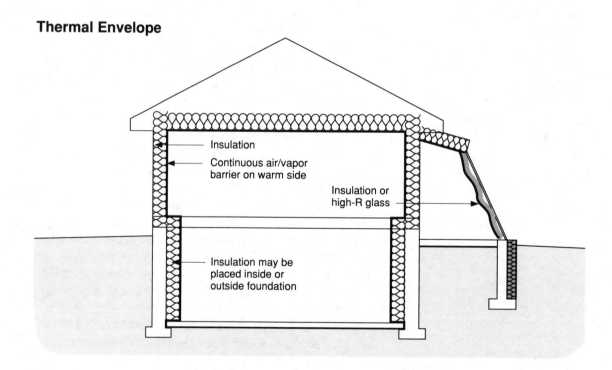

Insulation

Continuous air/vapor barrier on warm side

Insulation or high-R glass

Insulation may be placed inside or outside foundation

UNDERSTANDING R-VALUES

The Conduction Equation

Heat flow (either loss or gain) through a building surface is described by the conduction equation (see the illustration at right)

$$H = A \times \Delta T / R$$

where: H = British thermal units (Btu) per hour through the surface
A = area of surface in square feet
ΔT = difference in temperature in F°
R = R-value of the surface

HEAT CONDUCTION

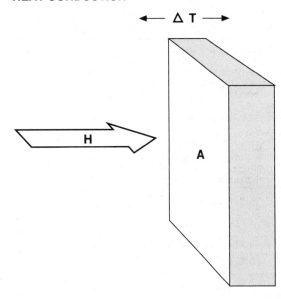

Adding R-values

Most building surfaces consist of numerous parts, each with differing thickness and R-value. The illustration below shows how R-values add up through the construction. Component R-values are from the tables and charts later in this chapter.

ADDING THE R-VALUES OF A WALL

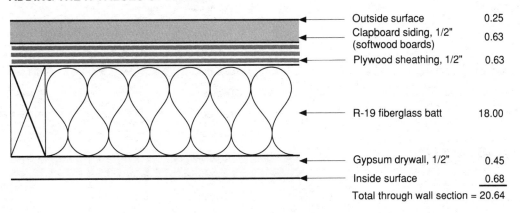

Outside surface	0.25
Clapboard siding, 1/2" (softwood boards)	0.63
Plywood sheathing, 1/2"	0.63
R-19 fiberglass batt	18.00
Gypsum drywall, 1/2"	0.45
Inside surface	0.68
Total through wall section =	20.64

R-VALUES OF BUILDING MATERIALS

Building Material	R-Value per Inch	R-Value for Unit
Boards and Panels		
Gypsum drywall	0.90	
Hardboard		
Medium density	1.37	
High-density underlay	1.22	
High-density tempered	1.00	
Laminated paperboard	2.00	
Particleboard		
Low density	1.41	
Medium density	1.06	
High density	0.85	
Underlayment	1.31	
Plywood, fir	1.25	
Vegetable fiber sheathing	2.64	
Wood		
Hardwood	0.90	
Softwood	1.25	
Flooring		
Carpet with fibrous pad		2.08
Carpet with rubber pad		1.23
Cork tile, 1/8"		0.28
Terrazzo		0.08
Tile - linoleum, vinyl, 1/8"		0.05
Wood		
Hardwood	0.90	
Softwood	1.25	

Building Material	R-Value per Inch	R-Value for Unit
Masonry		
Cement mortar	0.20	
Gypsum fiber concrete	0.60	
Sand and gravel or stone		
aggregate	0.09	
Lightweight aggregate		
120 lb/cu ft	0.18–0.09	
80 lb/cu ft	0.40–0.29	
40 lb/cu ft	1.08–0.90	
20 lb/cu ft	1.43	
Perlite, expanded	1.08	
Stucco	0.20	
Masonry Units		
Brick		
90 lb/cu ft	0.20	
120 lb/cu ft	0.11	
Clay tile, hollow		
One-cell, 3"		0.80
One-cell, 4"		1.11
Two-cell, 6"		1.52
Two-cell, 8"		1.85
Two-cell, 10"		2.22
Three-cell, 12"		2.50

Building Material	R-Value per Inch	R-Value for Unit	Building Material	R-Value per Inch	R-Value for Unit
Concrete block, normal weight			Sand aggregate	0.18	
8" empty cores		1.11–0.97	Vermiculite aggregate	0.59	
8" perlite cores		2.0	**Roofing**		
8" vermiculite cores		1.92–1.37	Asbestos-cement shingle		0.21
12" empty cores		1.23	Asphalt roll (90 lb)		0.15
Concrete block, medium weight			Asphalt shingle		0.44
8" empty cores		1.71–1.28	Built-up asphalt, 3/8"		0.33
8" perlite cores		3.7–2.3	Slate, 1/2"		0.05
8" vermiculite cores		3.3	Wood shingles (not furred)		0.94
8" expanded polystyrene beads		3.2	**Siding**		
Concrete block, lightweight			Shingles		
6" empty cores		1.93–1.65	Asbestos-cement		0.21
6" perlite cores		4.2	Wood, 16" (7" exposure)		0.87
6" vermiculite cores		3.0	Wood		
8" empty cores		3.2–1.90	Drop 1" x 8"		0.79
8" perlite cores		6.8–4.4	Bevel 1/2" x 8"		0.81
8" vermiculite cores		5.3–3.9	Bevel 3/4" x 10"		1.05
8" expanded polystyrene beads		4.8	Aluminum or steel		
12" empty cores		2.6–2.3	Hollow		0.61
12" perlite cores		9.2–6.3	With 3/8" backer		1.82
12" vermiculite cores		5.8	With 3/8" backer and foil		2.96
Stone	0.08				
Metals	Negligible				
Plasters					
Cement, sand aggregate	0.20				
Gypsum					
Lightweight aggregate	0.64				
Perlite aggregate	0.67				

Source: *1989 Fundamentals Handbook* (Atlanta: American Society of Heating, Refrigerating, and Air-conditioning Engineers, 1989).
Note: Cell means core of hollow masonry unit.

R-VALUES OF SURFACES AND TRAPPED AIR SPACES

R-VALUES OF SURFACES

Surface	Heat Flow	Type of Surface	
		Nonreflective	Foil-Faced
Still Air			
Horizontal	Upward	0.61	1.32
45° slope	Upward	0.62	1.37
Vertical	Horizontal	0.68	1.70
45° slope	Down	0.76	2.22
Horizontal	Down	0.92	4.55
Moving Air, Any Position			
7.5 mph wind	Any	0.25	–
15 mph wind	Any	0.17	–

Source: *Handbook of Fundamentals* (Atlanta: American Society of Heating, Refrigerating, and Air-conditioning Engineers, 1977).

R-VALUES OF TRAPPED AIR SPACES

Orientation	Thickness, inches	Heat Flow	Season	Type of Surface	
				Nonreflective	One Foil-Faced
Horizontal	3/4	Upward	Winter	0.87	2.23
	3/4	Downward	Winter	1.02	3.55
	4	Upward	Winter	0.94	2.73
	4	Downward	Winter	1.23	8.94
45° slope	3/4	Upward	Winter	0.94	2.78
	3/4	Downward	Summer	0.84	3.24
	4	Upward	Winter	0.96	3.00
	4	Downward	Summer	0.90	4.36
Vertical	3/4	Outward	Winter	1.01	3.48
	3/4	Inward	Summer	0.84	3.28
	4	Outward	Winter	1.01	3.45
	4	Inward	Summer	0.91	3.44

Source: *Handbook of Fundamentals* (Atlanta: American Society of Heating, Refrigerating, and Air-conditioning Engineers, 1977).

R-VALUES OF INSULATION MATERIALS

Insulation Material	R-Value per Inch	R-Value for Unit
Blankets and Batts		
Fiberglass		
3-1/2"		11
3-5/8"		13
6-1/2"		19
7"		22
9"		30
13"		38
Rock wool		
3-1/2"		11
3-5/8"		13
6-1/2"		19
7"		22
Loose Fill		
Cellulose	2.8 - 3.7	
Fiberglass 0.7 lb/cu ft	2.2	
Fiberglass 2.0 lb/cu ft	4.0	
Perlite	2.8	
Rock wool	3.1	
Sawdust or shavings	2.2	
Vermiculite	2.2	
Wood fiber	3.3	
Rigid Board		
Molded polystyrene 0.9 lb/cu ft	3.6	
Molded polystyrene 1.6 lb/cu ft	4.4	
Extruded polystyrene 1.9 lb/cu ft	5.0	

Insulation Material	R-Value per Inch	R-Value for Unit
Expanded rubber 4.5 lb/cu ft	4.6	
Fiberglass 4 - 9 lb/cu ft	4.0 - 4.4	
Perlite	2.6	
Phenolic	8.0	
Urethane	5.6 - 6.3	
Isocyanurate	5.6 - 6.3	
Sprayed or Foamed Fill		
Cellulose	3.0 - 4.0	
Urethane	5.6 - 6.2	

Source: *Handbook of Fundamentals* (Atlanta: American Society of Heating, Refrigerating, and Air-conditioning Engineers, 1977).

CHARACTERISTICS OF INSULATION MATERIALS

Type of Insulation	Rated R/Inch (range)	Maximum Service Temp, °F	Use As Vapor Barrier?	Resistance to			
				Water Absorption	Moisture Damage	Direct Sunlight	Fire
Roll, Batt, or Blanket							
Fiberglass	3.0 - 3.8	180	unfaced: P	G	E	E	G
Rock wool	3.0 - 3.8	>500	unfaced: P	G	E	E	E
Loose Fill							
Fiberglass	2.2 - 4.0	180	P	G	E	E	G
Rock wool	2.8 - 3.7	>500	P	G	E	E	E
Cellulose	2.8 - 3.7	180	P	P	P	G	F
Perlite	2.5 - 4.0	200	P	F	G	E	G
Treated perlite	3.4	200	P	G	G	E	G
Rigid Board							
Expanded polystyrene	3.6 - 4.4	165	P	P	G	P	P
Extruded polystyrene	5.0	165	G	E	E	P	P
Phenolic	6.5 - 8.3	300	G	G	E	E	P
Urethane, isocyanurate	5.6 - 6.3	200	G	G	E	P	P
Fiberglass board	3.8 - 4.8	180	F	G	E	E	G
Sprayed in Place							
Cellulose	3.0 - 4.0	180	G	P	F	G	F
Foamed in Place							
Urethane	5.6 - 6.8	165	E	E	E	P	P

Note: E = excellent; G = good; F = fair; P = poor.
Source: *Rodale Product Testing Report - Insulation Materials* (Emmaus, Pa: Rodale Press, 1982).

COMPRESSING FIBERGLASS

The R-value per inch of fiberglass insulation increases with density. Therefore, when a fiberglass batt is compressed into a smaller cavity, the total R-value of the batt decreases, but the R-value per inch increases, making the oversize batt a better performer than the batt designated for the space.

The chart below shows the actual R-values of nominal R-11, R-19, R-30, and R-38 batts when squeezed into various wall cavity thicknesses. For example, an R-19 batt (actually 6 inches thick and R-3.17/inch) rates R-18 and R-3.27/inch in a 2x6 (5-1/2-inch-thick) wall, and R-13 and R-3.71/inch in a 2x4 (3-1/2-inch-thick) wall.

R-Values of Compressed Batts

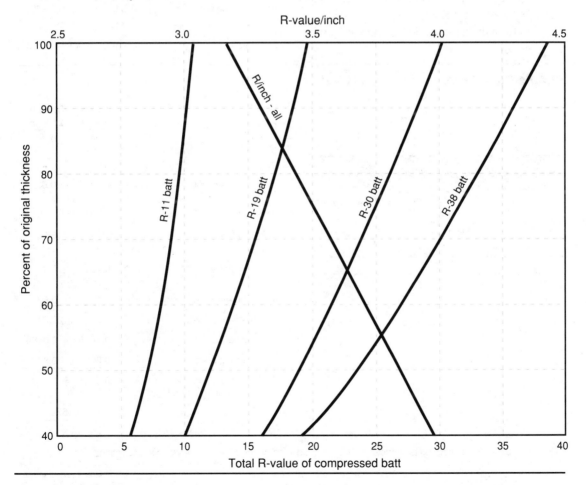

R-VALUES OF WALLS, ROOFS, AND FLOORS

In the pages that follow, a variety of walls, roofs, ceilings, and floors are shown, along with their computed average R-values. Note that the average R-values of the assemblies are often less than the R-values of the insulations alone. This is due to the greater conductivity of the framing. For this reason, staggered framing is sometimes used.

The average R-values listed were computed as shown in the example below, using component R-values from the tables and charts in the previous pages.

Wherever foam is shown, an average R-value of 6.0/inch is assumed. Correction factors for other R-values are listed in the table below.

CORRECTION FACTORS FOR RIGID FOAM INSULATIONS

Type of Foam	R/Inch	Correction Factor/Inch
Expanded polystyrene	4.0	-2.0
Extruded polystyrene	5.0	-1.0
Urethane/isocyanurate	6.0	0.0
Phenolic	8.0	+2.0

Sample Calculation of Average R-Value

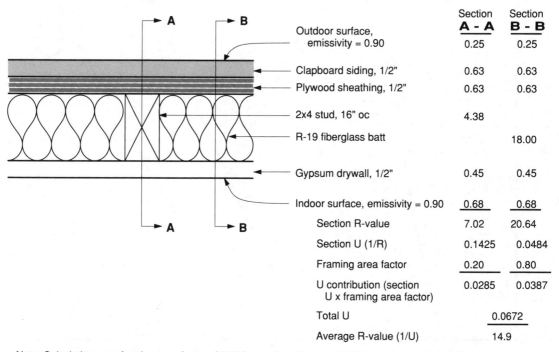

	Section A - A	Section B - B
Outdoor surface, emissivity = 0.90	0.25	0.25
Clapboard siding, 1/2"	0.63	0.63
Plywood sheathing, 1/2"	0.63	0.63
2x4 stud, 16" oc	4.38	
R-19 fiberglass batt		18.00
Gypsum drywall, 1/2"	0.45	0.45
Indoor surface, emissivity = 0.90	0.68	0.68
Section R-value	7.02	20.64
Section U (1/R)	0.1425	0.0484
Framing area factor	0.20	0.80
U contribution (section U x framing area factor)	0.0285	0.0387
Total U		0.0672
Average R-value (1/U)		14.9

Note: Calculation uses framing area factor of 0.20 for stud sections and 0.80 for between stud sections.

Frame Walls - Interior Insulation

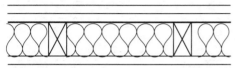

2x4 studs / R-11 batt   **R-11.6**

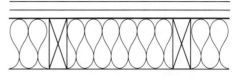

2x6 studs / R-19 batt   **R-17.2**

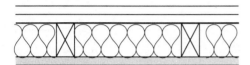

Blown cavity plus 1" foam   **R-19.9**

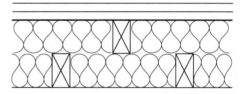

Double 2x4 studs / R-11 batts   **R-20.7**

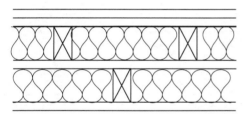

Blown cavity plus R-11 batt   **R-22.8**

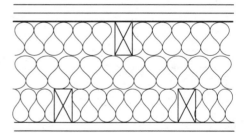

2x4 studs / R-11 batts   **R-31.7**

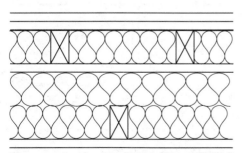

Blown cavity plus two R-11 batts   **R-33.8**

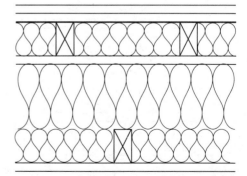

Blown cavity plus R-11 & 22 batts   **R-44.8**

Frame Walls - Exterior Insulation

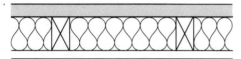

2x4 studs / R-11 batt   **R-15.6**
plus 3/4" foam sheathing

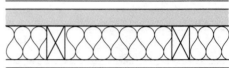

2x4 studs / R-11 batt   **R-17.1**
plus 1" foam sheathing

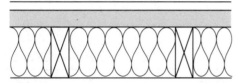

2x6 studs / R-11 batt   **R-22.6**
plus 1" foam sheathing

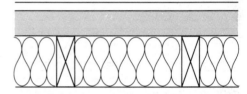

2x6 studs / R-11 batt   **R-25.6**
plus 1-1/2" foam sheathing

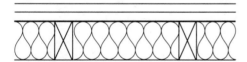

Full 4" cavity blown   **R-13.4**
with cellulose

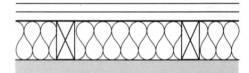

Full 4" cavity blown with   **R-19.4**
cellulose, plus 1" foam

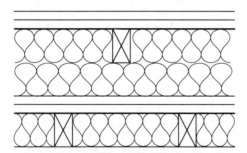

Full 4" cavity blown with cellulose   **R-34.8**
plus two R-11 batts

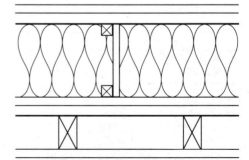

Unfilled wall cavity, plus 11"   **R-34.8**
plywood truss with R-38 batt

Masonry Walls

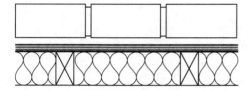

Brick veneer with 2x4
studs and R-11 batt   **R-12.1**

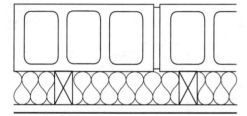

8" masonry block with   **R-12.8**
2x4 studs and R-11 batt

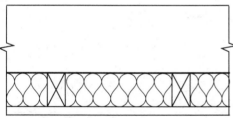

8" solid masonry with   **R-12.3**
2x4 studs and R-11 batt

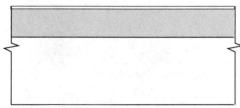

8" solid masonry with 4"   **R-18.9**
expanded polystyrene
foam and stucco finish

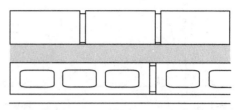

Brick, 2" foam, 4" block,   **R-15.3**
and strapping

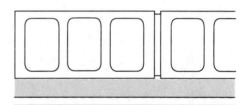

8" masonry block and 2" foam   **R-14.4**

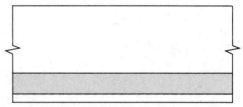

8" solid masonry and 3" foam   **R-21.4**

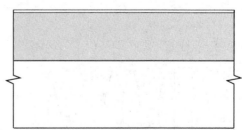

8" solid masonry with 6"   **R-26.9**
expanded polystyrene
foam and stucco finish

Attic Floors

R-11 batt **R-9.8**

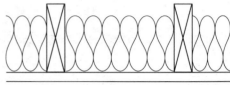

R-19 batt **R-16.5**

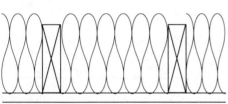

R-30 batt **R-25.4**

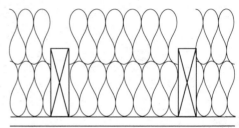

Two R-19 batts **R-35.5**

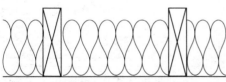

6" cellulose **R-16.5**

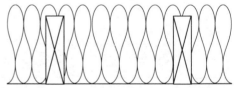

9" cellulose **R-26.1**

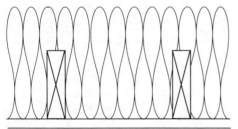

12" cellulose **R-35.7**

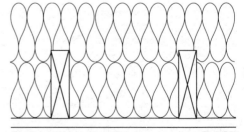

R-19 batt plus 6" cellulose **R-35.7**

Roofs and Ceilings

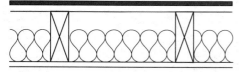

Vented ceiling with R-11 batt   **R-9.8**

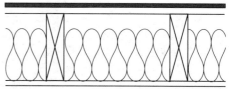

Vented ceiling with R-19 batt   **R-16.5**

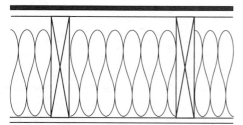

Vented ceiling with R-30 batt   **R-25.4**

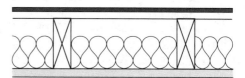

Vented ceiling with R-11
batt and 3/4" foam   **R-14.3**

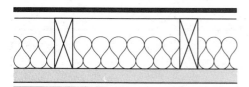

Vented ceiling with R-11
batt and 1-1/2" foam   **R-18.8**

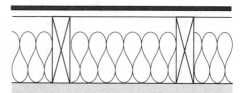

Vented ceiling with R-19
batt and 1-1/2" foam   **R-25.5**

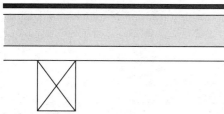

Cathedral ceiling with 1-1/2"
roofers and 3-1/2" foam   **R-25.3**

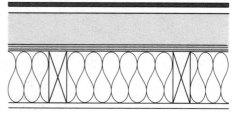

Unvented roof with R-19
batt and 3-1/2" foam   **R-17.2**

Floors

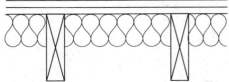

R-11 batt over crawl   **R-13.9**
or basement

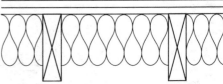

R-19 batt over crawl   **R-20.9**
or basement

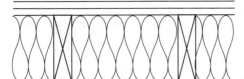

R-30 batt over crawl   **R-30.6**
or basement

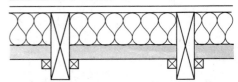

Open to air beneath   **R-21.8**
R-11 batt and 1-1/2" foam

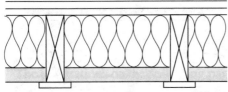

Open to air beneath   **R-28.8**
R-19 batt and 1-1/2" foam

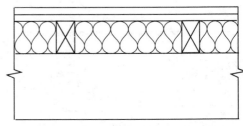

2x4 sleepers and   **adds R-10.4**
R-11 batt over slab   **to R of slab**

3/4" sleepers and   **adds R-3.7**
3/4" foam over slab   **to R of slab**

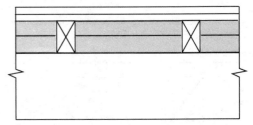

2x4 sleepers and   **adds R-14.2**
3-1/2" foam over slab   **to R of slab**

AIR/VAPOR BARRIERS

Humidity and Condensation

Unfortunately, stopping infiltration is not as simple as wrapping a building in plastic. Such an approach might lead to moisture problems within the building and its materials. To understand why, we need to understand the behavior of moisture in air.

The illustration below (a form of the psychrometric chart) describes the behavior of the gaseous form of water, *water vapor,* in ordinary air.

The horizontal scale is air temperature; the vertical scale is humidity, expressed as pounds of water vapor in a pound of air. The moisture condition of air can always be described by a single point on the chart. Point A, for example, represents air at 70°F and a certain moisture content.

One of the properties of a mixture of air and water vapor is saturation: Like a sponge, air can hold only a limited amount of water vapor before becoming saturated. The warmer the air,

Psychrometric Chart

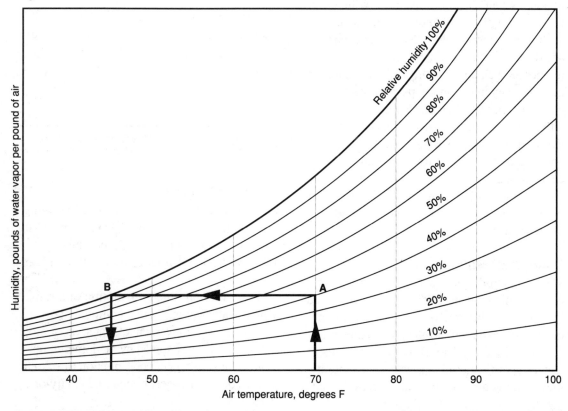

the more water vapor it can hold. This phenomenon is demonstrated by the chart's uppermost curve, known as the *saturation curve*. Air at any point on the saturation curve contains 100 percent of the possible water vapor for that temperature. Air anywhere else (such as at point A) contains a percentage of the saturation amount. This percentage is known as the *relative humidity* of the air. Thus, the air at point A is characterized as having a temperature of 70°F and relative humidity of 40 percent.

The air at point A is typical of the air in a building in winter. To illustrate the potentially damaging behavior of moist air in winter, let the air at point A flow into a cooler space such as an unheated attic. No water vapor is added or removed as the air flows; only the temperature changes as the air loses heat to its new surroundings. On the psychrometric chart, this parcel of air simply moves horizontally to the left until it strikes the saturation curve (its *dew point*) at point B. As the air cools further, it slides down the saturation curve, forcing water vapor out as liquid (water) or solid (ice), which is typically deposited on the coldest solid surface, the underside of the roof sheathing.

Sources of Moisture

The air in buildings in winter is often said to be *dry,* which means it has low relative humidity. Actually, interior winter air usually contains more water vapor than the outside air.

The table (above right) lists the amounts of water vapor (in liquid equivalents) evaporated into building air from typical sources, including the activities of a family of four.

SOURCES OF WATER VAPOR

Source	Quarts per Day
Construction materials first year	40
Standing water in basement	30
Damp basement or crawl space	25
Greenhouse connected to house	25
Humidifier - large	20
Drying 1 cord of firewood	16
Clothes dryer vented to inside	13
Respiration/perspiration - 4 people	4.7
Clothes washing	2.1
Unvented gas range	1.3
Cooking without lids	1.0
Houseplants - average number	0.5
Dish washing	0.5
Floor mopping	0.4
Showering/bathing	0.3

Source: *Walls, Windows and Roofs for the Canadian Climate* (Ottawa: National Research Council of Canada, 1973).

Moisture Transfer

As shown in the illustration on the following page, water vapor can move from warm living spaces to cooler spaces (wall, ceiling, and floor cavities and attics) in two different ways:

1. *convection* - the bulk motion of air, of which water vapor is but a component

2. *diffusion* - movement of water vapor molecules alone as they spread by random motion from an area of higher concentration (inside) to an area of lower concentration (the cavity or outdoors)

Permeance of Building Materials

Material placed on the warm side of a building surface to retard diffusion of water vapor is called a *vapor barrier*. Material intended to retard convection is called an *air barrier*. Material which accomplishes both is termed an air/vapor barrier. A material qualifies as a vapor barrier if its permeance is 1.0 perm or less.

Material	Permeance, perm
Building Materials	
Brick, 4"	0.8 - 1.1
Concrete, 8"	0.4
Gypsum drywall, 1/2"	38
Plaster (unpainted), 3/4"	15 - 20
Plywood, interior glue, 1/2"	1.0
Plywood, exterior glue, 1/2"	0.4
Softwood, 3/4"	2.9
Building Papers (air barriers)	
15 lb asphalt-impregnated felt	5.6
Spun-bonded plastic fiber	25 - 94
Insulation	
Loose fill, blanket and batt	>50
Expanded polystyrene	2.0 - 5.8
Extruded polystyrene	1.2
Paint	
Oil or alkyd, 3 coats	0.3 - 1.0
Latex VB paint, 2 coats	0.3 - 0.6
Films	
Aluminum foil, 1 mil (0.001")	0.0
Polyethylene, 4 mil	0.08

Source: *Handbook of Fundamentals* (Atlanta: American Society of Heating, Refrigerating and Air-conditioning Engineers, 1977).

Transport of Water Vapor

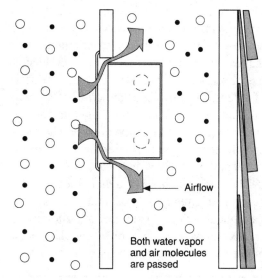

Airflow

Both water vapor and air molecules are passed

CONVECTION THROUGH HOLES

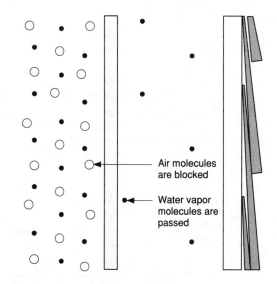

Air molecules are blocked

Water vapor molecules are passed

DIFFUSION THROUGH SURFACE

The Airtight Drywall Approach

The time-honored approach to controlling moisture condensation in building cavities has been the installation of a vapor barrier on the warm side. Vapor diffusion theory indicates that a low-perm barrier such as polyethylene on the warm side, coupled with an outside surface of at least five times greater permeability, will limit diffusion into the cavity and guarantee escape of that moisture to the outside. The method calls for lapping of barrier joints at framing and repair of all holes and tears.

This approach is still taught in most building books and required by most building inspectors, but an increasing number of building scientists believe that vapor diffusion plays but a minor role in moisture problems. They believe that the winter buoyancy (stack effect) of warm inside air is so powerful that air continually invades the building basement and lower walls and escapes upward and outward through ceiling and upper-wall openings, carrying 10 to 100 times as much moisture as the diffusion process alone. The air-barrier approach (also descriptively termed the airtight drywall approach, or ADA) stops airflow into and out of the building at the warm surfaces through careful detailing and gasketing at top and bottom of drywall panels and framing joints. A polyethylene membrane may also be used to block airflow, but latex vapor-barrier paint is commonly employed to retard vapor diffusion. The illustration below shows a typical ADA gasketing detail at a floor/exterior wall intersection.

ADA Gasketing

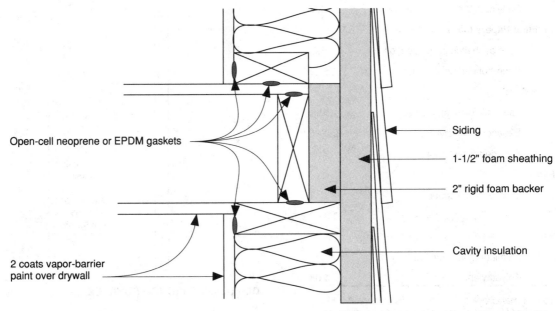

Open-cell neoprene or EPDM gaskets

2 coats vapor-barrier paint over drywall

Siding

1-1/2" foam sheathing

2" rigid foam backer

Cavity insulation

CAULKS AND WEATHER STRIPS

Airflow in Winter

New houses are, or should be, built with careful attention to air/vapor barriers - continuous barriers designed to stop air leakage between the building interior and building cavities and/or the outside.

Most houses built prior to 1980, however, are replete with unintentional cracks and holes. In fact, all of the seemingly miniscule openings in the average pre-1980 dwelling add up to an open area of approximately 2 square feet. Because of these leaks, infiltration (intrusion of cold outside air) accounts for between 25 and 50 percent of winter heating bills.

Although infiltration is most noticeable when cold winter wind pushes air around doors and windows, the major fraction of leakage occurs steadily and almost without notice as warm buoyant inside air finds its way out of the top of the building, to be replaced by cold outside air flowing in at lower levels (illustration at right).

A Heat Leak Hit List

The leaks in our "hot-air balloons" are not obvious until we look for them. The illustration and accompanying table on the following pages catalog 39 separate air/heat leaks in the normal residential building shell, along with their typical leakage areas in square inches.

Whether designing, building new, or weatherizing an older home, the illustration and table serve as a useful heat leak audit, pointing to the possible heat leaks and their relative importance.

Winter Airflow

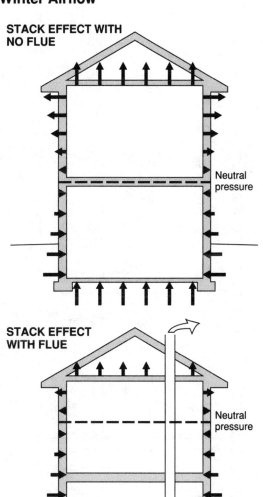

STACK EFFECT WITH NO FLUE

Neutral pressure

STACK EFFECT WITH FLUE

Neutral pressure

Heat Leaks in the Home

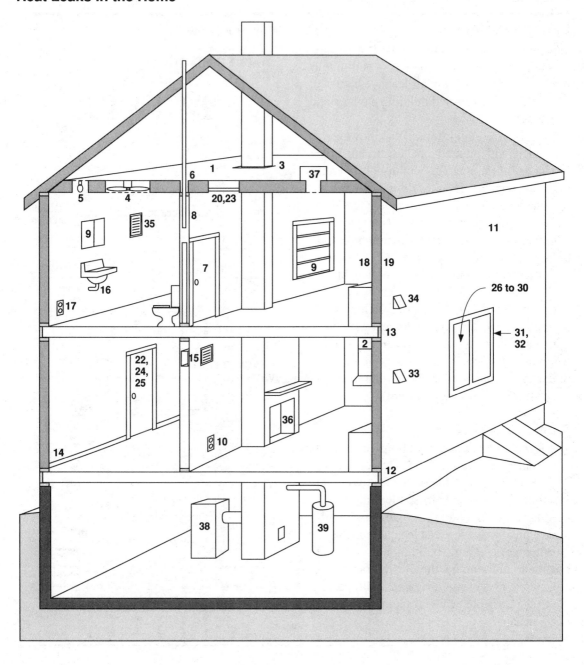

AIR LEAKAGE IN BUILDINGS

Heat Leak	Area, square inches
Ceiling	
1. General per 100 sq ft	0.05
2. Dropped ceiling, per 100 sq ft	
No plastic vapor barrier	78
With plastic vapor barrier	8
3. Chimney framing	12
Sealed	1
4. Whole-house fan, closed louvers	8
Covered with tight box	0.6
5. Lighting fixtures	
Recessed	4
Surface	0.3
6. Pipe or duct	1
Caulked at ceiling	0.2
Interior Walls	
7. Pocket door	5
8. Pipe or duct in wall	2
9. Recessed cabinet	0.8
10. Electric outlet or switch	0.2
With cover plate gasket	0.03
Exterior Walls	
11. General per 100 sq ft	0.8
12. Sill on masonry	65
Caulked	13
13. Band or box sill	65
Caulked	13
14. Floor/wall joint	27
Baseboard caulked	7
15. Duct in wall	9
16. Pipe in wall	2
17. Electric outlet or switch	0.2
With cover plate gasket	0.05
18. Poly vapor barrier (deduct)	-30
19. Styrofoam sheathing (deduct)	-15
Doors	
20. Attic fold-down	17
Weather-stripped	8
Insulated cover	2
21. Patio sliding	16
22. Entrance	8
Weather-stripped	6
Magnetic seal	4

Heat Leak	Area, square inches
23. Attic hatch	6
Weather-stripped	3
24. Air-lock entry (deduct)	-4
25. Storm door (deduct)	-3
Windows (weather-stripped)	
26. Double-hung	0.8
27. Horizontal slider	0.6
28. Awning	0.2
29. Casement	0.2
30. Fixed	0.2
Door and Window Frames	
31. Masonry wall	2
Caulked	0.4
32. Wood wall	0.6
Caulked	0.1
Vents	
33. Range, damper open	9
Damper closed	2
34. Dryer, damper open	4
Damper closed	1
35. Bathroom, damper open	3
Damper closed	1
Fireplace	
36. Fireplace, damper open	54
Avg damper closed	9
Tight damper	5
Stove insert	2
Heat and Hot Water	
37. Ducts in unheated space	56
Caulked and taped	28
38. Furnace	
With retention head burner	12
With stack damper	12
With both of above	9
39. Gas/oil boiler or water heater	8

Source: *Cataloguing Air Leakage Components in Houses* (Princeton, New Jersey: Princeton University Center for Energy and Environmental Studies, 1984).

Caulks and Caulking

With caulks, the old adage "you get what you pay for" holds true. The table below describes the characteristics of common caulks.

The best-performing caulks are, unfortunately, either not commonly available at hardware stores or very difficult to apply.

Oil and vinyl latex are such poor performers that they are not recommended for permanent sealing applications.

For the average homeowner, that leaves acrylic latex, silicon acrylic latex, and silicone.

Silicone is the best for joints with very large movement. It cannot be painted, however, and doesn't adhere well to plastics, pressure-treated wood, or unprimed wood. Paintable silicones still don't hold paint as well as the latexes.

Silicon acrylic latex is the best compromise for the homeowner. It stretches well, holds paint, adheres to almost anything, and is easy to apply.

CHARACTERISTICS OF CAULKS

Type of Caulk	Relative Cost	Life years	Ease of Application	Drying Time to Skin	Cure	Adhe- sion	Max- imum Gap	Max- imum Stretch	Problem Notes
Oil	Low	1 - 3	G	1 day	Months	F	1/4"	1%	6, 7
Vinyl latex	Low	3 - 5	E	1/2 hour	1 week	F	1/4"	2%	6, 7, 8
Acrylic latex	Low	5 - 10	E	1/2 hour	1 week	G	1/4"	2%	6, 9
Silicon acrylic latex	Medium	10 - 20	G	2 hours	1 week	G	1/2"	10%	2
Butyl rubber	Medium	5 - 10	F	1 day	6 months	G	1/2"	5%	6
Polysulfide	Medium	20 +	P	1 day	1 month	E	3/4"	25%	4, 5, 6, 7
Polyurethane	High	20 +	P	1 day	1 month	E	3/4"	25%	
Silicone	High	20 +	G	1 hour	1 week	G	1"	50%	1, 2, 3, 7
Urethane foam	High	10 - 20	P	1/2 hour	1 day	G	2"	1%	4, 6

Notes: E = excellent; G = good; F = fair; P = poor.
1. Can't be painted.
2. Won't adhere well to plastics.
3. Won't adhere to treated lumber.
4. Application temperature over 60°F.
5. Not available at hardware stores.
6. Degrades in direct sunlight.
7. Requires priming on porous surfaces.
8. Interior use only.
9. Requires priming over metal.

PROFESSIONAL CAULK APPLICATION GUIDE

Application	Most Suitable Caulk or Sealant
Wood to wood	All acrylic sealants, provided gap is less than 3/8 inch
	Silicone sealants for gaps greater than 3/8 inch
Wood to masonry	Butyl rubber for gaps up to 1/2 inch
	Polysulfide for gaps up to 3/4 inch and for damp conditions
Wood to metal	All acrylics for gaps up to 3/8 inch
	Butyl rubber for gaps up to 1/2 inch
	Silicone sealant for gaps up to 1 inch and for great movement
	Urethane foam sealant for gaps greater than 1 inch, but not for large gap movement
Masonry to masonry	Butyl rubber for gaps up to 1/2 inch
	Polysulfide for gaps ranging from 3/8 to 3/4 inch
	Grouts and special water-stop fillers for gaps over 3/4 inch
	Polysulfide for large gap movement
Polyethylene	Acoustic sealant (not an adhesive; must be supported mechanically with staples or other support)
High temperatures	Silicone sealants (effective at fairly high temperatures, but their temperature limit should not be exceeded)
	Room-temperature-vulcanized (RTV) silicone sealant (suitable for high temperatures, but requires specialized application procedures)
	High-temperature stove sealants or muffler cement
High humidity	Most silicones (particularly if containing mildew retardants)

Source: *Air Sealing Homes for Energy Conservation* (Ottawa: Energy, Mines and Resources Canada, 1984).

Caulking and Sealing Technique

Successful sealing requires three steps:

1. Select the best caulk or sealant for the application (consult the table on the previous page).

2. Prepare the surfaces exactly as recommended by the sealant manufacturer. Note that brands can vary in formulation and surface preparation.

3. Apply the sealant with the proper techniques, as shown in the illustrations below.

Surprisingly, caulk joints fail more often from use of too much rather than too little sealant. For narrow joints, push rather than drag the tip. For wide joints, first install flexible backer rod, in order to limit the depth of the joint.

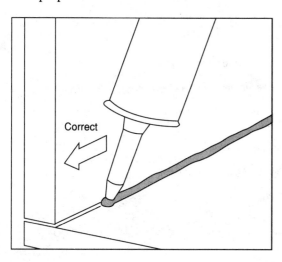

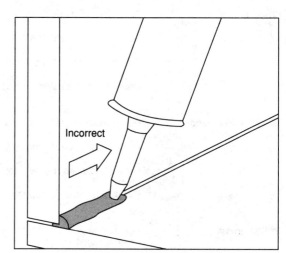

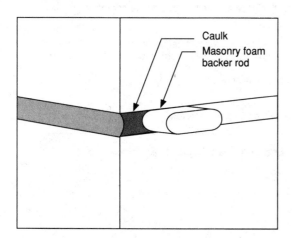

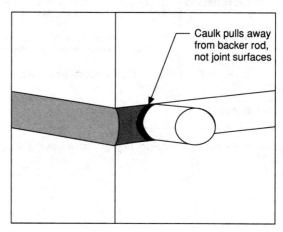

Sill and Baseboard

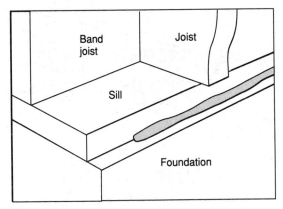

Band joist

Joist

Sill

Foundation

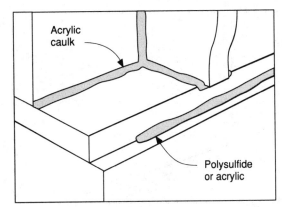

Acrylic caulk

Polysulfide or acrylic

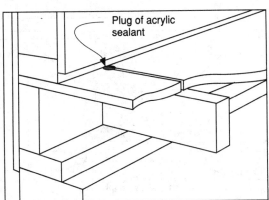

Plug of acrylic sealant

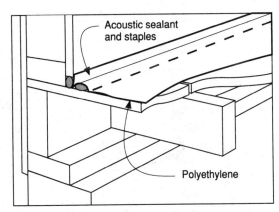

Acoustic sealant and staples

Polyethylene

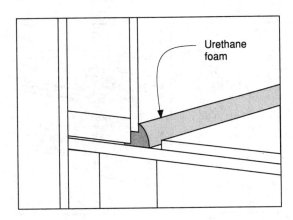

Urethane foam

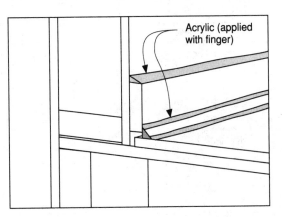

Acrylic (applied with finger)

Floor and Wall Penetrations

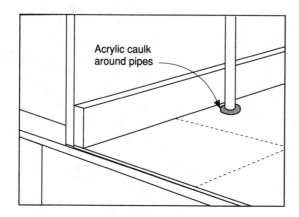

Acrylic caulk around pipes

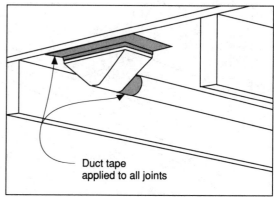

Duct tape applied to all joints

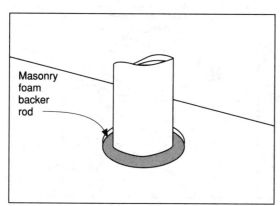

Masonry foam backer rod

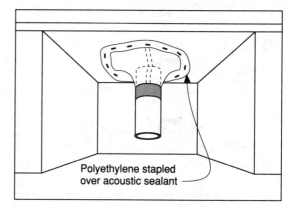

Polyethylene stapled over acoustic sealant

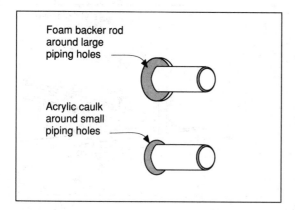

Foam backer rod around large piping holes

Acrylic caulk around small piping holes

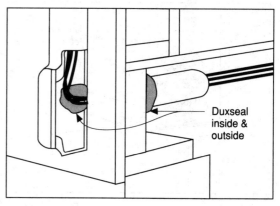

Duxseal inside & outside

Ceiling

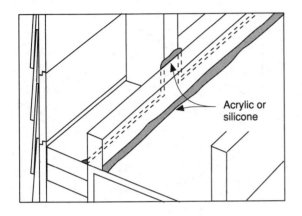

Acrylic or silicone

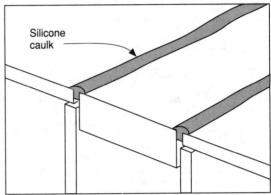

Silicone caulk

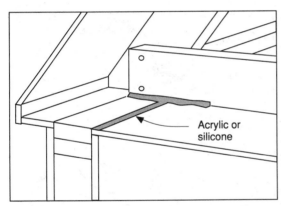

Acrylic or silicone

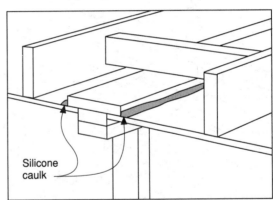

Silicone caulk

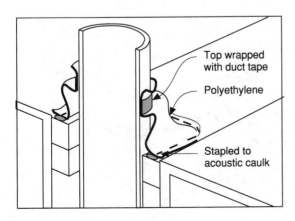

Top wrapped with duct tape

Polyethylene

Stapled to acoustic caulk

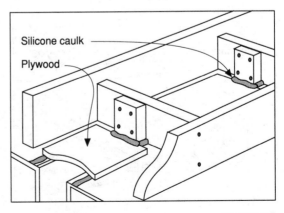

Silicone caulk

Plywood

(continued)

Ceiling—*Continued*

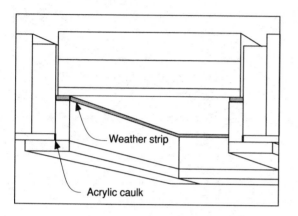

Weather strip

Acrylic caulk

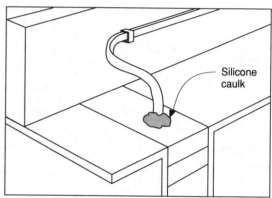

Silicone caulk

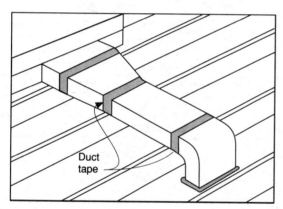

Duct tape

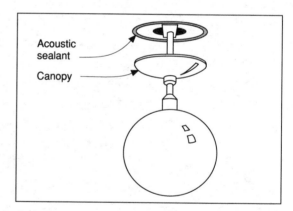

Acoustic sealant

Canopy

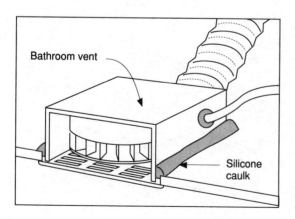

Bathroom vent

Silicone caulk

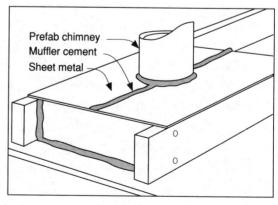

Prefab chimney
Muffler cement
Sheet metal

Weather Strips

While caulks are designed to seal joints having limited movement, weather strips are intended to seal operable joints such as those around doors and windows. The table below describes the best of the common household strips. Note that weather strips should *never be painted*.

THE BEST OF THE WEATHER STRIPS COMPARED

Type of Weather Strip		Relative Cost	Comments
	Closed-cell foam on attachment strip	Medium	Seals well on doors and windows; installed by nailing attachment strip to door or window jamb
			Operation of door or window may be hindered if installed too tight
			Not for sliding windows and doors
	Closed-cell rubber	Low	Effective when under compression (doors, hinged windows, clamped attic hatches, and so forth)
			Does not work with large gap variation
			Installation simple but relies on effective surface adhesion
	Hollow and foam-filled tube	Low to medium	Good if properly installed
			Does not require significant pressure to create seal
	Plastic V-shape	Low to medium	Effective and versatile for use throughout house
			Strips may be supplied with self-adhesive backing
			Protection against failure of adhesive backing supplied by stapling through back edge

Window Frame and Sash

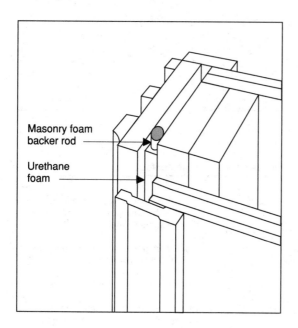

Masonry foam
backer rod

Urethane
foam

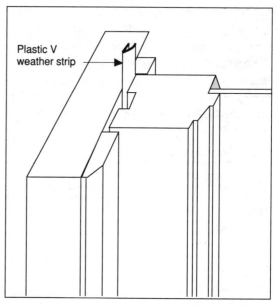

Plastic V
weather strip

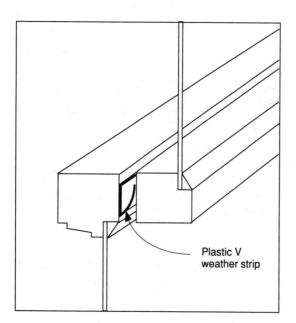

Plastic V
weather strip

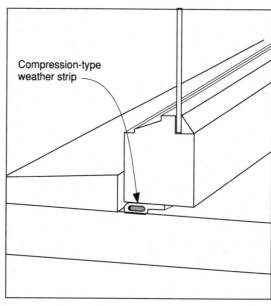

Compression-type
weather strip

Door Jambs

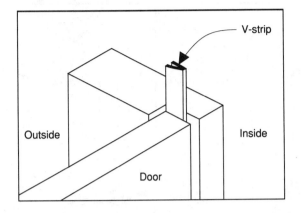

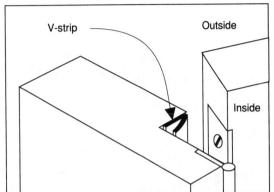

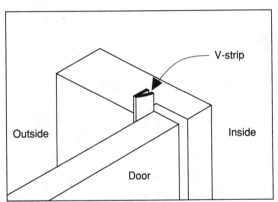

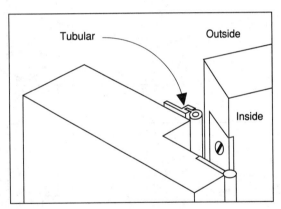

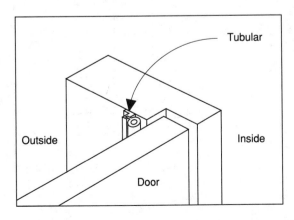

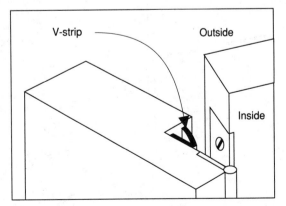

Door Sweeps and Thresholds

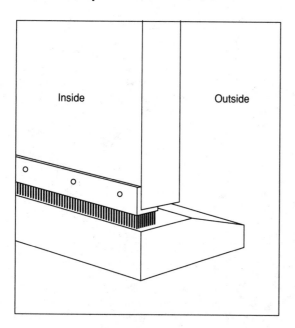

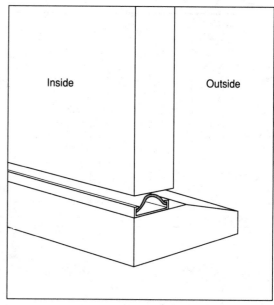

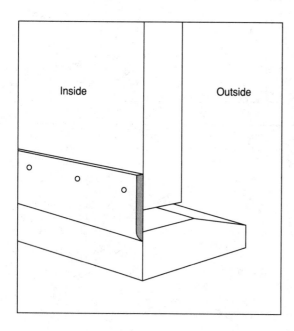

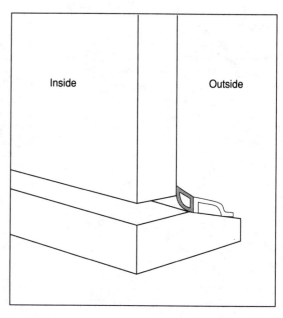

14

Floors, Walls, and Ceilings

Floors, walls, and ceilings are the surfaces we live with. It's important to select the best material and to install it properly to achieve long-term performance.

This chapter starts at the bottom, with *carpets and resilient floors.* It shows you the differences between carpet materials and weaves, and which are best for which room. Quality homes have always contained an abundance of *ceramic tile.* All of the standard shapes and sizes of mosaic, quarry, paver, and wall tiles are illustrated. A table also lists and explains the use of the four types of tile-setting adhesives and mortars.

Hardwood floors also are hallmarks of quality homes. Today you can buy unfinished or prefinished hardwood flooring in three styles: strip, plank, and parquet. This section shows you interesting parquet patterns, how to estimate coverage, and how to install wood flooring over concrete or wood subfloors.

Gypsum wallboard is one of the most popular of all home-owner projects. In this section you'll find the variety of wall-boards, how to estimate all of the materials needed to finish a room, and techniques from a professional manual.

A section on solid *wood paneling* shows the patterns commonly available, how to estimate quantities, and installation.

Suspended ceilings are inexpensive solutions for hiding basement pipes and wires or for lowering high ceilings. Step-by-step procedures for installing this engineered system are provided.

Most people blame the manufacturer when *paints and other finishes* fail. But more often, the problem is that the wrong finish was used. This section will help you avoid that mistake and find out what went wrong the first time.

Finishing touches include simple ways to install *shelving* and the complete collection of standard *wood moldings* available through your lumber dealer.

CARPET AND RESILIENT FLOORS

Carpet

Most wall-to-wall carpeting is produced by looping yarns through a coarse-fiber backing, binding the backs of the loops with latex, then applying a second backing for strength and dimensional stability. Finally, the loops may be left uncut for a tough, nubby surface or cut for a soft, plush surface.

The quality of carpeting is determined mostly by its *face weight,* defined as ounces of yarn (pile) per square yard.

Installation

There are two basic carpet installation methods:

Padded and stretched carpeting is stretched over a separate pad and mechanically fastened at joints and the perimeter. Soft foam pads are inexpensive and give the carpet a soft, luxurious feel. The more expensive jute and felt pads give better support and dimensional stability and make the carpet last longer.

Glued-down carpets are usually used in areas such as offices and stores, where carpets are subjected to heavily loaded wheel traffic. They are usually glued down with carpet adhesive with a pad. This minimizes destructive flexing of the backing and prevents rippling.

Fiber Materials

Manufacturers prefer to specify the trade names of their yarns rather than chemical type, since they all claim to have variations with superior qualities. However, the general characteristics are as shown below.

CARPET MATERIALS

Fiber	Advantages	Disadvantages
Acrylic (rarely used)	Resembles wool	Not very tough Attracts oily dirt
Nylon (most common)	Very tough Resists dirt Resembles wool Low static buildup	None
Polyester deep pilings	Soft and luxurious	Less resilient Attracts oily dirt
Polypropylene indoor-outdoor	Waterproof Resists fading Resists stains Easy to clean	Crushes easily
Wool	Durable Easy to clean Feels good Easily dyed	Most expensive

CARPET TYPES

Type of Weave	**Characteristics and Best Uses**
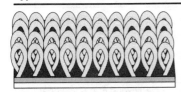	Even-height, tightly spaced uncut loops. Texture is hard and pebbly. Hard-wearing and easy to clean. Ideal for offices and high-traffic areas.
	Uneven height in patterns. Tightly spaced uncut loops. Texture is hard and pebbly. Hard-wearing and easy to clean. Ideal for offices and high-traffic areas.
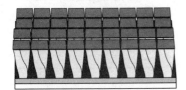	Evenly cut yarns with minimal twist. Extremely soft, velvety texture. Vacuuming and footprints appear as different colors, depending on light conditions. Ideal for formal rooms with light traffic, such as living rooms and bedrooms.
	Evenly cut yarns with tight twist. Extremely soft, velvety texture. Vacuuming and footprints appear as different colors, depending on light conditions. Ideal for formal rooms with light traffic, such as living rooms and bedrooms.
	Combination of both plush and level-loop. Hides dirt fairly well. Ideal for family rooms and children's bedrooms.
	Cut, tightly twisted yarns that twist upon themselves. Texture is rough. Hides dirt extremely well and is nearly as tough as level-loop. Ideal for entries, family rooms, and children's bedrooms.

Resilient Flooring

No-wax cushion vinyl has eliminated all competition in the resilient-flooring market. Available as 12-foot-wide rolls and as 9 x 9 and 12 x 12-inch tiles, cushion vinyl consists of a shiny (no-wax) clear vinyl coating, colored vinyl substrate, high-density vinyl foam, and a felt back. The roll material may be loose-laid, with double-sided tape or joint adhesive only at seams, or it may be fully glued down. Tiles are meant to be glued down. Some are self-sticking; others are applied over vinyl adhesive.

The primary measure of quality and longevity is the thickness of the clear no-wax wear layer. When the wear layer is worn through, the flooring has to be replaced or periodically treated with a vinyl dressing or wax.

All resilient flooring must be applied over a solid, smooth base. If the base is not smooth, the reflective vinyl will appear wavy. If the base is not solid (plywood with a missing knot, for example), women's heels or other heavily loaded objects may punch through.

The table at right describes adequate bases for resilient flooring.

BASES FOR RESILIENT FLOORING

Existing Floor	Cover or Repairs
Plywood subfloor, not rated as underlayment	Hardboard, particle board, or plywood underlayment
Plywood rated as underlayment	None needed
Single-layer board subfloor	Plywood or particle board underlayment, 5/8 inch minimum
Subfloor plus finish floor of strips less than 3 inches wide	Replace damaged strips; renail loose spots; sand smooth
Subfloor plus finish floor of strips more than 3 inches wide	Hardboard, particle board, or plywood underlayment
Concrete	None needed, but clean thoroughly by degreasing and wire brushing

CERAMIC TILE

Ceramic tiles for flooring and walls are available in four forms, all except wall tile in both glazed and unglazed finishes:

Mosaic tile is premounted and spaced on fabric backing. Nominal inch dimensions of standard tile are 1 x 1, 2 x 1, and 2 x 2, all 1/4 inch thick with 1/16-inch grout joints. Nonstandard, but common, tiles are 1-inch and 2-inch hexagons.

Quarry tile is formed by extrusion from clay or shale. Units are available with nominal inch dimensions of 3 x 3, 4 x 4, 6 x 3, 6 x 6, 8 x 4, and 8 x 8, all by 1/2 inch, plus 6 x 6 and 8 x 4 by 3/4 inch. Nonstandard tiles are common in hexagonal and Spanish patterns.

Paver tile is formed by compressing clay dust. Nominal inch dimensions are 4 x 4, 6 x 6, and 8 x 4, all by 3/8 inch, plus 4 x 4, 6 x 6, and 8 x 4, all by 1/2 inch. Hexagons and Spanish tiles also are available.

Wall tile (glazed only) is intended for decorative interior applications. Nominal inch

dimensions include 4-1/4 x 4-1/4, 6 x 4-1/4, and 6 x 6, all 5/16 inch thick, as well as hexagons, octagons, and Spanish patterns.

Mortars and Adhesives

A wide variety of materials is available for setting tiles. The choice of material depends on the application (wet or dry, freezing or not) and the skill level of the tile-setter. The table below and the illustration on the following page compare the most common materials.

Grout

Grout (the material filling the joints) may be the same as the tile-setting material, except that organic adhesive may be used only for tile-setting. In addition to the materials below, there is *latex grout* - essentially dry-set grout with liquid latex, which gives it more flexibility and water resistance.

TILE-SETTING MATERIALS

Material	Form	Bed, inches	Advantages	Disadvantages
Organic adhesive	Ready-mix mastic	1/16	Easy application; low cost; flexible bond	Interior-only suitability; resistance to immersion
Epoxy mortar	2 or 3 parts mixed at site	1/16–1/8	Excellent resistance to water and chemicals; very strong bond	Limited working time; difficult application
Dry-set mortar	Dry mix of portland cement, sand, and additives	1/4	Immersion resistance; freeze-resistance	Requirement of being kept moist for 3 days before grouting
Portland cement mortar	Portland cement, sand, and water mixed at site	3/4 walls; 1-1/4 floors	Allowance for slight leveling of uneven surfaces	Presoaking of tiles required; metal lath reinforcement recommended

Mortars and Adhesives for Tile-Setting

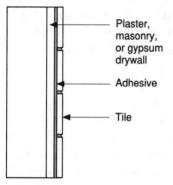

Plaster,
masonry,
or gypsum
drywall

Adhesive

Tile

ORGANIC ADHESIVE

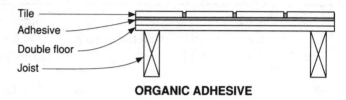

Tile

Adhesive

Double floor

Joist

ORGANIC ADHESIVE

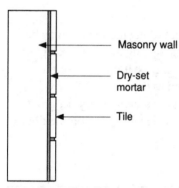

Masonry wall

Dry-set
mortar

Tile

DRY-SET MORTAR

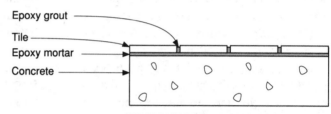

Epoxy grout

Tile

Epoxy mortar

Concrete

EPOXY MORTAR & GROUT

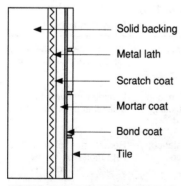

Solid backing

Metal lath

Scratch coat

Mortar coat

Bond coat

Tile

PORTLAND CEMENT MORTAR

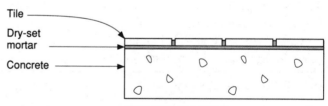

Tile

Dry-set
mortar

Concrete

DRY-SET MORTAR

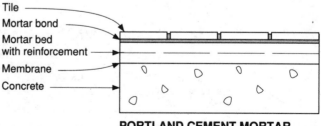

Tile

Mortar bond

Mortar bed
with reinforcement

Membrane

Concrete

PORTLAND CEMENT MORTAR

Standard Tile Trim Units

The ceramic tile industry has standardized not only basic tile shapes and dimensions, but the dimensions of various trim and accessory units as well. The illustrations below show most of the standardized trim units for mosaic, quarry, paver, and wall tiles. For illustrations of all units, contact the Tile Council of America.

In the illustrations, tile dimensions are listed as length x height.

Mosaic Tile Trim (all 1/4" thick)

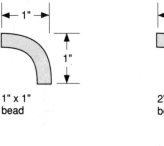

1" x 1"
bead

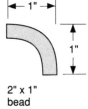

2" x 1"
bead

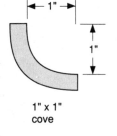

1" x 1"
cove

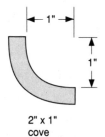

2" x 1"
cove

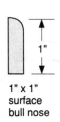

1" x 1"
surface
bull nose

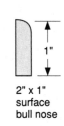

2" x 1"
surface
bull nose

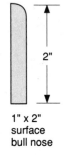

1" x 2"
surface
bull nose

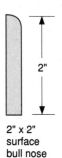

2" x 2"
surface
bull nose

Quarry Tile Trim

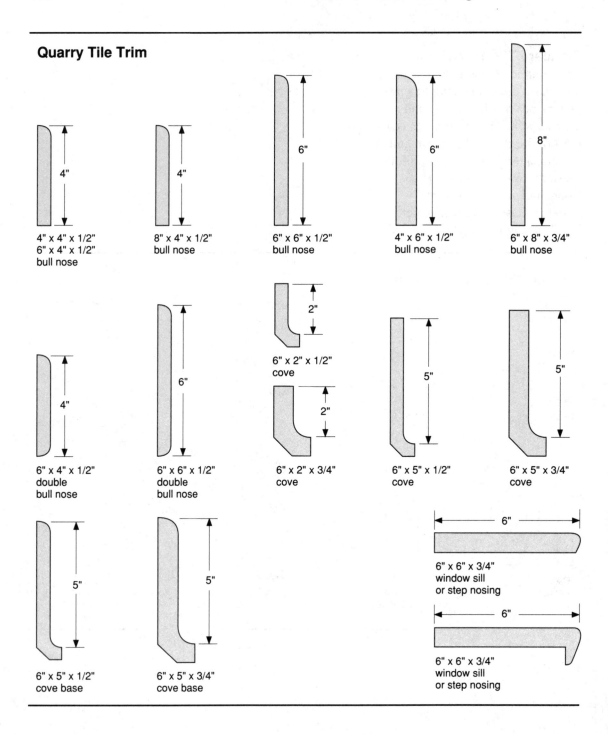

4" x 4" x 1/2"
6" x 4" x 1/2"
bull nose

8" x 4" x 1/2"
bull nose

6" x 6" x 1/2"
bull nose

4" x 6" x 1/2"
bull nose

6" x 8" x 3/4"
bull nose

6" x 4" x 1/2"
double
bull nose

6" x 6" x 1/2"
double
bull nose

6" x 2" x 1/2"
cove

6" x 2" x 3/4"
cove

6" x 5" x 1/2"
cove

6" x 5" x 3/4"
cove

6" x 5" x 1/2"
cove base

6" x 5" x 3/4"
cove base

6" x 6" x 3/4"
window sill
or step nosing

6" x 6" x 3/4"
window sill
or step nosing

Paver Tile Trim (1/2" thick unless noted)

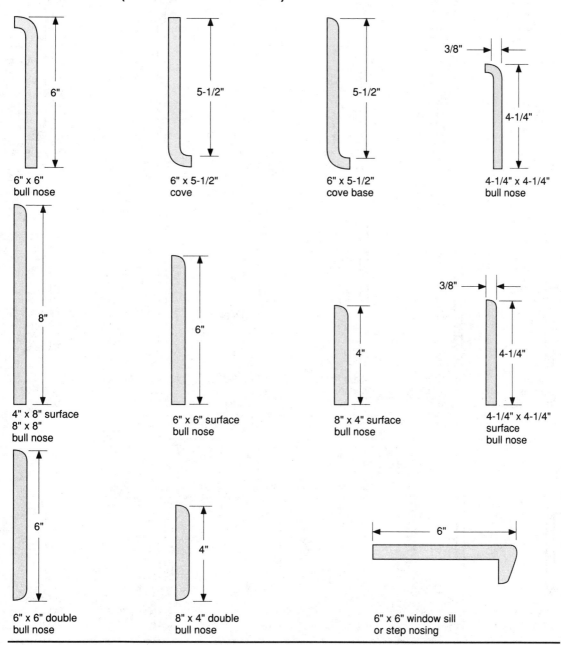

6" x 6"
bull nose

6" x 5-1/2"
cove

6" x 5-1/2"
cove base

4-1/4" x 4-1/4"
bull nose

4" x 8" surface
8" x 8"
bull nose

6" x 6" surface
bull nose

8" x 4" surface
bull nose

4-1/4" x 4-1/4"
surface
bull nose

6" x 6" double
bull nose

8" x 4" double
bull nose

6" x 6" window sill
or step nosing

Wall Tile Trim (thickness 5/16")

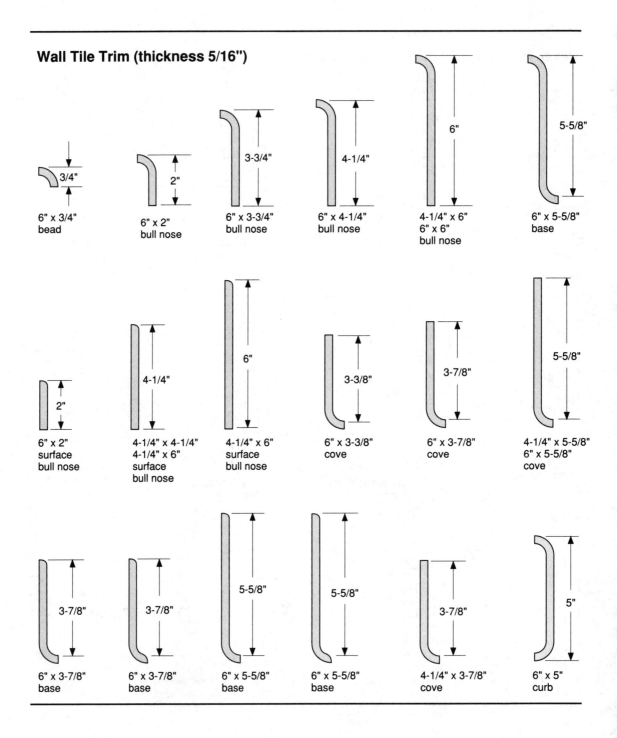

3/4"

6" x 3/4"
bead

2"

6" x 2"
bull nose

3-3/4"

6" x 3-3/4"
bull nose

4-1/4"

6" x 4-1/4"
bull nose

6"

4-1/4" x 6"
6" x 6"
bull nose

5-5/8"

6" x 5-5/8"
base

2"

6" x 2"
surface
bull nose

4-1/4"

4-1/4" x 4-1/4"
4-1/4" x 6"
surface
bull nose

6"

4-1/4" x 6"
surface
bull nose

3-3/8"

6" x 3-3/8"
cove

3-7/8"

6" x 3-7/8"
cove

5-5/8"

4-1/4" x 5-5/8"
6" x 5-5/8"
cove

3-7/8"

6" x 3-7/8"
base

3-7/8"

6" x 3-7/8"
base

5-5/8"

6" x 5-5/8"
base

5-5/8"

6" x 5-5/8"
base

3-7/8"

4-1/4" x 3-7/8"
cove

5"

6" x 5"
curb

HARDWOOD FLOORS

Hardwood flooring is available in three styles: strip, plank, and parquet.

Strip flooring is tongue-and-grooved on all four edges, making for a very secure installation without visible nails. Lengths vary by grade. The table below shows the standard cross-sectional sizes (thickness x width) and the board feet to order to cover a specified floor area. The table assumes a 5 percent cutting-waste factor.

Plank flooring is the same as strip flooring, except widths range from 3 to 8 inches. Because of its greater width, plank flooring is subject to more swelling and shrinkage with change in humidity. Standard planks should be installed with a gap the thickness of a putty knife. Laminated plank is also available in three-ply construction.

Since plank flooring is sold by the square footage of its face dimensions, simply multiply the floor area by 1.05 to allow for waste.

ESTIMATING HARDWOOD STRIP FLOORING (board feet, assuming 5% waste)

Floor Area square feet	Cross-Sectional Sizes, inches						
	3/4 x 1-1/2	3/4 x 2-1/4	3/4 x 3-1/4	1/2 x 1-1/2	1/2 x 2	3/8 x 1-1/2	3/8 x 2
5	8	7	6	7	7	7	7
10	16	14	13	14	13	14	13
20	31	28	26	28	26	28	26
30	47	42	39	42	39	42	39
40	62	55	52	55	52	55	52
50	78	69	65	69	65	69	65
60	93	83	77	83	78	83	78
70	109	97	90	97	91	97	91
80	124	111	103	111	104	111	104
90	140	125	116	125	117	125	117
100	155	138	129	138	130	138	130
200	310	277	258	277	260	277	260
300	465	415	387	415	390	415	390
400	620	553	516	553	520	553	520
500	775	691	645	691	650	691	650

Source: Oak Flooring Institute, Memphis, Tenn, 1986.

Parquet flooring consists of either slats (precisely milled short strips) or blocks (strips preassembled into square units). Slats are usually square edged. Blocks may be square edged or, more commonly, tongue-and-grooved on all four sides.

Most blocks are 5/16 inch thick. Face dimensions vary with manufacturer but are usually 6 x 6, 8 x 8, 9 x 9, 10 x 10, or 12 x 12 inches.

The illustration below shows a few of the dozens of parquet patterns which can be either purchased as blocks or assembled from strips.

Parquet Patterns

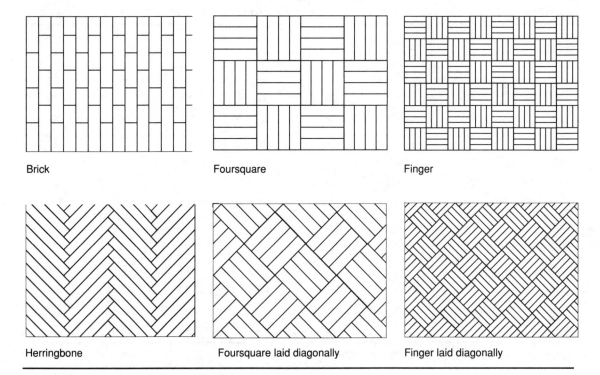

Brick   Foursquare   Finger

Herringbone   Foursquare laid diagonally   Finger laid diagonally

Installation

Solid wood flooring is subject to dimensional change when its moisture content changes. Make sure the flooring never gets wet, is stored in a dry location, and is allowed to equilibrate for at least 5 days at the installation site.

When flooring is installed over a *slab with plywood subfloor* (illustration at right), the slab must be dry and must not be located below grade. Tape a 1-square-foot piece of clear polyethylene to the slab for 24 hours; if no condensation appears, the slab is dry.

Apply cold cut-back asphalt mastic with a fine-tooth, 100-square-foot-per-gallon trowel. After it dries 1-1/2 hours minimum, unroll 6-mil polyethylene over the entire area with 6-inch overlaps. Walk on the entire surface to make sure the mastic makes contact. (Small bubbles may be punctured.) Nail 3/4-inch sheathing or underlayment grade plywood to the slab, leaving 3/4-inch spaces at walls and 1/4 inch between panels. Nail or cement finish flooring to the plywood per manufacturer's instructions.

Strip flooring may also be installed over a *slab and wood sleepers*. Lay pressure-treated 2 x 4 x 18 to 48-inch random-cut sleepers 12 inches on-center in asphalt mastic (enough for 100 percent contact). Stagger and overlap the sleepers at least 4 inches. Leave 3/4 inch at walls. Cover with polyethylene, lapping joints at sleepers.

With conventional wood joist construction over a basement or *crawl space,* outside cross-ventilation must be provided. Crawl space earth must also be covered with polyethylene. The subfloor should be either 5/8-inch or thicker performance-rated exterior or underlayment plywood or 3/4-inch square-edge, group 1, dense softwood boards laid diagonally on joists with

SLAB WITH PLYWOOD SUBFLOOR

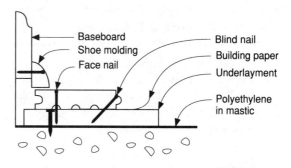

SLAB AND WOOD SLEEPERS

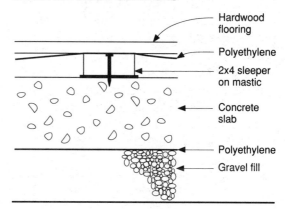

CRAWL SPACE

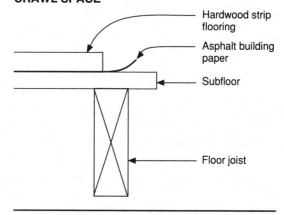

1/4-inch spaces. Plywood must be laid with face veneer across joists.

GYPSUM WALLBOARD

Gypsum Wallboard Types

Gypsum wallboard, due to its low cost, ease of application, ease of finishing, and superior fire and acoustic properties, is the most common wall material. The table below shows the readily available types and sizes of gypsum wallboard and their applications.

Fasteners

Various specialized fasteners have been developed specifically for gypsum drywall (see the table on the following page). The appropriate type depends not on the type of drywall, but on the type of substrate being fastened to. The drywall screw has proven so technically superior in wood-to-wood applications, as well, that it is now available in lengths to 3 inches.

Fastening Schedule

The illustration on the following page shows maximum fastener spacing. Select fastener lengths for adequate framing penetration:

- ring shank nails   3/4" into wood
- type W screws   5/8" into wood
- type S screws   3/8" into metal
- type G screws   1/2" into gypsum

GYPSUM DRYWALL PRODUCT TYPES

Type	Application	Edge Types	Thickness inches	Lengths feet
Regular	Usual wall and ceiling applications in dry locations where special fire rating not required	Tapered; square; rounded	1/4 3/8 1/2 5/8	6 to 12 6 to 16 6 to 16 6 to 16
Foil-backed	Same as regular, but with aluminum foil back face suitable as vapor barrier	Tapered; rounded; square	3/8 1/2 5/8	6 to 16 6 to 16 6 to 16
Water-resistant (type W)	Moist areas such as bathrooms; as base for ceramic tile	Tapered	1/2 5/8	8,11,12 8,11,12
Fire-rated (type X)	Walls and ceilings with increased fire rating	Tapered; rounded; square	1/2 5/8	6 to 16 6 to 16

FASTENERS FOR GYPSUM WALLBOARD

Type	Common Length	Application	Base Penetration
RING SHANK NAIL	1-1/4"	Single layer of wallboard to wood framing or furring	3/4"
TYPE W SCREW	1-1/4"	Wallboard to wood framing or furring	5/8"
TYPE S SCREW	1"	Wallboard to sheet metal studs and furring	3/8"
TYPE G SCREW	1-5/8"	Wallboard to gypsum wallboard or wood framing	1/2"

Gypsum Drywall Fastening

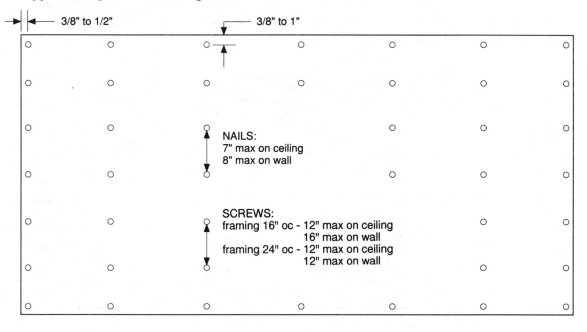

3/8" to 1/2"

3/8" to 1"

NAILS:
7" max on ceiling
8" max on wall

SCREWS:
framing 16" oc - 12" max on ceiling
                16" max on wall
framing 24" oc - 12" max on ceiling
                12" max on wall

Estimating Drywall Materials

The illustration at right and table below allow simple estimation of the materials required for the walls of a normal room with ceiling height up to 8 feet:

1. Measure the room perimeter in feet.
2. Divide the perimeter by 4.
3. Deduct for each

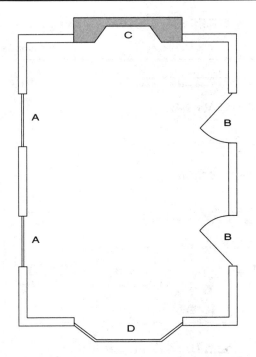

- window (A)   1/4 sheet
- door (B)   1/3 sheet
- fireplace (C)   1/2 sheet
- patio door or bay window (D)   2/3 sheet

4. Round up to the next whole sheet.
5. Find in the table below the quantity of nails or screws, gallons of ready-mix joint compound, and feet of joint tape.

DRYWALL SUPPLY ESTIMATOR

Sheets Drywall	Nails, pounds 1-1/4 inch	Nails, pounds 1-5/8 inch	Screws, pounds 1-1/4 inch	Screws, pounds 1-5/8 inch	Ready-Mix Compound, gal	Joint Tape, feet
8	1.0	1.4	0.8	1.0	2	95
10	1.3	1.7	1.0	1.3	2	118
12	1.5	2.0	1.2	1.6	3	142
14	1.8	2.4	1.4	1.8	3	166
16	2.0	2.7	1.6	2.1	4	190
18	2.3	3.1	1.8	2.4	4	213
20	2.6	3.4	2.0	2.6	4	237
22	2.8	3.7	2.3	2.9	5	260
24	3.1	4.1	2.5	3.1	5	285

Floating Corner Technique

The cracking of ceiling/wall joints and room corners is minimized by a technique known as *floating*. Drywall is always applied to the ceiling first, then the wall panels are shoved up tightly beneath.

Where the ceiling/wall joint runs parallel to the ceiling framing (see the illustration at top left), the ceiling panel is nailed or screwed at its edge, and the wall panel is not.

Where the ceiling/wall joint is perpendicular to the ceiling framing (see the illustration at top right) neither edge is fastened.

At room corner joints (see the illustration at bottom right), the edge of the underlying, or first-applied, sheet floats.

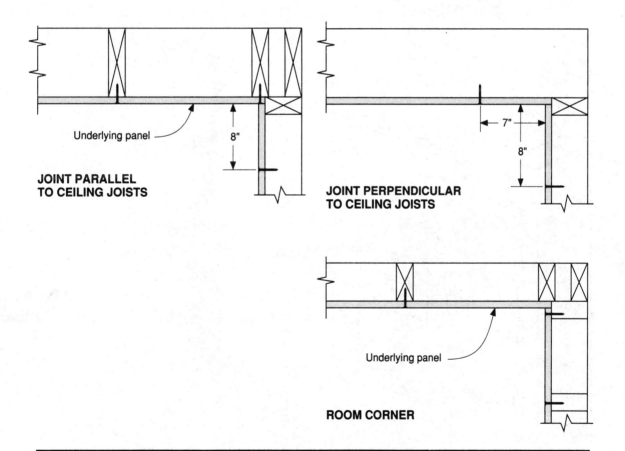

Underlying panel

8"

**JOINT PARALLEL
TO CEILING JOISTS**

7"

8"

**JOINT PERPENDICULAR
TO CEILING JOISTS**

Underlying panel

ROOM CORNER

Drywall Joint Taping

To achieve a perfectly smooth finish surface, the joints between sheets of gypsum drywall are taped and covered with joint compound. Tapered-edge boards have the two longer edges tapered to facilitate the process. Drywall should be applied either horizontally or vertically, in order to minimize the number of cut, or square-edge, joints, which require a much wider application of compound. The illustration below shows the normal three-coat process.

1. Fill the tapered area with compound, using a 6-inch trowel. Apply reinforcing tape, and press it into the compound so that no more than 1/32 inch of compound remains under the tape edges. Remove the excess compound.

2. After thorough drying, apply a second coat about 8 inches wide. At square-edge joints, 12 to 16 inches may be required.

3. After the second coat has dried, lightly sand or wipe it with a damp sponge to remove ridges. Apply a 12-inch-wide final coat with a 12-inch trowel. Lightly sand before painting.

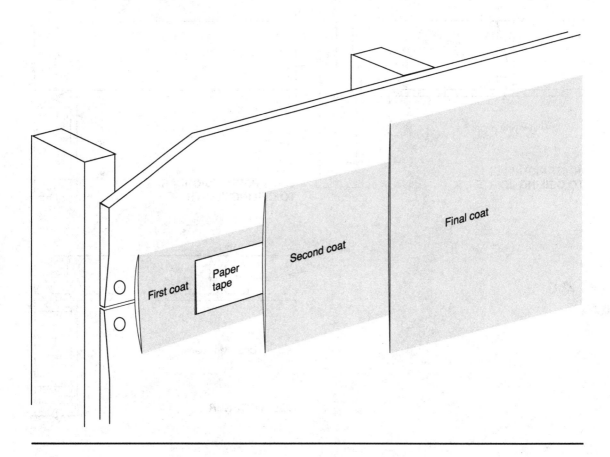

Metal Drywall Trim Accessories

The illustration below shows a variety of metal trim accessories designed to protect exposed drywall edges from mechanical damage and produce a neat finish with minimal effort or skill.

The trim is of lightweight perforated steel, which cuts easily with tin snips. The best results on corner and L-beads are obtained when the trim is bedded and nailed into joint compound, then finished in the same way as paper tape.

Metal Drywall Trim Accessories

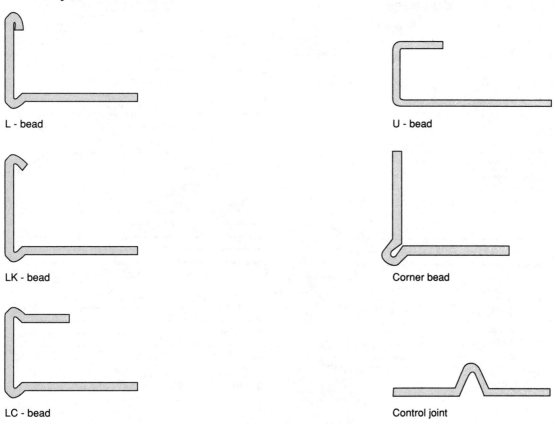

L - bead

U - bead

LK - bead

Corner bead

LC - bead

Control joint

Achieving Fire Ratings

Walls with 45-minute and 1-hour fire ratings are often required in residential construction. For example, the Uniform Building Code requires a 1-hour rating for the wall separating a dwelling from an attached garage and, unless that wall extends to the roof, a 45-minute rating for the garage ceiling. Commercial buildings may require even greater fire ratings.

Gypsum drywall is ideally suited to fire-rated construction because the material of which it is formed contains chemically bound water in the ratio of 1 quart to 10 pounds of drywall. The temperature of the drywall and the framing that it protects cannot rise much above 212°F until all of the water has been converted to steam and driven off.

The illustration below shows drywall constructions that achieve fire ratings from 1 to 4 hours. The framing can be either wood or steel.

Fire-Rated Constructions

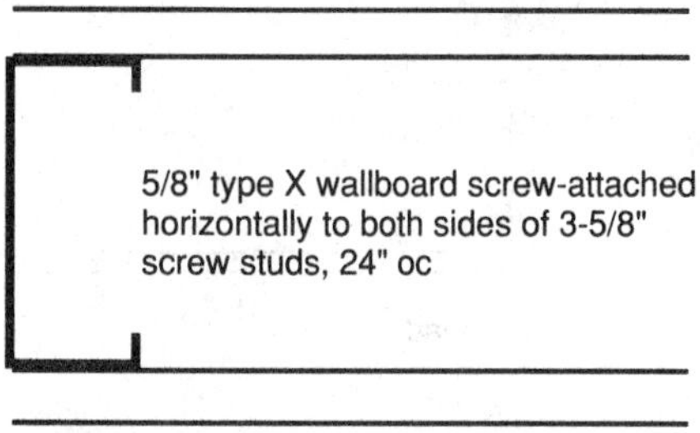

5/8" type X wallboard screw-attached horizontally to both sides of 3-5/8" screw studs, 24" oc

1-HOUR RATING
GA No. WP 1200

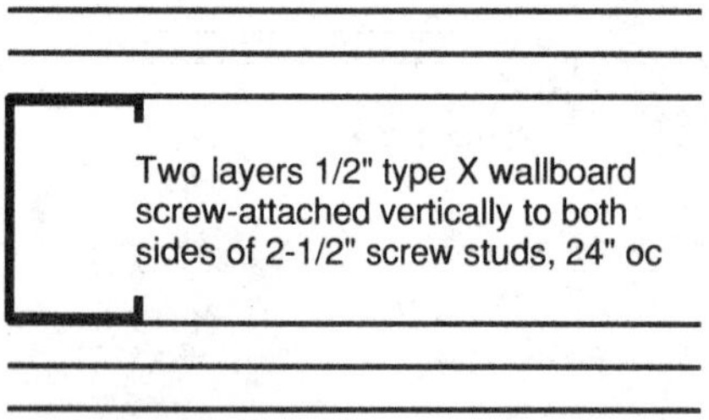

Three layers 1/2" type X wallboard each side of 1-5/8" screw studs, 24" oc. First layers vertical, face layer horizontal.

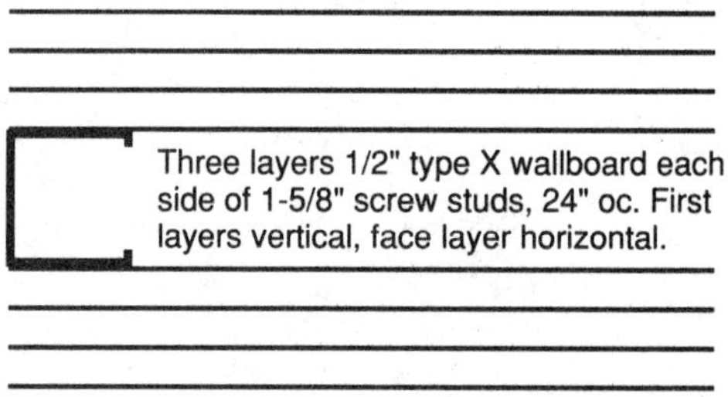

3-HOUR RATING
GA No. WP 2921

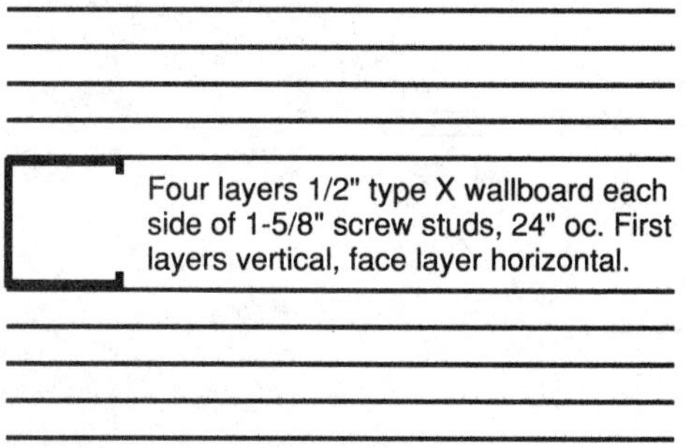

Two layers 1/2" type X wallboard screw-attached vertically to both sides of 2-1/2" screw studs, 24" oc

2-HOUR RATING
GA No. WP 1615

Four layers 1/2" type X wallboard each side of 1-5/8" screw studs, 24" oc. First layers vertical, face layer horizontal.

4-HOUR RATING
GA No. WP 2970

WOOD PANELING

Nothing beats the warmth of real wood paneling, whether it be of irreplaceable old walnut
or inexpensive knotty pine. The table below,
from the Western Wood Products Association, allows estimation of the board feet required to cover walls:

1. Calculate the gross area by multiplying
ceiling height by room perimeter, in feet.
2. Deduct the areas of windows and doors
to get the net area.
3. Multiply the net area by the appropriate
area factor in the table below, and add 10 percent for waste, to get board feet.

WOOD PANELING COVERAGE

Paneling Style	Nominal Size, inches	Width, inches Total	Face	Area Factor
SQUARE-EDGE BOARD	1 x 4	3-1/2	3-1/2	1.14
	1 x 6	5-1/2	5-1/2	1.09
	1 x 8	7-1/4	7-1/4	1.10
	1 x 10	9-1/4	9-1/4	1.08
	1 x 12	11-1/4	11-1/4	1.07
TONGUE-AND-GROOVE	1 x 4	3-3/8	3-1/8	1.28
	1 x 6	5-3/8	5-1/8	1.17
	1 x 8	7-1/8	6-7/8	1.16
	1 x 10	9-1/8	8-7/8	1.13
	1 x 12	11-1/8	10-7/8	1.10
PROFILE PATTERN (various)	1 x 6	5-7/16	5-1/16	1.19
	1 x 8	7-1/8	6-3/4	1.19
	1 x 10	9-1/8	8-3/4	1.14
	1 x 12	11-1/8	10-3/4	1.12
V-JOINT RUSTIC	1 x 6	5-3/8	5	1.20
	1 x 8	7-1/8	6-3/4	1.19
	1 x 10	9-1/8	8-3/4	1.14
	1 x 12	11-1/8	10-3/4	1.12
CHANNEL RUSTIC	1 x 6	5-3/8	4-7/8	1.23
	1 x 8	7-1/8	6-5/8	1.21
	1 x 10	9-1/8	8-5/8	1.16
	1 x 12	11-1/8	10-5/8	1.13

Wood paneling can be applied horizontally, vertically, diagonally, or in any combination thereof. It is particularly simple over exposed studs (see the illustration below). Only vertical installation requires the addition of either horizontal blocking or 1x4 strapping, spaced 36 inches on-center maximum.

Over a masonry wall (see right), 1x4 strapping should be fastened with either masonry nails or construction adhesive 36 inches on-center maximum. If the masonry wall is below grade, first apply a 6-mil vapor barrier, stapling it to the sill plate. Then install 1x4 furring, 36 inches on-center maximum, with masonry nails. An alternative is to install a stud wall and then wire and insulate between the studs before paneling.

Paneling over Masonry

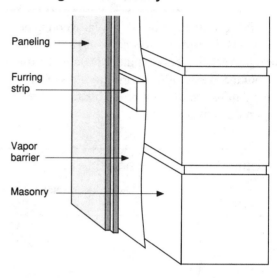

Paneling Patterns

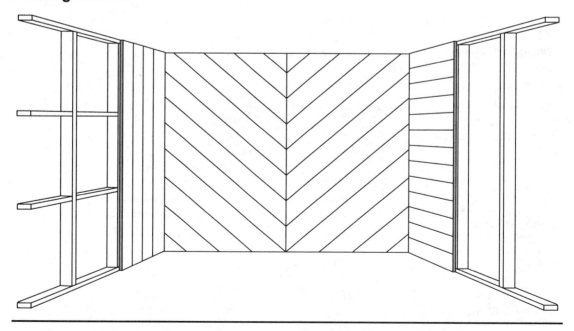

Trim

If the paneling has relief, baseboards will look better installed flush than installed over the paneling. Flush baseboards may be installed over furring strips, as in the illustration below, or over 2x4 blocking between the studs. There should be sufficient blocking to catch not only the baseboard, but the bottom of the paneling as well.

The illustrations at right apply to either door or window trim.

If the paneling is applied directly to the existing wall surface, simply remove the exist-ing door and window trim and replace them with square-edge strips of the same thickness and species as the paneling.

If the paneling is furred out, use jamb extenders to bring the jambs out to the finish wall surface, and apply casing over the joint.

Electrical switch and receptacle boxes can be left in place with convenient extension collars, available from electrical suppliers and larger hardware stores.

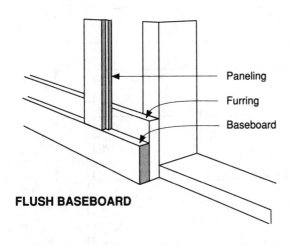

FLUSH BASEBOARD

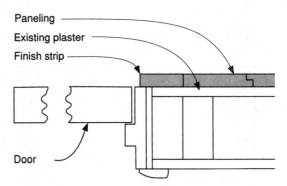

PANELING OVER EXISTING PLASTER

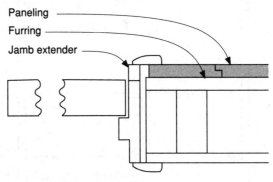

PANELING OVER FURRING

SUSPENDED CEILINGS

A suspended ceiling is a simple and inexpensive way to achieve a level ceiling beneath an existing ceiling or roof, whether flat or not. A strong point is the convenience of "dropping in" fluorescent fixtures at any point. A weakness is an office look, not appropriate to residences except in utilitarian rooms such as laundries, game rooms, or workshops.

Install a suspended ceiling as follows:

1. Plot the dimensions of the room on a large piece of 8 x 8-inch graph paper. Let each square represent either 3 or 6 inches. On the plot, draw a 2 x 4-foot grid so that the spaces next to the perimeter are symmetric and as large as possible (see the illustration below).

Laying Out a Ceiling Grid

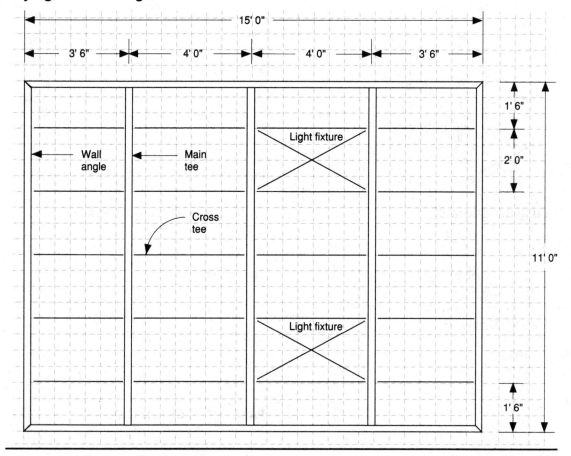

2. If the original ceiling is flat and level, measure down to the level of the new ceiling (allowing 6 inches minimum for recessed light fixtures) and snap a chalk line around the perimeter of the room. If the ceiling is not level, establish a level perimeter with a line level. Double-check the level by running the line level around the perimeter in the opposite direction and comparing.

3. Fasten sections of *wall angle* with the bottom flanges at the level of the chalk line. Nail them into studs if possible. Don't be shy about poking holes in the wall to find stud locations—as long as the holes are above the suspended-ceiling level.

4. Stretch strings along the positions of the main tees. Use a plumb bob or level to mark the locations where the string intersects the ceiling joists overhead. Again, don't be afraid to poke holes to find the joist locations: They'll never be seen.

5. Cut suspension wires 12 inches longer than the distance between the old ceiling or ceiling joists and the wall angle, and fasten a wire at each marked intersection. After fastening, make a right-angle bend in the wire at the level of the stretched string.

6. (See the illustration at top right.) Determine, from the layout sketch, the distance from the wall to the first intersection with a cross tee (1 foot 6 inches in the layout example on the previous page). Measure this distance from the end of a main tee, and select the slot just beyond. From the slot, measure back toward the original end, and mark. Cut the main tee 1/8 inch shorter than the mark, which allows for the thickness of the wall angle.

Cutting Main Tee

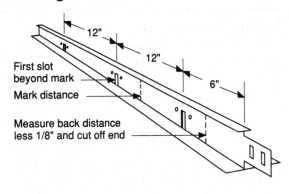

First slot beyond mark

Mark distance

Measure back distance less 1/8" and cut off end

7. Install all of the main tees, repeating step 6 each time. Main tees are 12 feet long, but splice fittings are available if required.

8. Install the cross tees by snapping the end tongues into the main tee slots. Cross tees at the perimeter will have to be cut to fit.

9. Install the wiring for any recessed light fixtures.

10. Drop in the ceiling panels, cutting perimeter panels as necessary with a utility knife and straightedge.

PAINTS AND OTHER FINISHES

Exterior Finishes

When it comes to finishing newly applied siding, there are six basic choices: no finish at all, water-repellent preservatives, semitransparent stains, paint, solid-color stains, and transparent coatings.

No finish—It should be noted that paints and stains do not necessarily preserve wood. If protected from excessive running water, such as by a moderate to large roof overhang, wood beveled siding and shingles may last a century or more. More specifically, paint and stain preserve an appearance. The decision is a tough one because, once painted, exterior wood requires some sort of recoating every 5 to 10 years at best.

Water-repellent preservatives—These keep water out of wood and thus reduce the mechanical stresses of swelling and shrinking. The addition of a preservative chemical prevents mildew and dry rot in horizontal members such as deck joists and window sills. Preservatives usually have no color of their own but darken natural wood somewhat. The very large drawback is that they need to be reapplied every few years—an unlikely prospect for the average home.

Semitransparent stains—They are most often water-repellent preservatives with light pigmentation. In addition to supplying the positive qualities of a repellent and preservative, the stains penetrate the wood surface without forming a skin. Thus, they impart color but cannot fail by peeling. These treatments must be applied over unpainted wood. Service life is greatest when applied to rough-sawn or already weathered natural wood, due to greater absorption.

Paint—For the best overall job of protecting wood outdoors, paint forms a protective coating. Oil-based paint seals out moisture better, but latex paint is more flexible and less likely to fail by cracking and peeling. Latex paint has become increasingly popular, especially with amateur painters, because it is easier to apply, dries quicker than oil-based paint, and cleans up with water. Cleanup for oil paints requires mineral spirits. A proper application of latex or oil paint lasts as long as 10 years.

The disadvantage of paint and other film-forming finishes, such as solid stains, is that recoating causes a buildup that eventually has to be laboriously removed.

Solid-color stains—Like paint, solid-color stains completely obscure the natural color and grain of the wood. They can be latex or oil-based. They have much more pigment than other stains. As a result, they form a film like paint and can even peel as paint does. Solid-color stains can be used over old paint or stain, and in fact, these stains are quite similar to thinned paint. The film layers can build up, but because the stain is thinner than paint, the surface can be recoated many more times before it needs to be stripped. Solid-color stains are not recommended for areas that will be walked on.

Transparent coatings—Forget urethanes, shellacs, and varnishes for exterior use! No matter what the manufacturer claims, the finish will crack and peel within a few years, requiring total removal, or at least extensive sanding before recoating.

PERFORMANCE OF EXTERIOR COATINGS

Type of Exterior Wood Surface	Water-Repellent Preservative		Semitransparent Stain		Paints and Solid Stains	
	Suitability	Life, years	Suitability	Life, years	Suitability	Life, years
Siding						
Cedar and Redwood						
Smooth (vertical grain)	Good	1-2	Fair	2-4	Good	4-6
Rough-sawn or weathered	Good	2-3	Excellent	5-8	Fair	3-5
Pine, Fir, Spruce, etc.						
Smooth (flat grain)	Good	1-2	Poor	2-3	Fair	3-5
Rough (flat grain)	Good	2-3	Good	4-7	Fair	3-5
Shingles						
Sawn	Good	2-3	Excellent	4-8	Fair	3-5
Split	Good	1-2	Excellent	4-8	NA	NA
Plywood						
(Douglas fir & southern pine)						
Sanded	Poor	1-2	Fair	2-4	Fair	3-5
Rough-sawn	Poor	2-3	Good	4-8	Fair	3-5
Medium-density overlay	NA	NA	NA	NA	Excellent	6-8
Plywood (cedar and redwood)						
Sanded	Poor	1-2	Fair	2-4	Fair	3-5
Rough-sawn	Poor	2-3	Excellent	5-8	Fair	3-5
Hardboard (medium density)						
Smooth						
Unfinished	NA	NA	NA	NA	Good	4-6
Preprimed	NA	NA	NA	NA	Good	4-6
Textured						
Unfinished	NA	NA	NA	NA	Good	4-6
Preprimed	NA	NA	NA	NA	Good	4-6
Millwork (usually pine)						
Windows, doors, trim	Good	NA	Fair	2-3	Good	3-6
Decking						
New (smooth)	Good	1-2	Fair	2-3	Poor	2-3
Weathered (rough)	Good	2-3	Good	3-6	Poor	2-3
Glued Laminated Members						
Smooth	Good	1-2	Fair	3-4	Fair	3-4
Rough	Good	2-3	Good	6-8	Fair	3-4
Waferboard	NA	NA	Poor	1-3	Fair	2-4

Source: *Wood Handbook* (Washington, DC: United States Department of Agriculture, 1987).

Note: NA = not applicable.

Interior Paints

Practically speaking, there are three types of interior primer and two types of interior paint:

Primer-sealer is designed to prevent bleed-through of wood resin contained in knots and pitch pockets. It's a good idea to apply at least two coats of primer-sealer to knots in wood before painting. Since primer-sealers are generally white, they cannot be used under clear finishes.

Latex primer is the best first coat over gypsum wallboard, plaster, and concrete. It adheres well to any surface except untreated wood (it raises the wood grain) or gloss oil paint. It can be applied over gloss oil paint, provided the surface is sanded to give the latex a grip.

Alkyd primer is just the opposite of latex primer: Alkyd primer is the best first coat over raw wood but should not be used on wallboard or masonry.

Latex paint is the most popular interior finish because it cleans up with water (warm, soapy water works best). It can be applied over gypsum wallboard, plaster, and masonry. Wood should be primed with alkyd primer first. Latex also adheres to latex and flat oils. Avoid gloss oils and alkyds other than primers.

Alkyd paint has nearly replaced natural oil paint. It has the same good qualities as natural oil such as linseed, but it is nearly odor free. Alkyd can be applied over any other paint or bare wood. It should not be applied to bare gypsum wallboard, plaster, or masonry.

The table below compares the advantages and disadvantages of the interior paints.

INTERIOR PRIMERS AND PAINTS

Materials	Advantages	Disadvantages
Primer-sealer	Stops bleeding of knots Dries extremely quickly	Stains white; must be painted Requires alcohol for cleanup
Latex primer	Cleans up with water Dries quickly unless humid Relatively odor free	Not good on wood
Alkyd primer	Best wood primer Suitable primer for all paint types	Not good on gypsum wallboard Requires paint thinner for cleanup
Latex paint	Cleans up with water Dries quickly Relatively odor free Inexpensive	Requires primer over wood Not as tough as alkyd Doesn't adhere to gloss finishes Wets and loosens wallpaper
Alkyd paint	Provides tough finish Relatively odor-free Adheres to all other paints	Dries relatively slowly Requires paint thinner for cleanup Requires primer over masonry, plaster, and gypsum wallboard

Diagnosing Paint Failures

We blame the manufacturer when paint fails prematurely. Ninety percent of paint failures, however, are due to either moisture problems or inadequate preparation of the surface. Use the table below to find out what went wrong and what can be done to correct the problem.

PAINT FAILURES

Symptom		Possible Causes	Suggested Cures
Alligatoring	Top layer only	Second coat applied before first coat dry	Remove to bare wood
	To wood	Paint too thick; lost flexibility	Remove to bare wood
Blistering	Top layer only	Paint applied in hot sun Surface oily or dirty Latex over heavily chalked oil paint	Paint on cloudy days Wash with detergent and rinse
	To wood	Moisture driving paint off (from roof leak, ice dam, impervious sheathing, bath or kitchen humidity, gutter leak)	Eliminate moisture source Strap under siding Wedge under siding
Checking of plywood		Expansion, contraction, and delamination of plywood veneers	Sand plywood and paint Replace plywood
Cracking	Top layer only	Inflexible paint	Scrape and wire-brush Use latex paint
	To wood	Inflexible paint	Remove to bare wood Use primer and latex
Flaking off masonry		Inadequate preparation of masonry	Wire-brush loose paint Apply masonry conditioner Apply 2 topcoats exterior latex
Mildew /mold	Inside	Cold wall or ceiling due to no insulation Lack of air movement	Treat with bleach, insulate, leave closet doors open
	Outside	Warm, humid outside air with no direct sunlight	Scrub with 1/2 cup bleach/gallon water, rinse, prime Use mildew-resistant paint
Peeling inside		Inadequate moisture storage of sheathing; rain wicking under siding	Install fiberboard or wood sheathing Strap under siding
Rust stains		Steel siding nails	Sand all rust, set nails 1/8", prime, fill with caulk, paint
Wrinkling and sagging		Too cold application (below 50°F) Second coat applied too soon Too much paint	Sand smooth, paint over

Clear Floor Finishes

Finishing wood floors is one of the most controversial subjects in house building and remodeling. All floor experts agree that there is not yet a wood finish that stands up to heavy foot traffic and retains its original beauty for more than a few years.

Many preservationists favor the oldest approach of all - nothing! Wood is tough and resilient and, except for stains which penetrate deeply, can be cleaned with water, detergent, and ammonia (except oak, which ammonia will turn black).

If you do decide to finish, however, you have two basic choices:

Penetrating finishes sink into the wood without forming a separate surface skin. These include linseed oil, tung oil, and various proprietary penetrating resins. The basic approach to long-term care is periodic cleaning and simple reapplication.

Surface finishes form hard "eggshell" skins on top of the wood. Length of life is primarily a function of the number of coats applied initially. Four coats of urethane last twice as long as two coats. Once the eggshell finish has been penetrated, water easily penetrates and begins to stain the wood. Refinishing usually requires complete stripping of the old finish.

CLEAR FLOOR FINISHES

Finish	Advantages	Disadvantages
No finish	Never requires removal	Most difficult to keep clean Doesn't highlight wood grain Doesn't control splintering

Penetrating
  Tung oil
  Linseed oil
  Resin oils
	Easy to renew
	Cannot be damaged by puncture
	Simple to apply
		Darkens wood
		Not abrasion resistant
		Needs to be retreated often (once per year)

Surface
  Lacquer
  Shellac
  Spirit varnish
	Quick drying
	Can be refurbished without removal
		Difficult to apply (dries too rapidly)
		Stained by water
		Too brittle for softwoods (pine, fir)

Alkyd varnish	Compatible with other undercoats Simple to apply Durable	Slow drying
Polyurethane	Easy to apply (brush, rag, sponge) Resistant to water and alcohol Very high gloss possible	Damage requires complete removal Too brittle for application over softwood Can be applied only over bare wood
Two-part (urea-formaldehyde base)	Toughest finish of all	Requires professional application Excessive wood lost in removal

SHELVING

No residence or office has enough shelving. Adjustable shelving (shown on the following page) is the most convenient and flexible. Fixed shelving, however, is lower in cost and, if well executed, looks more finished.

Below is a variety of simple fixed shelving requiring minimal skills and tools.

For fixed shelving, carefully consider the heights of the items to be shelved. Book heights, for example, range from 8 inches for small paperbacks to more than 15 inches for oversize coffee-table editions. You'll probably get at least one additional useful shelf by carefully measuring your needs.

Fixed Shelves

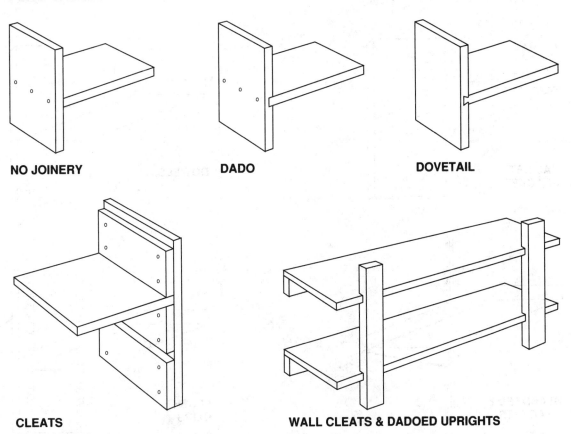

NO JOINERY

DADO

DOVETAIL

CLEATS

WALL CLEATS & DADOED UPRIGHTS

Adjustable Shelves

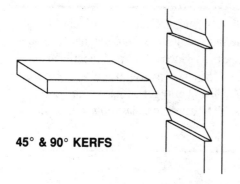

45° & 90° KERFS

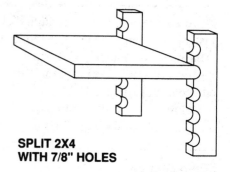

**SPLIT 2X4
WITH 7/8" HOLES**

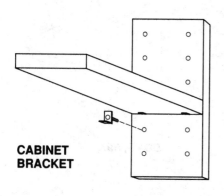

**CABINET
BRACKET**

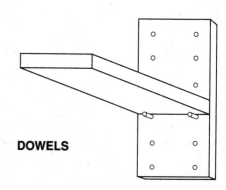

DOWELS

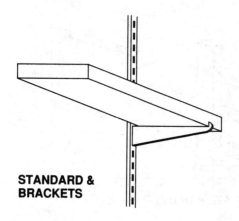

**STANDARD &
BRACKETS**

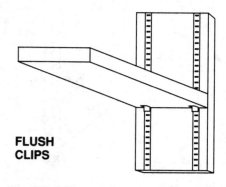

**FLUSH
CLIPS**

WOOD MOLDINGS

Wood moldings are strips of wood milled with plane (flat) or curved surfaces continuous over their lengths. The name *molding* derives from the fact that they are so perfectly smooth, they appear to be molded of plaster. In fact, they were first intended to be painted to resemble the stone carvings found on classical Greek and Roman buildings.

The purposes of moldings include decoration (such as edging for paneling), protection (such as chair rails), and concealment (such as base and cove moldings):

Crown, bed, and some cove moldings are milled from thin stock but installed at an angle at the wall/ceiling intersection. This allows the molding to be wide, yet follow irregularities in the wall and ceiling.

Quarter rounds are designed to both cover and reinforce the joint between paneling and a frame.

Base shoes are used alone or on top of regular board stock to cover the wall/floor joint and to protect the base of the wall.

Astragals cover panel joints and sometimes simply add decorative relief.

Screen moldings are used to fasten and cover the edges of screening material in door and window frames.

Casings visually frame windows and doors and cover the joint between the door or window frame and the plaster or drywall finish. They are hollow on the back side in order to still lie flat despite possible cupping.

Brick moldings provide casings for doors and windows in masonry walls.

Drip caps are designed to shed water over doors and windows but have been mostly replaced by metal flashings.

Stops are used on both door and window frames to stop or constrain the motion of the door or sash.

Panel strips and mullion casings are used to join large sections of paneling or to join several window units together.

The most common species of wood used in moldings is ponderosa pine. Other species used in significant quantities are sugar pine, fir, larch, cedar, and hemlock. Clear wood in long lengths is increasingly more expensive, so manufacturers have turned to finger jointing to eliminate defects. As large wood moldings become more expensive, plaster moldings have become more available, particularly for restoration work.

The Wood Moulding and Millwork Producers Association (WM) promulgates standard molding patterns. The illustrations that follow contain full-scale sections of the most popular patterns of the *WM Series,* the latest (1977) standard. Regional millwork distributors generally offer selections of these patterns, customized to local demand.

CROWNS

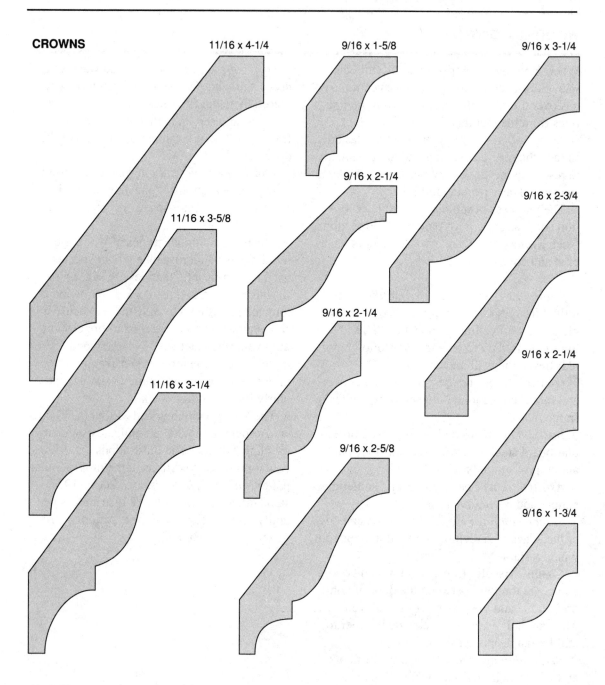

11/16 x 4-1/4

9/16 x 1-5/8

9/16 x 3-1/4

11/16 x 3-5/8

9/16 x 2-1/4

9/16 x 2-3/4

11/16 x 3-1/4

9/16 x 2-1/4

9/16 x 2-1/4

9/16 x 2-5/8

9/16 x 1-3/4

Note: Dimensions are in inches.

BEDS

COVES

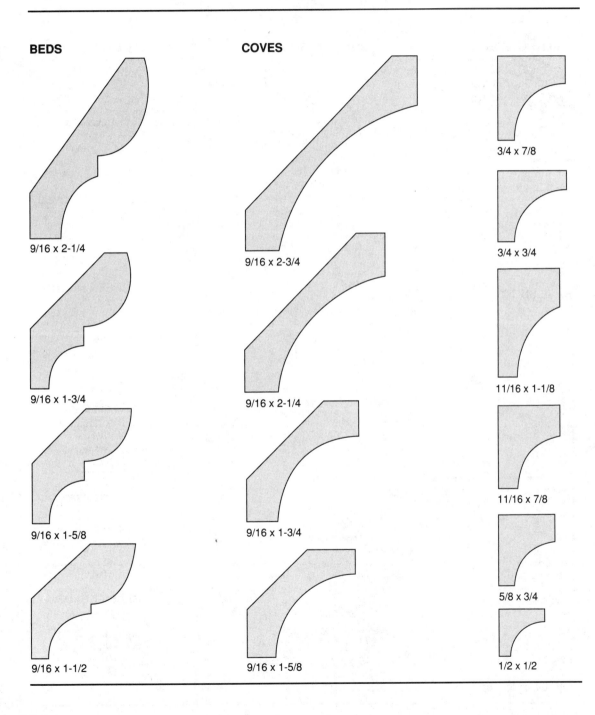

9/16 x 2-1/4

9/16 x 1-3/4

9/16 x 1-5/8

9/16 x 1-1/2

9/16 x 2-3/4

9/16 x 2-1/4

9/16 x 1-3/4

9/16 x 1-5/8

3/4 x 7/8

3/4 x 3/4

11/16 x 1-1/8

11/16 x 7/8

5/8 x 3/4

1/2 x 1/2

HALF ROUNDS

1/2 x 1

3/8 x 11/16

5/16 x 5/8

1/4 x 1/2

BASE SHOES

1/2 x 3/4

7/16 x 3/4

7/16 x 11/16

1/2 x 3/4

QUARTER ROUNDS

1-1/16 x 1-1/16

11/16 x 1-3/8

3/4 x 3/4

11/16 x 11/16

5/8 x 5/8   1/4 x 1/4

1/2 x 1/2   3/8 x 3/8

FLAT ASTRAGALS

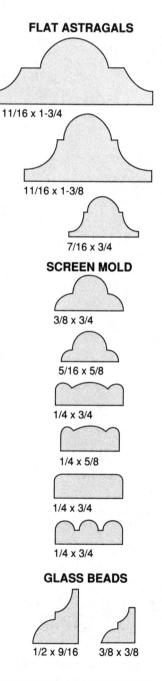

11/16 x 1-3/4

11/16 x 1-3/8

7/16 x 3/4

SCREEN MOLD

3/8 x 3/4

5/16 x 5/8

1/4 x 3/4

1/4 x 5/8

1/4 x 3/4

1/4 x 3/4

GLASS BEADS

1/2 x 9/16   3/8 x 3/8

Note: Dimensions are in inches.

CASING

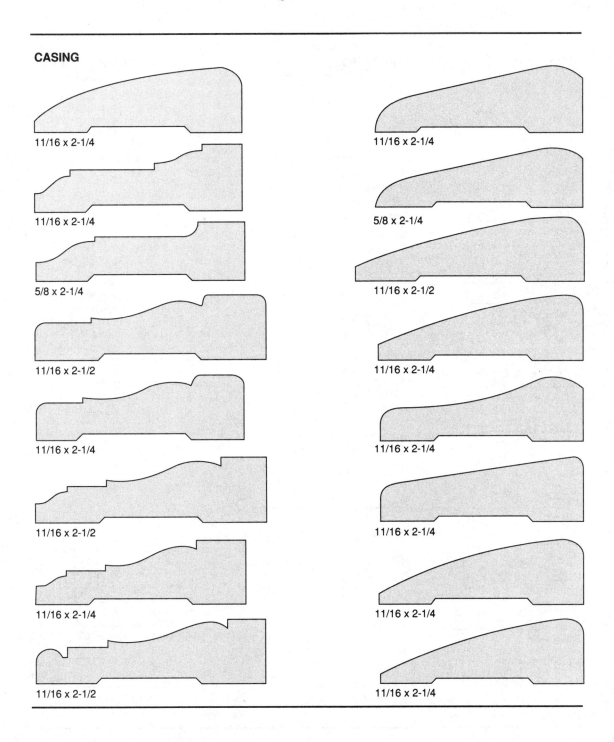

11/16 x 2-1/4

11/16 x 2-1/4

5/8 x 2-1/4

11/16 x 2-1/2

11/16 x 2-1/4

11/16 x 2-1/2

11/16 x 2-1/4

11/16 x 2-1/2

11/16 x 2-1/4

5/8 x 2-1/4

11/16 x 2-1/2

11/16 x 2-1/4

11/16 x 2-1/4

11/16 x 2-1/4

11/16 x 2-1/4

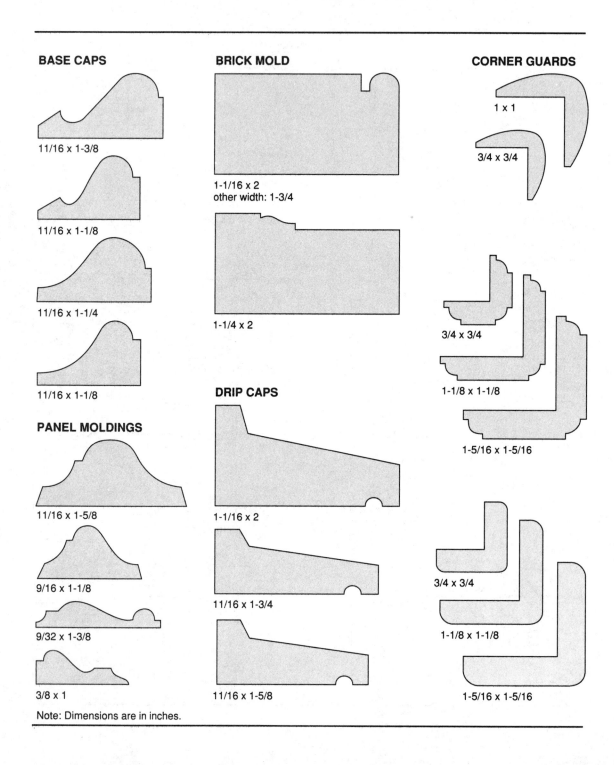

BASE CAPS

11/16 x 1-3/8

11/16 x 1-1/8

11/16 x 1-1/4

11/16 x 1-1/8

PANEL MOLDINGS

11/16 x 1-5/8

9/16 x 1-1/8

9/32 x 1-3/8

3/8 x 1

BRICK MOLD

1-1/16 x 2
other width: 1-3/4

1-1/4 x 2

DRIP CAPS

1-1/16 x 2

11/16 x 1-3/4

11/16 x 1-5/8

CORNER GUARDS

1 x 1

3/4 x 3/4

3/4 x 3/4

1-1/8 x 1-1/8

1-5/16 x 1-5/16

3/4 x 3/4

1-1/8 x 1-1/8

1-5/16 x 1-5/16

Note: Dimensions are in inches.

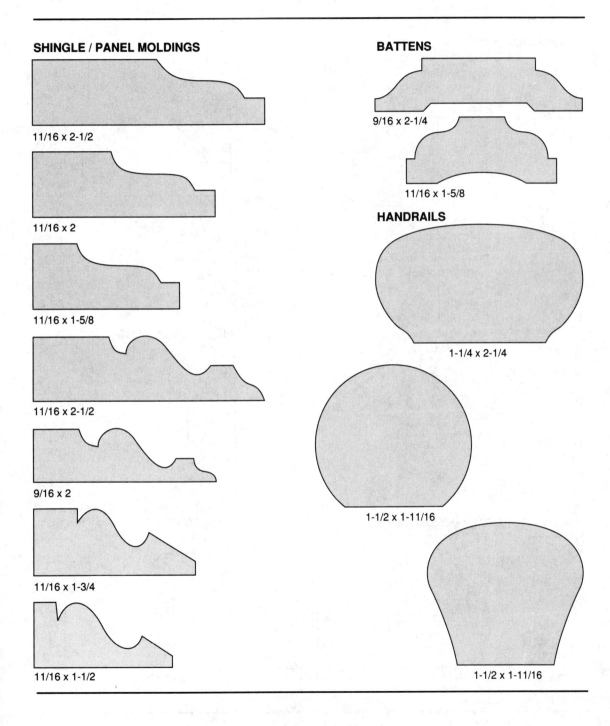

SHINGLE / PANEL MOLDINGS

11/16 x 2-1/2

11/16 x 2

11/16 x 1-5/8

11/16 x 2-1/2

9/16 x 2

11/16 x 1-3/4

11/16 x 1-1/2

BATTENS

9/16 x 2-1/4

11/16 x 1-5/8

HANDRAILS

1-1/4 x 2-1/4

1-1/2 x 1-11/16

1-1/2 x 1-11/16

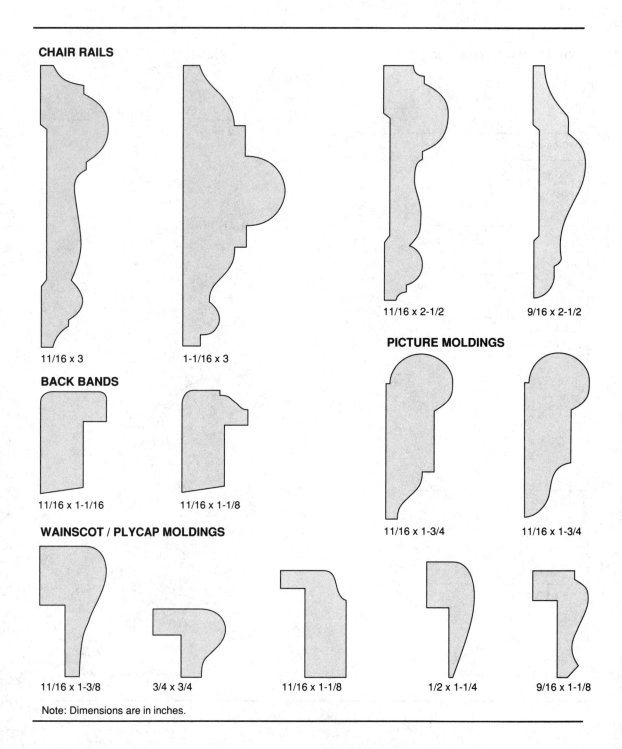

CHAIR RAILS

11/16 x 3

1-1/16 x 3

11/16 x 2-1/2

9/16 x 2-1/2

BACK BANDS

11/16 x 1-1/16

11/16 x 1-1/8

PICTURE MOLDINGS

11/16 x 1-3/4

11/16 x 1-3/4

WAINSCOT / PLYCAP MOLDINGS

11/16 x 1-3/8

3/4 x 3/4

11/16 x 1-1/8

1/2 x 1-1/4

9/16 x 1-1/8

Note: Dimensions are in inches.

CASINGS

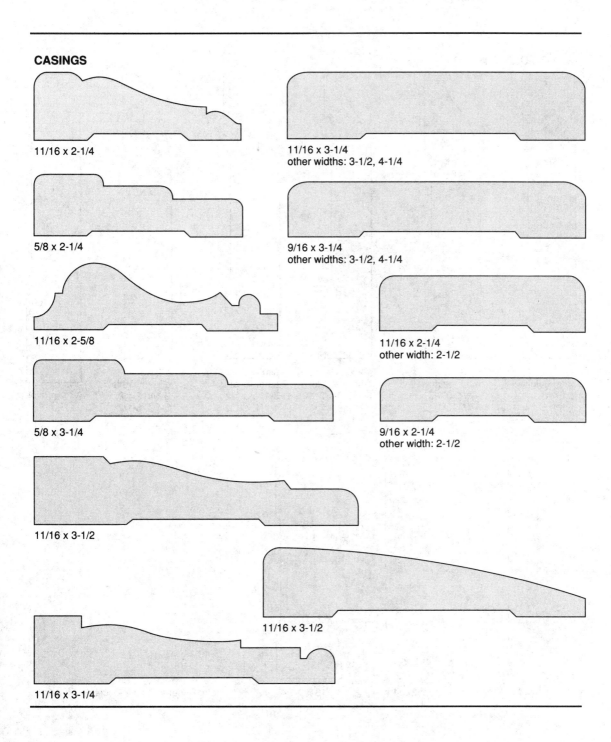

11/16 x 2-1/4

11/16 x 3-1/4
other widths: 3-1/2, 4-1/4

5/8 x 2-1/4

9/16 x 3-1/4
other widths: 3-1/2, 4-1/4

11/16 x 2-5/8

11/16 x 2-1/4
other width: 2-1/2

5/8 x 3-1/4

9/16 x 2-1/4
other width: 2-1/2

11/16 x 3-1/2

11/16 x 3-1/2

11/16 x 3-1/4

BASE MOLDINGS

9/16 x 3
other heights:
3-1/4, 3-1/2,
4-1/4, 5-1/4

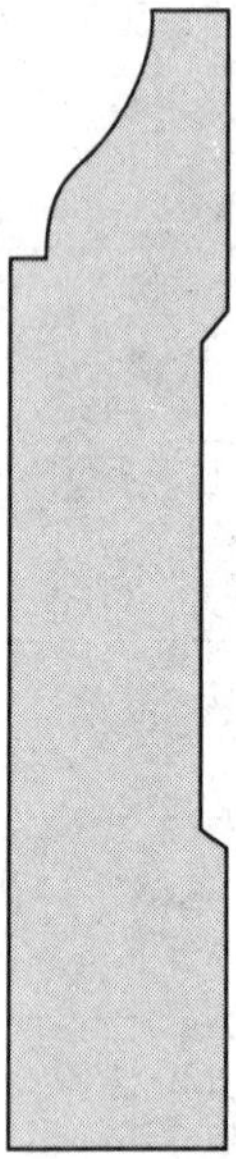

9/16 x 3
other heights:
3-1/4, 3-1/2,
4-1/4, 5-1/4

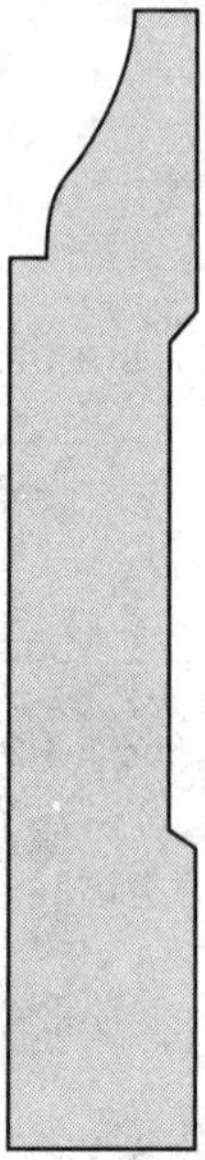

1/2 x 3
other heights:
3-1/4, 3-1/2

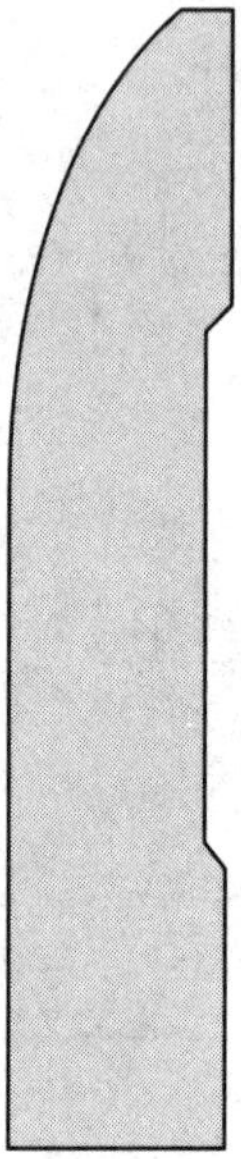

9/16 x 3
other heights:
3-1/4, 3-1/2,
4-1/4, 4-1/2

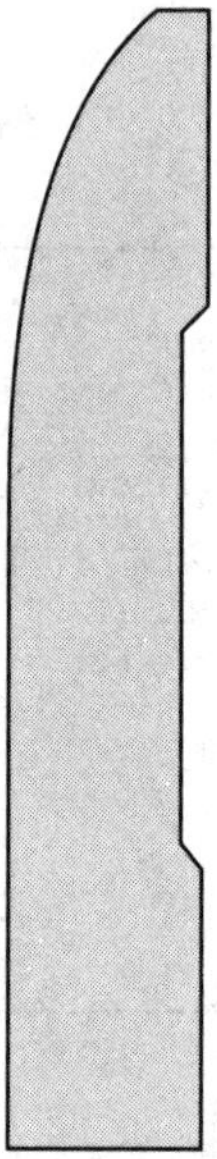

1/2 x 3
other heights:
2-1/4, 2-1/2,
3-1/4, 3-1/2

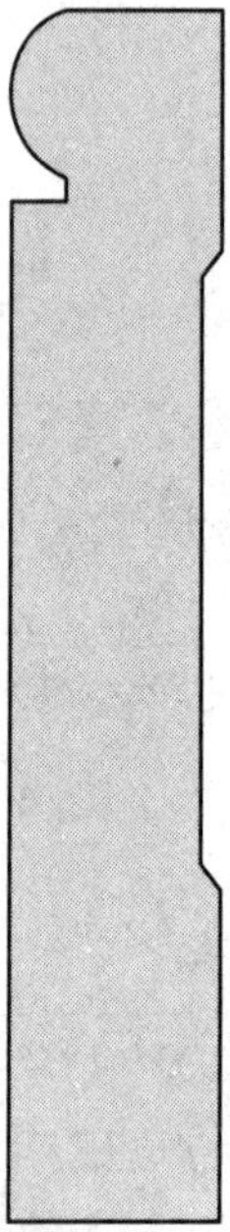

9/16 x 3-1/4

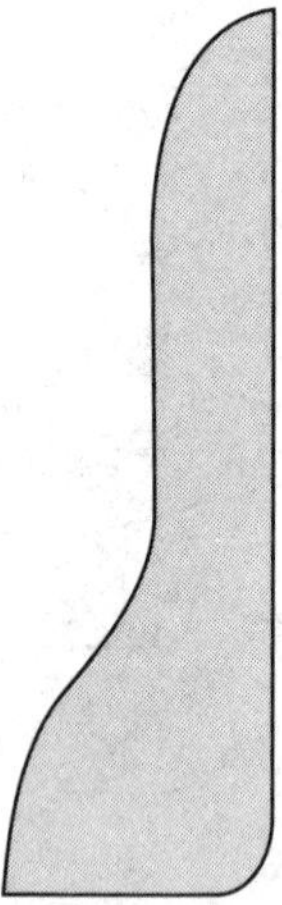

11/16 x 2-1/4

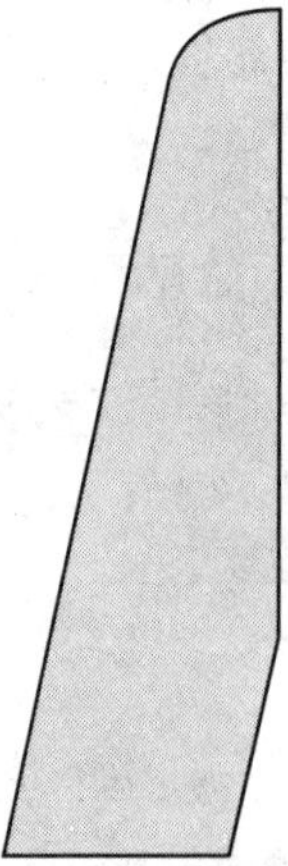

11/16 x 2-1/4

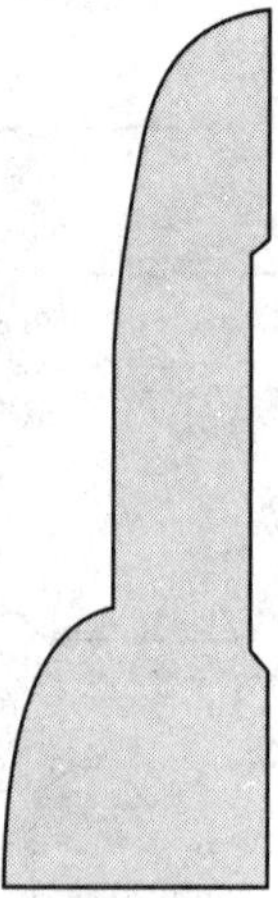

11/16 x 2-1/4

Note: Dimensions are in inches.

STOPS

7/16 x 1-3/8
other widths: 7/8, 1-1/8,
1-1/4, 1-5/8, 1-3/4, 2-1/4

7/16 x 1-3/8
other widths: 3/4, 7/8, 1-1/8,
1-1/4, 1-5/8, 1-3/4, 2-1/4

7/16 x 1-3/8
other widths: 3/4, 7/8, 1-1/8,
1-1/4, 1-5/8, 1-3/4, 2-1/4

3/8 x 1-3/8
other widths: 3/4, 7/8, 1-1/8,
1-1/4, 1-5/8, 1-3/4, 2-1/4

3/8 x 1-3/8
other widths: 3/4, 7/8, 1-1/8,
1-1/4, 1-5/8, 1-3/4, 2-1/4

3/8 x 1-3/8
other widths: 3/4, 7/8, 1-1/8,
1-1/4, 1-5/8, 1-3/4, 2-1/4

7/16 x 1-3/8
other widths: 3/4, 7/8, 1-1/8,
1-1/4, 1-5/8, 1-3/4, 2-1/4

7/16 x specified widths

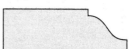

7/16 x 1-3/8
other widths: 3/4, 7/8, 1-1/8,
1-1/4, 1-5/8, 1-3/4, 2-1/4

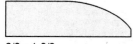

3/8 x 1-3/8
other widths: 3/4, 7/8, 1-1/8,
1-1/4, 1-5/8, 1-3/4, 2-1/4

3/8 x specified widths

3/8 x 1-3/8
other widths: 3/4, 7/8, 1-1/8,
1-1/4, 1-5/8, 1-3/4, 2-1/4

PANEL STRIPS/MULLION CASINGS

3/8 x 1-3/4, 2, 2-1/4

3/8 x 1-3/4, 2, 2-1/4

3/8 x 1-3/4, 2, 2-1/4

3/8 x 1-3/4, 2, 2-1/4

3/8 x 1-3/4, 2, 2-1/4

3/8 x 1-3/4, 2, 2-1/4

3/8 x 1-3/4, 2, 2-1/4

RABBETED STOOLS

11/16 x 2-1/4, 2-1/2, 2-3/4, 3-1/4

1-1/16 x 2-3/4, 3-1/4, 3-5/8

11/16 x 2-1/4, 2-1/2, 2-3/4, 3-1/4

1-1/16 x 2-3/4, 3-1/4, 3-5/8

FLAT STOOLS

11/16 x specified widths

Note: Dimensions are in inches.

Decorating with Wood Moldings

A variety of decorative effects can be achieved by applying simple moldings to building surfaces. Walls, ceilings, and doors are all candidates. In addition, stand-alone items such as screens, fences, and trellises can be constructed of S4S structural frames and lattice in-fill. On this and the next page are ideas from the Wood Moulding and Millwork Producers Association.

Cabinet Doors

Bar and Counter Fronts

Interior and Exterior Doors

15   Heating

Americans have learned that energy supplies can no longer be taken for granted. No longer is the heating system a neglected, mysterious object in the nether regions of our homes that we try not to think about. Winter comfort requires that we have the right type of heating system, properly sized to the heat load of the house.

This chapter begins with a simple form for calculating *heat loads,* both for the coldest day of winter and for the entire winter.

Next, you'll learn how the various *heating systems* work and the pros and cons of each.

Most of us have considered switching fuels at some time. Many of us supplement our main heating systems with wood stoves. *"Comparing Fuels and Efficiencies"* gives you a simple method for comparing the cost per useful Btu of all fuels and a chart that shows the percentage fuel savings from installing a more efficient system.

If you are going to have a fireplace anyway, why not make it as fuel-efficient as possible? *"Fireplaces"* contains detailed, energy-efficient plans for both standard and Rumford fireplaces.

For others, a wood stove is a more practical alternative to the fireplace. *"Wood Stove Installation"* shows the fire code requirements for stove and stovepipe clearances. It also shows how to vent a wood stove into an existing fireplace.

If you don't already have a sound and safe masonry flue, the section on *metal prefabricated chimneys* will be helpful, showing every detail of six typical installations.

Finally, we provide you with a *checklist of code requirements* relating to heating.

HEAT LOADS

Two types of heat load (rate of heat loss) are

design heat load, the rate of heat loss in British thermal units (Btu) per hour when the outside temperature is at the *design minimum temperature* for the site; used to size the heating system

annual heat load, the total heat loss in Btu for the entire heating season; used to estimate the annual heating bill

Use the work sheet on the following page to estimate both of these loads for your home. An example work sheet also follows.

Heat Load Work Sheet Instructions

Line 1. Use line 1 if you have an unheated attic. Get the R-value from chapter 13, or use 3.0 if the attic is uninsulated.

Line 2. Use line 2 if the ceiling is the underside of the roof. Get the R-value from chapter 13, or use 3.0 if the roof is uninsulated.

Line(s) 3. Get the wall R-values from chapter 13, or use 4.0 if the wall is uninsulated. Use a different line for each different section of wall. Subtract window and door areas.

Line(s) 4. The area of most exterior doors is 20 square feet. Use R-2 for a solid wood door, R-3 for a wood door plus storm door, and R-6 for an insulated door.

Line(s) 5. The window area is the area of the sash. Use R-1 for single glazing, R-2 for double glazing, R-3 for triple and Low-E glazing, and R-4 for Heat Mirror.

Line 6. Use this line if your home, or a portion of it, sits on piers or a ventilated crawl space. Get the floor R-values from chapter 13, or use 5.0 if the floor is uninsulated.

Line 7. Use this line if your home sits on a concrete slab. Use R-20 if the slab is uninsulated. Add the insulation R-value (from chapter 13) if it is insulated.

Line 8. Use this line if your home has a basement. Use R-5 if the foundation is uninsulated. Add the insulation R-value (from chapter 13) if it is insulated.

Line 9. For air changes per hour, use 1.5 for a drafty house, 0.75 for a typical 10 to 30-year-old house, 0.50 for an average new house, and 0.25 for a new "energy-efficient" house. Heated volume is 8x heated floor area.

Line 10. Add up all of the numbers appearing in the right-hand column above this line.

Line 11. First enter the sum from line 10. Next enter 65 minus the design minimum temperature ($DMT_{97.5\%}$) from the table of cities following the example work sheet. Multiply the entries and enter the result in the right column. Your heating contractor can use this result to size your heating system.

Line 12. Enter the sum from line 10. Next find heating degree-days, base 65°F, (HDD_{65}) from the table of cities following the example work sheet. Multiply the entries. The result is the total annual heat loss.

To estimate fuel requirements, divide this number by 100,000 for gallons of oil; 70,000 per hundred cubic feet of gas; 3,410 for kilowatt-hours (kwhr) of electric-resistance heat; and 6,830 for an electric heat pump.

HDD_{65} is used with the assumption that the house requires heat when the daily average outdoor temperature drops below 65°F. If your house retains internal heat gains well, the heat may not come on until the average temperature drops to 55°F, for example. In that case, use HDD_{55} from the same table instead.

Work Sheet for Design Minimum and Annual Heat Loads

Heat Loss	Calculations			Results
1. Ceiling under attic	_____ sq ft ÷	_____ ceiling R-value	=	_____
2. Cathedral ceiling	_____ sq ft ÷	_____ ceiling R-value	=	_____
3. Exterior wall #1	_____ sq ft ÷	_____ wall R-value	=	_____
Exterior wall #2	_____ sq ft ÷	_____ wall R-value	=	_____
Exterior wall #3	_____ sq ft ÷	_____ wall R-value	=	_____
Exterior wall #4	_____ sq ft ÷	_____ wall R-value	=	_____
Exterior wall #5	_____ sq ft ÷	_____ wall R-value	=	_____
Exterior wall #6	_____ sq ft ÷	_____ wall R-value	=	_____
4. Exterior door #1	_____ sq ft ÷	_____ door R-value	=	_____
Exterior door #2	_____ sq ft ÷	_____ door R-value	=	_____
Exterior door #3	_____ sq ft ÷	_____ door R-value	=	_____
5. Window type #1	_____ sq ft ÷	_____ window R-value	=	_____
Window type #2	_____ sq ft ÷	_____ window R-value	=	_____
Window type #3	_____ sq ft ÷	_____ window R-value	=	_____
6. Floor over crawl	_____ sq ft ÷	_____ floor R-value	=	_____
7. Slab-on-grade	_____ sq ft ÷	_____ total slab R-value	=	_____
8. Foundation wall	_____ sq ft ÷	_____ total wall R-value	=	_____
9. Air changes per hour	_____ × 0.018 × heated volume	_____ cu ft	=	_____
10. Sum of results from all lines above			=	_____
11. **Design heat load:** line 10	_____ ×	_____ (65°F - DMT)	=	_____ Btu/hr
12. **Annual heat load:** line 10	_____ × 24 ×	_____ HDD_{65}	=	_____ Btu/yr

Sample Heat Load Calculations

The next page contains an example work sheet showing the calculations for design heat load and annual heat load for the small house in Boston, Massachusetts, shown at right.

The house is deliberately kept simple in order to clarify the calculations. Many homes will have more than one type of exterior wall, foundation, or window, and they will require multiple entries for these items.

Line 1. The ceiling measures 30×40 feet, so its area is 1,200 square feet. Chapter 13 gives an R-value of 35.5 for its two R-19 batts.

Line 3. After the areas of windows and doors are deducted, the remaining area of exterior wall is 996 square feet. The 2x6 wall with R-19 batts has an R-value of 17.2.

Line 4. The first exterior door is solid wood with storm and has a combined R-value of 3.0. The second has an insulated core and R-value of 6.0.

Line 5. All of the windows are single glazed with storm windows (total R-value of 2.0). The total area of window sash is 84 square feet.

Line 6. The house sits on a crawl space that is ventilated in winter and is insulated with R-19 batts between the joists. Chapter 13 gives this type of floor an R-value of 20.9.

Line 9. The house is a recently built tract home, so its air change rate is about 0.50 changes per hour. The heated volume is the floor area times the ceiling height, 8 feet.

Line 10. The sum of all of the results in the right column is 287.5.

House for Heat Load Example

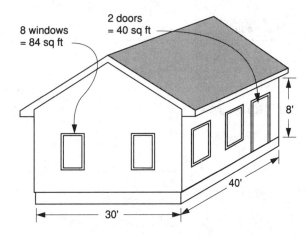

8 windows = 84 sq ft

2 doors = 40 sq ft

8'

40'

30'

Volume of heated space = 9,600 cubic feet
Air exchange rate = 0.5 changes per hour

Line 11. From the table following the example work sheet, $DMT_{97.5\%}$ is 6°F for Boston. The result for this line is a design heat load at 6°F of 16,963 Btu per hour.

Line 12. From the same table, for Boston the $HDD_{65} = 5,621$. The result of multiplying the three numbers on line 12 is the annual heat load: 38,785,000 Btu.

If the house were heated with oil, the approximate winter fuel consumption would be the annual heat load divided by 100,000, or 388 gallons of oil.

Example Work Sheet for Design Minimum and Annual Heat Loads

Heat Loss	Calculations		Results
1. Ceiling under attic	1,200 sq ft ÷ 35.5	ceiling R-value =	33.8
2. Cathedral ceiling	_____ sq ft ÷ _____	ceiling R-value =	_____
3. Exterior wall #1	996 sq ft ÷ 17.2	wall R-value =	57.9
Exterior wall #2	_____ sq ft ÷ _____	wall R-value =	_____
Exterior wall #3	_____ sq ft ÷ _____	wall R-value =	_____
Exterior wall #4	_____ sq ft ÷ _____	wall R-value =	_____
Exterior wall #5	_____ sq ft ÷ _____	wall R-value =	_____
Exterior wall #6	_____ sq ft ÷ _____	wall R-value =	_____
4. Exterior door #1	20 sq ft ÷ 3.0	door R-value =	6.7
Exterior door #2	20 sq ft ÷ 6.0	door R-value =	3.3
Exterior door #3	_____ sq ft ÷ _____	door R-value =	_____
5. Window type #1	84 sq ft ÷ 2.0	window R-value =	42.0
Window type #2	_____ sq ft ÷ _____	window R-value =	_____
Window type #3	_____ sq ft ÷ _____	window R-value =	_____
6. Floor over crawl	1,200 sq ft ÷ 20.9	floor R-value =	57.4
7. Slab-on-grade	_____ sq ft ÷ _____	total slab R-value =	_____
8. Foundation wall	_____ sq ft ÷ _____	total wall R-value =	_____
9. Air changes per hour	0.50 × 0.018 × heated volume 9,600 cu ft =		86.4
10. Sum of results from all lines above		=	287.5

11. **Design heat load:** line 10 ___287.5___ × ___59___ (65°F - DMT) = ___16,963___ Btu/hr

12. **Annual heat load:** line 10 ___287.5___ × 24 × ___5,621___ HDD$_{65}$ = ___38,785,000___ Btu/yr

DESIGN MINIMUM TEMPERATURE AND HEATING DEGREE-DAYS FOR US AND CANADIAN CITIES

City	$DMT_{97.5\%}$	HDD_{50}	HDD_{55}	HDD_{60}	HDD_{65}
Alabama, Birmingham	19	702	1,240	1,961	2,844
Arizona, Phoenix	31	187	459	919	1,552
Arkansas, Little Rock	19	984	1,617	2,414	3,354
California, Los Angeles	43	64	299	849	1,819
Colorado, Denver	-2	2,592	3,588	4,733	6,016
Connecticut, Hartford	1	2,971	3,948	5,075	6,350
Delaware, Wilmington	12	1,978	2,829	3,818	4,940
DC, Washington	12	2,004	2,869	3,864	5,010
Florida, Orlando	33	39	126	348	733
Georgia, Atlanta	18	758	1,362	2,150	3,095
Idaho, Boise	4	2,420	3,395	4,536	5,833
Illinois, Chicago	-4	2,954	3,881	4,940	6,127
Indiana, Indianapolis	0	2511	3,403	4,421	5,577
Iowa, Des Moines	-7	3491	4,435	5,510	6,710
Kansas, Topeka	3	2325	3,175	4,137	5,243
Kentucky, Louisville	8	1816	2,625	3,563	4,645
Louisiana, Baton Rouge	25	232	530	1,006	1,670
Maine, Portland	-5	3648	4,758	6,039	7,498
Massachusetts, Boston	6	2,374	3,300	4,381	5,621
Michigan, Detroit	4	2,931	3,890	4,986	6,228
Minnesota, Duluth	-19	5,560	6,792	8,189	9,756
Mississippi, Jackson	21	471	898	1,506	2,300
Missouri, St Louis	4	1,961	2,762	3,686	4,750
Montana, Great Falls	-20	3,761	4,893	6,191	7,652
Nebraska, North Platte	-6	3,326	4,325	5,473	6,743
Nevada, Las Vegas	23	631	1,129	1,788	2,601
New Hampshire, Concord	-11	3,653	4,726	5,954	7,360

City	DMT$_{97.5\%}$	HDD$_{50}$	HDD$_{55}$	HDD$_{60}$	HDD$_{65}$
New Jersey, Newark	11	2,056	2,908	3,905	5,034
New Mexico, Albuquerque	14	1,497	2,292	3,216	4,292
New York, Syracuse	-2	3,215	4,218	5,366	6,678
North Carolina, Greensboro	14	1,202	1,916	2,797	3,825
North Dakota, Bismarck	-24	5,235	6,364	7,627	9,044
Ohio, Columbus	2	2,524	3,438	4,491	5,702
Oklahoma, Tulsa	12	1,245	1,910	2,731	3,680
Oregon, Salem	21	1,265	2,220	3,411	4,852
Pennsylvania, Pittsburgh	5	2,635	3,574	4,669	5,930
Rhode Island, Providence	6	2,566	3,543	4,669	5,972
South Carolina, Columbia	20	590	1,094	1,772	2,598
South Dakota, Rapid City	-9	3,681	4,749	5,965	7,324
Tennessee, Nashville	12	1,195	1,874	2,720	3,696
Texas, Dallas	19	505	943	1,543	2,290
Utah, Salt Lake City	5	2,648	3,612	4,725	5,983
Vermont, Burlington	-12	4,142	5,230	6,464	7,876
Virginia, Richmond	14	1,296	2,021	2,909	3,939
Washington, Seattle	28	1,386	2,393	3,662	5,185
West Virginia, Charleston	9	1,726	2,540	3,488	4,590
Wisconsin, Madison	-9	4,086	5,143	6,352	7,730
Wyoming, Casper	-11	3,723	4,850	6,131	7,555
Alberta, Edmunton	96[1]	6,317	7,563	9,016	10,650
British Columbia, Vancouver	51[1]	1,791	2,781	4,041	5,588
Manitoba, Winnipeg	95[1]	6,925	8,062	9,338	10,790
Nova Scotia, Halifax	57[1]	3,457	4,500	5,746	7,211
Ontario, Ottawa	83[1]	4,912	5,965	7,158	8,529
Quebec, Normandin		7,037	8,308	9,762	11,376

[1] Design temperature difference (same as 65°F - DMT) is used in Canada.

HEATING SYSTEMS

Hydronic Boiler

The hydronic, or forced-hot-water, system heats water in a gas or oil boiler and circulates it through loops of pipe to distribute heat to separate heating zones.

Annual fuel utilization efficiencies (AFUEs) are similar to those of furnaces: about 85 percent for high-efficiency models and about 90 percent for condensing boilers.

Pros include even heating, ease of zoning with separate thermostats, and small, easy-to-conceal pipes.

Cons include no air cleaning or humidification and no sharing of ducts with air-conditioning.

Hydronic Boiler

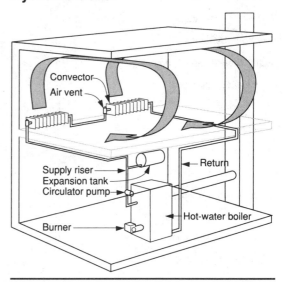

Steam Boiler

In a steam system, steam is produced in a steam-rated oil or gas boiler, circulated through insulated pipes to room radiators, and condensed in the radiator, giving up its heat of vaporization. The condensed water then drains back to the boiler for reheating.

Steam boilers are now only for replacement, since the system is difficult to control.

Pros include longevity and heat distribution by both radiation and convection.

Cons include difficult control and expensive installation.

Steam Boiler

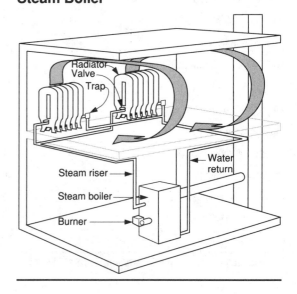

Warm-Air Furnace

The first furnaces distributed heat by natural convection of warm, buoyant air through large ducts. The modern warm-air furnace is a great improvement. Heat is produced by clean and efficient combustion of gas or oil, and the warm air is distributed evenly throughout the building by a blower, supply and return ducts, and registers.

AFUEs average about 84 percent for high-efficiency furnaces and 93 percent for condensing furnaces.

Pros include circulation and filtration of air, humidification and dehumidification, possible integration with heat exchangers, and ductwork that can be shared with air-conditioning.

Cons include bulky, hard-to-conceal ducts, noise at high air velocity, and sound transmission between rooms.

Coal/Wood Stove

Solid-fuel (wood or coal) stoves heat both by radiation to the immediate surroundings and by natural convection of warmed air.

Pros include inexpensive fuel and radiant heat that helps to define social spaces.

Cons are smoke and pollution, dangerous creosote deposits in the chimney, no thermostatic control, and messy wood and ashes.

Warm-Air Furnace

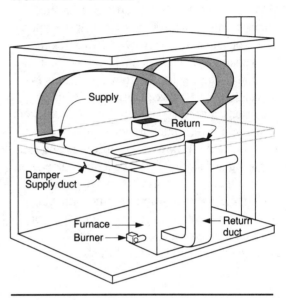

Coal/Wood Stove

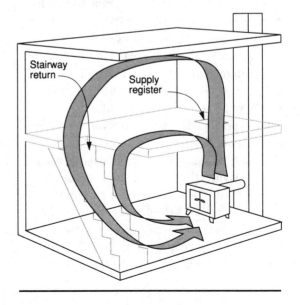

Electric-Resistance Baseboards

Electric-resistance baseboards convert electricity to heat with 100 percent efficiency; no heat goes up a flue. However, 100 percent conversion efficiency is misleading. The generation of the electricity from fossil fuels is typically only 40 percent efficient, making electricity a very expensive "fuel." Exceptions you could investigate include regionally lower cost due to hydropower and utility incentives for using and storing lower-cost off-peak power.

Pros include the least expensive installation cost of all, room-by-room thermostatic control, and freedom from pipes and ducts. Cons include the highest operating cost and a lack of air cleaning and humidification.

Electric Radiant Panels

Radiant panels—typically installed under a drywall ceiling or as part of a dropped ceiling—heat by emitting infrared (heat) radiation. When the radiation strikes a surface (floor, furnishings, or people), it changes to sensible heat.

Proponents of radiant heat claim it to be the most natural and comfortable of all heat distribution systems, since it emulates that most natural of all heating systems, the sun.

If the point of heating is to achieve comfort, then electric radiant heat is more efficient than electric-resistance heating because it heats the occupants directly, rather than the air around them. This allows lower indoor air temperatures, which lowers the heating bill.

Pros include a high degree of comfort and a low installation cost. Cons include a high operating cost relative to fossil fuels.

Electric Baseboards

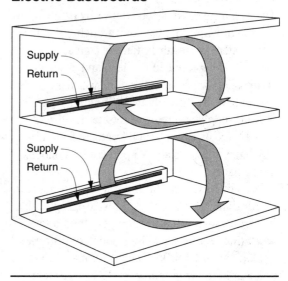

Radiant Panels

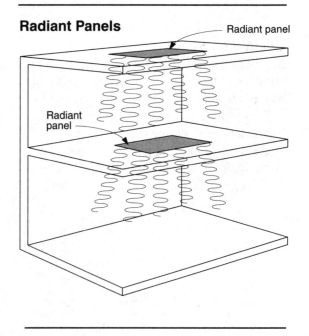

Heat Pump

Heat pumps operate on the same principle as refrigerators. By compressing and expanding a gas (the refrigerant), they reverse nature and pump heat from a cooler "source" to a warmer "sink." By reversing the pump, you can cool, as well as heat, a house.

Most heat pumps extract heat from outside air. These air-source heat pumps are cost-effective only in warm regions where air temperature rarely dips below 45°F. For more northern areas, consider more expensive but more efficient water-source and ground-source heat pumps.

Pros include circulation and filtration of air, humidification and dehumidification, and combined heating and air conditioning.

The single disadvantage is the relatively high installation cost—especially for water-source and ground-source systems.

Direct-Vent Heater

The direct-vent heater takes its name from the fact that, due to small size and high efficiency, its combustion gases are cool enough to be safely vented directly through the wall. Both propane and kerosene versions are available. Air return and supply are directly into and out of the unit, eliminating duct work, as well. The savings due to elimination of chimney, ducts, and pipes result in a relatively low installation cost.

Pros include both a low installation cost and a low operating cost.

Cons are a small output (about 50,000 Btu/h maximum) and a lack of ducts, limiting direct-vent heaters to relatively small, open spaces and superinsulated buildings.

Heat Pump

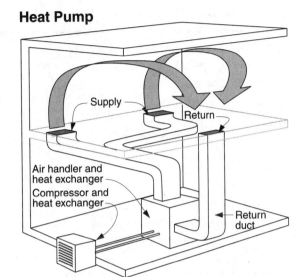

Direct-Vent Heater

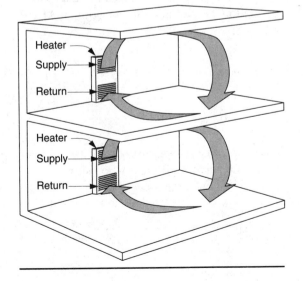

COMPARING FUELS AND EFFICIENCIES

The only way to compare heating costs is on an apple-to-apple basis, in this case cost per Btu of useful heat:

$$\text{Cost per Btu} = \frac{100 \times P}{F \times AFUE}$$

where

F = Btu content of a unit of fuel
P = price per unit of fuel, $
$AFUE$ = annual fuel utilization efficiency, percent

The table at right lists the Btu contents of fuels in their customary quantities. Coal is also sold by the ton. Natural gas is also sold by the 100 cubic feet and the nearly equivalent therm (100,000 Btu content).

Example: What is the cost of 1 million (10^6) Btu of useful heat from #2 fuel oil at $1.10 per gallon, burned in a furnace with an AFUE rating of 82 percent?

$$\text{Cost per Btu} = \frac{100 \times \$1.10}{139,000 \times 82}$$
$$= \$9.65 \times 10^{-6}$$
Cost per 10^6 Btu = $9.65

Example: What is the cost per 10^6 Btu of heat from a 60 percent-efficient wood stove burning red oak at $95 per cord?

$$\text{Cost per } 10^6 \text{ Btu} = \frac{10^6 \times 100 \times \$95}{24.0 \times 10^6 \times 60}$$
$$= \$6.60$$

ENERGY CONTENT OF FUELS

Fuel		Btu per	Unit
Coal	Anthracite	15,000	lb
	Bituminous	13,000	lb
	Cannel	11,000	lb
	Lignite	11,000	lb
Electricity		3,412	kwhr
Gas	Natural	1,030	cu ft
	Propane	91,600	gal
Oil	#1 (kerosene)	134,000	gal
	#2 (residential)	139,000	gal
	#4	150,000	gal
	#6	153,000	gal
Wood	Ash	25.0×10^6	cord
	Beech	28.0×10^6	cord
	Birch, white	23.4×10^6	cord
	Birch, yellow	25.8×10^6	cord
	Fir, Douglas	21.4×10^6	cord
	Hemlock	18.5×10^6	cord
	Hickory	30.6×10^6	cord
	Maple, red	24.0×10^6	cord
	Maple, sugar	29.0×10^6	cord
	Oak, red	24.0×10^6	cord
	Oak, white	30.6×10^6	cord
	Pine, pitch	22.8×10^6	cord
	Pine, white	15.8×10^6	cord
	Poplar	17.4×10^6	cord
	Spruce	17.5×10^6	cord
	Tamarack	23.1×10^6	cord

Savings from Increased Efficiency

The cost-per-Btu formula on the previous page can also be used to calculate the savings realized from increasing heating-system efficiency. The table below is more convenient, however, because it shows the reduction in annual heating bill as a percentage of the present bill.

Example: What savings will be realized on a heating bill of $700 by installing a new gas furnace if the AFUE is increased from 65 to 85 percent?

From the table below, the annual savings will be 23.5 percent of $700, or $164.50.

Example: How long will it take for a flame retention head burner replacement to pay back its cost if the burner increases the estimated annual efficiency of a boiler from 60 to 75 percent? The present fuel bill is $1,200 per year, and the new burner costs $400.

From the table below, the annual savings will be 20 percent of $1,200, or $240. The $400 cost will be paid back in $400÷$240 = 1.7 years. This is equivalent to receiving 60 percent interest on the $400 investment.

PERCENTAGE FUEL SAVINGS FROM INCREASED ANNUAL EFFICIENCY

From AFUE, %	To AFUE, %									
	55	60	65	70	75	80	85	90	95	100
40	27.3	33.3	38.5	42.9	46.7	50.0	52.9	55.6	57.9	60.0
45	18.2	25.0	30.8	35.7	40.0	43.8	47.1	50.0	52.6	55.0
50	09.1	16.7	23.1	28.6	33.3	37.5	41.2	44.4	47.4	50.0
55	00.0	08.3	15.4	21.4	26.7	31.3	35.3	38.9	42.1	45.0
60		00.0	07.7	14.3	20.0	25.0	29.4	33.3	36.8	40.0
65			00.0	07.1	13.3	18.8	23.5	27.8	31.6	35.1
70				00.0	06.7	12.5	17.6	22.2	26.3	30.0
75					00.0	06.3	11.8	16.7	21.1	25.0
80						00.0	05.9	11.1	15.8	20.0
85							00.0	05.6	10.5	15.0
90								00.0	05.3	10.0
95									00.0	05.0

FIREPLACES

Standard Fireplace with Exterior Air

The modified standard fireplace design, illustrated on the following page, offers two key advantages: It uses outside air for combustion, and it cuts down on infiltration of cold air caused by replacement of warm air used for combustion. Both factors mean greater efficiency. To reduce infiltration when the fireplace is not in operation, tight-fitting dampers, louvers, and glass screens should be used.

The intake is located in an outside wall or in the back of the fireplace and requires a screened closable louver, preferably one that can be operated from inside the house.

To ensure sufficient air, the passageway should have a cross-sectional area of at least 55 square inches. The insulated passageway can be built into the base of the fireplace assembly or channeled between joists in the floor.

The inlet brings the outside air into the firebox. A damper is required for volume and direction control. The damper is located in the front of the firebox, and although its dimensions may vary, most openings will be about 4-1/2 × 13 inches.

The air intake pit, located directly below the inlet, should have the same dimensions and should be about 13 inches deep.

CONVENTIONAL FIREPLACE DIMENSIONS, inches

| | Finished Fireplace Opening | | | | | | Rough Brickwork | | | | Flue Size[1] | Steel Angle |
A	B	C	D	E	F	G	H	I	J	K	L x M	N
24	24	16	11	14	18	8-4/4	32	21	19	10	8 x 12	A-36
26	24	16	13	14	18	8-3/4	34	21	21	11	8 x 12	A-36
28	24	16	15	14	18	8-3/4	36	21	21	12	8 x 12	A-36
30	29	16	17	14	23	8-3/4	38	21	24	13	12 x 12	A-42
32	29	16	19	14	23	8-3/4	40	21	24	14	12 x 12	A-42
36	29	16	23	14	23	8-3/4	44	21	27	16	12 x 12	A-48
40	29	16	27	14	23	8-3/4	48	21	29	16	12 x 12	A-48
42	32	16	29	16	24	8-3/4	50	21	32	17	12 x 16	B-54
48	32	18	33	16	24	8-3/4	56	23	37	20	12 x 16	B-60
54	37	20	37	16	29	13	68	25	45	26	16 x 16	B-66
60	40	22	42	18	30	13	72	27	45	26	16 x 20	B-72
72	40	22	54	18	30	13	84	27	56	32	20 x 20	C-84

[1] Flue size assumes chimney height of ≥ 25 feet.
Source: *Residential Fireplace Design* (Reston, Va: Brick Institute of America, 1993).

Standard Fireplace with Exterior Air

SECTION

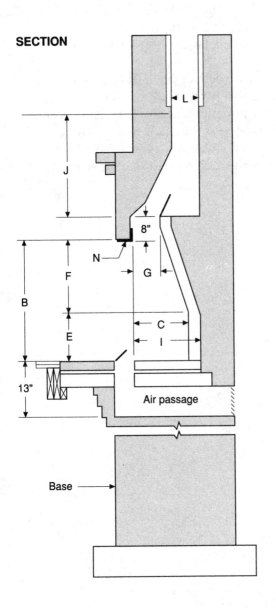

PLAN VIEW

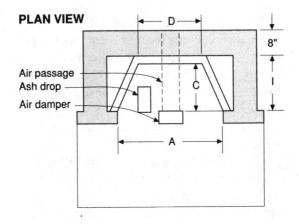

FRONT ELEVATION

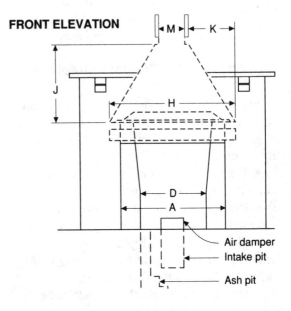

Rumford Fireplace with Exterior Air

Most of the heat that can be gotten from a fireplace is radiated heat. There is a particular class of fireplace design, featuring shallow fireboxes with obliquely flared sides and backs, that produces the maximum amount of radiated heat. This design is called the *Rumford fireplace*. The design rules below and the illustration on the following page provide details and dimensions for a Rumford fireplace and guidance on how it should be designed and built.

If you combine this data with the information on using outside air for combustion and draft, and glass screens, you can achieve the maximum in energy-efficient fireplace design. A Rumford fireplace, with exterior air supply and glass screens for the opening, will provide the most radiated heat, the least infiltration, and the best all-around energy performance to be found in true (noninsert) fireplaces.

Design Rules for Rumford Fireplaces

- The width of the firebox (D) must equal the depth (C).
- The vertical portion of the firebox (E) must equal the width (D).
- The thickness of the firebox (I minus C) should be at least 2-1/4 inches.
- The area of the fireplace opening (A × B) must not exceed ten times the flue opening area.
- The width of the fireplace opening (A) and its height (B) should each be two to three times the depth of the firebox (C).
- The opening height (B) should not be larger than the width (A).

- The throat (G) should be not less than 3 nor more than 4 inches.
- The centerline of the throat must align with the center of the firebox.
- The smoke shelf (R) should be 4 inches wide.
- The width of the lintel (O) should be not less than 4 nor more than 5 inches.
- The vertical distance from lintel to throat (P) must be at least 12 inches.
- A flat plate damper is required at the throat and must open toward the smoke shelf.

Rumford Fireplace with Exterior Air

SECTION

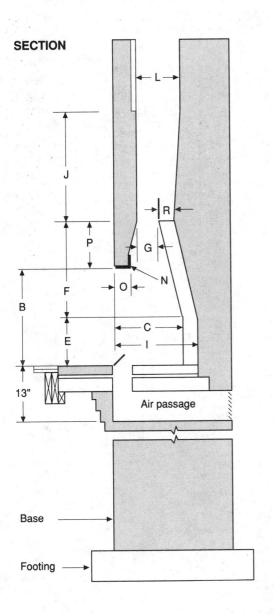

Base

Footing

PLAN VIEW

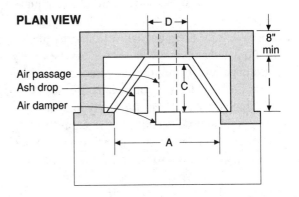

8" min

Air passage
Ash drop
Air damper

FRONT ELEVATION

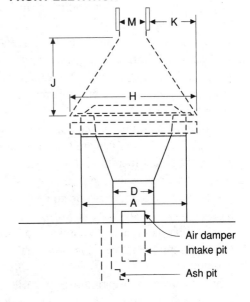

Air damper
Intake pit
Ash pit

WOOD STOVE INSTALLATION

Clearances

Wood and coal stoves are classified as circulating, radiant, or cookstove. A circulating stove has two walls: an inner wall surrounding the firebox, and an outer wall. Heated air rises, drawing cool air from the floor into the space between the inner and outer walls. Because the air circulates, clearances for this type of stove are less than those needed for a radiant stove.

Much of the heat from the single-walled radiant stove is in the form of infrared radiation. Clearances are greater because the infrared energy radiates into combustible materials and changes their composition, lowering the temperature at which they can spontaneously combust. Since the changes are not always evident, proper clearances should be maintained to prevent the possibility of fire.

Stovepipes ordinarily run cooler than the stoves to which they are connected. However, in case of a chimney fire, the stovepipe may temporarily become much hotter than the stove. Clearances are therefore specified for stovepipes as well.

Combustible materials include anything that can burn. Examples are the wood box, magazine racks, furniture, draperies, and wood paneling. Even a gypsum-drywalled or plastered wall is combustible since the wood studs behind the surface can burn.

The illustration at right shows minimum side and rear clearances for radiant stoves and single-walled stovepipes, both with and without protection of combustible surfaces.

The tables on the the next pages list more detailed specifications.

**RADIANT STOVE,
NO REAR PROTECTION**

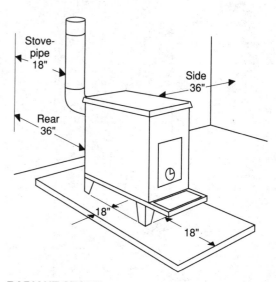

**RADIANT STOVE,
REAR PROTECTION**

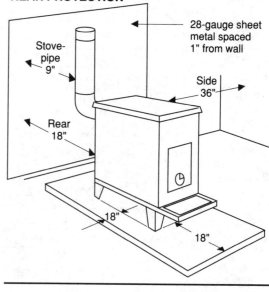

MINIMUM CLEARANCES FOR WOOD-BURNING STOVES WITH NO PROTECTION

Surface	Radiant	Circulating	Cookstove Lined Firepot	Cookstove Unlined Firepot	Stovepipe	UL-Listed Stoves
Ceiling	36"	36"	30"	30"	18"	Install following directions
Front	36"	24"	Cooking room	Cooking room	18"	
Side	36"	12"	Firing side 24" Opposite side 18"	Firing side 36" Opposite side 18"	18"	
Rear	36"	12"	24"	36"	18"	

MINIMUM CLEARANCES FOR WOOD-BURNING STOVES WITH ADDED PROTECTION

Minimum Protection Covering All Combustible Surfaces	Where the Required Clearance with No Protection Is:								
	36 inches			18 inches			12 inches		
	Above	Sides	Rear	Above	Sides	Rear	Above	Sides	Rear
28-gauge sheet metal spaced 1" from wall and 1" from floor	18"	12"	12"	9"	6"	6"	6"	4"	4"
22-gauge sheet metal on 1" mineral fiber batts reinforced with wire or equivalent, spaced 1" from wall and 1" from floor	18"	12"	12"	4"	3"	3"	2"	2"	2"

FLOOR CLEARANCES AND RECOMMENDED FLOOR COVERINGS

Floor Clearance	Suitable Protective Covering
More than 18"	24-gauge sheet metal
6" to 18"	24-gauge sheet metal over 1/4" layer of asbestos millboard or equivalent
Less than 6"	4" hollow masonry laid to provide air circulation through masonry layer, with 24-gauge sheet metal covering masonry
Listed stoves	Install according to manufacturer's directions

Chimney/Roof Clearances

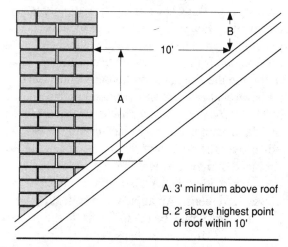

A. 3' minimum above roof

B. 2' above highest point of roof within 10'

Source of all tables: *Recommended Standards for the Installation of Woodburning Stoves* (Augusta, Maine: Maine State Fire Marshal's Office, 1979).

Installing a Wood Stove in a Fireplace

Except for the modern prefabricated fireplace insert, wood stoves are far more efficient than fireplaces. The reasons are these:

• A wood stove radiates heat in all directions, rather than from the opening only.

• A wood stove convects warm air to the room from all of its surfaces, rather than up the chimney.

• A wood stove draws the minimum amount of air required for combustion, rather than 10 to 100 times that required.

The illustration at right shows three ways to connect a wood stove to a fireplace.

Above the fireplace provides the best draft, produces the least mess, and is the safest. If there is not already a clay-lined thimble installed, call a mason or stove specialist to do the conversion.

Into the throat provides nearly as good a draft, provided the sheet metal throat plate is tight fitting. For safety, make sure the throat plate is fastened securely. An advantage of this method is that a small stove can fit partially into the fireplace, particularly if the outlet is vertical rather than horizontal as shown.

Through a cover panel provides the least draft and results in messy black creosote deposits over the entire interior surface of the fireplace. It may, however, be the simplest option, provided the fireplace already has glass fire screens. An approved panel, such as a stove mat, may be cut to fit inside the frame of the fireplace screens.

Fireplace Installations

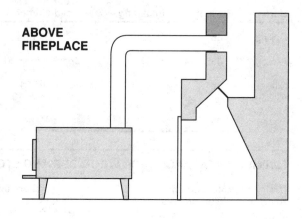

ABOVE FIREPLACE

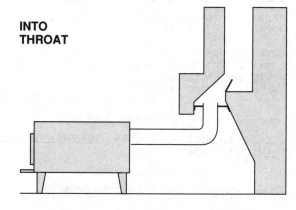

INTO THROAT

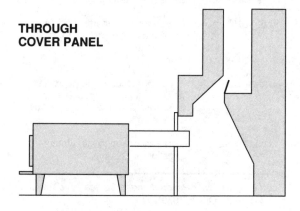

THROUGH COVER PANEL

Stovepipe Installation

Proper installation of the stovepipe connecting the stove to the chimney is important for both safety and performance:

• The stovepipe should be as short as possible. Horizontal runs should be no longer than 75 percent of the vertical chimney height above the connecting thimble.

• The stovepipe should be as straight as possible. No more than two 90-degree bends should be used. Additional bends could cause creosote to collect in the stovepipe or chimney, block flue gas flow, and increase the potential for fire.

• Use a stovepipe that has a diameter as large as the collar where the pipe joins the stove.

• Horizontal runs should rise 1/4 inch per foot, with the highest point at the thimble.

• When joining stovepipe, overlap the joint at least 2 inches, with the crimped end pointing down to prevent creosote leaks. Secure each joint with three sheet metal screws. A fireproof sealant (furnace cement) may be used as well.

• All pipe joints should fit snugly, including connections with the stove and thimble. The pipe should not project into the chimney and hinder the draft.

• The stovepipe should not pass through ceilings. Factory-built, listed, all-fuel chimney should be used for passing through ceilings. Follow the manufacturer's directions.

• A stovepipe passing through walls must be supported and spaced by vented thimbles of three times the diameter of the pipe. An alternative is a section of factory-built, listed, all-fuel chimney installed as directed.

Connecting to Chimneys

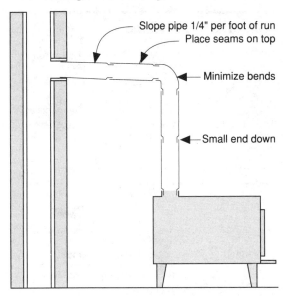

Running through Combustible Walls

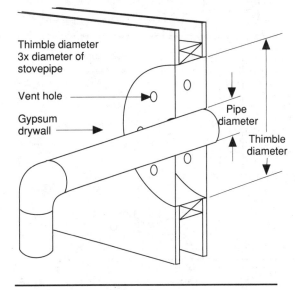

METAL PREFABRICATED CHIMNEYS

Offset to Avoid Ridge

Through Attic Insulation

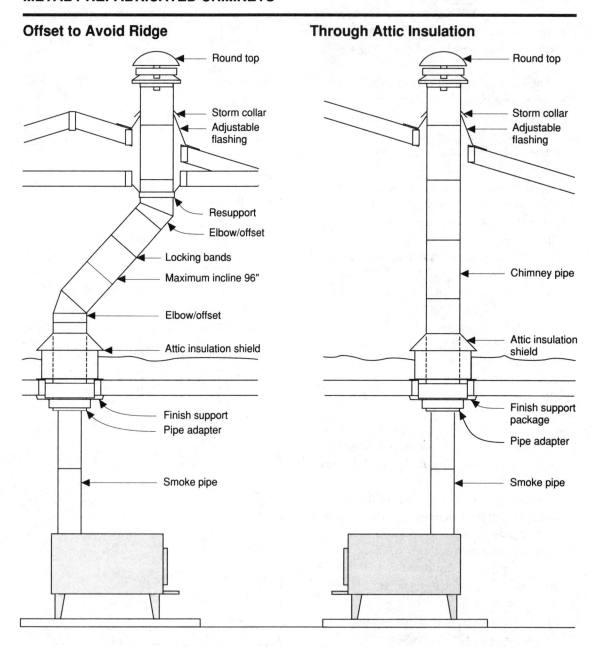

Offset to Avoid Ridge:
- Round top
- Storm collar
- Adjustable flashing
- Resupport
- Elbow/offset
- Locking bands
- Maximum incline 96"
- Elbow/offset
- Attic insulation shield
- Finish support
- Pipe adapter
- Smoke pipe

Through Attic Insulation:
- Round top
- Storm collar
- Adjustable flashing
- Chimney pipe
- Attic insulation shield
- Finish support package
- Pipe adapter
- Smoke pipe

Through Occupied Space Above

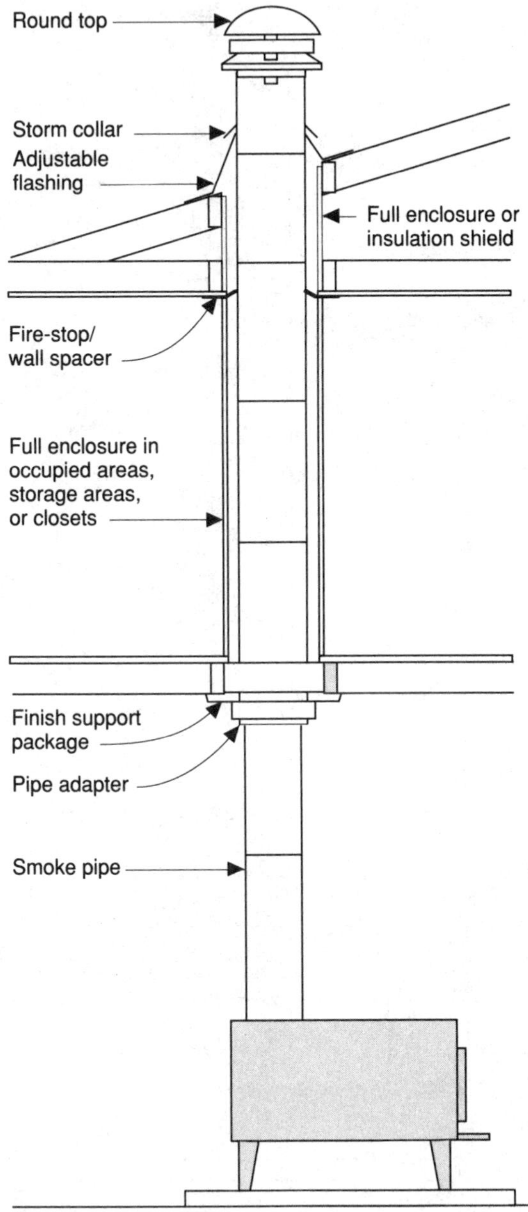

Round top

Storm collar
Adjustable flashing

Full enclosure or insulation shield

Fire-stop/ wall spacer

Full enclosure in occupied areas, storage areas, or closets

Finish support package

Pipe adapter

Smoke pipe

Through Steep-Pitch Roof

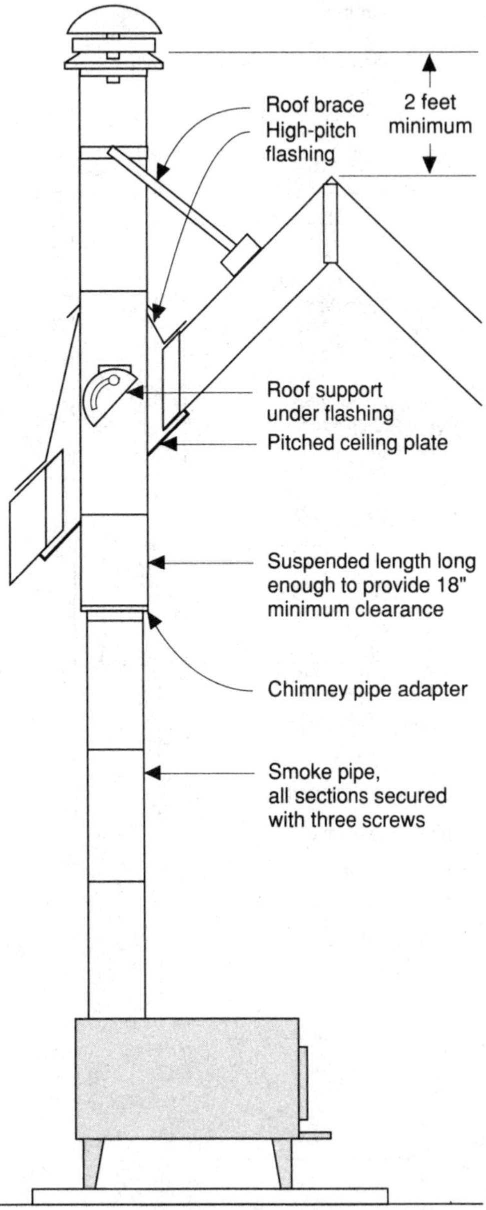

Roof brace
High-pitch flashing

2 feet minimum

Roof support under flashing

Pitched ceiling plate

Suspended length long enough to provide 18" minimum clearance

Chimney pipe adapter

Smoke pipe, all sections secured with three screws

Tee-Supported Outside Chimney

Fireplace Chimney

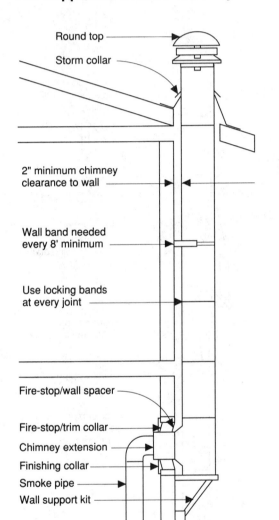

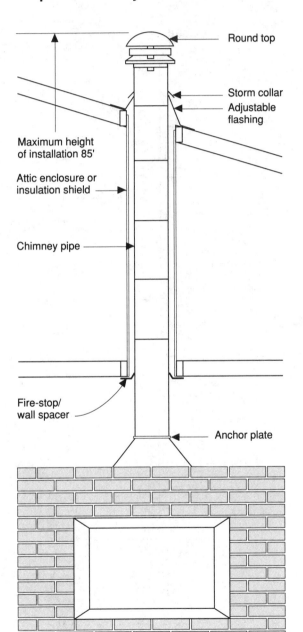

Round top

Storm collar

2" minimum chimney clearance to wall

Wall band needed every 8' minimum

Use locking bands at every joint

Fire-stop/wall spacer

Fire-stop/trim collar

Chimney extension

Finishing collar

Smoke pipe

Wall support kit

Round top

Storm collar

Adjustable flashing

Maximum height of installation 85'

Attic enclosure or insulation shield

Chimney pipe

Fire-stop/ wall spacer

Anchor plate

CHECKLIST OF CODE REQUIREMENTS

The following is a partial list of requirements from the 1995 Council of American Building Officials (CABO) *One and Two Family Dwelling Code*. Consult the publication for the full text and additional provisions.

Installations

(1307.3) Heating appliances in garage:
- protected from impact by automobiles
- igniters and switches ≥ 18" above floor

(1401.4) Equipment room installation:
- access to remove largest piece
- door ≥ 20" wide
- 30" wide × 30" high minimum clear space at control side with door of equipment room open

(1401.5.1) Permanent outlet and light controlled by switch at opening

(1401.6) Crawl space installation:
- access to remove equipment
- passage ≥ 20" wide × ≥ 30" high to controls

(1401.6.1) Ground clearance:
- suspended equipment ≥ 6" above ground
- supported equipment on level and on masonry ≥ 3" above ground

(1402.1) Fuel-burning warm-air furnaces:
- not in rooms designed to be closets
- in bedrooms or bathrooms, in sealed enclosures so that combustion air not taken from living space, except for direct-vent furnaces

(1602.2) Vented wall furnace location:
- ≥ 6" from adjoining walls in corners
- > 18" below overhead projections

- inlet/outlet ≥ 12" from door swing

(1602.3) Vented wall furnace installation:
- wall thickness to meet manufacturer's specifications
- no ducts to be attached

Combustion Air–General

(2001.1.1) For buildings considered to be of unusually tight construction:
- defined as buildings with continuous wall/ceiling vapor barriers, weatherstripped doors and windows, and caulked building joints and penetrations
- combustion air to be from outside
- combustion air not to be from areas containing combustible vapors
- effect of other fans on combustion air supply to be considered

All Combustion Air from Inside

(2002.1) Required volume of equipment space:
- > 50 cu ft per 1,000 Btu/h input rating
- connecting rooms with no doors may be considered part of space

(2002.2) Confined space (does not meet 2002.1):
- two permanent openings to adjacent spaces
- top opening ≤ 12" from top of enclosure
- bottom opening ≤ 12" from floor
- each opening minimum of 1 sq in per 1,000 Btu/h input rating of all appliances in space, but not less than 100 sq in

(2002.3) If volume is adequate but building is of unusually tight construction, combustion air to be from outside

All Combustion Air from Outside

(2003.1) Outdoor air openings:
- two permanent openings to outside
- top opening ≤ 12" from top of enclosure
- bottom opening ≤ 12" from bottom
- outside duct may not serve both openings
- top duct either level or sloping upward

(2003.2) Size of openings:
- vertical ducts 1 sq in/4,000 Btu/h minimum
- horizontal ducts 1 sq in/2,000 Btu/h minimum
- minimum duct dimension 3"

Vents

(2104.2.2) Natural-draft appliance:
- vent to terminate ≥ 5' above highest appliance outlet
- gas wall furnace vent to terminate ≥ 12' above bottom of furnace

(2104.2.5) Direct-vent appliance:
- vent terminals for ≤ 50,000 Btu/h appliance ≥ 9" from building openings
- vent terminals for > 50,000 Btu/h appliance ≥ 12" from building openings
- all vent terminals ≥ 12" above grade

(2104.3.2) Multiple-appliance vents:
- appliances to be on same floor
- inlets offset from each other
- natural-draft appliance not to be connected to positive-pressure mechanical-draft system
- solid fuel (wood, coal) appliances not to share vent with other-fuel appliances
- smaller of two connectors to enter at highest level possible

Chimney and Vent Connectors

(2103.4.2) Floor/wall/ceiling penetrations:
- chimney and vent connectors not to pass through floor/wall/ceiling unless listed or passed through devices listed for purpose
- single-wall metal pipe passed through ventilated thimble of ≥ 4" larger diameter

(2103.4.3) Connector length:
- horizontal run of uninsulated connector ≤ 75% of height of chimney above connector
- horizontal run of listed connector ≤ 100% of height of chimney above connector

(2103.4.4) Connector size:
- ≥ flue collar of appliance
- except by manufacturer's instructions and approved by building code official

(2103.4.7) Appliance not to be connected to fireplace flue unless fireplace opening or fireplace flue is sealed below connection

Masonry Chimneys

(2105.3) Masonry chimney connectors:
- entry to chimney ≥ 6" above bottom, except if 6" not available, install capped T in connector next to chimney for cleanout
- end of connector flush with inside of liner
- thimbles firmly cemented into masonry

(2105.4) Size of masonry chimneys:
- effective area of natural-draft chimney flue to be ≥ area of single appliance connector
- effective area of natural-draft chimney flue serving more than one appliance ≥ area of largest appliance connector plus 50% of areas of additional appliance connectors

16   Cooling

The object of this chapter is to help you achieve cool comfort for the lowest possible cost. You'll find there are numerous techniques you can employ before resorting to mechanical air-conditioning. To understand how and why these other techniques work, the chapter begins by explaining the relationship between *cooling and comfort.*

Inexpensive *attic radiant barriers* have been shown to result in dramatic cooling savings in southern states.

You've always known that *window shading* cuts down on solar gain in summer, but this chapter shows you exactly how much you will save using different types of shades and different widths of roof overhang.

"Moving Air with Fans" shows you how to select and size the right type of fan for your cooling needs, from a small oscillating fan in the kitchen, to an old-fashioned but surprisingly effective ceiling fan, to large window and whole-house fans.

If you live in a dry climate, chances are excellent that you can cool air by 20 degrees or more with an economical *evaporative cooler.* Here you'll find a table listing the evaporative cooling potential for your area and a formula for estimating the size of cooler for your home.

If all of the above fail to cool you sufficiently, use the work sheet provided for calculating the size of air conditioner needed to cool either a single room or the whole house.

COOLING AND COMFORT

Many people think of temperature as the only variable affecting their comfort. In fact, as the human comfort chart below illustrates, humidity, air movement, and radiation all strongly affect our comfort:

Relative humidity affects the rate of evaporation from the skin. Since evaporation of perspiration absorbs heat, it lowers the temperature of the skin. Thus, at high temperatures, we are more comfortable at lower relative humidities.

Air movement across the body increases heat loss from the body and allows us to feel comfortable at higher temperatures. In a 5 mph breeze, for example, the average person would feel comfortable at up to 87°F at 60 percent relative humidity, and up to 93°F at 20 percent relative humidity.

Radiation adds heat to the body. The chart below shows that radiation of 150 Btu/square foot-hour (approximately the level of midday winter solar radiation) allows the average person, dressed in normal clothing, to be comfortable down to a temperature of 45°F!

The Human Comfort Zone

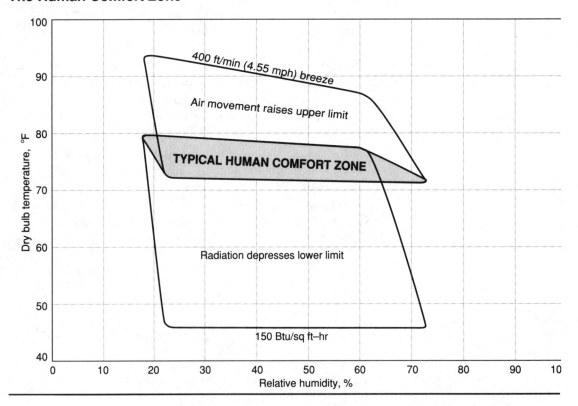

ATTIC RADIANT BARRIERS

A radiant barrier is a layer of aluminum foil installed in an interior space to block radiant heat transfer from a heat-radiating surface (such as a hot roof) to a heat-absorbing surface (such as conventional attic insulation). Radiant barriers are used predominately in attics in southern states, where they have proved effective in lowering cooling bills.

A radiant barrier is not insulation. It does not block *conducted heat,* as insulation does. You can pick up a Styrofoam cup of hot coffee because the Styrofoam is retarding the conduction of heat from the inside of the cup to the outside. But you'd have a rough time drinking hot coffee from an aluminum beer can. The heat would conduct right through.

Radiant heat consists of rays of energy that travel through air space and don't become heat until they strike an object. An aluminum barrier works because it is capable of reflecting 95 to 98 percent of the radiant heat that strikes it. But since aluminum is a good heat conductor, there must be an air space between the radiating surface and the surface to be shielded from radiant energy. As a result, there are two ways to effectively install a radiant barrier in an attic:

Under the sheathing is the simplest and cheapest method for new construction. The barrier, usually consisting of aluminum foil backed by a tear-resistant material, is rolled out on top of the rafters before the sheathing is nailed down.

Attached to the bottom of the rafters is the easiest, most effective retrofitting method. Simply staple the material to the underside of the rafters.

Whichever method you use, install the barrier with the foil facing down into the attic space. At first, the foil would work equally well facing up or down. However, dust would collect on upward-facing foil, preventing the barrier from working.

This dust problem also makes it ineffective to install the barrier over attic floor insulation, facing up.

Note that small tears or gaps in the radiant barrier will not significantly reduce overall performance.

Attic Section with Three Possible Locations for Radiant Barrier

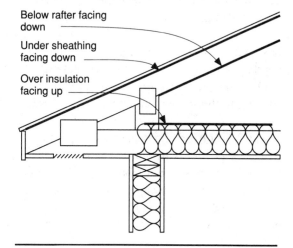

Below rafter facing down

Under sheathing facing down

Over insulation facing up

WINDOW SHADING

A high percentage of summer heat gain comes through windows as solar radiation. Surprisingly, much of the radiation is diffuse or reflected, and not directly from the sun. As a result, it pays to shade all windows, regardless of the direction they face. The single exception is passive solar glazings in northern states, which should have calculated overhangs as shown in chapter 17.

The tables in this section are based on preliminary research performed at the Florida Solar Energy Center for 30° north latitude. Summer solar gain is not very sensitive to latitude, however, so the results are applicable throughout the southern United States.

Roof Overhangs

The illustration and table at right show the percentage reductions in heat gain through the entire cooling season for 4 × 4-foot windows when shaded by continuous overhangs of various widths, W. The reductions are seen to be nearly independent of window orientation.

Of course, there are reasons other than solar shading for building overhangs:

• A 2-foot overhang protects siding and windows and doors from the weathering effects of rain. In addition, windows can be left open without worrying about sudden showers.

• A 6 to 8-foot overhang is common for porch roofs, which are very common in the South.

• A ten-foot overhang could be provided by a carport or vine covered patio.

Roof Overhang

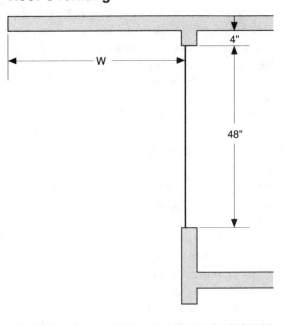

PERCENTAGE SUMMER HEAT GAIN REDUCTION FROM ROOF OVERHANGS

Window Facing	Width of Overhang (W), feet					
	1	2	3	4	6	10
North	16	32	44	54	66	78
East	14	32	47	58	72	84
South	17	35	47	56	67	79
West	15	32	47	58	71	83
Avg	16	33	46	57	69	81

Source: *Comparison of Window Shading Strategies for Heat Gain Prevention* (Cape Canaveral, Fla: Florida Solar Energy Center, 1984).

Window Awnings

The illustration at right shows the variable slope of opaque window awnings. All should have side panels. Both the awnings and windows are 4 × 4 feet, so at a 90° slope, the awning would completely cover the window. The table below the illustration shows the cooling-season reductions in heat gain through single-glazed windows at different awning slopes.

As the slope increases, visibility from the window decreases until, at 90° slope, nothing can be seen. The table indicates that much of the shading benefit is achieved with a slope of only 30°, at which point the top half of the view is blocked.

Canvas awnings can be rolled up, but the other shading options would be difficult or impossible to adjust in different seasons. We therefore need to know for each the net effect, or cooling-season savings less reduction in winter solar gain.

Both cooling and heating bills vary with latitude. As latitude increases, cooling bills decrease and heating bills increase. Net savings from fixed shading devices therefore vary with latitude. In the South, net savings are positive; in the North they can be negative.

The table at right shows how strongly net savings vary. Savings in Miami (virtually zero winter heating load) are nearly three times those in Jacksonville, only 4.5° to the north. For example, permanent charcoal-colored screens over 100 square feet of uniformly distributed glazing would save 1,000 kwhr in Miami, but only 350 kwhr in Jacksonville.

Shading devices above 30° north latitude should be of the adjustable variety in order to maximize both summer and winter savings.

Awning with Sides

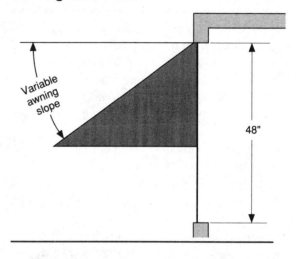

NET SAVINGS FROM SHADING, kwhr/sq ft-yr

Window Facing	Tinted Glass	Silver Film	Charcoal Screen	Awning No Side	Awning & Sides
Jacksonville, 30.5° N Latitude					
North	1.5	3.6	3.8	2.9	4.2
East	1.5	3.4	3.6	4.1	4.5
South	0.4	0.9	1.0	0.4	0.9
West	2.3	5.4	5.6	5.6	6.8
Avg	1.4	3.3	3.5	3.3	4.1
Miami, 26° N Latitude					
North	3.1	7.4	7.7	5.8	8.6
East	4.5	10.5	11.0	10.2	13.1
South	4.0	9.4	9.8	8.2	11.3
West	4.7	10.9	11.4	10.3	13.5
Avg	4.1	9.6	10.0	8.6	11.6

Source: *Comparison of Window Shading Strategies for Heat Gain Prevention* (Cape Canaveral, Fla: Florida Solar Energy Center, 1984).

MOVING AIR WITH FANS

During the cooling season in hot, humid regions of the United States, air-conditioning costs are the largest part of the monthly utility bill. In Florida, for example, air-conditioning accounts for about 30 percent of the total annual electric bill in an all-electric home. To a large extent, fans can reduce the need for air-conditioning by raising the upper limit of the comfort zone.

There are three generic types of fans in common use:

- *air-circulating,* which cool by air motion
- *whole-house,* which cool by air exchange
- *attic vent,* which lower attic temperatures

Air-Circulating Fans

Air-circulating fans include oscillating, box, and ceiling fans. The breeze they create can easily allow air conditioner setbacks of 5 to 10 F°, lowering the air-conditioning bill by 40 to 80 percent.

Portable *oscillating fans* are best at cooling small areas, such as people sitting at desks or working in the kitchen. Larger *box fans* can be placed in doorways to move large volumes of air between rooms. The quietest and most efficient by far, however, are *ceiling fans.* A ceiling fixture can usually be easily converted to a ceiling fan plus light. The illustration below shows required clearances and proper fan size.

Sizing Ceiling Fans

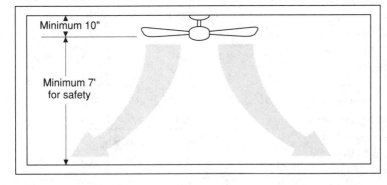

MINIMUM FAN SIZE

Largest Room Dimension, feet	Diameter of Fan, inches
12 or less	36
12 to 16	48
16 to 18	52
Over 18	2 fans

Whole-House Fans

As the illustration below shows, a centrally located whole-house fan can pull air from every room in which there is an open window. The total open window area should be three times the fan intake area. Similarly, the total attic vent area should be three times the fan intake, assuming screened vents.

The fan should be sized to replace one-third of the house air every minute. For example, a 1,500-square-foot house with 8-foot ceiling contains 12,000 cubic feet. The fan should therefore be rated at $1/3 \times 12,000 = 4,000$ cubic feet per minute (cfm). If the rating is for "free air," increase the cfm by 20 percent.

Attic Whole-House Fan

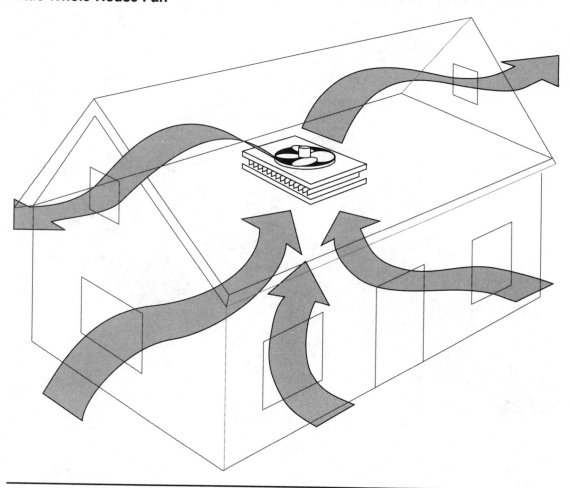

Window Fan as a Whole-House Fan

Whole-house fans are effective, but they are also expensive, difficult to retrofit, and power hungry (500 to 700 watts). Operated properly, a window fan can be nearly as effective:

• Locate the specific areas you wish to cool, such as the two bedrooms below.

• Place the fan to blow *out* in a room far from the cooled rooms.

• Open windows in the rooms to be cooled and all intervening doors.

• Close all other windows.

Get ready to enjoy a cooling breeze through your bedroom without the noise of a fan.

Air Exchange with a Window Fan

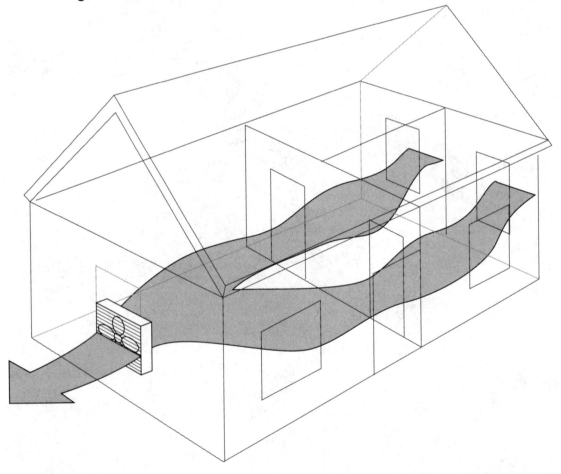

Attic Vent Fans

Powered attic vent fans ventilate the attic but not the house. Research has shown that attics with recommended natural ventilation and insulation levels do not need attic vent fans. Such vent fans typically cost more to operate than they save in cooling bills. Instead, install the recommended R-value of attic insulation (see chapter 13) with an attic radiant barrier as described earlier in this chapter. For effective natural attic ventilation, use continuous soffit vents at the eaves plus ridge vents at the peak.

The cooling effect of a breeze depends on its velocity. Air velocity can be approximated if you know the total area through which the breeze is blowing (such as total open area of inlet windows) and the fan rating in cfm.

The chart below shows ranges of fan cfm versus fan size for the fans discussed in this chapter, as well as their approximate efficiencies in cfm per watt of electrical consumption. Note that ceiling fans are at least three times more efficient than any other type.

Fan Size and Efficiency

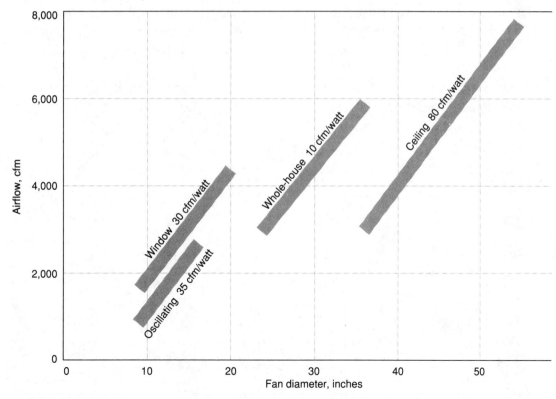

EVAPORATIVE COOLERS

When water evaporates, it absorbs heat. This is why you feel so cool when you emerge from swimming on a dry, breezy day.

Evaporative coolers (also known as swamp coolers) utilize this phenomenon to lower air temperature. As the illustration below shows, hot, dry air blown through a water-soaked pad emerges as humid, but much cooler, air. The temperature drop can be predicted:

Temperature drop $= E \times (DB - WB)$
where
   E   = cooler efficiency, percent
   DB = intake air dry bulb temperature
   WB = intake air wet bulb temperature

As a rule of thumb, evaporative coolers are recommended wherever the temperature drop (DB - WB) is more than 20F° and the cooled air would be below 79°F. The table on the following page lists these criteria for selected cities. Cities which approximately meet both criteria appear in *italics*.

To size an evaporative cooler

1. Compute the volume of house air.
2. Find the recommended minutes per air change for your location.
3. Divide the house volume by minutes to find the recommended cooler capacity in cfm.
4. If your home is very energy efficient, divide cfm by 2; if not insulated, multiply by 2.

How Evaporative Coolers Work

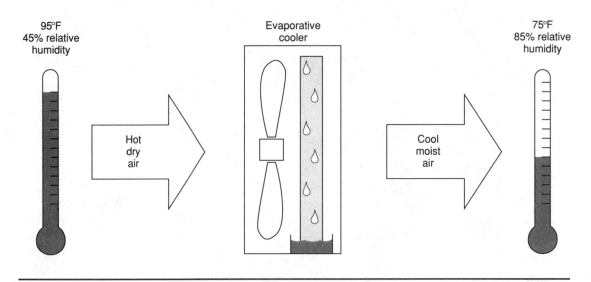

POTENTIAL FOR EVAPORATIVE COOLING

Location	DB °F	WB °F	Temperature Drop, F°[1]	Cooled Temp, °F	Minutes per Air Change	Location	DB °F	WB °F	Temperature Drop, F°[1]	Cooled Temp, °F	Minutes per Air Change
AL, Birmingham	96	74	18	78	NR	NV, Las Vegas	108	66	34	74	3
AZ, Phoenix	109	71	30	79	2	NH, Concord	90	72	14	76	NR
AR, Little Rock	99	76	18	81	NR	NJ, Newark	94	74	16	78	NR
CA, Los Angeles	93	70	18	75	2	NM, Albuquerque	96	61	28	68	3
CO, Denver	93	59	27	66	4	NY, Syracuse	90	73	14	76	NR
CT, Hartford	91	74	14	77	NR	NC, Greensboro	93	74	15	78	NR
DE, Wilmington	92	74	14	78	NR	ND, Bismarck	95	68	22	73	3
DC, Washington	93	75	14	79	NR	OH, Columbus	92	73	15	77	NR
FL, Orlando	94	76	14	80	NR	OK, Tulsa	101	74	22	79	1.3
GA, Atlanta	94	74	16	78	NR	OR, Portland	89	68	17	72	NR
ID, Boise	96	65	25	71	4	PA, Pittsburgh	91	72	15	76	NR
IL, Chicago	94	75	15	79	NR	RI, Providence	89	73	13	76	NR
IN, Indianapolis	92	74	14	78	NR	SC, Columbia	97	76	17	80	NR
IA, Des Moines	94	75	15	79	NR	SD, Rapid City	95	66	23	72	3
KS, Topeka	99	75	20	79	2	TN, Nashville	97	75	18	79	NR
KY, Louisville	95	74	17	78	NR	TX, Dallas	102	75	22	80	2
LA, Baton Rouge	95	77	14	81	NR	UT, Salt Lake City	97	62	28	69	4
ME, Portland	87	72	12	75	NR	VT, Burlington	88	72	13	75	NR
MD, Baltimore	9	77	12	80	NR	VA, Richmond	95	76	15	80	NR
MA, Boston	91	73	14	77	NR	WA, Seattle	85	68	14	71	NR
MI, Detroit	91	73	14	77	NR	WV, Charleston	92	74	14	78	NR
MN, Duluth	85	70	12	73	NR	WI, Madison	91	74	14	77	NR
MS, Jackson	97	76	17	80	NR	WY, Casper	92	58	27	65	4
MO, St Louis	98	75	18	80	NR						
MT, Great Falls	91	60	25	66	3						
NE, North Platte	97	69	22	75	3						

Note: NR means not recommended for this location. Cities with good evaporative cooling potential appear in *italics*.
[1] Temperature drop assumes a cooler efficiency of 80%.

AIR-CONDITIONING

If all else fails to cool you into the comfort zone, you have no choice but to air-condition. Air conditioners, powerful but expensive tools, lower humidity as well as temperature.

What you need first is an estimate of your *peak cooling load,* the number of Btu per hour that need to be removed under the worst conditions of the cooling season. The work sheet and tables that follow allow you to find that load, whether you are cooling just a bedroom or the entire house, no matter where you live in the United States.

You may wish to photocopy the work sheet so that you will be able to calculate the peak cooling load for a second room.

Carefully read the instructions for each line before entering any numbers. An example calculation follows the work sheet.

When you are done, look for an air conditioner with a *rated cooling capacity* that closely matches your peak cooling load. The appliance efficiency label lists the capacity as well as the energy efficiency ratio (EER), the ratio of Btu of cooling to watts consumed.

Cooling Factors
(for line 9 of air conditioner work sheet)

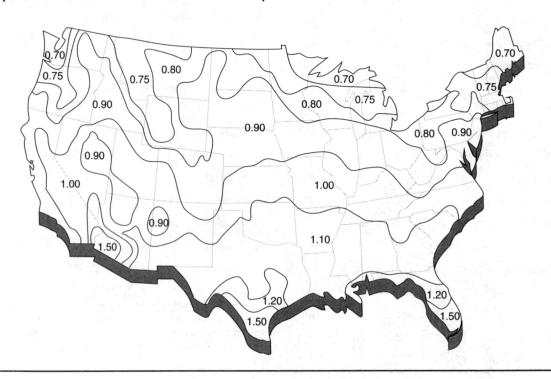

Instructions for Work Sheet

Lines 1 and 2. Use line 1 if you have a ventilated attic; otherwise use line 2. Find the shading factor in column 1 of table 1. The insulation factor is 0.8 × the nominal R-value of the insulation. (See chapter 13 for insulation R-values.) Use 2.4 if there is no insulation.

Line 3. Follow the same instructions as for lines 1 and 2. Exterior walls are those facing the outside. Include doors but not windows.

Line 4. Interior walls are those that separate the cooled space from unconditioned spaces. If you are cooling the entire house, there will be no interior walls. The insulation factor is 0.8 × the nominal R-value of the wall insulation, or 2.4 if there is none.

Line 5. Get the floor factor from table 2; the insulation factor is as in line 4.

Line 6. Enter the total floor area of the cooled space. Estimate air changes per hour as 0.4 for the tightest possible house to 1.3 for a drafty one.

Line 7. Calculate window areas as height × width of sash (frames holding the glass). Get the glazing factors from table 3.

Line 8. Get the shading factors from table 1 and the glazing factors from table 3.

Line 9. Add the results from lines 1 to 8 and multiply by the cooling factor from the map on the previous page.

Line 10. Multiply your average monthly spring or fall kilowatt-hours (get them from your utility bills) by 1.4. For the average home the result should be about 600.

Line 11. Give the average number of people occupying the cooled space during the hot months.

Line 12. Add lines 9, 10, and 11, and multiply by the mass factor from table 4.

TABLE 1. SHADING FACTORS

	Roof, Wall, Ceiling	Windows
Unshaded areas	1.00	1.00
Fully shaded areas	0.70	0.20
Partially shaded by awning, overhang, small trees	0.90	0.65
Shaded inside by shades, drapes, films		0.45

TABLE 2. FLOOR FACTORS

Floor Above	Factor
Open crawl space	1.0
Closed crawl space	0.0
Full basement	0.0
Unconditioned room	0.9
Ground (slab-on-grade)	0.1

TABLE 3. GLAZING FACTORS

Type of Glazing	Line 7	Line 8
Single-glazed window	1.0	1.0
Double-glazed window	0.5	0.8
Triple-glazed window	0.33	0.65

TABLE 4. THERMAL MASS FACTORS

Building Construction	Factor
Light wood frame	1.00
Solid masonry or wood frame with exterior masonry veneer	0.90
Wood frame with masonry interior walls, floors, or other mass	0.80
Earth-sheltered walls and roof	0.50

Work Sheet for Sizing Air Conditioners

Heat Source	Calculations	Results

1. Roof over ventilated attic   _____ sq ft × 44 × _____ shading factor ÷ _____ insulation factor = _____

2. Cathedral ceiling or   _____ sq ft × 48 × _____ shading factor ÷ _____ insulation factor = _____
roof over unventilated attic

3. Exterior wall facing
   north   _____ sq ft × 18 × _____ shading factor ÷ _____ insulation factor = _____

   east   _____ sq ft × 28 × _____ shading factor ÷ _____ insulation factor = _____

   south   _____ sq ft × 24 × _____ shading factor ÷ _____ insulation factor = _____

   west   _____ sq ft × 28 × _____ shading factor ÷ _____ insulation factor = _____

4. Interior walls facing   _____ sq ft × 12 ÷ _____ insulation factor =   _____
unconditioned rooms

5. Floors over unconditioned   _____ sq ft × 20 × _____ floor factor ÷ _____ insulation factor = _____
spaces

6. Infiltration: area of living space _____ sq ft × _____ air changes per hour × 1.6 =   _____

7. Window conduction   _____ sq ft × 16 × _____ glazing factor =   _____

8. Window solar gain
   north   _____ sq ft × 16 × _____ shading factor ÷ _____ glazing factor = _____

   east, south, southeast   _____ sq ft × 80 × _____ shading factor ÷ _____ glazing factor = _____

   west, southwest, northwest   _____ sq ft × 140 × _____ shading factor ÷ _____ glazing factor = _____

   northeast   _____ sq ft × 50 × _____ shading factor ÷ _____ glazing factor = _____

9. Sum of lines 1 - 8   _____ × _____ cooling factor from map =   _____

10. Utility gain   _____ watts being consumed in space × 3.4 =   _____

11. People gain   _____ number of people in space × 600 =   _____

12. Peak cooling load, Btu/hr: sum of lines 9 - 11 × _____ thermal mass factor =   _____

Sample Air Conditioner Calculation

The next page contains a completed form showing the calculations for the required capacity of an air conditioner for the small house in Boston, Massachusetts, shown at right.

Line 1. The ceiling measures 30×40 feet, so its area is 1,200 square feet. The roof, however, is pitched $30°$, so its area is 1,386 square feet. (If you don't know how to do this calculation, see the relationships between the sides of a triangle in chapter 22.) The unshaded roof has a shading factor of 1.00, from column 1 of table 1. The insulation factor is $0.8 \times$ the nominal R-value of 38.

Line 3. The north and south walls are each 320 square feet, less the window area of 21 square feet, or 299 square feet. The east and west walls measure 219 square feet by the same process. The west wall is fully shaded, so its shading factor (table 1) is 0.70. The rest are unshaded, so their shading factors are 1.00. The nominal R-values of 19 are multiplied by 0.8 to get insulation factors of 15.2.

Line 4. The entire house is air-conditioned, so the interior walls have no effect.

Line 5. The house sits on a vented (open) crawl space, so the floor factor is 1.0.

Line 6. There are 0.50 air changes per hour.

Line 7. The glazing factor, from column 1 of table 3, for double-glazed windows is 0.5.

Line 8. Both east and south windows are entered on one line. The shading factors are the same as for the walls in line 3. The glazing factor of 0.8 is found in column 2 of table 3.

House for Air Conditioner Example

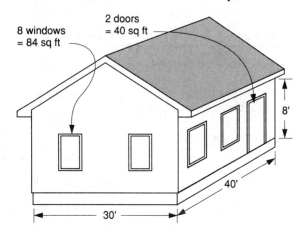

Air exchange rate = 0.5 changes per hour
Ceiling R-38; walls R-19; floor R-19 over vented crawl
21 sq ft of windows facing N, S, E, W
West wall fully shaded; other walls and roof unshaded
Spring and fall utility bills average 350 kwhr
3 occupants in summer

Line 9. The sum of the results column for all of the lines above is 13,921. Boston's cooling factor of 0.75 is found from the map at the beginning of this section.

Line 10. The electric utility bills for the spring and fall months show an average consumption of 350 kilowatt-hours per month.

Line 12. There are three occupants of the home during the cooling season.

Line 12. The thermal mass factor for a light wood frame house (1.00) is found in table 4. The sum of lines 9 through 11 is 13,906, so the peak cooling load is 13,906 Btu per hour. This is a small load for a house and could be satisfied by two 7,000 Btu/hr room air conditioners.

Example Work Sheet for Sizing Air Conditioners

Heat Source	Calculations	Results

1. Roof over ventilated attic $\underline{1,386}$ sq ft × 44 × $\underline{1.00}$ shading factor ÷ $\underline{30.4}$ insulation factor = $\underline{2,006}$

2. Cathedral ceiling or roof over unventilated attic _____ sq ft × 48 × _____ shading factor ÷ _____ insulation factor = _____

3. Exterior wall facing

north $\quad\underline{299}$ sq ft × 18 × $\underline{1.00}$ shading factor ÷ $\underline{15.2}$ insulation factor = $\underline{354}$

east $\quad\underline{219}$ sq ft × 28 × $\underline{1.00}$ shading factor ÷ $\underline{15.2}$ insulation factor = $\underline{403}$

south $\quad\underline{299}$ sq ft × 24 × $\underline{1.00}$ shading factor ÷ $\underline{15.2}$ insulation factor = $\underline{472}$

west $\quad\underline{219}$ sq ft × 28 × $\underline{0.70}$ shading factor ÷ $\underline{15.2}$ insulation factor = $\underline{282}$

4. Interior walls facing unconditioned rooms _____ sq ft × 12 + _____ insulation factor = _____

5. Floors over unconditioned spaces $\underline{1,200}$ sq ft × 20 × $\underline{1.0}$ floor factor + $\underline{15.2}$ insulation factor = $\underline{1,579}$

6. Infiltration: area of living space $\underline{1,200}$ sq ft × $\underline{0.50}$ air changes per hour × 1.6 = $\underline{960}$

7. Window conduction $\underline{84}$ sq ft × 16 × $\underline{0.5}$ glazing factor = $\underline{672}$

8. Window solar gain

north $\quad\underline{21}$ sq ft × 16 × $\underline{1.00}$ shading factor ÷ $\underline{0.8}$ glazing factor = $\underline{420}$

east, south, southeast $\quad\underline{42}$ sq ft × 80 × $\underline{1.00}$ shading factor ÷ $\underline{0.8}$ glazing factor = $\underline{4,200}$

west, southwest, northwest $\quad\underline{21}$ sq ft × 140 × $\underline{0.70}$ shading factor ÷ $\underline{0.8}$ glazing factor = $\underline{2,573}$

northeast $\quad$ _____ sq ft × 50 × _____ shading factor + _____ glazing factor = _____

9. Sum of lines 1 - 8 $\quad\underline{13,921}$ × $\underline{0.75}$ cooling factor from map = $\underline{10,440}$

10. Utility gain $\underline{350}$ × 1.4 = $\underline{490}$ watts being consumed in space × 3.4 = $\underline{1,666}$

11. People gain $\quad\underline{3}$ number of people in space × 600 = $\underline{1,800}$

12. Peak cooling load, Btu/hr: sum of lines 9 - 11 × $\underline{1.00}$ thermal mass factor = $\underline{13,906}$

17   Passive Solar Heating

Since man constructed the first rudimentary shelter, he has always instinctively utilized free energy from the sun. *What is passive solar* today? What are the techniques and materials, and what can we expect in terms of lowered fuel bills?

After several decades of experimentation, builders and homeowners alike have settled on two basic *passive solar system types:* the direct gain or solar-tempering approach, and the greenhouse.

If you are considering building a solar-tempered home or thinking of adding a solar greenhouse, you'll need to know the *performance factors* that spell the difference between radiant warmth and cold disappointment.

A simple *solar design procedure* guides you through the selection of an appropriate solar goal for your climate, the colors and materials you should use, and the amounts and areas of heat-absorbing construction materials you should provide for 24-hour comfort.

WHAT IS PASSIVE SOLAR?

There are many techniques for reducing energy consumption in buildings. Techniques such as increasing insulation, caulking and weather-stripping, and using high-performance windows and doors have been described in previous chapters. These energy conservation techniques are primarily buffers against cold climates, reducing the rate of escape of interior heat.

There are, however, techniques that capture free energy from the sun, reducing, and in some cases eliminating, the need for conventional central heating sources. These solar techniques can be roughly divided into two categories: *active* and *passive* solar heating. Passive techniques rely upon the interrelationship of heat from the sun, mass of the building, and siting; with these they capture, store, and release solar energy. With little if any increased cost, and with no noisy equipment to maintain, passive solar has become the technique of choice.

During the last 15 to 20 years, designers have learned a lot about the actual performance of passive solar structures. This chapter contains a condensation of that knowledge in the forms of simple graphs, tables, and design procedures that allow the design of near-optimum residential passive solar buildings anywhere in North America.

Passive solar, of course, is no substitute for standard energy conservation techniques. An underlying assumption for all that follows is a very high level of energy conservation. State and local energy standards should be considered the minimum for any passive solar design.

Guidelines for energy conservation in conjunction with passive solar design include the following:

- Insulate walls, roofs, and floors one step beyond the local norm, i.e., if the code calls for R-19 walls, make them R-25.
- Select triple-glazed, Low-E, or Heat Mirror windows, and insulated doors.
- Reduce air infiltration through the use of continuous air/vapor barriers and caulking and weather-stripping of all openings.
- Reduce the area of windows and doors on the north side of the building.
- Orient the building and openings to maximize the effects of cooling summer breezes and minimize the effects of winter winds.
- Utilize landscape elements to provide summer shade and to block winter winds.
- If natural deciduous shading is not possible, provide overhangs and projections to shade glazings during the cooling season.
- Ventilate roofs and attics to avoid condensation damage and summer overheating.

PASSIVE SOLAR SYSTEM TYPES

Direct Gain

Of all the passive solar types, direct gain is
the easiest to understand, since it is a simple
variation from an ordinary house with south-
facing windows. A direct-gain design is one
in which the solar radiation directly enters and
heats the living spaces. The building itself is a
solar collector.

In the heating season during daylight hours,
sunlight enters through south-facing windows,
patio doors, clerestories, or skylights. The radiation strikes and is absorbed by floors,
walls, ceilings, and furnishings. As anyone
who has ever been in a south-facing room in
winter realizes, some of the heat is transferred
to the air immediately, warming the room.
Some of the heat is absorbed into the structure
and objects in the room, to be released slowly
during the night, filling some of the overnight
heating requirement. In extreme designs, increased surface absorptivities and increased
amounts of mass allow storage of several
days' heating supply.

Although ceilings can be designed to store
heat, common direct-gain storage materials
are most easily incorporated into floors and
walls, which frequently serve a structural purpose as well (a masonry bearing wall, for example). Two very simple but effective storage
masses that can be incorporated into any
home are a masonry, tile, or slate floor, and
walls with double layers of gypsum drywall.

Open floor plans are recommended, to allow distribution of the released heat throughout the house by natural air circulation.

Direct Gain

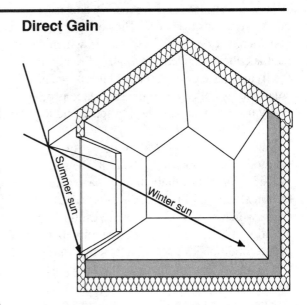

Sun Space or Greenhouse

Attached sun spaces are frequently constructed as extensions to homes. They are generally
considered secondary use spaces, in which
heat is either collected and vented directly to
the living space or is stored for later use. The
solar energy collected by the sun space is generally used to heat both the sun space and the
adjacent living space.

Sun spaces are designed for one of two basic modes of operation. In the first, the sun
space is *isolated* from the living space (illustration at top right) by an insulated wall and
doors which may be closed. As a result of this
isolation, the sun space is not treated as part
of the conditioned space, and its temperature
is allowed to fluctuate beyond the range of
human comfort. In the second case, the sun
space is *integrated* with the living space
(illustration at bottom right), and its temperature is controlled with auxiliary heat or heat
from the main living areas.

Integration is desirable when the space is
primarily a living space. Isolation is desirable
when the sun space is used primarily as a
greenhouse, generating more water vapor than
the house can safely absorb without causing
condensation and mildew.

Sun space glazings are usually tilted for
maximum light penetration and collection, but
the tilt results in increased summer heat gain.
Two solutions are deciduous shade trees to
block direct summer sun and ventilating windows and doors left open during the summer.

In northern areas sun space glazings should
either be of a high-performance type (Low-E
or Heat Mirror) or covered at night with a
form of movable window insulation.

Isolated Sun Space

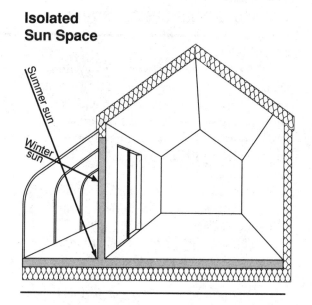

Integrated Sun Space

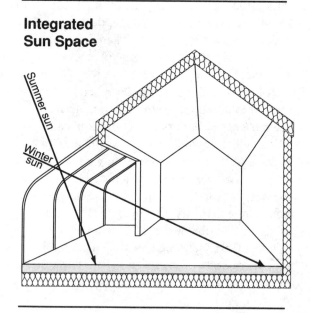

PERFORMANCE FACTORS

How well a passive system works depends upon 12 factors. Five of these factors are incorporated into a simple design procedure at the end of this chapter. Those factors are

- system type
- available daily radiation
- local climate
- area of south glazing
- thermal mass size and placement

A sixth factor, building insulation level, is assumed to be very high. The remaining factors are assumed by the design procedure to be the best possible:

- solar access (clear view of the sun)
- orientation of glazing
- tilt of glazing
- shading by trees and overhangs
- absorptivity of interior surfaces
- heat capacity of thermal mass

Solar Access

The amount of solar heat that reaches glazing is affected by its orientation to the sun and the fraction of energy blocked by permanent outside objects such as trees and buildings. Optimum solar access permits no shading between the hours of 8 AM and 4 PM from September 21 through March 21. These hours and days are represented by the shaded area in the chart on the following page.

In planning glazing, you can plot solar access on a sun chart, a map of the sky viewed from the location of the glazing. In the pages that follow, you'll find sun charts for latitudes in the United States and southern Canada. Your latitude can be found on page 38.

On the charts, the vertical axis represents the angle of the sun above the horizon; the horizontal axis represents the direction east or west of true south. To plot obstructions on your sun chart, you need to determine their azimuths and altitudes.

To find azimuth, first find true south as described on pages 37 and 38. Then, simply use a protractor and a piece of string as shown in the illustration on the next page. Standing at the proposed location of the glazing, point the zero on the protractor to true south and the string toward the object. The string will lie on the azimuth reading.

To find altitude, sight the top of an object along the straight edge of the protractor, and read altitude where the string crosses the outside of the scale. Take as many readings as you need to plot the outlines of all obstructing objects on the sun chart, as shown in the "Optimum Solar Access" and "Marginal Solar Access" examples on page 481.

Sky Chart for 40° North Latitude

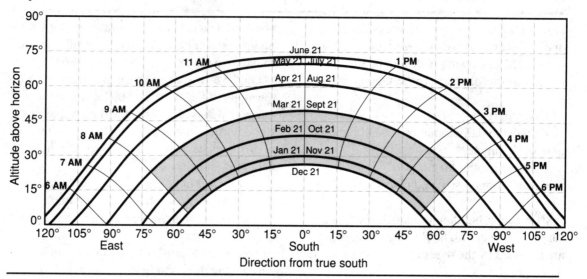

Finding Altitude

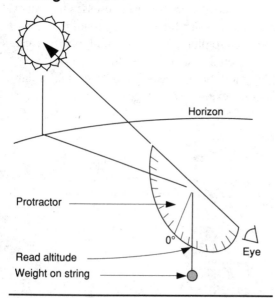

Finding Azimuth

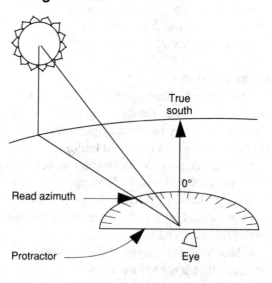

Optimum Solar Access

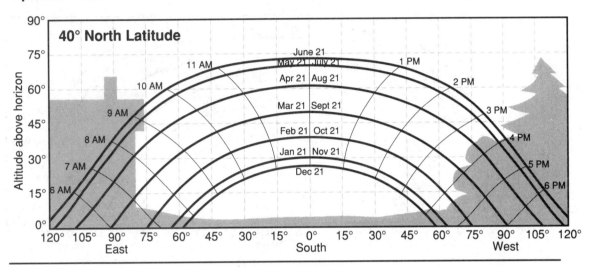

The illustration above shows a near-optimum solar access. The illustration at bottom shows a marginal level of solar access, caused by shrubs and trees between 3 and 4 PM and a neighboring building. The chart shows that the glazing will receive less than 50 percent of the possible clear-day radiation. More shading will make the site unsuitable for solar design.

Marginal Solar Access

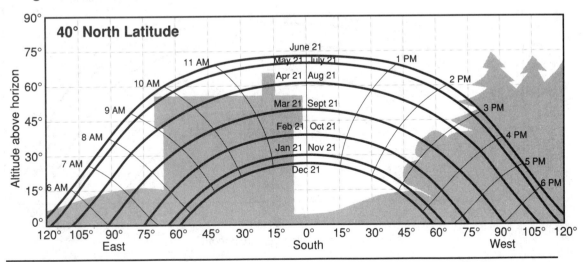

28° North Latitude

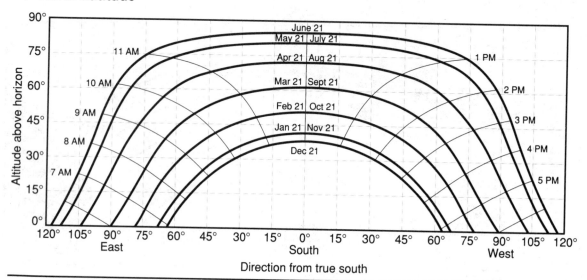

32° North Latitude

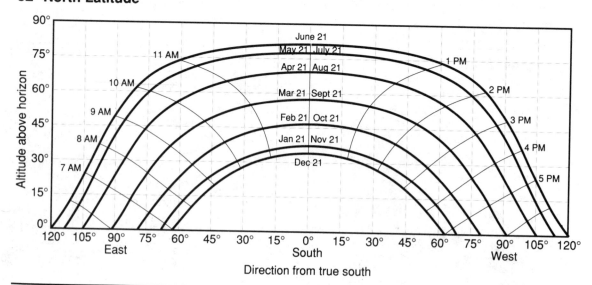

36° North Latitude

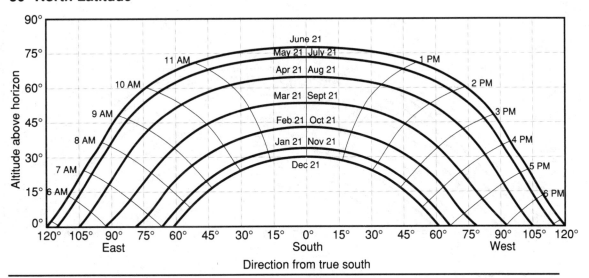

40° North Latitude

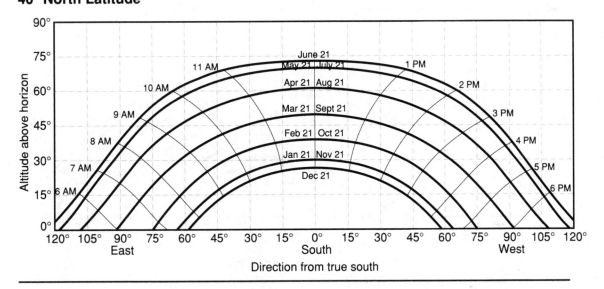

44° North Latitude

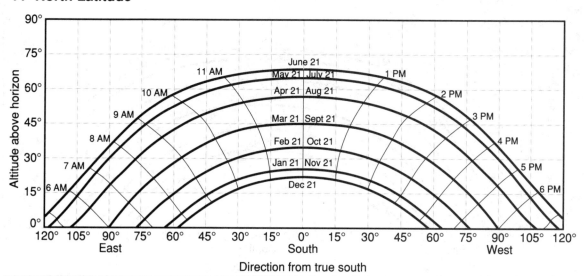

48° North Latitude

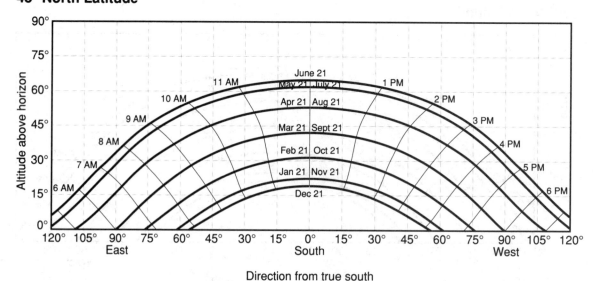

Glazing Orientation

A key circumstance that makes passive solar heating possible is the fact that the sun is lower in the southern sky in the winter than it is in the summer. As the chart below shows, this means that south-facing windows will receive more solar heat in winter than in summer, while for north, east, and west exposures, the reverse is true.

This is why passive solar buildings usually have 50 to 100 percent of their windows on the south wall and virtually no glazing on the north wall. Nonsolar houses usually have approximately 25 percent of their glazing on each of the four sides.

While the chart below gives clear-day radiation at 40 degrees north latitude, the principle applies throughout the northern hemisphere. The Btu will change, but the curves will remain the same.

Clear-Day Radiation at 40° North Latitude

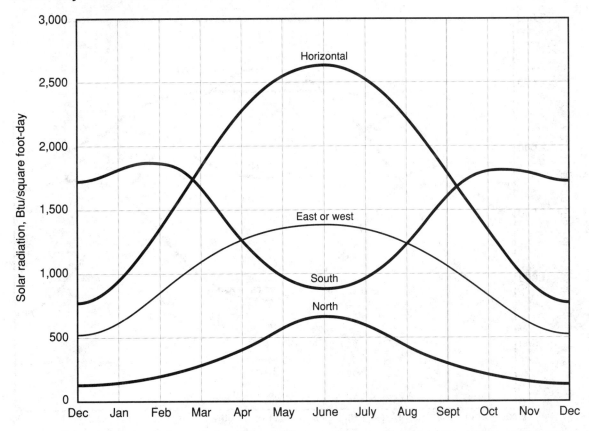

Glazing Tilt

Over the years, there has been controversy over the proper tilt for solar glazing. As the chart below shows, tilted glazing receives slightly more direct radiation than vertical glazing. But vertical surfaces receive more total radiation if the foreground is covered with snow. The net effect is that tilt has little effect upon performance. Moreover, tilted glazing has some practical disadvantages as compared with vertical glazing. Tilted glazing is more difficult to seal and is more prone to leak. Codes require tempered or safety glass. And as the chart also shows, tilted glazings receive more summer radiation and therefore require shading.

Still, tilted glazing usually is favored for true plant-growing greenhouses because direct sunlight is needed throughout the structure.

Radiation on Tilted South-Facing Glazings at 40° North Latitude

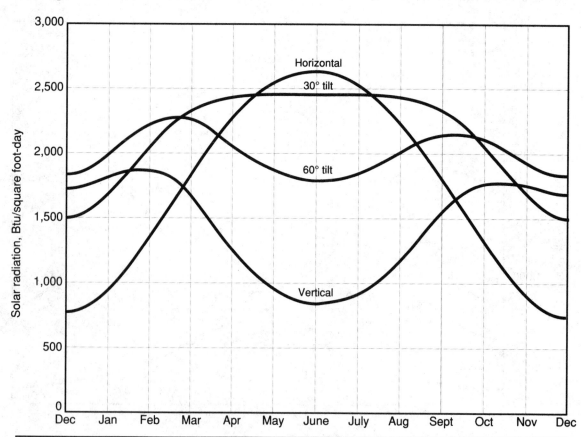

Shading

As explained in chapter 16, south vertical glazings must be shaded to prevent summer overheating. You can use a variety of awnings and inside or outside shades and shutters. But the most practical, attractive, and maintenance-free ways to provide shade are deciduous trees and roof overhangs.

Save existing shade trees and incorporate them into the siting of the building. If trees prove insufficient, roof overhangs can be built. Use the formula below to find the proper overhang projection from the site latitude and height of eaves above the window sill. Select 100 percent shade May 10 to August 1 if you have no trees.

$$OH = H \times F$$

where

OH	= overhang's horizontal projection
H	= height from eaves to window sill
F	= factor found in the table below

Example: What is the required horizontal projection of eaves (beyond the glazing plane) to fully shade, on June 21, a south-facing window located at 40° north latitude if the window sill is 7 feet 2 inches lower than the eaves?

H	= 7' 2" = 86"
F	= 0.29
OH	= 86" × 0.29
	= 25"

Roof overhangs are not practical for shading windows facing east or west or tilted south-facing windows.

South Roof Overhang

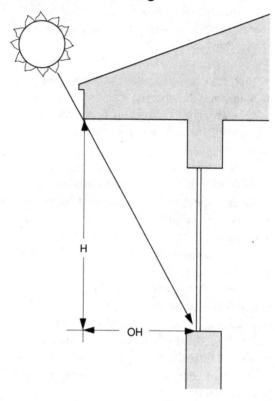

FACTORS FOR SOUTH-FACING OVERHANGS

° North Latitude	For 100% Shade June 21 Only	May 10 - Aug 1
28	0.09	0.18
32	0.16	0.25
36	0.22	0.33
40	0.29	0.40
44	0.37	0.50
48	0.45	0.59

Solar Absorptance

Once through the glazing, transmitted solar energy striking an interior surface can do one of two things:

   1. be reflected to another room surface or back through the glazing
   2. be absorbed (converted from radiant energy to sensible heat) by the surface

   The fraction not reflected, but converted to sensible heat, is termed the *absorptance* of the surface. The proper absorptance of room surfaces is more complicated than one might at first assume. If all light were absorbed, room lighting conditions would be poor, with all of the light coming from a single direction. If little of the light were absorbed in the first several reflections, much of the radiation would escape back through the windows. Light absorbed by lightweight surfaces results in heated air and little thermal storage.
   Solar designers recommend the following rules of thumb:

   • Lightweight objects should be light in color to avoid overheating of the room air, promote more even light distribution, and reflect radiation onto more massive surfaces.
   • Surfaces of massive objects should be dark in color and be placed to receive direct sunlight in order to efficiently collect and store heat.
   • Ceilings should be white, and deep rooms should have light-colored back walls to diffuse light more evenly.
   • Masonry floors should not be covered with wall-to-wall carpeting.

Example: A room has both wood-paneled and brick walls. Make the paneled walls light in color and the brick walls dark.

Example: A combined living/kitchen space has both wood and slate floors. You may carpet the wood floor, but leave the slate floor uncovered.

SOLAR ABSORPTANCE OF SURFACES

Material	Solar Absorptance
Flat black paint	0.95
Water	0.94
Dark gray slate	0.89
Dark brown paint	0.88
Dark blue-gray paint	0.88
Brown concrete	0.85
Medium brown paint	0.84
Medium light brown paint	0.80
Light gray oil paint	0.75
Red oil paint	0.74
Red brick	0.70
Concrete	0.65
Medium yellow paint	0.57
Medium blue paint	0.51
Light green paint	0.47
White semigloss paint	0.30
White gloss paint	0.25
White plaster	0.07

Heat Storage Capacity of Building Materials

Thermal mass is the amount of heat absorbed by a material as its temperature rises, expressed as Btu/F°. The *specific heat* of a material is its thermal mass per pound. Note that the specific heat of water is exactly 1.00, the highest of all natural materials.

As shown in the illustration at right, the benefit of building thermal mass into a house is that the house becomes less responsive to extremes in outdoor temperature. This means the house won't get too hot during the day or cold at night. The wider the outdoor temperature swing, the more beneficial the mass.

The table at right shows that there is little reason to turn to exotic materials. Masonry floors and walls, exposed wood, and extra-thick gypsum drywall are all effective storage masses. Water has been included in the table and the design procedures at the end of the chapter, in case you wish to go supersolar.

You have already seen how material and color can affect the efficiency of thermal mass. A third efficiency factor is thickness. When it comes to thermal mass, thicker is not necessarily better. If a wall is too thick, it remains too cool to become a heat source. If too thin, it warms too quickly and returns its heat before evening, when it is needed.

Optimum thicknesses for common heat storage materials are

- adobe   8 to 12 inches
- brick   10 to 14 inches
- concrete   12 to 18 inches
- water   6 or more inches

Temperature Swing and Thermal Mass

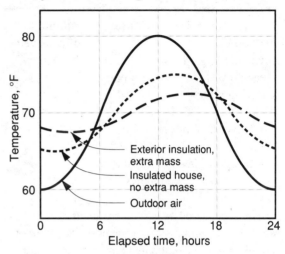

HEAT CAPACITIES OF MATERIALS

Material	Specific Heat Btu/lb-F°	Density lb/cu ft	Heat Capacity Btu/cu ft-F°
Brick	0.20	123	24.6
Cement	0.16	120	19.2
Concrete	0.22	144	31.7
Glass	0.18	154	27.7
Gypsum drywall	0.26	78	20.2
Iron	0.12	450	54.0
Limestone	0.22	103	22.4
Marble	0.21	162	34.0
Sand	0.19	94.6	18.1
Steel	0.12	489	58.7
Water	1.00	62.4	62.4
White oak	0.57	47	26.8
White pine	0.67	27	18.1

A SOLAR DESIGN PROCEDURE

The tables in this section provide a simple method for the preliminary design of passive solar buildings. Initial sizing of both window area and storage mass can be quickly achieved knowing only

- location of the site
- floor area
- insulation level of the building
- R-value of the windows

The procedure lets you determine the area of south-facing windows, amount and placement of storage mass, and estimated reduction in winter heating bills.

Glazed Area and Fuel Saving

Extensive computer simulations have been performed by researchers at Los Alamos Laboratory for passive solar homes in the cities listed in the table on the following two pages. Interpretation of the table is simple:

A south-facing window area of between (X1) and (X2) percent of the floor area can be expected to reduce the winter fuel bill of a home by (Y1) to (Y2) percent.

The lower window area and fuel-saving percentages correspond roughly to ordinary homes with ordinary window areas, with the exception that the windows are all on the south wall. The higher set of figures gives the maximum recommended target percentages for the location.

The fuel saving depends on the type of window: double-glazed, triple-glazed or equivalent R-3 glazing such as Low-E, quadruple-glazed or equivalent R-4 glazing such as Heat Mirror, or double-glazed with R-9 night insulation. Values for double-glazed and double-glazed with R-9 night insulation systems were derived from the computer simulations; triple-glazed and quadruple-glazed values were interpolated assuming the use of R-1 and R-2 night window insulation over double-glazed windows.

Example: A south-facing window area of 15 to 29 percent of the floor area can be expected to reduce the winter fuel bill of a home with no south-facing glass in Boston, Massachusetts, by 17 to 25 percent if the glazing is double glazed, and 26 to 39 percent if the glazing is triple glazed or Low-E glass.

Assumed Heat Loss

The table assumes a building heat loss of 6 Btu per degree-day per square foot of floor area. This corresponds roughly to the current energy standard for northern states: R-19 walls, R-38 ceilings, double glazing, and an infiltration rate of 3/4 air change per hour. If your home's insulation levels are different, the table values may be adjusted proportionally. For example, if your home has R-28 walls, R-57 ceilings, triple glazing, and 1/2 air change per hour (one and a half times as much insulation and two-thirds as much heat loss as the standard), you can reduce your window areas by one-third and still achieve the same percentage fuel saving.

GLAZED AREA AND SOLAR SAVINGS

City	South-Facing Window Area as % of Floor Area (X1) - (X2)	Resulting Fuel Savings for Window Types, range of percent			
		Double-Glazed (Y1) - (Y2)	Triple-Glazed (Y1) - (Y2)	Quadruple-Glazed (Y1) - (Y2)	Double-Glazed Plus R-9 (Y1) - (Y2)
Alabama, Birmingham	9 - 18	22 - 37	27 - 45	28 - 48	34 - 58
Arizona, Phoenix	6 - 12	37 - 60	41 - 66	43 - 68	48 - 75
Arkansas, Little Rock	10 - 19	23 - 38	28 - 47	30 - 51	37 - 62
California, Los Angeles	5 - 9	36 - 58	39 - 63	40 - 64	44 - 72
Colorado, Denver	12 - 23	27 - 43	34 - 54	38 - 60	47 - 74
Connecticut, Hartford	17 - 35	14 - 19	24 - 36	28 - 43	40 - 64
Delaware, Wilmington	15 - 29	19 - 30	26 - 42	30 - 48	39 - 63
DC, Washington	12 - 23	18 - 28	25 - 40	28 - 46	37 - 61
Florida, Orlando	3 - 6	30 - 52	33 - 56	34 - 58	37 - 63
Georgia, Atlanta	8 - 17	22 - 36	26 - 44	28 - 48	34 - 58
Idaho, Boise	14 - 28	27 - 38	35 - 50	38 - 56	48 - 71
Illinois, Chicago	17 - 35	17 - 23	27 - 39	31 - 47	43 - 67
Indiana, Indianapolis	14 - 28	15 - 21	23 - 35	27 - 42	37 - 60
Iowa, Des Moines	21 - 43	19 - 25	30 - 44	36 - 52	50 - 75
Kansas, Topeka	14 - 28	24 - 35	32 - 48	35 - 54	45 - 71
Kentucky, Louisville	13 - 27	18 - 27	24 - 39	27 - 44	35 - 59
Louisiana, Baton Rouge	6 - 12	26 - 43	29 - 49	30 - 52	34 - 59
Maine, Portland	17 - 34	14 - 17	25 - 36	31 - 45	45 - 69
Massachusetts, Boston	15 - 29	17 - 25	26 - 39	29 - 46	40 - 64
Michigan, Detroit	17 - 34	13 - 17	23 - 33	26 - 41	39 - 61
Minnesota, Duluth	25 - 50	Not rec	24 - 33	32 - 43	50 - 70
Mississippi, Jackson	8 - 15	24 - 40	28 - 47	29 - 50	34 - 59
Missouri, St Louis	15 - 29	21 - 33	28 - 45	32 - 50	41 - 65
Montana, Great Falls	18 - 37	23 - 28	35 - 46	41 - 54	56 - 77
Nebraska, North Platte	17 - 34	25 - 36	34 - 51	39 - 58	50 - 76
Nevada, Las Vegas	9 - 18	35 - 56	40 - 63	42 - 66	48 - 75

Passive Solar Heating

City	South-Facing Window Area as % of Floor Area (X1) - (X2)	Resulting Fuel Savings for Window Types, range of percent			
		Double-Glazed (Y1) - (Y2)	Triple-Glazed (Y1) - (Y2)	Quadruple-Glazed (Y1) - (Y2)	Double-Glazed Plus R-9 (Y1) - (Y2)
New Hampshire, Concord	17 - 34	13 - 15	25 - 35	30 - 44	45 - 68
New Jersey, Newark	13 - 25	19 - 29	26 - 42	30 - 48	39 - 64
New Mexico, Albuquerque	11 - 22	29 - 47	35 - 57	38 - 61	46 - 73
New York, Syracuse	19 - 38	Not rec	20 - 29	25 - 37	37 - 59
North Carolina, Greensboro	10 - 20	23 - 37	28 - 47	31 - 51	37 - 63
North Dakota, Bismarck	25 - 50	Not rec	27 - 36	35 - 47	56 - 77
Ohio, Columbus	14 - 28	13 - 18	21 - 32	25 - 39	35 - 57
Oklahoma, Tulsa	11 - 22	24 - 38	30 - 49	33 - 54	41 - 67
Oregon, Salem	12 - 24	21 - 32	27 - 33	30 - 47	37 - 59
Pennsylvania, Pittsburgh	14 - 28	12 - 16	20 - 30	23 - 37	33 - 55
Rhode Island, Providence	15 - 30	17 - 24	26 - 40	29 - 46	40 - 64
South Carolina, Columbia	8 - 17	25 - 41	29 - 48	31 - 52	36 - 61
South Dakota, Pierre	22 - 43	21 - 33	35 - 44	41 - 54	58 - 80
Tennessee, Nashville	10 - 21	19 - 30	24 - 39	27 - 44	33 - 55
Texas, Dallas	8 - 17	27 - 44	31 - 51	33 - 55	38 - 64
Utah, Salt Lake City	13 - 26	27 - 39	31 - 51	33 - 57	48 - 72
Vermont, Burlington	22 - 43	Not rec	23 - 33	29 - 42	46 - 68
Virginia, Richmond	11 - 22	21 - 34	27 - 44	30 - 49	37 - 61
Washington, Seattle	11 - 22	21 - 30	28 - 41	31 - 46	39 - 59
West Virginia, Charleston	13 - 25	16 - 24	22 - 35	25 - 40	32 - 54
Wisconsin, Madison	20 - 40	15 - 17	28 - 38	34 - 48	51 - 74
Wyoming, Casper	13 - 26	27 - 39	38 - 53	41 - 60	53 - 78
Alberta, Edmunton	25 - 50	Not rec	26 - 34	34 - 44	54 - 72
British Columbia, Vancouver	13 - 26	20 - 28	27 - 40	31 - 45	40 - 60
Manitoba, Winnipeg	25 - 50	Not rec	26 - 34	34 - 45	54 - 74
Nova Scotia, Dartmouth	14 - 28	17 - 24	27 - 41	32 - 49	45 - 70
Ontario, Ottawa	25 - 50	Not rec	28 - 37	36 - 49	59 - 80
Quebec, Normandin	25 - 50	Not rec	26 - 35	34 - 45	54 - 74

Source: *Passive Solar Design Handbook, vol 3* (Washington, DC: United States Department of Energy, 1980).
Note: Not rec = not recommended at this site.

Sizing and Placing Storage Mass

All buildings have mass in their floors, walls, ceilings, and furnishings. If they didn't, they'd all overheat on sunny days. The table below shows the approximate areas of south-facing windows in average-insulated and well-insulated wood frame homes before overheating occurs.

SOUTH WINDOW AREA LIMITS
(window area as percentage of floor area)

Degree-Days[1]	Average Jan Temp, °F	Average House[2]	Well-Insulated House[3]
4,000	40	11	6
5,000	30	13	6
6,000	25	13	7
7,000	20	14	7

[1] See chapter 15 for heating degree-days, base 65°F.
[2] R-11 walls, R-19 ceiling, double-glazed windows.
[3] R-25 walls, R-38 ceiling, triple-glazed windows.

Example: With no additional mass, what is the maximum allowable area of south-facing window for a 1,800-square-foot home in Boston (6,000 DD_{65}) constructed with R-25 walls, R-49 ceiling, and R-3 windows?

The energy efficiency of the home is close to the well-insulated house in the table, so the appropriate percentage is 7, and the maximum glazed area is $0.07 \times 1,800 = 126$ square feet.

The example above was a typical nonsolar home. For the higher fuel savings listed in the table on the previous page, however, a much greater window area is suggested. For example, the suggested window percentage for Boston is 15 to 29 percent, or two to four times the percentage in the example. Such a home will require additional storage mass.

Adding Mass

In the pages that follow, five distinctly different storage mass patterns are shown. For each, an accompanying table specifies the material, thickness, and surface area of mass required for each square foot of glazing in excess of the norm.

Example: Assuming you wish to achieve 25 percent solar savings for the home in Boston, you will need about 15 percent of the floor area in south glazing. You therefore need additional thermal mass to compensate for 15 percent - 7 percent = 8 percent of the floor area, or 144 square feet of glazing. Using pattern 1 ("Floor or Wall in Direct Sunlight") and assuming a bare 6-inch concrete slab as the mass, you'll find you need 3×144 square feet = 432 square feet of slab in direct sunlight.

The Mass Patterns

Pattern 1 corresponds to a home with a masonry floor. The floor may be a concrete slab-on-grade, or it may consist of a masonry veneer over a wood or concrete base.

Pattern 2 might represent a home with one or more exposed masonry walls or a home with an extra-thick plaster or drywall ceiling.

Pattern 3 might occur when a building is remodeled, exposing an interior masonry party wall. It could also represent a large masonry fireplace.

Pattern 4 typically represents a solar greenhouse with a masonry rear wall.

Pattern 5 is not very common but accounts for any massive structure in direct sunlight that does not reach the ceiling. A masonry planter/room divider would fall into pattern 5.

Pattern 1:
Floor or Wall in Direct Sun

Pattern 1 is defined as storage mass that has one surface exposed to the living space and a back surface that is insulated. The exposed surface is further defined as being in direct sun for at least 6 hours a day. Architecturally, this pattern combined with pattern 2 is useful for direct-gain passive solar rooms.

The mass can be either a directly irradiated floor slab, as shown, or a directly irradiated outside wall (inside walls are considered in pattern 3). As with patterns 2 and 3, the mass element is one-sided; that is, heat moves into and out of the mass from the same surface.

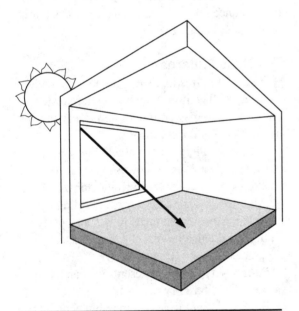

MASS SIZING FOR FLOOR OR WALL IN DIRECT SUN

Material Thickness	Sq Ft of Mass per Sq Ft of Window				
	Concrete	Brick	Gypsum	Oak	Pine
1/2"	—	—	76	—	—
1"	14	17	38	17	21
1-1/2"	—	—	26	—	—
2"	7	8	20	10	12
3"	5	6	—	10	12
4"	4	5	—	11	12
6"	3	5	—	11	13
8"	3	5	—	11	13

Example: The design procedure has resulted in 100 square feet of south-facing glass in a room with 200 square feet of floor and 380 square feet of windowless wall. Does a 4-inch concrete slab provide enough thermal mass to prevent overheating?

According to the table, you should provide 4 square feet of 4-inch concrete slab for each square foot of glazing. That would require 400 square feet of slab. Increasing the slab thickness to 8 inches would still require 300 square feet of slab. Therefore you must either reduce the glazed area or add further mass, using one of the other four patterns.

Pattern 2:
Floor, Wall, or Ceiling in Indirect Sun

The mass in pattern 2 is like that in pattern 1, that is, the mass is one-sided and insulated on the back side. The distinction here is that the mass is receiving not direct radiation, but reflected sun.

In a simple direct-gain space, some of the mass will be of pattern 1 (a floor slab near the solar glazing, for example), and some mass will be of pattern 2 (the ceiling, for example). Much of the mass in such a space will be directly irradiated some of the time and indirectly irradiated the rest of the day. In these cases, an interpolation between pattern 1 and pattern 2 must be carried out, as described in pattern 1.

MASS SIZING FOR FLOOR, WALL, OR CEILING IN INDIRECT SUN

Material Thickness	Sq Ft of Mass per Sq Ft of Window				
	Concrete	Brick	Gypsum	Oak	Pine
1/2"	—	—	114	—	—
1"	25	30	57	28	36
1-1/2"	—	—	39	—	—
2"	12	15	31	17	21
3"	8	11	—	17	20
4"	7	9	—	19	21
6"	5	9	—	19	22
8"	5	10	—	19	22

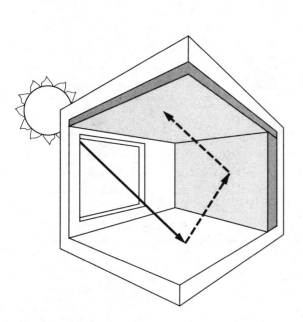

Example: You have decided to use an 8-inch concrete slab for the room described in mass pattern 1. This leaves 33 square feet of glazing to provide mass for. How many square feet of 8-inch brick wall will be required to provide the mass?

According to the table above, 10 square feet of 8-inch brick wall is required to balance each square foot of glazing. You therefore need 330 square feet of brick wall. Since the total wall area is 380 square feet, this is a practical solution.

Pattern 3:
Floor, Wall, or Ceiling Remote from Sun

As in patterns 1 and 2, the storage mass in this pattern is one-sided. The difference is that the mass receives neither direct radiation nor reflected radiation. It is instead heated by the room air that is warmed as a result of solar gains elsewhere in the building.

This pattern is useful for storage materials in spaces deeper within a passive building, away from the rooms receiving solar gains. However, the solar-heated air must reach the remote mass either by natural or forced air circulation.

Reasonable judgment is required here - a hallway open to a south room could be included, a back room totally closed off from the solar-heated space should be excluded.

MASS SIZING FOR FLOOR, WALL, OR CEILING REMOTE FROM SUN

Material Thickness	Sq Ft of Mass per Sq Ft of Window				
	Concrete	Brick	Gypsum	Oak	Pine
1/2"	—	—	114	—	—
1"	27	32	57	32	39
1-1/2"	—	—	42	—	—
2"	17	20	35	24	27
3"	15	17	—	26	28
4"	14	17	—	24	30
6"	14	18	—	28	31
8"	15	19	—	28	31

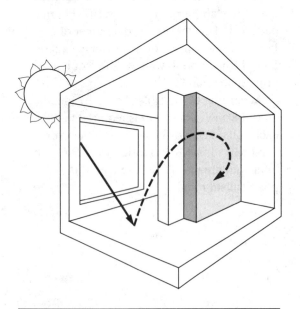

Example: Your remodeling plan calls for removing half of a wood-framed gypsum wall to open the south-facing kitchen to the living room. The remaining wall has an 80-square-foot fireplace of 8-inch brick on the living-room side. You plan to add a south-facing window in the kitchen. How many square feet of window will the mass of the fireplace balance?

According to the table above, the fireplace alone will account for only 4 square feet of window. Obviously, you must look elsewhere for thermal mass.

Pattern 4:
Full Masonry Wall or Water Wall in Direct Sun

Pattern 4 is defined as a floor-to-ceiling wall of massive material that receives direct sun on one side and is exposed to the living space on the other side. In other words, the sunlit side is isolated from the living space.

This pattern is useful for isolated sun spaces and greenhouses. The storage wall may have high and low vents or be unvented, as shown, without affecting the values in the table.

The performance of the wall improves with thickness up to about 18 inches but is not very sensitive to variations in thickness within normal buildable ranges. For brick walls, higher-density bricks (with water absorption of less than 6 percent) are recommended over bricks of lower density. Note that the mass surface area refers to the area of the sunlit side only.

MASS SIZING FOR MASS WALL OR WATER WALL IN DIRECT SUN

Material and Thickness	Sq Ft of Mass Surface per Sq Ft of Window
8" thick brick	1
12" thick concrete	1
8" thick water wall	1

Example: You are considering adding an attached solar greenhouse. The primary purpose of the greenhouse will be growing plants. The greenhouse structure should therefore be isolated to avoid excess humidity in the living space.

As the table above shows, 1 square foot of 8-inch brick, 12-inch concrete, or 8-inch water wall (water containers) for each square foot of glazing will conveniently provide all of the required thermal mass.

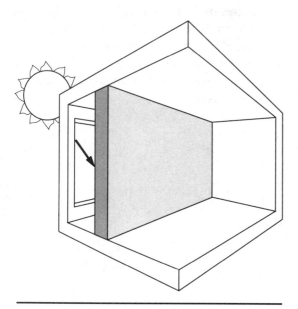

Pattern 5:
Partial Masonry Wall or Water Containers in Direct Sun

Similar to pattern 4, mass in this pattern is sunlit on one side and exposed to the living space on the other side. The distinction is that there is free air circulation around this mass material so that heat may be gained by the living space from either side of the partial wall or from all sides of the water containers.

This pattern may represent a freestanding masonry wall or a series of water containers.

The mass is assumed to be in full sun for at least 6 hours. As with pattern 4, the wall thicknesses listed are not very sensitive to variations, and the wall surface area listed is for one side of the wall only. Water containers are listed in the table at right as gallons per square foot of glazing.

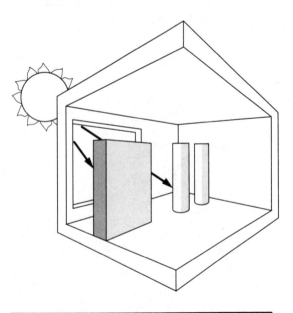

MASS SIZING FOR PARTIAL MASS WALL OR WATER CONTAINERS IN DIRECT SUN

Material and Thickness	Sq Ft of Storage Mass Surface per Sq Ft of Window
8" thick brick	2
6" thick concrete	2
Water containers	7 gallons per sq ft of window area

Example: You plan to add a greenhouse with 160 square feet of south-facing glazing. Unlike the example in pattern 4, however, you plan to use the space for living rather than growing. You'd like the greenhouse to be open to the adjacent kitchen. Will a 3-foot-high by 20-foot-long room divider constructed of 8-inch brick provide sufficient thermal mass?

The area of the room divider exposed to direct sunlight is 60 square feet. According to the table above, the room divider alone will account for only 30 square feet of glazing. You will probably need to employ a pattern 1 brick floor as well.

18 Lighting

Proper lighting around the home is important for safety, for reading, for working, for atmosphere, and for the long-term health of your eyes. In order to understand why different light sources and intensities are recommended for different applications, you need to understand the relationships between *light and seeing*.

"Lamps" shows you the great variety of bulb shapes and bases available today. A table on lamps lists the color characteristics and efficiencies of more than 30 incandescent, fluorescent, and high-intensity-discharge lamps for use in and around the home.

Perhaps the most useful section in this chapter is *"Lighting Guidelines for the Home"*—specific lighting recommendations for 28 applications in the home.

LIGHT AND SEEING

Light Units

The relationship between lighting units is displayed in the illustration at right. A point light source with a strength (candlepower) of 1 candela results in an *illuminance* of 1 *footcandle*, or 1 lumen per square foot. Since a sphere of radius 1 foot has a surface area of 12.57 square feet, the total light output is 12.57 *lumens*.

Illumination levels are usually given in footcandles, although lumens per square foot is equivalent. Total lamp output is always given in lumens.

The intensity of light falling on a surface is the illuminance. The intensity of light given off or reflected by a surface is its *luminance*. For non-light-emitting surfaces,

Luminance = illuminance × *reflectivity*

Example: What is the luminance of a surface of reflectivity 0.50 when illuminated at an intensity of 100 footcandles?

Luminance = illuminance × reflectivity
= 100 footcandles × 0.50
= 50 footcandles

Visual Acuity

As shown in the illustration at right, the human eye can detect light from nearly an entire half sphere (radius 90°). The ability to discriminate among small details, however, is limited to a radius of about 1°, the *central field*. The area surrounding the central field is the *surround*.

The ability of the eye to discern the small details of a task within the central field is determined by four factors:

Light Measurements

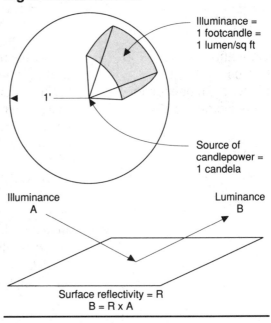

Illuminance = 1 footcandle = 1 lumen/sq ft

Source of candlepower = 1 candela

Illuminance
A

Luminance
B

Surface reflectivity = R
B = R x A

The Visual Field

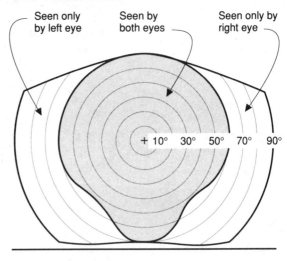

Seen only by left eye

Seen by both eyes

Seen only by right eye

+ 10°   30°   50°   70°   90°

- size of the field
- contrast between detail and background
- time (eye fatigue)
- brightness of the task

The size of a task can be magnified by a lens. In the case of printed material, contrast is maximized by printing black on white. Luminance is a function of both illumination and task reflectance. Deficiencies in size, contrast, and time can all be compensated to a degree by an increase in illumination.

Glare

An area anywhere within the visual field that has sufficient luminance to cause either discomfort or a reduction in visual acuity is an area of *glare*. Direct glare is illumination direct from a light source, such as a window or exposed lamp. Reflected glare is light which has been reflected from a shiny or glossy surface, usually within the task area.

For optimum visual comfort, the luminance of the task should be slightly greater than the luminance of the surround. The table below lists maximum recommended luminance ratios. Since the response of the eye is logarithmic, the ratios are deceptive. A ratio of 10 to 1 is only 3-1/2 camera f-stops.

Compared Areas	Maximum Ratio
Task to adjacent area	3 to 1
Task to remote dark surface	10 to 1
Task to remote light surface	0.1 to 1
Window to adjacent wall	20 to 1
Task to any area within visual field	40 to 1

Color

Most objects emit no light; they simply reflect incident light received from a light source or luminaire. The perceived color of an object is determined by the color (energy content at different wavelengths) of the light source and the reflectivity of the object at the corresponding wavelengths.

Objects which reflect all wavelengths equally are termed *white,* and light sources which emit light of all frequencies are termed *white lights.* Not all wavelengths are emitted equally, however, even from white lights. For incandescent sources, the wavelength of most intense emission increases with the temperature of the source.

A measure of peak emission wavelength is the color-correlated temperature of the source. Direct light from the sun is at a color temperature of about 6,000°Kelvin (K). Due to selective absorption and scattering in the atmosphere, the sun's color temperature falls to about 3,500°K an hour before sunset. The color-correlated temperatures of incandescent sources range from about 2,400 to 3,100°K. Most fluorescent lamps have color-correlated temperatures between 3,000°K (warm white) and 4,200°K (cool white).

Incandescent lamps emit light at all wavelengths. Fluorescent and other gas-discharge lamps emit light only at specific wavelengths. Color rendition may therefore be different between incandescent and gas-discharge lamps even though their color-correlated temperatures may be the same. The color-rendering index (CRI) compares lamp output with natural daylight at all frequencies, with a CRI of 100 indicating a perfect match.

LAMPS

Light Sources

Lamps fall into one of three broad categories, depending on the way in which they convert electricity to light:

Incandescent lamps emit light from filaments heated to incandescence by an electric current. Efficiencies range from about 10 to 20 lumens per watt. Quartz-halogen lamps achieve higher efficiencies, of about 15 to 24 lumens per watt, through the use of higher-temperature filaments. Strong points include low initial cost, wide range of available output for the same base style and bulb size, and ability to concentrate the light beam. Weak points are high operating costs and relatively short lives.

Fluorescent lamps emit light from phosphor coatings stimulated by high-voltage discharge through the mercury-vapor-filled bulbs. Rare earth phosphor additives modify color output. Ballasts are required to limit the arc current. Excluding the tiniest lamps, efficiencies range from about 70 to 90 lumens per watt, including ballast power. Strong points include long lamp life, low operating cost, and low operating temperature. Weak points include size, color rendition, and 60 Hz hum and flicker.

High-intensity-discharge (HID) lamps emit light directly from electric arcs through metal vapor. Color rendition is inappropriate for residential applications, but efficiencies range up to 144 lumens per watt.

Efficiencies of Light Sources

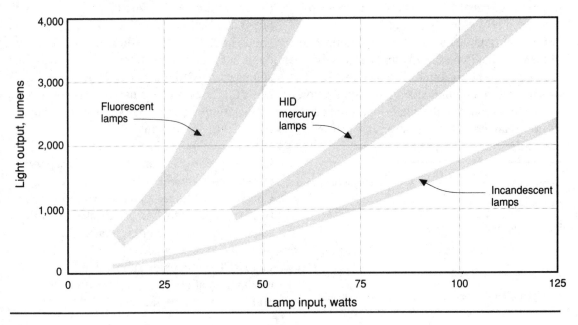

CHARACTERISTICS OF COMMON LAMPS (Sylvania)

Type	Lamp	Color Temp, °K	CRI[1]	Lumens	Watts	Lumens per Watt
Incandescent	A19, frosted, 25 watt	2,900		357	25	14
	A19, frosted, 40 watt	2,900		460	40	12
	A19, frosted, 60 watt	2,950		890	60	15
	A19, frosted, 75 watt	3,000		1,210	75	16
	A19, frosted, 100 watt	3,050		2,850	150	18
	12V quartz-halogen, 20 watt	2,900		300	19.4	15
	12V quartz-halogen, 45 watt	3,100		980	44.5	22
Fluorescent	Twin tube 27K	2,700	81	600	9	67
	Incandescent/fluorescent	2,750	89	1,110	30	37
	Deluxe warm white	2,950	74	1,550	30	52
	Warm white	3,000	52	2,360	30	79
	Designer 3000K	3,000	67	3,300	40	83
	Royal white	3,000	80	2,400	30	80
	Octron 3100K	3,100	75	3,650	40	91
	White	3,450	57	1,900	30	63
	Octron 3500K	3,500	75	3,650	40	91
	Natural white	3,600	86	3,050	55	55
	Designer 4100K	4,100	67	8,800	95	93
	Octron 4100K	4,100	75	3,650	40	91
	Deluxe cool white	4,100	89	2,100	40	53
	Lite white	4,150	48	4,300	60	72
	Cool white	4,200	62	3,150	40	79
	Design 50	5,000	90	1,610	30	54
	Daylight	6,300	76	1,900	30	63
High-intensity discharge	Mercury, Warmtone	3,300	52	3,700	100	37
	Mercury, Brite White Deluxe	4,000	45	3,650	100	37
	Mercury, clear	5,900	22	3,380	100	34
	Metal halide, clear	3,200	65	6,800	100	68
	Metal halide, coated	3,900	70	16,000	250	64
	High-pressure sodium, Lumalox	2,000	22	8,850	100	89
	High-pressure sodium, Unalox	1,900	20	11,700	150	78

Source: *Large Lamp Ordering Guide* (Danvers, Mass: Sylvania Lighting Center, 1986).
[1] Color-rendering index.

Lamp Bases

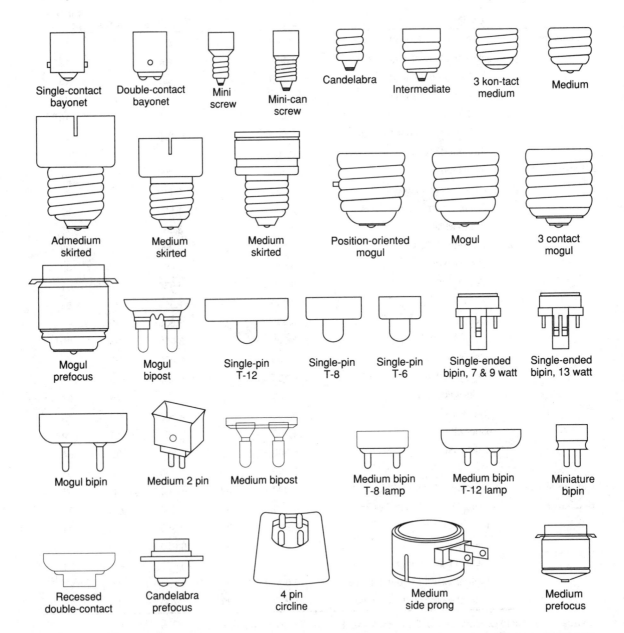

Single-contact
bayonet

Double-contact
bayonet

Mini
screw

Mini-can
screw

Candelabra

Intermediate

3 kon-tact
medium

Medium

Admedium
skirted

Medium
skirted

Medium
skirted

Position-oriented
mogul

Mogul

3 contact
mogul

Mogul
prefocus

Mogul
bipost

Single-pin
T-12

Single-pin
T-8

Single-pin
T-6

Single-ended
bipin, 7 & 9 watt

Single-ended
bipin, 13 watt

Mogul bipin

Medium 2 pin

Medium bipost

Medium bipin
T-8 lamp

Medium bipin
T-12 lamp

Miniature
bipin

Recessed
double-contact

Candelabra
prefocus

4 pin
circline

Medium
side prong

Medium
prefocus

Incandescent Lamp Bulb Shapes

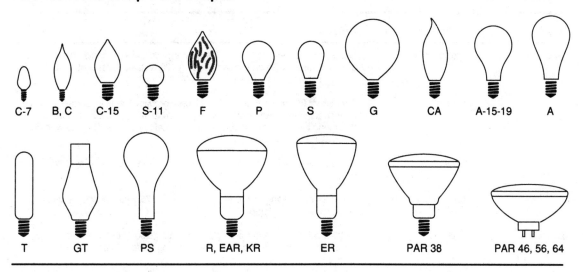

C-7 B, C C-15 S-11 F P S G CA A-15-19 A

T GT PS R, EAR, KR ER PAR 38 PAR 46, 56, 64

Fluorescent Lamp Bulb Shapes

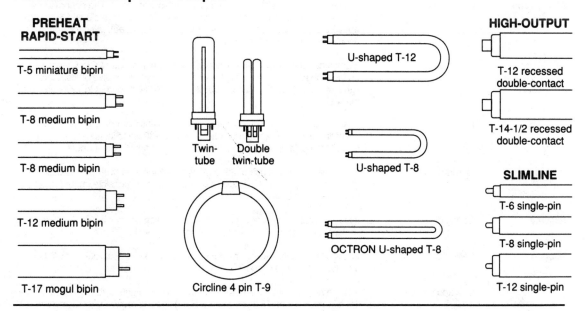

PREHEAT RAPID-START

T-5 miniature bipin

T-8 medium bipin

T-8 medium bipin

T-12 medium bipin

T-17 mogul bipin

Twin-tube Double twin-tube

Circline 4 pin T-9

U-shaped T-12

U-shaped T-8

OCTRON U-shaped T-8

HIGH-OUTPUT

T-12 recessed double-contact

T-14-1/2 recessed double-contact

SLIMLINE

T-6 single-pin

T-8 single-pin

T-12 single-pin

LIGHTING GUIDELINES FOR THE HOME

Area	Application	Guidelines
Bathroom	Small mirrors	Use 75-watt incandescents or 20-watt warm white fluorescents on each side of the mirror about 30 inches apart. Install a 100-watt incandescent or 40-watt fluorescent ceiling fixture as well.
	Large mirrors	For mirrors 36 inches or wider, install three or four 60-watt incandescents in a 22-inch-minimum-width fixture, or install a 36 to 48-inch diffused fluorescent fixture along the top of the mirror. For a theatrical look, install 15 to 25-watt G bulbs along the top and sides.
	Shower light	Use a 60-watt incandescent in a wet-location ceiling fixture. Also, check your local code.
	Toilet compartment	Install either a ceiling or wall fixture with a 60 to 75-watt incandescent or a 30 to 40-watt fluorescent lamp.
Bedroom	General	Install a ceiling fixture or track lighting of wattage sufficient to provide 10 footcandles of uniform lighting. Install small ceiling lights in large closets.
	Reading in bed	Provide an individual 75 to 100-watt incandescent or 22 to 32-watt fluorescent lamp, with the bottom of the shade at eye level and 22 inches to the side of the center of the book. As an option, headboard track lighting should provide one 20 to 50-watt incandescent R bulb for each person, mounted 30 inches above mattress level.
Dining room	Chandelier	Provide a total of 200 to 300 watts of incandescent lamps. The bottom of the chandelier should be at least 12 inches narrower than the table and 30 inches above the surface. If the room ceiling is higher than 8 feet, raise the chandelier 3 inches for each additional foot.
	Buffet server	Flank a buffet with sconces 60 inches above the floor. Use 25 to 60-watt shielded incandescent bulbs or 15 to 25-watt unshielded decorative bulbs. Track lights of 50 to 75 watts may be substituted 2 to 3 feet apart, 9 to 12 inches from the wall.
Entrance	Foyer	In areas of less than 75 square feet, use 100 watts incandescent or 25 to 40 watts fluorescent. For larger areas use 150 to 180 watts incandescent. Consider wall lamps flanking a mirror, a pendant over a table, or a chandelier.
	Outside	Flank the door with a pair of 25 to 60-watt incandescent wall fixtures 66 inches above standing level at the door. If only one fixture is possible, mount it at the lock side of the door.
Hallway	Ceiling or wall	Install at least one fixture every 10 feet. Recessed or track accent lighting for wall art is acceptable.
Kitchen	Ceiling	Install either incandescent or fluorescent ceiling fixtures sufficient for 50 footcandles of general lighting.
	Sink and range	Install over the front edge of the counter two downlights with 75-watt reflective flood lamps spaced 18 inches apart. Range hoods require 60-watt incandescents.

LIGHTING GUIDELINES—*Continued*

Area	Application	Guidelines
	Under cabinet	Mount as close to the cabinet front as possible 8 watts per foot of fluorescent fixtures. Cover at least two-thirds of the total counter length.
	Island counter	Light an island the same as a sink.
	Dinette	Install a pendant or track lighting of 120 watts incandescent or 32 to 40 watts fluorescent over the table or counter.
Living & family	General	Install a combination of accent, wall-grazing, and wall-washing track lighting sufficient to achieve 10 footcandles average.
	Reading	Light the same as for reading in bed.
	Music stand	Install one 75-watt reflective or parabolic reflector flood lamp in a recessed or track fixture 12 inches to the left and 24 inches in front of the music. Two fixtures 30 inches apart are even better.
	Television	Provide low-level lighting to avoid reflections from the screen.
	Game table	Install one recessed 150-watt incandescent or 40-watt fluorescent fixture over each half of the table. For a card or pool table, mount a single 100-watt shaded pendant 36 inches above the center of the table.
	Bar	Install 25 to 50-watt recessed or track reflector bulbs 16 to 24 inches apart over the bar.
Outside	Vegetation	Illuminate trees and bushes with spotlights mounted high on walls or from ground level. Do not allow light to shine at neighboring houses.
Stairs		Provide ceiling or wall fixtures at both top and bottom. Control them from both locations with three-way switches.
Study	Desk	Position one 150-watt incandescent or 32-watt fluorescent lamp with the bottom of the shade 15 inches above the desk and 12 inches from the front edge.
Track	Accent	Ceiling-mounted fixtures should be aimed at a 30-degree angle from the vertical to prevent light from shining in anyone's eyes and to avoid reflections on the illuminated objects. Usually, one fixture is required for each object being accented. To locate the ceiling fixture, the horizontal distance from the wall should be 57 percent of the vertical distance from the center of the object to the ceiling.
	Wall grazing	For dramatic shadows on surfaces such as draperies, stone, or brick, mount the track 6 to 12 inches from the wall with fixtures spaced the same distance.
	Wall washing	For nontextured surfaces, mount the track 2 to 3 feet from the wall for ceilings up to 9 feet high. Use proportional spacing for higher ceilings. Space the fixtures by the same distance.
Workshop	Workbench	Suspend a double 48-inch fluorescent fixture over the front edge of the bench and 48 inches above its surface.

Source: *Lighting Your Life* (Chicago: American Home Lighting Institute, 1986).

19   Sound

The *quality of sound* in our homes has a great effect on the quality of our lives. When we play music or engage in conversation, we want to hear clearly and hear well. On the other hand, when junior is playing what teenagers call music or when the neighbors upstairs are fighting, we don't want to hear that at all.

In this chapter you'll see how to achieve quality sound in a room and what to do about a room that either rings or is acoustically flat.

You'll also find guidelines for acceptable levels of *sound transmission* between rooms and between floors. Tables of *sound transmission classes of walls* and *sound transmission classes and impact insulation classes of floor-ceilings* show specific construction techniques for reducing sound transmission to almost any level.

QUALITY OF SOUND

Sound is the sensation produced when pressure waves in the acoustic range of frequencies strike the eardrum. A typical young person can hear sound over the range of 20 to 20,000 cycles per second, or hertz (Hz).

Sound quality involves source intensity, absorption, echo, standing-wave amplification, and reverberation time.

Intensity

Sound intensity is measured in decibels (db). The 0 db level is defined as a sound energy level of 10^{-16} watts per square centimeter and corresponds roughly to the smallest sound detectable by humans. A 10 db increase in intensity is an increase of 10 times in the power of the sound wave and is perceived as a doubling of volume. The table at right lists characteristic sound intensities.

Absorption

The fraction of energy absorbed by a surface is its *absorption coefficient*, α. Gypsum drywall is highly reflective ($\alpha = 0.05$ at 500 Hz), while heavy carpet is highly absorptive ($\alpha = 0.57$ at 500 Hz). One unit of absorption is equivalent to an area of one square foot of perfect ($\alpha = 1.00$) absorption.

High absorption does not reduce the intensity of sound received directly from a source but reduces the buildup of reflected sound. It thus reduces the total intensity of sound in a space.

Absorption is measured at 125, 250, 500, 1,000, 2,000, and 4,000 Hz. In order to give a material a single absorption figure, the noise-reduction coefficient (NRC) is defined as the

average of coefficients at 250, 500, 1,000, and 2,000 Hz. The table on the following page lists typical NRCs for home surfaces.

SOUND INTENSITIES

Intensity, db	Sound Source
0	Threshold of hearing
10	Rustling leaves
20	Rural background
30	Bedroom conversation
40	Living-room conversation
50	Large office activity
60	Face-to-face conversation
70	Auto interior at 55 mph
80	Face-to-face shouting
90	Downtown traffic
100	Table saw
110	Symphony orchestra maximum
120	Elevated train from platform
130	Threshold of pain (rock concert)
140	Jet engine

Absorption and Reflection

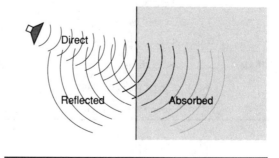

TYPICAL NOISE-REDUCTION COEFFICIENTS (NRC)

Material		Absorption Coefficient at (hertz)				
		250	500	1,000	2,000	NRC
Brick	Bare	0.03	0.03	0.04	0.05	0.04
	Painted	0.01	0.02	0.02	0.02	0.02
Carpet	On slab	0.06	0.14	0.37	0.60	0.29
	On foam pad	0.24	0.57	0.69	0.71	0.55
Concrete Block	Bare	0.44	0.31	0.29	0.39	0.36
	Painted	0.05	0.06	0.07	0.09	0.07
Fabric	10 oz velour hung straight	0.04	0.11	0.17	0.24	0.14
	14 oz velour pleated double	0.31	0.49	0.75	0.70	0.56
Floor	Bare concrete	0.01	0.02	0.02	0.02	0.02
	Resilient on concrete	0.03	0.03	0.03	0.03	0.03
	Bare wood	0.11	0.10	0.07	0.06	0.09
	Parquet on concrete	0.04	0.07	0.06	0.06	0.06
Furniture	Bare wood	0.04	0.05	0.07	0.07	0.06
	Metal	0.04	0.05	0.07	0.07	0.06
	Upholstered with plastic	0.45	0.50	0.55	0.50	0.50
	Upholstered with fabric	0.37	0.56	0.67	0.61	0.55
Glass	Plate	0.06	0.04	0.03	0.02	0.04
	Double strength	0.25	0.18	0.12	0.07	0.16
Wall	1/2" gypsum drywall	0.10	0.05	0.04	0.07	0.07
	Marble	0.01	0.01	0.01	0.02	0.01
	Glazed tile	0.01	0.01	0.01	0.02	0.01
	Plywood paneling	0.22	0.17	0.09	0.10	0.15
Plaster	Smooth finish on brick	0.02	0.02	0.03	0.04	0.03
	Smooth finish on lath	0.03	0.04	0.05	0.04	0.04
	Rough finish on lath	0.02	0.03	0.04	0.04	0.04

Echo

An echo is a sound reflection. To be perceived as a distinct sound, the reflected wave must arrive at least 1/17 second after the direct wave. Since sound travels at about 1,000 feet per second, an echo must have traveled at least 60 feet further than the direct sound (see illustration at right). Thus, echoes occur only in rooms more than 30 feet long. Echoes are stronger when the reflecting surface is highly reflective and is concave toward the listener (focuses the reflected wave toward the listener).

Standing Waves

If the dimensions of a room are a multiple of the wavelength of a sound (illustration at right), the sound wave is reinforced by reflection. Frequencies are thus selectively amplified, distorting the total sound. The problem is made worse by reflective room surfaces, parallel room surfaces, and room dimensions in simple ratios, such as 1/4, 1/3, and so forth.

Reverberation Time

The length of time required for the intensity of sound in a space to fall by 60 db is the *reverberation time* of the space. A room with too short a reverberation time is acoustically dead; too long a reverberation time confuses sounds. Reverberation time can be calculated as

$$T = \frac{V}{20 A}$$

where

T = reverberation time in seconds
V = volume of the room in cubic feet
A = total absorption of the room

Echo in a Room

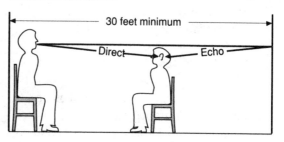

Examples of Standing Waves

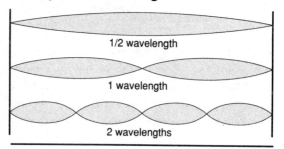

Acceptable Room Reverberation Times

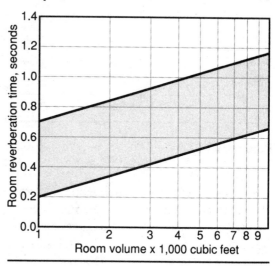

Calculation of Room Absorption and Reverberation Time

Calculation of the total absorption in a room allows determination of the room's reverberation time in seconds:

1. Calculate the areas of all surfaces, including the room furnishings.
2. Multiply each surface area by its 500 Hz absorption coefficient.
3. Sum the products to obtain the total absorption.
4. Calculate reverberation time, $T = V \div 20A$.

Example: What is the 500 Hz reverberation time of a $12 \times 15 \times 8$-foot room having a bare wood floor, gypsum drywall ceiling, smooth plaster walls, two 3×5-foot windows covered by pleated velour drapes, and 40 square feet (floor area) of fabric-upholstered furniture?

Surface	Area, sq ft	α at 500 Hz	Product
Floor	140	0.10	14.0
Ceiling	180	0.05	9.0
Wall	402	0.04	16.1
Drapes	30	0.49	14.7
Furniture	40	0.56	22.4

Total room absorption, A = 76.2

$$\text{Reverberation time} = \frac{1{,}440}{20 \times 76.2} = 0.94 \text{ seconds}$$

The volume of the room in question is $12 \times 15 \times 8$ feet = 1,440 cubic feet. Referring to the graph of acceptable reverberation times on the previous page, you find 0.3 to 0.8 seconds for a room of this volume. Thus, the computed reverberation time falls outside the acceptable range. Carpeting the floor (carpet on foam pad, $\alpha = 0.57$) would increase total room absorption to 142 and bring the reverberation time down to an acceptable 0.51 seconds.

SOUND TRANSMISSION

Sound can be transmitted through a wall, floor, or ceiling in three ways, as seen in the illustration at right:

Leaks or openings allow sound to propagate as though the wall, floor, or ceiling didn't exist. Examples are space under a door, back-to-back electrical fixtures, rough-in holes for pipes, and heating ducts connecting rooms.

Airborne sounds set a building surface into motion, causing it to reradiate to the other side, much like the two heads of a drum.

Impact of an object falling on a floor causes the attached ceiling to radiate sound to the room below.

Walls and floor-ceilings are rated by their abilities to reduce sound transmission. The reduction is measured in decibels, where 10 db corresponds to a factor-of-10 difference in sound energy and a factor-of-2 difference in loudness. The table at right shows the effects on hearing of various levels of sound reduction between spaces.

Walls are rated by *sound transmission class* (STC), roughly the sound reduction at 500 Hz, but taking into consideration frequencies from 125 to 4,000 Hz as well. Floor-ceilings are rated by both STC and *impact insulation class* (IIC), a measure of the noise heard when objects are dropped on the floor above.

The table at right, from the US Department of Housing and Urban Development (HUD) Minimum Property Standards, serves as a guide to minimum acceptable STCs and IICs for residential construction. Walls and ceilings with higher ratings are desirable in a quality home, especially one with children.

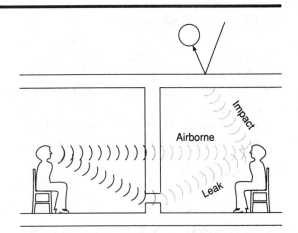

EFFECTIVENESS OF SOUND REDUCTION

db	Effect on Hearing
25	Little effect; normal speech heard clearly
30	Loud speech understood fairly well
35	Loud speech audible but not understood
40	Loud speech heard as a murmur
45	Loud speech a strain to hear
50	Loud speech not heard at all

RECOMMENDED RESIDENTIAL STC & IIC

STC	IIC	Building Surface
45	—	Wall separating living space from living space or public space, such as corridor
50	—	Wall separating living space from commercial space or high-noise service space, such as boiler or mechanical room
45	45	Floor-ceiling separating living space from living space or public space, such as corridor
50	50	Floor-ceiling separating living space from commercial space or high-noise service space

SOUND TRANSMISSION CLASSES OF WALLS

The STC rating of a wall can be increased by

- increasing the mass (doubling the drywall, for example)
- decoupling the surfaces (staggering studs,
for example)
- adding sound-absorbing material (filling
stud cavities with insulation, for example)

The table below shows STC ratings for walls.

SOUND TRANSMISSION CLASSES OF WALLS

Wall Type	STC	Description
	STC 42	5/8" fire-rated gypsum wallboard screw-attached horizontally to both sides of 3-5/8" screw studs, 24" oc. All wallboard joints staggered. Note that rating of 42 is below HUD minimum.
	STC 44	5/8" fire-rated gypsum wallboard screw-attached vertically to both sides of 3-5/8" screw studs, 24" oc. Second layer laminated vertically and screwed to one side only.
	STC 47	5/8" fire-rated gypsum wallboard screw-attached vertically to both sides of 3-5/8" screw studs, 24" oc. Second layer laminated vertically and screwed to one side only. 3-1/2" fiberglass or mineral wool in cavity.
	STC 49	First layer 5/8" fire-rated gypsum wallboard screw-attached vertically to both sides of 3-5/8" screw studs, 24" oc. Second layer laminated vertically to both sides.
	STC 56	First layer 5/8" fire-rated gypsum wallboard screw-attached vertically to both sides of 3-5/8" screw studs, 24" oc. Second layer laminated or screw-attached vertically to both sides. 3" fiberglass in cavity.

SOUND TRANSMISSION CLASSES OF WALLS—*Continued*

Wall Type	STC	Description
	STC 35	5/8" fire-rated gypsum wallboard nailed to both sides of 2x4 wood studs, 16" oc.
	STC 40	5/8" fire-rated gypsum wallboard. Base layer nail-applied to 2x4 wood studs, spaced 24" oc. Face layer nail-applied.
	STC 43	5/8" fire-rated gypsum wallboard. One side screw-applied to resilient furring channel, spaced 24" oc, on 2x4 studs spaced 16" oc. Other side nailed direct to studs.
	STC 46	5/8" fire-rated gypsum wallboard nailed on both sides to staggered 2x4 wood studs, 16" oc, on single 6" plate.
	STC 50	5/8" fire-rated gypsum wallboard. One side screw-applied to resilient furring channel, spaced 24" oc, on 2x4 studs spaced 16" oc. Other side nailed direct to studs. 3-1/2" fiberglass in stud cavity.
	STC 50	Two layers of 5/8" fire-rated gypsum wallboard. One side screw-applied to resilient furring channel, spaced 24" oc, on 2x4 studs spaced 16" oc. Other side nailed direct to studs.
	STC 51	Two layers 5/8" fire-rated gypsum wallboard nailed on both sides to staggered 2x4 wood studs, 16" oc, on single 6" plate.
	STC 58	5/8" fire-rated gypsum wallboard. Base layer applied vertically, nailed 6" oc. Face layer applied horizontally, nailed 8" oc. Nailed to double row of wood studs 16" oc on separate plates. 3-1/2" fiberglass in cavity.

Source: *Gypsum Wallboard Construction* (Charlotte, NC: Gold Bond Products, 1982).

SOUND TRANSMISSION CLASSES AND IMPACT INSULATION CLASSES OF FLOOR-CEILINGS

Both STC and IIC ratings of floor-ceilings are increased by the same three measures used in walls: mass, decoupling, and absorption materials. In addition, IICs are dramatically increased by installation of impact-absorbing flooring materials such as carpets and pads.

As with walls, ratings higher than the minimum HUD values listed earlier are very desirable in quality homes.

SOUND TRANSMISSION AND IMPACT INSULATION CLASSES OF FLOOR-CEILINGS

Floor-Ceiling Type	STC	IIC	Description
	STC 37	IIC 34	1/8" vinyl asbestos tile on 1/2" plywood underlayment, over 5/8" plywood subfloor on 2x joists at 16" oc. Ceiling 1/2" gypsum drywall nailed to joists.
	STC 37	IIC 56	1/4" foam rubber pad and 3/8" nylon carpet on 1/2" plywood underlayment, over 5/8" plywood subfloor on 2x joists at 16" oc. Ceiling 1/2" gypsum drywall nailed to joists.
	STC 46	IIC 44	0.075" vinyl sheet on 3/8" plywood underlayment, over 5/8" plywood subfloor on 2x joists at 16" oc with 3" fiberglass batts. Ceiling 5/8" gypsum drywall screwed to resilient channels.
	STC 48	IIC 45	1/16" vinyl sheet on19/32" T&G Sturd-I-Floor on 2x joists at 16" oc with 3" fiberglass batts. Ceiling 5/8" gypsum drywall screwed to resilient channels.
	STC 51	IIC 80	44 oz carpet and 40 oz pad on 1-1/8" T&G Sturd-I-Floor on 2x joists at 16" oc with 3" fiberglass batts. Ceiling 5/8" gypsum drywall nailed to separate joists.
	STC 52	IIC 78	44 oz carpet and 40 oz pad on 19/32" T&G Sturd-I-Floor, nailed to 2x3 sleepers, glued between joists to 1/2" insulation board, stapled to 1/2" plywood on 2x joists at 16" oc with 3" fiberglass batts. Ceiling 5/8" gypsum drywall screwed to resilient channels.

SOUND TRANSMISSION AND IMPACT INSULATION CLASSES—*Continued*

Floor-Ceiling Type	STC	IIC	Description
	STC 53	IIC 51	25/32" wood strip flooring nailed to 2x3 sleepers, glued between joists to 1/2" insulation board, stapled to 1/2" plywood on 2x joists at 16" oc with 3" fiberglass batts. Ceiling 5/8" gypsum drywall screwed to resilient channels.
	STC 53	IIC 45	25/32" wood strip flooring on 1/2" plywood subfloor on 2x joists at 16" oc with 3" fiberglass batts. Ceiling 5/8" gypsum drywall nailed to separate joists.
	STC 53	IIC 74	44 oz carpet and 40 oz pad on 1-5/8" of 75 pcf perlite/sand concrete over 5/8" plywood subfloor on 2x joists at 16" oc. Ceiling 5/8" gypsum drywall screwed to resilient channels.
	STC 54	IIC 50	25/32" wood strip flooring nailed to 2x3 sleepers, glued to 3" wide strips of 1/2" insulation board, nailed above the floor joists to 1/2" plywood subfloor on 2x joists at 16" oc. Ceiling 5/8" gypsum drywall nailed to separate joists. 3" & 1-1/2" fiberglass batts between joists and sleepers.
	STC 54	IIC 51	5/16" wood block flooring glued to 1/2" plywood underlayment, glued to 1/2" soundboard over 1/2" plywood subfloor on 2x joists at 16" oc. Ceiling 5/8" gypsum drywall screwed to resilient channels with 3" fiberglass between joists.
	STC 55	IIC 72	Carpet and pad on 1/2" plywood underlayment, glued to 1/2" soundboard over 5/8" subfloor on 2x joists at 16" oc. Ceiling 5/8" gypsum drywall screwed to resilient channels with 3" fiberglass between joists.
	STC 57	IIC 56	Vinyl flooring glued to 1/2" plywood underlayment over 1x3 furring strips between joists, on top of 1/2" soundboard, over 5/8" plywood subfloor on 2x joists at 16" oc. Ceiling 5/8" gypsum drywall screwed to resilient channels with 3" fiberglass between joists.
	STC 58	IIC 55	Vinyl flooring on 1/2" plywood underlayment, over 1/2" soundboard over 5/8" subfloor on 2x joists at 16" oc. Ceiling 5/8" gypsum drywall screwed to resilient channels with 3" fiberglass between joists.

Source: *Gypsum Wallboard Construction* (Charlotte, NC: Gold Bond Products, 1982).

20   Fasteners

Houses consist of thousands of pieces held together by what seems like millions of fasteners. What is the right type of fastener, how long should it be, how much will it hold, and how many should I use? These are the questions that must be answered every day on a construction project. And this chapter provides the much-needed answers.

"Nails" provides a field guide to 36 types of nails and 52 different applications. It shows you the relationship between pennyweight (d) and nail length. It lists the type, size, and number of nails to use in every step of residential construction. It even contains tables showing you exactly how much force you can expect a single nail to resist, in 33 species of wood.

"Wood Screws" shows how to drill just the right pilot hole for each size of screw. It also contains tables of allowable holding power for screws.

"A Variety of Screws and Bolts" contains illustrations of screws, bolts, screw heads, and washers.

Metal framing aids are a boon to both contractors and do-it-yourselfers, resulting in stronger fastening in less time than required using more traditional methods of nailing. Illustrated are dozens of these useful aids.

Finally, modern chemistry has produced nothing more amazing than the variety of *adhesives* you can buy at your local building supply store. Unfortunately, they all claim to be the best for every application. Hopefully, the adhesives guide will clear up the confusion.

NAILS

Sizes (common nails, actual size)

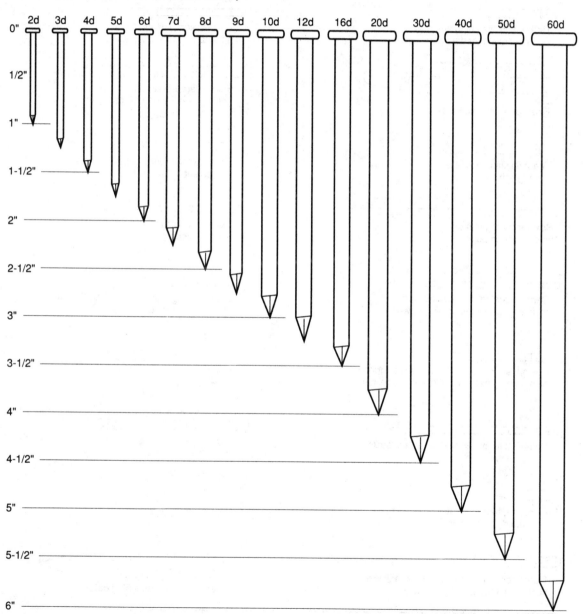

A Variety of Nails

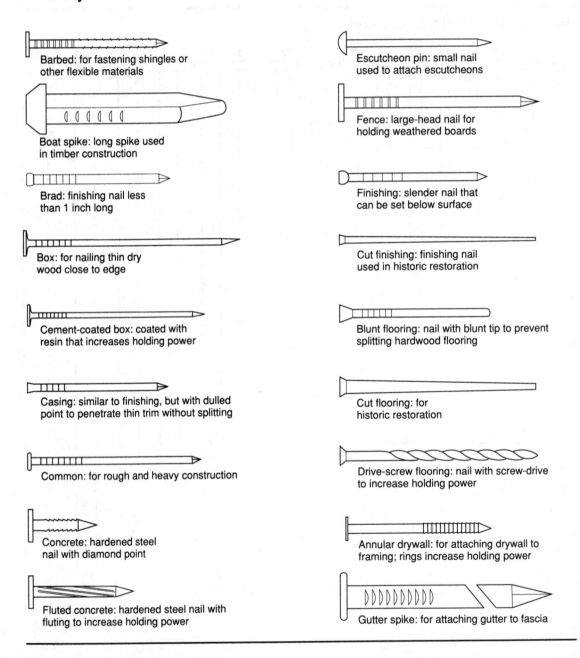

Barbed: for fastening shingles or other flexible materials

Boat spike: long spike used in timber construction

Brad: finishing nail less than 1 inch long

Box: for nailing thin dry wood close to edge

Cement-coated box: coated with resin that increases holding power

Casing: similar to finishing, but with dulled point to penetrate thin trim without splitting

Common: for rough and heavy construction

Concrete: hardened steel nail with diamond point

Fluted concrete: hardened steel nail with fluting to increase holding power

Escutcheon pin: small nail used to attach escutcheons

Fence: large-head nail for holding weathered boards

Finishing: slender nail that can be set below surface

Cut finishing: finishing nail used in historic restoration

Blunt flooring: nail with blunt tip to prevent splitting hardwood flooring

Cut flooring: for historic restoration

Drive-screw flooring: nail with screw-drive to increase holding power

Annular drywall: for attaching drywall to framing; rings increase holding power

Gutter spike: for attaching gutter to fascia

A Variety of Nails —*Continued*

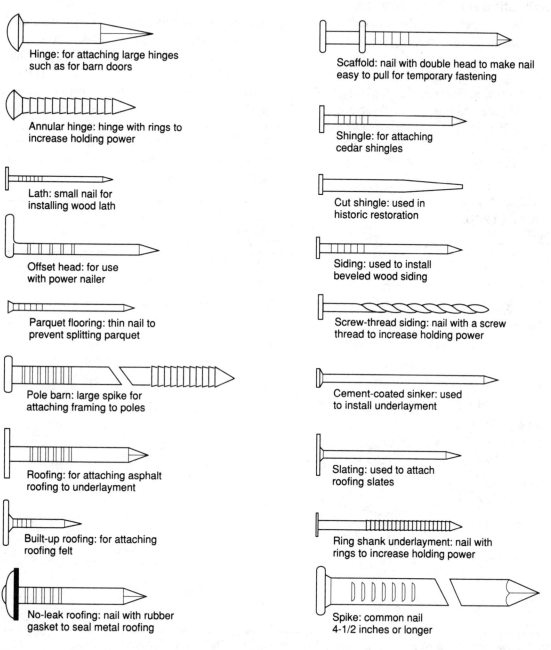

Hinge: for attaching large hinges
such as for barn doors

Annular hinge: hinge with rings to
increase holding power

Lath: small nail for
installing wood lath

Offset head: for use
with power nailer

Parquet flooring: thin nail to
prevent splitting parquet

Pole barn: large spike for
attaching framing to poles

Roofing: for attaching asphalt
roofing to underlayment

Built-up roofing: for attaching
roofing felt

No-leak roofing: nail with rubber
gasket to seal metal roofing

Scaffold: nail with double head to make nail
easy to pull for temporary fastening

Shingle: for attaching
cedar shingles

Cut shingle: used in
historic restoration

Siding: used to install
beveled wood siding

Screw-thread siding: nail with a screw
thread to increase holding power

Cement-coated sinker: used
to install underlayment

Slating: used to attach
roofing slates

Ring shank underlayment: nail with
rings to increase holding power

Spike: common nail
4-1/2 inches or longer

Estimating Nail Requirements

Use the table below and the residential nailing schedule on the next two pages to estimate your requirements, in pounds, for the most common residential building nails. Note that special coatings such as electroplating, etching, and glue add insignificantly to nail weight. Hot-dip galvanizing, however, can increase weight by 20 percent. You should also allow 5 percent for waste.

Example: How many pounds of nails are required to fasten headers to joists for a 24 × 40-foot floor? In the table on the next page, you can find three 16d common nails needed per connection. There are 33 joists, so you need 3 nails × 33 joists × 2 connections = 198 nails. In the table below, you can find 49 16d common nails per pound, so you need 4 pounds. Adding 20 percent for galvanized finish and 5 percent for waste, you should allow 5 pounds.

NAILS PER POUND

Length inches	Penny-weight (d)	Box	Casing	Common	Drywall	Finishing	Roofing	Siding	Spike
						Type of Nail			
1	2	—	—	876	—	1351	255	—	—
1-1/4	3	635	—	568	375	807	210	—	—
1-1/2	4	473	473	316	329	584	180	—	—
1-3/4	5	406	406	271	289	500	150	—	—
2	6	236	236	181	248	309	138	236	—
2-1/4	7	210	—	161	—	238	—	236	—
2-1/2	8	145	145	106	—	189	118	210	—
3	10	94	94	69	—	121	—	94	—
3-1/4	12	87	—	63	—	—	—	—	—
3-1/2	16	71	71	49	—	90	—	—	—
4	20	52	—	31	—	62	—	—	—
4-1/2	30	—	—	24	—	—	—	—	—
5	40	—	—	18	—	—	—	—	—
5-1/2	50	—	—	18	—	—	—	—	—
6	60	—	—	14	—	—	—	—	—
7	—	—	—	11	—	—	—	—	6
8	—	—	—	—	—	—	—	—	5
10	—	—	—	—	—	—	—	—	3
12	—	—	—	—	—	—	—	—	2.5

Source: *Keystone Steel & Wire Pocket Nail Guide* (Peoria, Ill: Keystone Steel & Wire Company, 1989).
Note: Sinkers up to 12d are 1/8" shorter than common nails of the same penny size.
Sinkers 16d or larger are 1/4" shorter than common nails of the same penny size.

NAILING SCHEDULE FOR LIGHT CONSTRUCTION

Application		Method	Number	Size	Type of Nail
Finish	Casings	Face-nail	—	—	Casing, 1-1/2" min into base
	Moldings	Face-nail	—	—	Finishing, 1-1/2" min into base
Floor	Board subfloor	Face-nail	2	8d	Common
	Plywood subfloor	Face-nail	6" oc	8d	Common
	2×6, T&G subfloor	Face-nail	2	16d	Common
	2×6, T&G finish floor	Blind-nail	1	16d	Casing
	Softwood finish	Face-nail	2/16" oc	8d	Common
	T&G hardwood strip	Blind-nail	16" oc	8d	Blunt flooring
	Underlayment	Face-nail	6" oc	1-1/4"	Ring shank underlayment
Frame	Header to joist	End-nail	3	16d	Common
	Header to sill	Toenail	16" oc	10d	Common
	Joist to sill	Toenail	2	10d	Common
	Bridging to joist	Toenail	2	8d	Common
	Ledger to beam	Face-nail	3/16" oc	16d	Common
	Sole plate to stud	End-nail	2	16d	Common
	Top plate to stud	End-nail	2	16d	Common
	Stud to sole plate	Toenail	4	8d	Common
	Sole plate to joist	Face-nail	16" oc	16d	Common
	Doubled studs	Face-nail	16" oc	10d	Common
	Double top plate	Face-nail	16" oc	10d	Common
	Double header	Face-nail	12" oc	12d	Common
	Ceiling joist to top plate	Toenail	3	8d	Common
	Overlapping joists	Face-nail	4	16d	Common
	Rafter to top plate	Toenail	2	8d	Common
	Rafter to ceiling joist	Face-nail	5	10d	Common

Application		Method	Number	Size	Type of Nail
	Rafter to hip or valley	Toenail	3	10d	Common
	Ridge board to rafter	End-nail	3	10d	Common
	Rafter to rafter	Toenail	4	8d	Common
	Collar tie (2") to rafter	Face-nail	2	12d	Common
	Collar tie (1") to rafter	Face-nail	3	8d	Common
	Let-in brace (1")	Face-nail	2 each stud	8d	Common
	Corner studs	Face-nail	12" oc	16d	Common
	Built-up beams (3 or 4)	Face-nail	32" oc each side	20d	Common
Paneling	3/4" wood	Blind-nail	24" oc	6d	Finishing
	Hardboard	Face-nail	8" oc	2"	Annular drywall
	Plywood	Face-nail	8" oc	3d	Finishing
	Gypsum drywall	Face-nail	6" oc	1-1/4"	Annular drywall
Roof	Asphalt, new	Face-nail	4	7/8"	Roofing
	Asphalt, reroof	Face-nail	4	1-3/4"	Roofing
	Wood shingle, new	Face-nail	2	4d	Shingle
	Wood shingle, reroof	Face-nail	2	6d	Shingle
Sheathing	3/8" plywood	Face-nail	6" oc	6d	Common
	1/2" & over plywood	Face-nail	6" oc	8d	Common
	1/2" fiberboard	Face-nail	3" oc	1-1/2"	Roofing
	3/4" fiberboard	Face-nail	3" oc	1-3/4"	Roofing
	3/4" boards	Face-nail	6" oc	8d	Common
	Foam board	Face-nail	12" oc	—	Roofing, 1/2" into base
Siding	Bevel and lap	Face-nail	16" oc	8d	Aluminum siding
	Drop and shiplap	Blind-nail	16" oc	8d	Finishing
	Plywood	Face-nail	8" oc	7d	Aluminum siding
	Hardboard	Face-nail	16" oc	2"	Hardboard siding
	Shingles	Face-nail	2	4d	Aluminum siding

Source: *Wood Frame House Construction* (Washington, DC: US Department of Agriculture, 1975).

Holding Power of Common Nails

The tables in this section show the withdrawal resistance (force required to pull straight out) and lateral resistance (sideways force to pull out) of common nails driven perpendicular to the grain. Assumptions include bright common nails, and seasoned wood that will remain dry or wet wood that will remain wet. Nails driven into unseasoned wood lose much of their holding power as the wood dries.

Holding power increases with wood density or specific gravity. The tables are grouped either by species group or by specific gravity. Use the table below to determine either.

Nail holding power can be increased by

- surface etching or coating
- annular or spiral threads
- clinching (bending the tips over)

Holding power is decreased by

- blunting the points
- end-nailing (does not resist withdrawal)
- toenailing (33% reduction)

WOOD SPECIES GROUPS AND OVEN-DRY SPECIFIC GRAVITIES

Species Group	Specific Gravity	Wood Species	Species Group	Specific Gravity	Wood Species
I	0.75	Hickory and pecan		0.43	Sitka spruce
	0.68	Beech		0.42	California redwood (closed grain)
	0.67	Oak, red and white		0.42	Hem-fir
	0.66	Birch, sweet and yellow		0.42	Ponderosa pine - sugar pine
II	0.55	Southern pine		0.42	Red pine
	0.54	Sweet gum and tupelo		0.42	Spruce-pine-fir
	0.51	Douglas fir - larch		0.41	Eastern spruce
	0.49	Ponderosa pine		0.40	Idaho white pine
III	0.48	Douglas fir, south	IV	0.41	Eastern cottonwood
	0.48	Southern cypress		0.40	Western white pine
	0.47	Mountain hemlock		0.39	Coast Sitka spruce
	0.46	Northern pine		0.38	Eastern white pine
	0.46	Yellow poplar		0.37	California redwood (open grain)
	0.45	Eastern hemlock - tamarack		0.36	Balsam fir
	0.44	Lodgepole pine		0.36	Engelmann spruce - alpine fir
	0.43	Eastern hemlock		0.35	Western cedars
				0.31	Northern white cedar

Fasteners

ALLOWABLE LATERAL LOADS FOR COMMON NAILS (pounds)

Species Group	Common Nail Size, d							
	6	8	10 & 12	16	20	30	40	60
I	77 (1.2)	97 (1.4)	116 (1.5)	133 (1.7)	172 (2.0)	192 (2.1)	218 (2.3)	275 (2.7)
II	63 (1.3)	78 (1.5)	94 (1.7)	108 (1.8)	139 (2.2)	155 (2.3)	176 (2.5)	223 (2.9)
III	51 (1.5)	64 (1.7)	77 (2.0)	88 (2.2)	114 (2.5)	127 (2.7)	144 (3.0)	182 (3.5)
IV	41 (1.6)	51 (1.9)	61 (2.1)	70 (2.3)	91 (2.7)	102 (2.9)	115 (3.2)	146 (3.7)

Source: *National Design Specification for Stress-Grade Lumber and Its Fastening* (Washington, DC: National Forest Products Association, 1971).
Note: Inches of penetration into member holding point shown in parentheses.

ALLOWABLE WITHDRAWAL LOADS FOR COMMON NAILS (pounds)

Specific Gravity [1]	Common Nail Size, d							
	6	8	10 & 12	16	20	30	40	60
0.75	76	88	99	109	129	139	151	177
0.68	59	69	78	85	101	109	118	138
0.66	55	64	72	79	94	101	110	128
0.62	47	55	62	68	80	86	94	110
0.55	35	41	46	50	59	64	70	81
0.51	29	34	38	42	49	53	58	67
0.47	24	27	31	34	40	43	47	55
0.45	21	25	28	30	36	39	42	49
0.43	19	22	25	27	32	35	38	44
0.41	17	19	22	24	29	31	33	39
0.39	15	17	19	21	25	27	29	34
0.37	13	15	17	19	22	24	26	30
0.33	10	11	13	14	17	18	19	23
0.31	8	10	11	12	14	15	17	19

Source: *National Design Specification for Wood Construction* (Washington, DC: National Forest Products Association, 1986).

Note: Loads are per inch of penetration into member holding point.
[1] Based on oven-dry weight and volume.

WOOD SCREWS

Wood-screw size is specified by diameter gauge (see illustration below) and by length, where length is the distance from tip to plane of the wood surface.

Screws are designed to draw two pieces together. For maximum effectiveness, the first piece is drilled out to the diameter of the body, while the receiving piece is drilled just large enough to prevent splitting. Drill sizes for pilot holes are shown in the table below.

Maximum wood-screw loads for different wood species are given on the following page.

Wood-Screw Pilot Holes

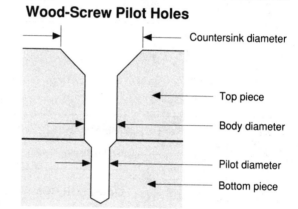

WOOD-SCREW PILOT HOLE DIMENSIONS

Hole	Screw Size									
	2	3	4	5	6	8	10	12	14	16
Body dia, inch	5/64	3/32	7/64	1/8	9/64	5/32	3/16	7/32	15/64	17/64
Pilot Drill: Softwood	#65	#58	1/16	5/64	5/64	3/32	7/64	1/8	11/64	3/16
Hardwood	#56	#54	5/64	3/32	3/32	7/64	1/8	5/32	7/32	15/64
Body drill	#42	#37	1/8	9/64	9/64	11/64	3/16	7/32	1/4	9/32
Countersink dia, inch	5/32	3/16	7/32	1/4	9/32	11/32	3/8	7/16	15/32	11/16

Wood-Screw Gauges (actual size)

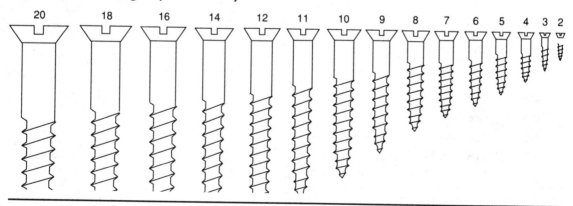

20 18 16 14 12 11 10 9 8 7 6 5 4 3 2

Fasteners

ALLOWABLE LATERAL LOADS FOR WOOD SCREWS (pounds)

Species Group	Wood-Screw Size							
	6	8	10	12	14	16	18	20
I	91 (0.97)	129 (1.15)	173 (1.33)	224 (1.51)	281 (1.69)	345 (1.88)	415 (2.06)	492 (2.24)
II	75 (0.97)	106 (1.15)	143 (1.33)	185 (1.51)	232 (1.69)	284 (1.88)	342 (2.06)	406 (2.24)
III	62 (0.97)	87 (1.15)	117 (1.33)	151 (1.51)	190 (1.69)	233 (1.88)	280 (2.06)	332 (2.24)
IV	48 (0.97)	68 (1.15)	91 (1.33)	118 (1.51)	148 (1.69)	181 (1.88)	218 (2.06)	258 (2.24)

Source: *National Design Specification for Stress-Grade Lumber and Its Fastening* (Washington, DC: National Forest Products Association, 1971).
Note: Inches of penetration into member holding point shown in parentheses.

ALLOWABLE WITHDRAWAL LOADS FOR WOOD SCREWS (pounds)

Specific Gravity [1,2]	Wood-Screw Size							
	6	8	10	12	14	16	18	20
0.75	220	262	304	345	387	428	470	511
0.68	181	215	250	284	318	352	386	420
0.66	171	203	235	267	299	332	364	396
0.62	151	179	207	236	264	293	321	349
0.55	119	141	163	186	208	230	253	275
0.51	102	121	140	160	179	198	217	236
0.47	87	103	119	136	152	168	184	201
0.45	79	94	109	124	139	154	169	184
0.43	72	86	100	113	127	141	154	168
0.41	66	78	91	103	116	128	140	153
0.39	60	71	82	93	105	116	127	138
0.37	54	64	74	84	94	104	114	124
0.33	43	51	59	67	75	83	91	99
0.31	38	45	52	59	66	73	80	87

Source: *National Design Specification for Wood Construction* (Washington, DC: National Forest Products Association, 1986).

Note: Load is per inch penetration into member holding point.
[1] Based on oven-dry weight and volume.
[2] Allowable loads assume 19% moisture content or lower.

A VARIETY OF SCREWS AND BOLTS

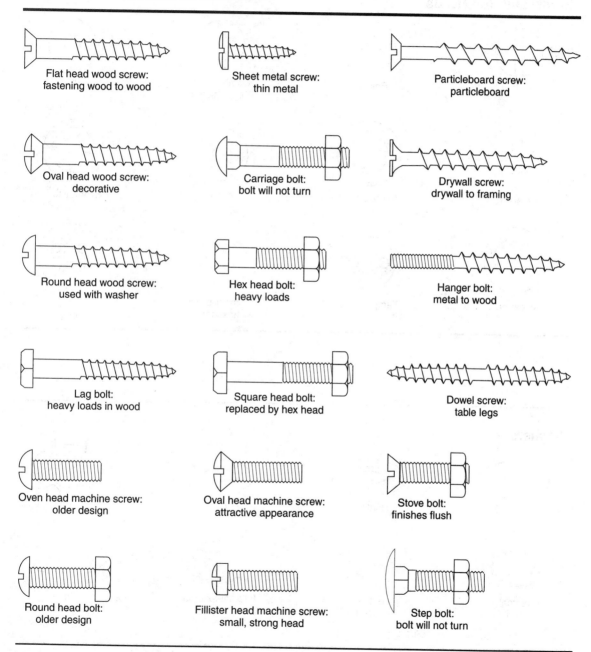

Flat head wood screw:
fastening wood to wood

Sheet metal screw:
thin metal

Particleboard screw:
particleboard

Oval head wood screw:
decorative

Carriage bolt:
bolt will not turn

Drywall screw:
drywall to framing

Round head wood screw:
used with washer

Hex head bolt:
heavy loads

Hanger bolt:
metal to wood

Lag bolt:
heavy loads in wood

Square head bolt:
replaced by hex head

Dowel screw:
table legs

Oven head machine screw:
older design

Oval head machine screw:
attractive appearance

Stove bolt:
finishes flush

Round head bolt:
older design

Fillister head machine screw:
small, strong head

Step bolt:
bolt will not turn

Screw and Bolt Heads

Slotted

Phillips

Combination
Phillips/slotted

Square

Frearson

Internal torx

Clutch

External torx

Tamper-proof

Tamper-proof

Tamper-proof
hexagon

Tamper-proof
torx

Washers

Flat USS

Flat SAE

Finish

Torque

Internal-
tooth

External-
tooth

Internal-external-
tooth

Split-
lock

METAL FRAMING AIDS

Metal framing aids do not represent an admission of incompetence on the part of a builder. They are an improvement over older techniques such as end-nailing and toenailing, let-in bracing, and splicing with wood cleats.

Use metal framing aids to

- separate wood from moisture
- strengthen butt and splice joints
- support rafter and beam ends
- brace walls against racking
- reinforce cutouts in studs and joists
- provide hurricane resistance

The following illustrations show many of the United Steel metal framing aids for residential construction. Your lumberyard is sure to carry this or another similar product line.

Foundation Anchors

FA1
Uplift load = 785 lb

PAHD42
Uplift load = 3,362 lb

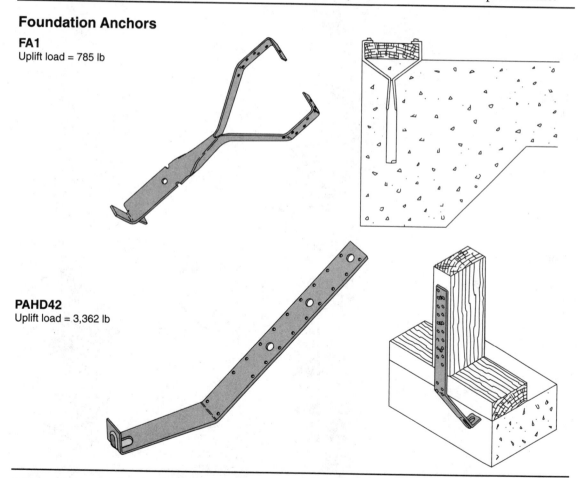

Source: *Full Line Catalog—Construction Hardware* (Montgomery, Minn: United Steel Products Company, 1996).

Post Anchors

WA44
Uplift load = 1,375 lb

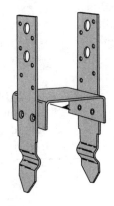
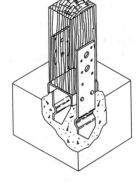

PA44
Uplift load = 380 lb
Bearing load = 5,135 lb

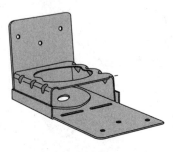
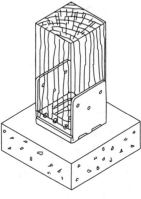

EBG44
Uplift load = 1,085 lb
Bearing load = 2,485 lb

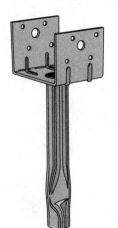
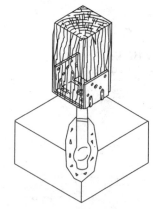

Source: *Full Line Catalog—Construction Hardware* (Montgomery, Minn: United Steel Products Company, 1996).

Post Beam Caps

PB44
Uplift load = 770 lb

PCM44
Uplift load = 1,050 lb

KCC44
Uplift load = 2,985 lb

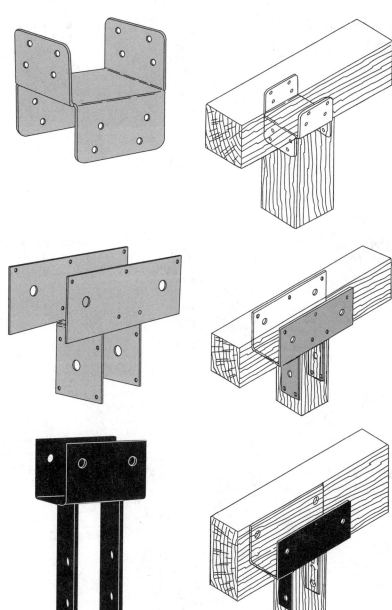

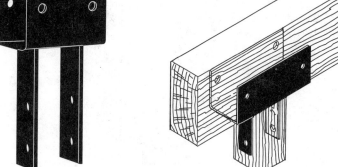

Source: *Full Line Catalog—Construction Hardware* (Montgomery, Minn: United Steel Products Company, 1996).

Face-Mount Hangers

JL26—2 x 6,8
Load = 590 lb

JL28—2 x 8,10
Load = 970 lb

SP26—2 x 6
Load = 805 lb

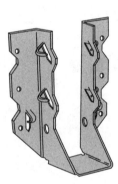

JESX—2 x 10,12,14
Load = 1,610 lb

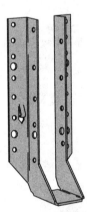

SU26-2—two 2 x 6,8
Load = 1,290 lb

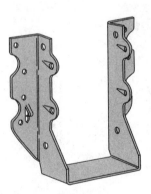

JT—three 2 x 8,10,12
Load = 2,720 lb

Source: *Full Line Catalog—Construction Hardware* (Montgomery, Minn: United Steel Products Company, 1996).

Face-Mount Hangers

S26/8
Light slope
Load = 945 lb

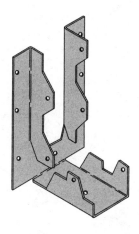

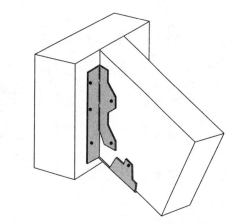

SKHL26—2 x 6,8
Skewed left
Skewed right = SKHR26
Load = 810 lb

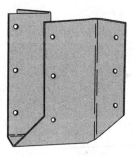

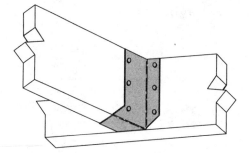

JH30—two 2 x 6,8,10
Load = 1,875 lb

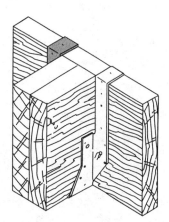

Source: *Full Line Catalog—Construction Hardware* (Montgomery, Minn: United Steel Products Company, 1996).

Top-Mount Hangers

KLB26
Load = 1,365 lb

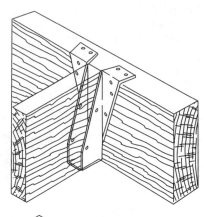

HDO28
Load = 1,960 lb

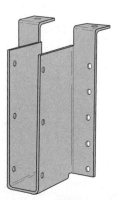
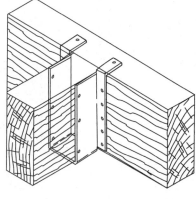

MPH210
Load = 2,545 lb

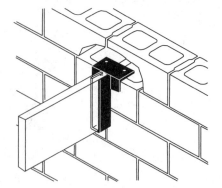

Source: *Full Line Catalog—Construction Hardware* (Montgomery, Minn: United Steel Products Company, 1996).

Beam Hangers

LBH5—5-1/8" laminated
Load = 7,980 lb

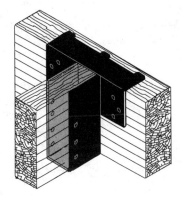

KHHB5—5-1/8" laminated
Load = 5,960 lb

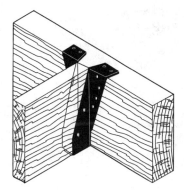

KLEG5—5-1/8" laminated
Load = 10,400 lb

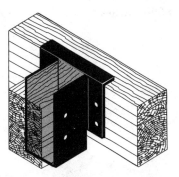

Source: *Full Line Catalog—Construction Hardware* (Montgomery, Minn: United Steel Products Company, 1996).

Wall Bracing

WBT – rolled edge
WBC – L-style

WBT   WBC

WB – X-style
RWB – Rolled

WB   RWB

Breaking RWB

S365 – Quick strip
S360 – Wrap-around

S365   S360

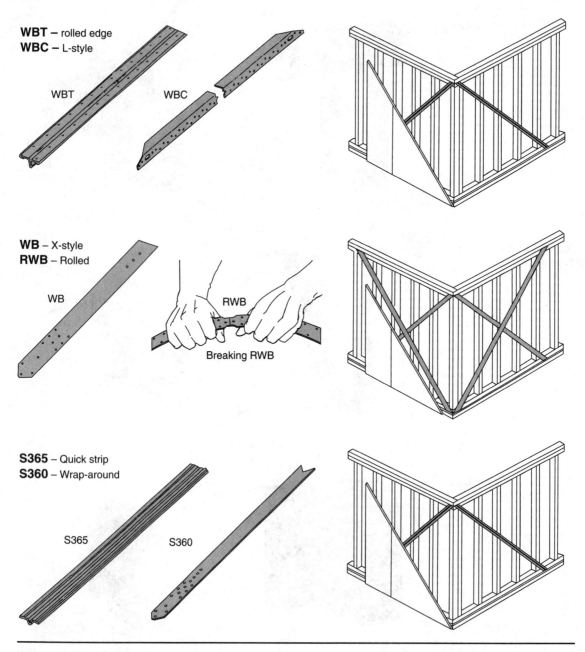

Source: *Full Line Catalog—Construction Hardware* (Montgomery, Minn: United Steel Products Company, 1996).

Angles and Straps

KSA

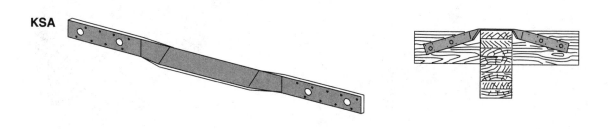

S345   **S320**

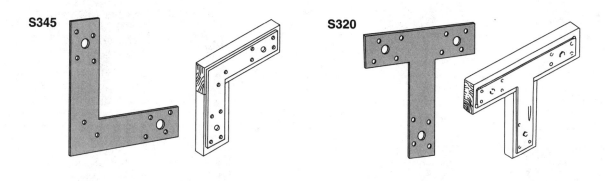

MPA1

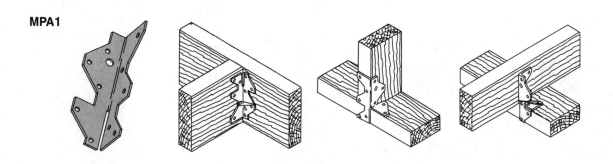

Source: *Full Line Catalog—Construction Hardware* (Montgomery, Minn: United Steel Products Company, 1996).

Rafter/Hurricane Ties

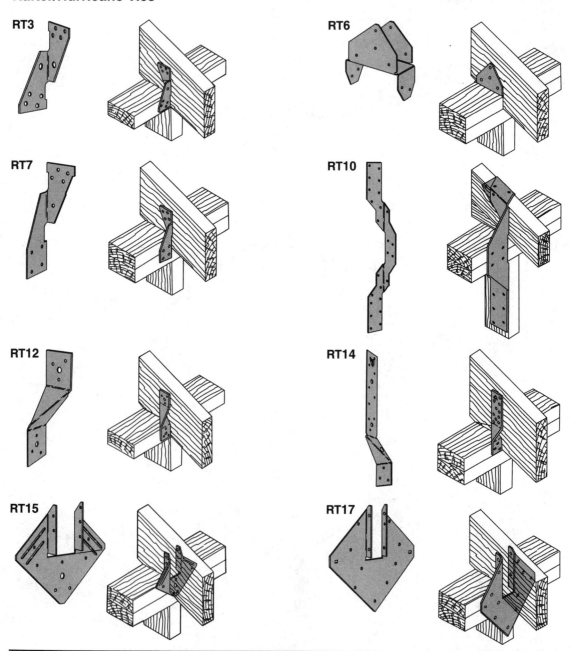

Source: *Full Line Catalog—Construction Hardware* (Montgomery, Minn: United Steel Products Company, 1996).

ADHESIVES

The adhesives described here and in the table on the next page are household glues, adhesives generally available at hardware stores. There are literally thousands of adhesives on the market, but 99 percent are variations on the following 11 generic products:

Aliphatic glue, otherwise known as carpenter's glue, is a water-based, yellow-drying wood glue. It is relatively fast setting and forms a strong bond with wood. Use it for woodworking projects except for furniture.

Catalyzed acrylic glue is a waterproof adhesive which comes in two parts. It sets up in about 1 minute and achieves full strength in about 15 minutes. It will bond most materials, including those with greasy surfaces.

Clear cement is also known as model or airplane cement. It works best on porous materials such as open-grain wood, fabric, and paper. It sets up extremely fast and dries to an almost invisible clarity.

Contact cement is unique in that it bonds after drying. Spread cement on both surfaces and leave it until dry to the touch. When the two coated surfaces are brought together, they bond instantly and permanently. The cement is not very strong, but it is water resistant, making it perfect for bonding sheets of plastic laminate to plywood or particleboard for kitchen and bathroom countertops.

Cyanoacrylate, also known as instant-setting glue, bonds almost instantly to any nonporous material. Some varieties will bond wood weakly, but none can be used on paper or foam. One problem is that they bond to skin extremely well.

Epoxy glue comes in two components measured and mixed at the time of application. The bond is extremely strong between nonporous materials, and it is water and chemical resistant. The chemicals are clear, but color and metallic fillers are sometimes added to make it appear like epoxy glass, plastic, steel, and so forth. Use it to bond metals and plastics.

Plastic resin glue is based on urea-formaldehyde and comes in a dry powder that is mixed with water at the time of application. It is an extremely water-resistant and strong wood glue. It requires clamping overnight, however, and dries to a red-brown color. It is excellent for gluing furniture.

Resorcinol glue, also known as waterproof glue, is a two-part mixture of liquid resin and powder catalyst. It is slow to cure but forms a strong and completely waterproof bond between porous surfaces, particularly wood. It is too difficult for most home applications.

Silicone rubber is primarily a caulk, but it is waterproof, clear, and extremely flexible, making it the best for bathroom applications. A drawback is that it can't be painted.

Urethane adhesive offers strength and flexibility. It requires overnight clamping and dries to an ugly, rubbery finish, but its bond strength to wood and its flexibility make it perfect for repairing indoor furniture, toys, and tools.

White wood glue is a low-cost, water-soluble, easy-to-use glue that bonds fairly well to porous materials such as wood, paper, and fabric. Strength of bond is only moderate, however, and it has little water resistance. It is good for children's projects.

ADHESIVES GUIDE

Adhesive Type	Set Time	Bond Strength Wood	Glass	Water Resistance Wood	Glass	Gap Filling	Color Dry
Aliphatic (carpenter's) glue	1 hr	E	P	P	P	E	Yellow
Catalyzed acrylic glue	15 min	E	E	G	G	F	Brown
Clear (airplane) cement	10 min	E	P	G	F	F	Clear
Contact cement	5 min	F	P	G	F	P	Brown
Cyanoacrylate (instant) glue	1 min	F	E	E	F	F	Clear
Epoxy glue, regular set	2 hr	F	E	E	E	E	Clear
Epoxy glue, quick set	5 min	F	E	E	E	E	Clear to white
Plastic resin	24 hr	E	P	E	P	F	Red to brown
Resorcinol (waterproof) glue	8 hr	E	P	E	P	G	–
Silicone rubber	8 hr	F	P	E	E	E	Clear or white
Urethane adhesive	8 hr	E	F	E	E	E	White
White wood glue	1 hr	G	P	P	P	E	White

Note: E = excellent; G = good; F = fair; P = poor.

21　Outdoors

Good residential design incorporates outdoor spaces into the overall plan. *Decks* can provide graceful transitions from enclosed living/dining areas to the outdoors.

Laying out is the first step in deck construction. Framing consists of connecting the posts, beams, and joists. High decks often require bracing of the posts as well as dealing with grade level changes. Visual interest can be increased by varying the pattern of decking.

Stairs provide access to ground-level spaces. Railings are required whenever a deck is more than 30 inches above the ground. Benches reduce the need for outdoor furniture and may be incorporated into the railing design or may be freestanding.

Overhead structures make a deck usable in hot, sunny climates and also heighten the formal definition of space.

No matter how ambitious your design, you will find that complex decks are really nothing more than combinations of simple deck designs, of which we provide two examples.

Fences, in conjunction with landscaping, are used to define the largest living spaces (lawns and gardens) as well as to enhance the overall appearance of a property.

Laying out a fence is similar to laying out a deck. Fence framing basically consists of connecting rails to posts in any of a variety of post/rail joints and providing for one or more gates. Most popular is the picket fence, of which we show a number of variations.

DECKS

Laying Out

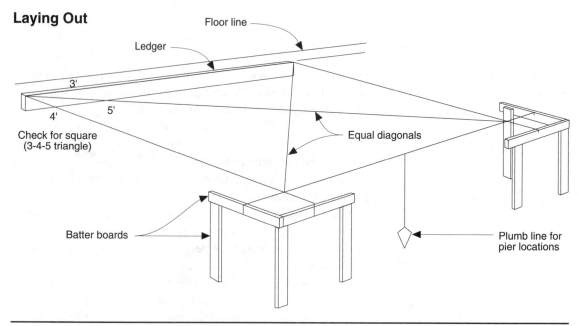

Setting Posts

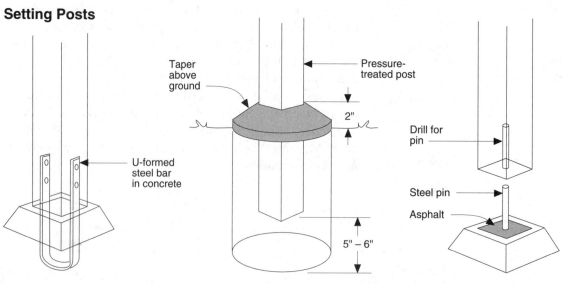

Framing

ATTACHING TO BUILDING

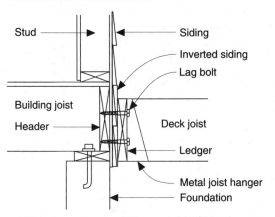

Stud

Siding

Inverted siding

Lag bolt

Building joist

Deck joist

Header

Ledger

Metal joist hanger

Foundation

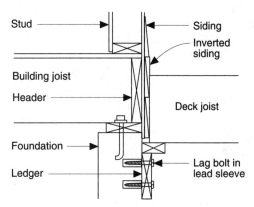

Stud

Siding

Inverted siding

Building joist

Header

Deck joist

Foundation

Ledger

Lag bolt in lead sleeve

POSTS AND BEAMS

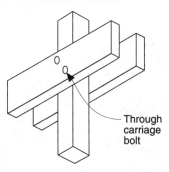

Through carriage bolt

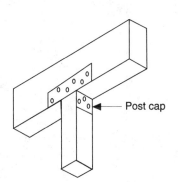

Post cap

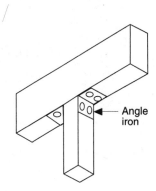

Angle iron

BEAMS AND JOISTS

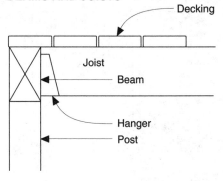

Decking

Joist

Beam

Hanger

Post

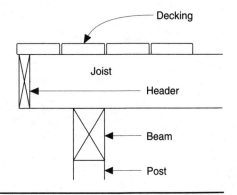

Decking

Joist

Header

Beam

Post

Bracing

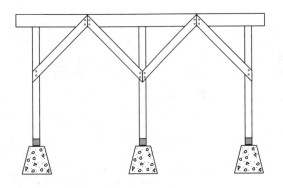

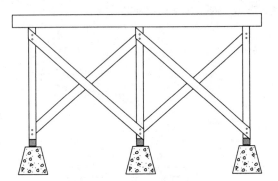

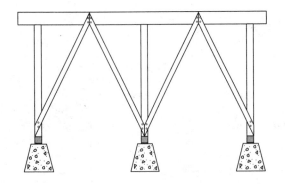

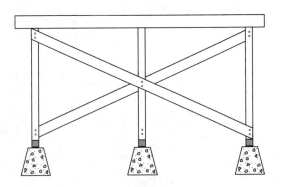

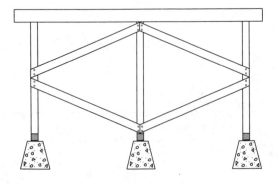

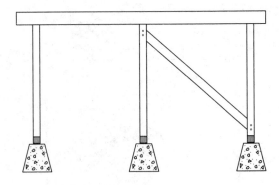

Level Changes

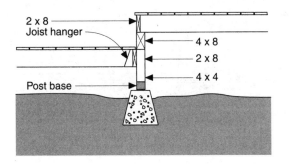
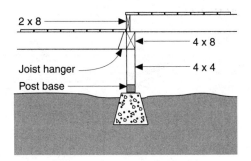

2 x 8
Joist hanger
4 x 8
2 x 8
4 x 4
Post base

2 x 8
4 x 8
Joist hanger
4 x 4
Post base

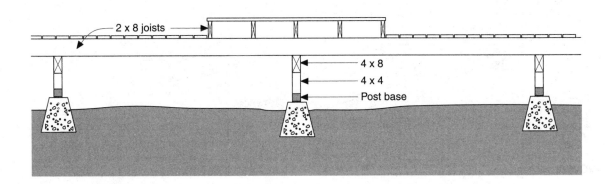

2 x 8 joists
4 x 8
4 x 4
Post base

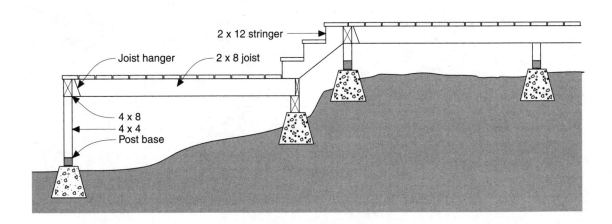

2 x 12 stringer
Joist hanger   2 x 8 joist
4 x 8
4 x 4
Post base

Decking

DIAGONAL

HERRINGBONE

CURVED

ANGLED

Stairs

FIGURING RISE AND RUN

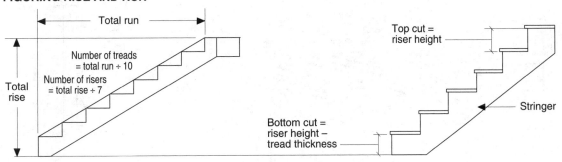

Total run

Number of treads
= total run ÷ 10

Number of risers
= total rise ÷ 7

Total rise

Top cut =
riser height

Bottom cut =
riser height –
tread thickness

Stringer

HUNG STRINGER

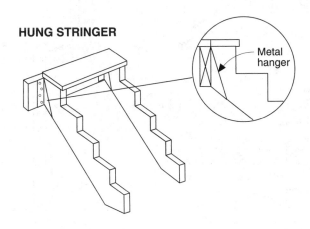

Metal hanger

BOLTED STRINGER

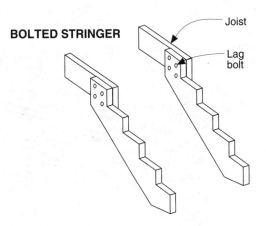

Joist

Lag bolt

OPEN TREAD

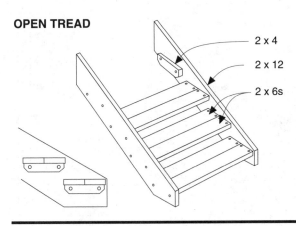

2 x 4

2 x 12

2 x 6s

CLOSED TREAD

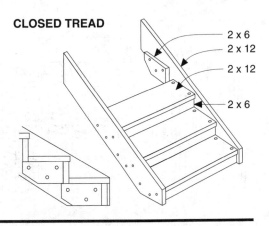

2 x 6

2 x 12

2 x 12

2 x 6

Railings

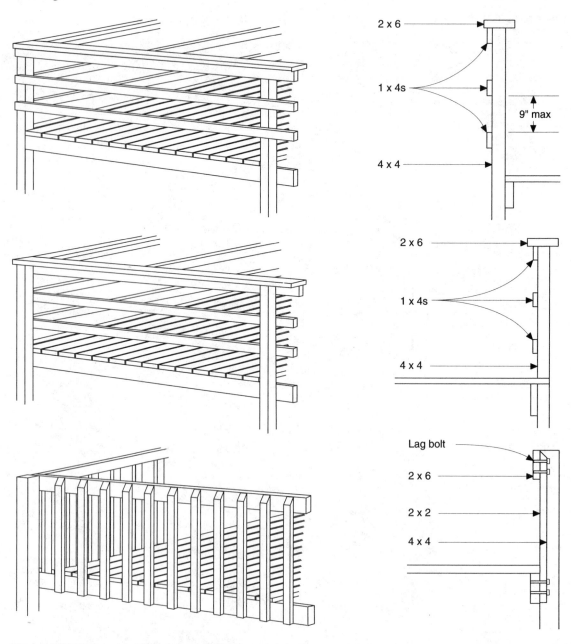

Benches

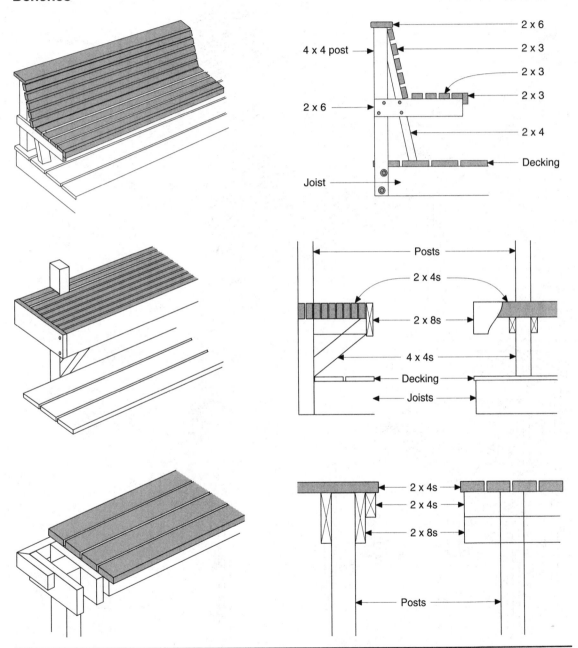

Overhead Structures

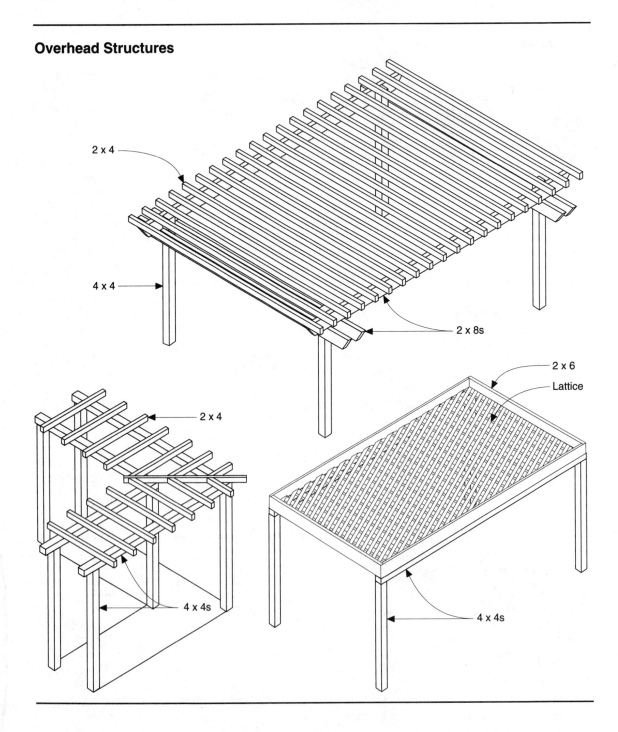

2 x 4

4 x 4

2 x 8s

2 x 6

Lattice

2 x 4

4 x 4s

4 x 4s

Simple Deck 1

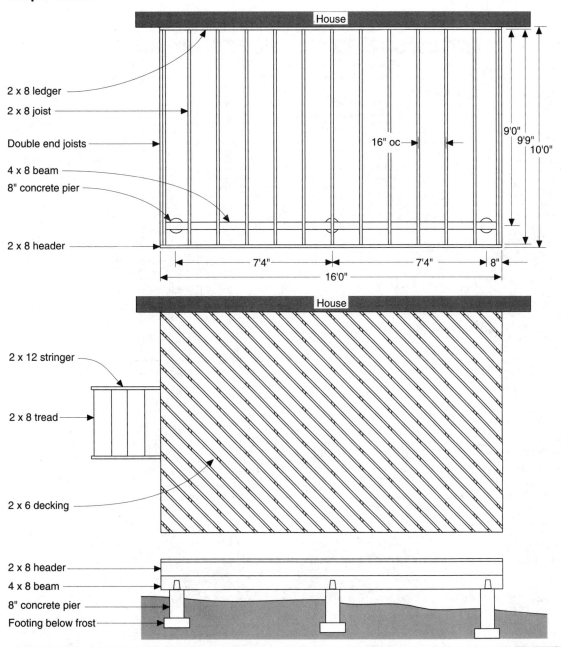

2 x 8 ledger

2 x 8 joist

Double end joists

4 x 8 beam

8" concrete pier

2 x 8 header

House

16" oc

9'0"
9'9"
10'0"

7'4"   7'4"   8"

16'0"

House

2 x 12 stringer

2 x 8 tread

2 x 6 decking

2 x 8 header

4 x 8 beam

8" concrete pier

Footing below frost

Simple Deck 2

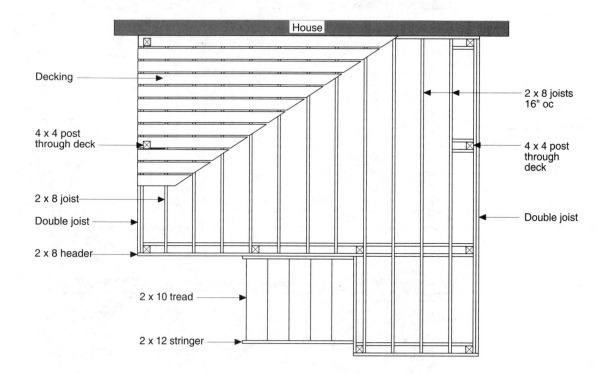

House

Decking

2 x 8 joists 16" oc

4 x 4 post through deck

4 x 4 post through deck

2 x 8 joist

Double joist

Double joist

2 x 8 header

2 x 10 tread

2 x 12 stringer

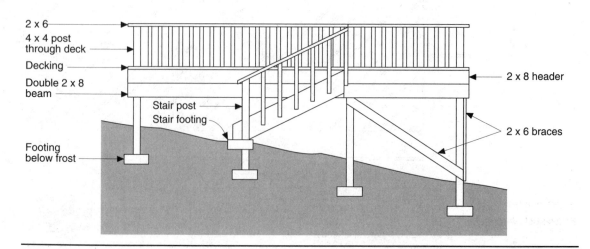

2 x 6

4 x 4 post through deck

Decking

Double 2 x 8 beam

2 x 8 header

Stair post

Stair footing

2 x 6 braces

Footing below frost

FENCES

Post/Rail Joints

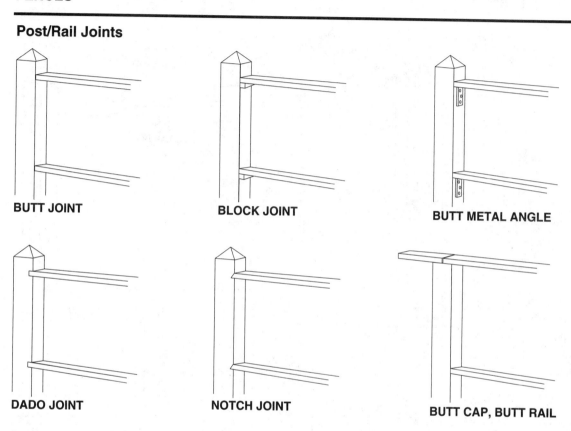

BUTT JOINT

BLOCK JOINT

BUTT METAL ANGLE

DADO JOINT

NOTCH JOINT

BUTT CAP, BUTT RAIL

Gates

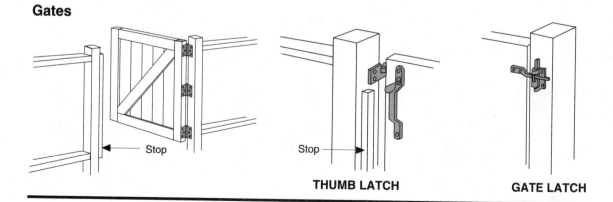

Stop

Stop

THUMB LATCH

GATE LATCH

Picket Fences

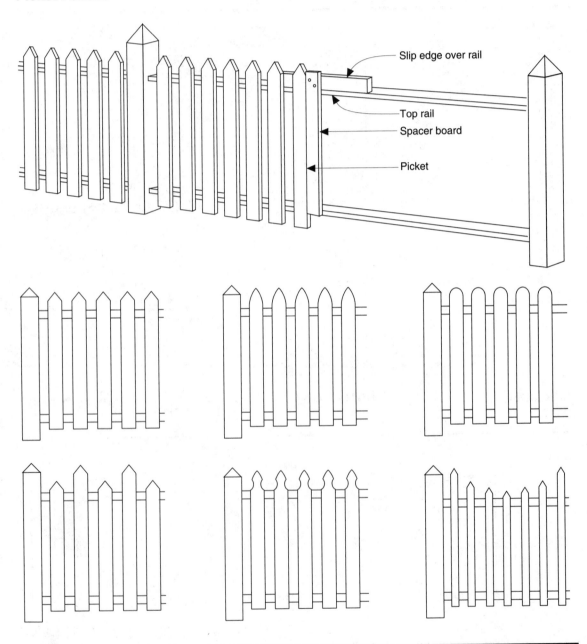

Slip edge over rail

Top rail

Spacer board

Picket

22

Finance and Measurement

Few building projects are paid for with cash; usually somebody must obtain a mortgage. This chapter contains a guide to *home mortgage types* that will let you in on what the banker and the real estate broker are talking about. The table of *interest on loans* will help you calculate the monthly payments.

Geometric figures and *trigonometry* may bring back bad memories, but sooner or later you'll find it useful or even necessary to compute the volume of a space, an area of carpeting, or the angle of a saw cut. This chapter contains the formulas for all of the shapes you could think of.

How many square feet are there in an acre? How many ounces in a kilogram? What is the decimal value of 3 feet 4-5/16 inches? The answers to these and dozens more building questions are all contained in *"Units and Conversions."*

Finally, how thick is 18 gauge? It depends on which gauge, as you'll see in *"Gauges Used in Building."*

HOME MORTGAGE TYPES

Loan Type	How Loan Works	Pros and Cons
Adjustable-rate (ARM)	Interest rate tied to published financial index such as prime lending rate. Most have annual and lifetime caps.	Interest rate lower initially. Interest and payments usually increase over time.
Assumable	Buyer takes over seller's mortgage.	Mortgage usually fixed-rate at below-market interest rate.
Balloon	Payments based on long term (usually 30 years), but entire principle due in short term (usually 3 to 5 years).	Refinancing required at unknown interest rate at time of balloon payment.
Buy-down	Seller pays part of interest for first years.	Loan more affordable for buyer whose income is expected to rise.
Fixed-rate (FRM)	Interest rate and monthly payments constant for life of loan (usually 30 years).	Interest rate usually higher. Equity grows slowly in early years.
Graduated-payment (GPM)	Payments increase for first few years, then remain constant. Interest rate may vary.	Loan more affordable for buyer whose income is expected to rise.
Growing-equity (GEM)	Payments increase annually, with increase applied to principle. Interest rate usually constant.	Buyer equity increases rapidly, but buyer must be able to make increased payments.
Interest-only	Entire payment goes toward interest only. Principle remains at original amount.	Requires lowest payments. Owner equity consists of down payment plus appreciation.
Owner-financed (seller take back)	Seller holds either first or second mortgage.	Interest rate may be higher than market rate.
Renegotiable (rollover)	Same as adjustable-rate mortgage, but interest rate adjusted less often.	Interest rate and monthly payment variable, but fixed for longer periods.
Reverse annuity	Lender makes monthly payments to borrower. Debt increases over time to maximum percentage of appraised value of property.	Provides monthly income to borrower but decreases equity.
Shared-appreciation (SAM)	Lender charges less interest in exchange for share of appreciation when property sold.	Makes property more affordable. Reduces owner's gain upon sale.
Wraparound (blended-rate)	Existing lower-interest loan combined with new additional loan for single loan with intermediate interest rate.	Decreases interest rate.
Zero-interest (no-interest)	No interest charged. Fixed monthly payments usually over short term.	Sale price usually inflated. One-time fee may be charged. No tax deduction allowed.

Source: Federal Trade Commission, Washington, DC.

INTEREST ON LOANS

Monthly Payment

The table below shows monthly payments for each $1,000 borrowed for fixed-interest rates from 0 to 18 percent and for periods from 1 to 40 years. Monthly payments at other interest rates can be accurately interpolated using the figures in the table.

Example: What are the monthly payments on a 10-year, $1,000 loan at 11 percent interest? From the table, the payments at 10 and 12 percent are $13.22 and $14.35. Since 11 percent is the average of 10 and 12 percent, the monthly payment is ($13.22 + $14.35) ÷ 2 = $13.78 for each $1,000 borrowed.

Total Interest Paid

Knowing the monthly payment, it is easy to find the total interest paid over the life of a loan. Total interest is the difference between the sum of the payments over life and the original amount.

Example: For a 30-year, 12 percent, $100,000 mortgage, what is the total interest paid? From the table, the monthly payment for each $1,000 is $10.29. Therefore, the total lifetime payment is 100 × 12 months × 30 years × $10.29 = $370,440. Since the original loan is for $100,000, the total interest paid is the difference: $270,440.

MONTHLY PAYMENTS FOR FIXED-RATE LOANS
(payment for each $1,000)

Rate %	1	2	3	4	5	10	15	20	30	40
0	83.33	41.67	27.78	20.83	16.67	8.33	5.56	4.17	2.78	2.08
1	83.79	42.10	28.21	21.26	17.09	8.76	5.98	4.60	3.22	2.53
2	84.24	42.54	28.64	21.70	17.53	9.20	6.44	5.06	3.70	3.03
3	84.69	42.98	29.08	22.13	17.97	9.66	6.91	5.55	4.22	3.58
4	85.15	43.42	29.52	22.58	18.42	10.12	7.40	6.06	4.77	4.18
5	85.61	43.88	29.98	23.03	18.88	10.61	7.91	6.60	5.37	4.83
6	86.07	44.33	30.43	23.49	19.34	11.11	8.44	7.17	6.00	5.51
7	86.53	44.78	30.88	23.95	19.81	11.62	8.99	7.76	6.66	6.22
8	86.99	45.23	31.34	24.42	20.28	12.14	9.56	8.37	7.34	6.96
9	87.46	45.69	31.80	24.89	20.76	12.67	10.15	9.00	8.05	7.72
10	87.92	46.15	32.27	25.37	21.25	13.22	10.75	9.66	8.78	8.50
12	88.85	47.08	33.22	26.34	22.25	14.35	12.01	11.02	10.29	10.09
14	89.79	48.02	34.18	27.33	23.27	15.53	13.32	12.44	11.85	11.72
16	90.74	48.97	35.16	28.35	24.32	16.76	14.69	13.92	13.45	13.36
18	91.68	49.93	36.16	29.38	25.40	18.02	16.11	15.44	15.08	15.02

GEOMETRIC FIGURES

Plane Shapes (two-dimensional)

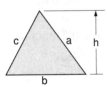

TRIANGLE
Area = $\dfrac{bh}{2}$
Perimeter = $a + b + c$

SQUARE
Area = ab
Perimeter = $4a$

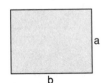

RECTANGLE
Area = ab
Perimeter = $2a + 2b$

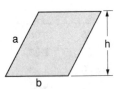

PARALLELOGRAM
Area = bh
Perimeter = $2a + 2b$

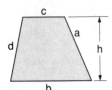

TRAPEZOID
Area = $\dfrac{h(b + c)}{2}$
Perimeter = $a + b + c + d$

N-SIDED POLYGON
Area = $\dfrac{N(ra)}{2}$
Perimeter = Na

CIRCLE
Area = πr^2
Perimeter = $2\pi r$

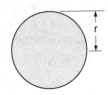

CIRCULAR SECTOR
Area = $\dfrac{\pi r}{360}$
Perimeter = $0.01745 r A°$

CIRCULAR SEGMENT
Area = $\dfrac{r^2}{2} \left(\dfrac{\pi A° - \sin A°}{180} \right)$
$c = 2r \sin \dfrac{A°}{2}$

ELLIPSE
Area = πab
Perimeter (approximate)
$= \pi \sqrt{2(a^2 + b^2)}$

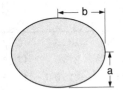

PARABOLA
Area = $\dfrac{2ab}{3}$
Perimeter = $b \left(1 + \dfrac{8a^2}{3b^2} \right)$

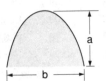

CIRCLE/SQUARE
Area = $0.2146a^2$
Perimeter = $2a + \dfrac{\pi a}{2}$

Solid Bodies (three-dimensional)

CUBE
Area = $6a^2$
Volume = a^3

CYLINDER
Area = $2\pi r^2 + 2\pi rh$
Volume = $\pi r^2 h$

RECTANGULAR PRISM
Area = $2ab + 2ac + 2bc$
Volume = abc

CONE
Area = $\pi r \sqrt{r^2 + h^3} + \pi r^2$
Volume = $\dfrac{\pi r^2 h}{3}$

SPHERE
Area = $4\pi r^2$
Volume = $\dfrac{4\pi r^3}{3}$

SPHERICAL SEGMENT
Area = $2\pi ra$
Volume = $\dfrac{\pi a^2 (3r - a)}{3}$

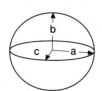

ELLIPSOID
Volume = $\dfrac{\pi abc}{3}$

PARABOLOID
Volume = $\dfrac{\pi r^2 h}{2}$

TRIGONOMETRY

RIGHT TRIANGLE

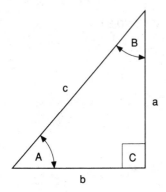

DEFINITION OF RATIOS OF SIDES (TRIG FUNCTIONS)

$\sin A = a/c$ $\sin B = b/c$ $\sin C = 1$

$\cos A = b/c$ $\cos B = a/c$ $\cos C = 0$

$\tan A = a/b$ $\tan B = b/a$ $\tan C = \infty$

PYTHAGOREAN THEOREM

$a^2 + b^2 = c^2$

$c = \sqrt{a^2 + b^2}$

ANY TRIANGLE

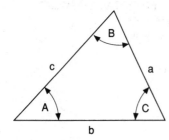

LAW OF COSINES

$a^2 = b^2 + c^2 - 2bc \cos A$ $\cos A = \dfrac{b^2 + c^2 - a^2}{2bc}$

$b^2 = a^2 + c^2 - 2ac \cos B$ $\cos B = \dfrac{a^2 + c^2 - b^2}{2ac}$

$c^2 = a^2 + b^2 - 2ab \cos C$ $\cos C = \dfrac{a^2 + b^2 - c^2}{2ab}$

LAW OF SINES

$$\frac{a}{\sin A} = \frac{b}{\sin B} = \frac{c}{\sin C}$$

$\dfrac{a}{b} = \dfrac{\sin A}{\sin B},$ etc.

Trigonometry Tables

Deg	Sin	Cos	Tan	Deg	Sin	Cos	Tan	Deg	Sin	Cos	Tan
1	.0175	.9998	.0175	31	.5150	.8572	.6009	61	.8746	.4848	1.8040
2	.0349	.9994	.0349	32	.5299	.8480	.6249	62	.8829	.4695	1.8807
3	.0523	.9986	.0524	33	.5446	.8387	.6494	63	.8910	.4540	1.9626
4	.0698	.9976	.0699	34	.5592	.8290	.6745	64	.8988	.4384	2.0503
5	.0872	.9962	.0875	35	.5736	.8192	.7002	65	.9063	.4226	2.1445
6	.1045	.9945	.1051	36	.5878	.8090	.7265	66	.9135	.4067	2.2460
7	.1219	.9925	.1228	37	.6018	.7986	.7536	67	.9205	.3907	2.3559
8	.1392	.9903	.1405	38	.6157	.7880	.7813	68	.9272	.3746	2.4751
9	.1564	.9877	.1584	39	.6293	.7771	.8098	69	.9336	.3584	2.6051
10	.1736	.9848	.1763	40	.6428	.7660	.8391	70	.9397	.3420	2.7475
11	.1908	.9816	.1944	41	.6561	.7547	.8693	71	.9455	.3256	2.9042
12	.2079	.9781	.2126	42	.6691	.7431	.9004	72	.9511	.3090	3.0777
13	.2250	.9744	.2309	43	.6820	.7314	.9325	73	.9563	.2924	3.2709
14	.2419	.9703	.2493	44	.6947	.7193	.9657	74	.9613	.2756	3.4874
15	.2588	.9659	.2679	45	.7071	.7071	1.0000	75	.9659	.2588	3.7321
16	.2756	.9613	.2867	46	.7193	.6947	1.0355	76	.9703	.2419	4.0108
17	2924	.9563	.3057	47	.7314	.6820	1.0724	77	.9744	.2250	4.3315
18	.3090	.9511	.3249	48	.7431	.6691	1.1106	78	.9781	.2079	4.7046
19	.3256	.9455	.3443	49	.7547	.6561	1.1504	79	.9816	.1908	5.1446
20	.3420	.9397	.3640	50	.7660	.6428	1.1918	80	.9848	.1736	5.6713
21	.3584	.9336	.3839	51	.7771	.6293	1.2349	81	.9877	.1564	6.3138
22	.3746	.9272	.4040	52	.7880	.6157	1.2799	82	.9903	.1392	7.1154
23	.3907	.9205	.4245	53	.7986	.6018	1.3270	83	.9925	.1219	8.1443
24	.4067	.9135	.4452	54	.8090	.5878	1.3764	84	.9945	.1045	9.5144
25	.4226	.9063	.4663	55	.8192	.5736	1.4281	85	.9962	.0872	11.4301
26	.4384	.8988	.4877	56	.8290	.5592	1.4826	86	.9976	.0698	14.3007
27	.4540	.8910	.5095	57	.8387	.5446	1.5399	87	.9986	.0523	19.0811
28	.4695	.8829	.5317	58	.8480	.5299	1.6003	88	.9994	.0349	28.6363
29	.4848	.8746	.5543	59	.8572	.5150	1.6643	89	.9998	.0175	57.2900
30	.5000	.8660	.5774	60	.8660	.5000	1.7321	90	1.000	.0000	∞

Note: Deg = degrees of angle; Sin = sine; Cos = cosine; Tan = tangent.

UNITS AND CONVERSIONS

Converting Inches to Feet

Very often in building calculations, exact measurements in feet and inches must be converted to feet and decimal fractions. For some people this is a simple matter; for those who find the process baffling, the table at right will help.

To make the conversion from inches to decimal fractions of a foot, find the decimal fraction for the number of whole inches (left half of table) and the decimal fraction for the partial inches (right half of table), and add them together.

Example 1: Find the decimal value of 13 feet 7-9/16 inches.

13'	= 13.0000'
7"	= 0.5833'
9/16"	= 0.0469'
Total	= 13.6302'

Example 2: Find the decimal value of the floor area of a rectangular room of width 10 feet 8-1/2 inches and length 12 feet 3-5/8 inches.

Width:	10'	=	10.0000'
	8"	=	.6667'
	1/2"	=	.0417'
Total width		=	10.7084'

Length:	12'	=	12.0000'
	3"	=	.2500'
	5/8"	=	.0521'
Total length		=	12.3021'

Area = width × length = 131.74 sq ft

INCHES-TO-FEET CONVERSION

Whole Inches	Fraction of Foot	Partial Inches	Fraction of Foot
1	.0833	1/16	.0052
2	.1667	1/8	.0104
3	.2500	3/16	.0156
4	.3333	1/4	.0208
5	.4167	5/16	.0260
6	.5000	3/8	.0313
7	.5833	7/16	.0365
8	.6667	1/2	.0417
9	.7500	9/16	.0469
10	.8333	5/8	.0521
11	.9167	11/16	.0573
12	1.0000	3/4	.0625
		13/16	.0677
		7/8	.0729
		15/16	.0781

Unit Conversions

Multiply	By	To Get	Multiply	By	To Get
Length			Mile	1,760	Yards
Centimeter	0.3937	Inches	Mile	1,609	Meters
Centimeter	10	Millimeters	Mile	1.609	Kilometers
Centimeter	0.01	Meters			
Inch	2.54	Centimeters	**Area**		
Inch	0.0833	Feet	Square centimeter	0.1550	Square inches
Inch	0.0278	Yards	Square centimeter	100	Square millimeters
Foot	30.48	Centimeters	Square centimeter	0.0001	Square meters
Foot	0.3048	Meters	Square inch	6.4516	Square centimeters
Foot	12	Inches	Square inch	0.0069	Square feet
Foot	0.3333	Yards	Square inch	7.72×10^{-4}	Square yards
Yard	91.44	Centimeters	Square foot	929	Square centimeters
Yard	0.9144	Meters	Square foot	0.0929	Square meters
Yard	36	Inches	Square foot	144	Square inches
Yard	3	Feet	Square foot	0.1111	Square yards
Meter	39.37	Inches	Square yard	8,361	Square centimeters
Meter	3.281	Feet	Square yard	0.8361	Square meters
Meter	1.094	Yards	Square yard	1,296	Square inches
Meter	100	Centimeters	Square yard	9	Square feet
Meter	0.001	Kilometers	Square meter	1,550	Square inches
Kilometer	3,281	Feet	Square meter	10.765	Square feet
Kilometer	1,094	Yards	Square meter	1.1968	Square yards
Kilometer	0.6214	Miles	Square meter	10,000	Square centimeters
Kilometer	1,000	Meters	Square meter	1.0×10^{-6}	Square kilometers
Mile	5,280	Feet	Square kilometer	1.076×10^{7}	Square feet

(continued)

Unit Conversions—*Continued*

Multiply	By	To Get
Square kilometer	1.197×10^6	Square yards
Square kilometer	0.3861	Square miles
Square kilometer	1.0×10^6	Square meters
Square mile	2.788×10^7	Square feet
Square mile	3.098×10^6	Square yards
Square mile	640	Acres
Square mile	2.590	Square kilometers

Multiply	By	To Get
Cubic yard	4.667×10^4	Cubic inches
Cubic yard	27	Cubic feet
Cubic meter	6.102×10^4	Cubic inches
Cubic meter	35.320	Cubic feet
Cubic meter	1.3093	Cubic yards
Cubic meter	1.0×10^6	Cubic centimeters
Cubic meter	1,000	Liters

Volume

Multiply	By	To Get
Cubic centimeter	0.0610	Cubic inches
Cubic centimeter	1,000	Cubic millimeters
Cubic centimeter	1.0×10^{-6}	Cubic meters
Cubic inch	16.387	Cubic centimeters
Cubic inch	5.787×10^{-4}	Cubic feet
Liter	0.2642	Gallons, US
Liter	1.0568	Quarts
Liter	1,000	Cubic centimeters
Gallon, US	0.0238	Barrels (42 gallons)
Gallon, US	4	Quarts
Gallon, US	231	Cubic inches
Gallon, US	3,785	Cubic centimeters
Cubic foot	2.832×10^4	Cubic centimeters
Cubic foot	0.0283	Cubic meters
Cubic foot	1,728	Cubic inches
Cubic foot	0.0370	Cubic yards
Cubic yard	7.646×10^5	Cubic centimeters
Cubic yard	0.7646	Cubic meters

Mass

Multiply	By	To Get
Pound	4.448	Newtons
Pound	32.17	Poundals
Ton, US short	2,000	Pounds
Ton, US long	2,240	Pounds
Ton, metric	2,205	Pounds
Ton, metric	1,000	Kilograms
Gram	0.0353	Ounces
Gram	2.205×10^{-3}	Pounds
Gram	0.001	Kilograms
Gram	15.432	Grains
Ounce	28.35	Grams
Ounce	0.0284	Kilograms
Ounce	0.0625	Pounds
Pound	453.6	Grams
Pound	0.4536	Kilograms
Pound	16	Ounces
Kilogram	35.28	Ounces
Kilogram	2.205	Pounds

Multiply	By	To Get
Kilogram	1,000	Grams
Ton, US	0.9070	Tons metric
Ton, US	907	Kilograms
Ton, US	2,000	Pounds
Ton, metric	1.102	Tons, US
Ton, metric	2,205	Pounds
Ton, metric	1,000	Kilograms

Multiply	By	To Get
Temperature		
Degree C	1.8	Degrees F
Degree F	0.5556	Degrees C
Degree K	1	Degrees C
Degree C	1.8 (°C+32)	Degrees F
Degree F	0.556 (°F-32)	Degrees C
Degree K	(°K-273)	Degrees C

Energy

Multiply	By	To Get
Erg	1.0×10^{-7}	Joules
Joule	1	Newton-meters
Joule	1.0×10^{7}	Ergs
Joule	0.2389	Calories
Joule	9.48×10^{-4}	British thermal units
Joule	0.7376	Foot-pounds
Calorie	3.97×10^{-3}	British thermal units
Btu/hour	0.293	Joules/second
Btu/hour	252	Calories/hour

Surveyor's Measure

Multiply	By	To Get
Link	7.92	Inches
Rod	16.5	Feet
Chain	4	Rods
Rood	40	Square rods
Acre	160	Square rods
Acre	43,560	Square feet
Square mile	640	Acres
Township	36	Square miles

Light

Multiply	By	To Get
Lux	1	Lumens/square meter
Lux	0.0929	Lumens/square foot
Lux	0.0929	Footcandles
Footcandle	10.76	Lux
Lumen/square foot	10.76	Lux

GAUGES USED IN BUILDING

GAUGE DIAMETERS (inches)

Wire Gauge	American Standard	US Standard	Brown & Sharpe
5/0	.4305	.4375	–
4/0	.3938	.4063	.4600
3/0	.3625	.3750	.4096
2/0	.3310	.3438	.3648
1/0	.3065	.3125	.3249
1	.2830	.2813	.2893
2	.2625	.2656	.2576
3	.2437	.2500	.2294
4	.2253	.2344	.2043
5	.2070	.2188	.1819
6	.1920	.2031	.1620
7	.1770	.1875	.1443
8	.1620	.1719	.1285
10	.1350	.1406	.1019
12	.1055	.1094	.0808
14	.0800	.0781	.0641
16	.0625	.0625	.0508
18	.0475	.0500	.0403
20	.0348	.0375	.0320
22	.0286	.0313	.0253
24	.0230	.0250	.0201
26	.0181	.0188	.0159
28	.0162	.0156	.0126
30	.0140	.0125	.0100

WEIGHTS OF METAL SHEET

American Standard Wire Gauge	Pounds Per Square Foot		
	Aluminum	Steel	Copper
5/0	5.92	17.58	19.95
4/0	5.41	16.08	18.24
3/0	4.98	14.80	16.79
2/0	4.55	13.52	15.34
1/0	4.21	12.52	14.20
1	3.89	11.56	13.11
2	3.61	10.72	12.16
3	3.35	9.95	11.29
4	3.10	9.20	10.44
5	2.85	8.45	9.59
6	2.64	7.84	8.90
7	2.43	7.23	8.20
8	2.23	6.62	7.51
10	1.86	5.51	6.25
12	1.45	4.31	4.89
14	1.10	3.27	3.71
16	0.86	2.55	2.90
18	0.65	1.94	2.20
20	0.48	1.42	1.61
22	0.39	1.17	1.33
24	0.32	0.94	1.07
26	0.25	0.74	0.84
28	0.22	0.66	0.75
30	0.19	0.57	0.65

Glossary

Absorptance: ratio of energy absorbed by a surface to the amount striking it.

Absorption: weight of water absorbed, expressed as percentage of dry weight; interception of radiant energy or sound waves.

Absorptivity: (see *absorptance*).

Accelerator: chemical added to concrete to speed setting.

Active solar collector: mechanical system for collecting solar heat.

Admixture: substance added to mortar to change its properties.

Aggregate: granular materials used in masonry.

Air barrier: material or surface designed to prevent passage of air, but not water vapor.

Air-entrained concrete: concrete containing microscopic air bubbles to make it less susceptible to freeze damage.

Alkyd: synthetic resin paint base. Alkyd resin has largely replaced linseed oil.

Alternating current (AC): electrical current which reverses direction regularly (60 hertz, or cycles per second, in the US).

Altitude: vertical angle of the sun above the horizon.

Ampacity: ampere-carrying capacity of a wire.

Ampere: unit of electrical current. Often abbreviated *amp*.

Anchor bolt: bolt set into a foundation to fasten it to the building sill.

Apron: vertical panel below the window sill.

Ash drop: opening in the floor of a fireplace for ash disposal.

Ashlar: one of several masonry patterns, consisting of large rectangular units of cut stone.

Asphalt: waterproof organic liquid used to waterproof.

Asphalt plastic cement: asphalt used to seal roofing materials together.

Awning: shading device mounted above a window.

Awning window: single window sash hinged at the top and swinging outward.

Azimuth angle: direction to the sun, usually measured from true north. Solar azimuth is measured east or west from true south.

Backer rod: foam rope used to fill large gaps before caulking.

Backfill: material used to fill excavation around a foundation.

Balloon frame: wood frame in which studs are continuous from the sill to the top plate of the top floor.

Baluster: vertical member under a railing.

Baseboard: horizontal molding at the base of a wall. Also known as mopboard.

Batten: thin molding of rectangular section used to cover a joint.

Bay window: window that projects from a wall.

Beam: horizontal structural member designed to support loads in bending.

Bearing wall: wall that supports a load from above.

Bevel cut: wood cut made at any angle other than 0° or 90°.

Bitumin: substance containing oil or coal-based compounds; asphalt.

Blocking: short member bracing between two longer framing members.

Boiler: central heating appliance that generates either hot water or steam.

Bond: strength of adhesion. Also one of several patterns in which masonry units may be laid.

Bow window: same as a bay window, except the projection approximates a circular arc.

Braced frame: heavy-timber frame braced in the corners by lighter members.

Branch circuit: one of several circuits in a building, originating at the service entrance panel and protected by a separate circuit breaker or fuse.

Brick: rectangular masonry unit hardened by firing in a kiln.

Brick mold: standard wood molding used as outside casing around doors and windows.

Brick veneer: brick facing over wood or masonry.

Bridging: bracing between floor joists to prevent twist.

British thermal unit (Btu): amount of heat required to raise the temperature of 1 pound of water by $1F°$.

Brown coat: next-to-last plaster or stucco coat.

Building code: rules adopted by a government for the regulation of building.

Built-up roof: roofing consisting of many alternating layers of asphalt and felt.

Bull nose: rounded masonry unit for use in corners.

Bundle: package of shingles.

Bus bar: rectangular metal (usually copper) bar for carrying large electrical current.

Butt: bottom edge of a roof shingle.

Butt joint: joint in which two members meet without overlap or miter.

Calcium chloride: concrete accelerator.

Cant strip: beveled strip around the perimeter of a roof.

Capillarity: movement of water through small gaps due to adhesion and surface tension.

Casement window: window hinged on the side and opening outward.

Casing: inside or outside molding which covers space between a window or door jamb and a wall.

Caulk: material used to fill building joints and cracks.

Cavity wall: masonry wall with a continuous space between the inside and outside bricks that acts as a capillary break.

Cellulose insulation: loose-fill insulation consisting of shredded and treated newspaper.

Ceramic mosaic: sheet of small ceramic tiles.

Chalk line: straight line made by snapping a taut string coated by colored chalk.

Check: cracks in the surface of wood resulting from drying (and shrinking) of the surface faster than of the interior.

Chimney: vertical tube for venting flue gases by natural convection.

Chimney fire: burning of creosote and other deposits within a chimney.

Circuit: two or more wires carrying electricity to lights, receptacles, switches, or appliances.

Circuit breaker: electromechanical device that opens when the current exceeds its rating.

Clapboard: board for overlapping as horizontal siding.

Closed valley: roof valley where the shingles extend in an unbroken line across the valley intersection.

Collar: preformed vent pipe flashing. Also the part of an appliance that connects to a stove or vent pipe.

Collar tie: rafter tie beam.

Column: structural member designed to carry a vertical load in compression.

Common nail: large-diameter nail for rough framing.

Compression: action of forces to squeeze or compact.

Concealed-nail roofing: method of applying asphalt roll roofing where all nails are in the underlying layer and the top layer is attached by asphalt cement only.

Concrete: hardened mixture of portland cement, sand, gravel, and water.

Condensation: process of water vapor turning to liquid water.

Conduction: transfer of heat through an opaque material. Also the transfer of electrons (current) through a material.

Conductor: wire intended to carry electric current.

Conduit: metal or plastic pipe that surrounds electrical wires and protects them from physical damage.

Control joint: groove in concrete to control the location of cracking.

Convection: heat transfer through either the natural or forced movement of air.

Corner bead: strip of metal designed to provide protection of a plaster or drywall corner.

Corner board: vertical board at a wall intersection for butting siding.

Corner bracing: diagonal boards, metal strips, or rigid panels used at building corners to prevent racking.

Cornice: top, projecting molding of the entablature.

Countersink: to sink a nail or screw below the surface.

Course: row of roofing or siding.

Cove molding (trim): popular molding (trim) for ceiling/wall intersections.

Coverage: minimum number of layers of roofing at any section.

Crawl space: space beneath a building not high enough for a person to stand in.

Cricket: small roof for diverting water.

Curing: process of hardening of concrete over time.

Cut-in brace: corner brace of framing lumber cut into studs.

Cutout: space between tabs in a roofing shingle.

Cycle: one complete reversal of electrical current and voltage. Cycles per second are called hertz.

Dado: rectangular groove cut across the grain of wood.

Damper: valve designed to control the flow of air or smoke.

Damp-proofing: treating a masonry surface to retard capillary action.

Dead load: load imposed on a structure by the weight of the building materials only.

Decay: deterioration of wood from attack by fungi or insects.

Decibel (db): logarithmic measure of sound intensity. An increase of 6 db is the same as doubling the sound pressure.

Deciduous plant: one that loses its leaves in winter.

Deck: roof surface to which roofing is applied.

Deflection: distance moved upon application of a specified load on a structural member.

Deflection ratio: ratio of clear span to deflection at design load.

Degree-day: difference between the average of daily high and low temperatures and a fixed temperature - usually 65°F.

Dew point: air temperature at which water vapor begins to condense as either water or ice (frost).

Diffuse radiation: solar energy received from a direction different from that of the sun.

Dimension lumber: framing lumber 2 to 5 inches in nominal thickness and up to 12 inches in nominal width.

Direct current (DC): electrical current that flows in a single direction.

Direct gain: heating system in which energy received from the sun directly enters and heats the living spaces.

Dormer: vertical window projecting from a roof. Gabled dormers have peaked roofs; shed dormers have shed roofs.

Double coverage: result of applying asphalt roll roofing with sufficient overlap to get a double layer. See *coverage*.

Double-hung window: window with vertically sliding upper and lower sash. If the upper sash is fixed, the window becomes single hung.

Double-strength glass: sheet glass of nominal thickness 0.125 inch.

Dovetail: flared mortise and tenon that form a locking joint.

Downspout: vertical section of pipe in a gutter system.

Draft: air pressure difference between the inside and outside of a chimney. Also the rate of flue gas or combustion airflow.

Drip cap: molding at the top of a window or door.

Drip edge: material designed to force water to drip from roof rakes and eaves.

Drywall: interior finish material in large sheets. Plywood paneling and gypsum drywall are two examples.

Drywall nail: special, ringed nail for fastening drywall.

Duct: enclosure for distributing heated or cooled air.

Duct tape: aluminized cloth tape used to seal duct joints.

Eaves: lower edge of a sloped roof.

Eaves flashing: flashing at eaves designed to prevent leaking from an ice dam.

Egress window: window whose clear dimensions are large enough that it can serve as a fire exit.

Elastic modulus: ratio of stress to strain in a material. Also known as the modulus of elasticity.

Elbow: right-angle bend in stovepipe.

Elevation: view of a vertical face of a building.

Ell: L-shaped pipe fitting.

Energy efficiency: percentage of energy in fuel that is converted to useful energy.

Envelope: collection of building surfaces that separate the building interior from the outside, i.e., roof, walls, floor, windows, and doors.

Exposed-nail roofing: mineral-surfaced roll roofing where nails are exposed.

Exposure: portion of roofing or siding material exposed after installation.

Extension jamb: addition to a door or window jamb to bring the jamb up to full wall thickness. Also known as a jamb extender.

Face: side of a masonry unit or wood panel intended to be exposed.

Face brick: brick intended to be used in an exposed surface.

Face nailing: nailing perpendicular to the face or surface. Also called direct nailing.

Fahrenheit: temperature scale defined by the freezing (32°) and boiling (212°) points of water.

Fascia: vertical flat board at a cornice. Also spelled *facia*.

Fastener: any device for connecting two members.

Felt: fibrous sheet material for roofing.

Finger joint: wood joint formed by interlocking fingers.

Finish coat: final coat of a material.

Finish grade: final ground level around a building.

Finish nail: thin nail intended to be driven flush or countersunk.

Fire-stop: framing member designed to block the spread of fire within a framing cavity.

Flashing: material used to prevent leaks at intersections and penetrations of a roof.

Float glass: glass formed by floating molten glass on molten tin.

Flue: passage in a chimney for the venting of flue gases or products of combustion.

Flue gases: mixture of air and the products of combustion.

Fluorescent light: lamp that emits light when an electric discharge excites a phosphor coating.

Footing: bottom section of a foundation that rests directly on the soil.

Forced-air system: heat transfer system using a blower.

Foundation: section of a building that transfers the building load to the earth.

Frame: assemblage of structural support members.

Frieze: middle section of an entablature, between the architrave below and the cornice above. In wood construction, the horizontal board between the top of the siding and the soffit. Also a decorative band near the top of a wall.

Frost heave: expansion of the earth due to freezing of interstitial water.

Frost line: maximum depth of freezing in the soil.

Furnace: appliance that generates hot air.

Furring: strip of wood that provides space for insulation or that levels an uneven surface.

Gable: upper, triangular portion of an end wall.

Gable roof: roof having gables at opposite ends, each equally pitched.

Gasket: elastic strip that forms a seal between two parts.

Girder: main supporting beam.

Girt: horizontal beam framed into the posts.

Glare: excessive contrast in lighting.

Glass: transparent material composed of silica (sand), soda (sodium carbonate), lime (calcium carbonate), and small quantities of other minerals.

Glass block: glass molded into hollow blocks that serve to support loads and pass light.

Glazing: glass or other transparent material used for windows.

Gloss paint: paint with a high percentage of resin that dries to a highly reflective finish.

Grade: level of the ground.

Ground: any metal object that is connected to and serves as the earth in an electrical system.

Grounded wire: wire in a circuit that is connected to the ground and serves to return current from the hot wire back to the ground. Identified by a white insulation jacket.

Ground fault interrupter: circuit breaker that trips on leakage of current.

Grounding wire: bare or green wire in a circuit that connects metal components, such as appliance cabinets, to the ground.

Grout: very thin mortar applied to masonry joints.

Gusset: flat plate used on either side of a wood joint to aid in connection.

Gutter: horizontal trough for collecting rain water from a roof.

Gypsum: calcium sulfate, a naturally occurring mineral.

Gypsum drywall: rigid paper-faced board made from hydrated gypsum and used as a substitute for plaster and lath.

Handrail: upper rail in a balustrade.

Hard water: water rich in calcium.

Hardwood: wood from a deciduous tree.

Head: top element in many structures.

Header: beam over a door or window for supporting the load from above. Also any beam which crosses and supports the ends of other beams. Also a brick placed to tie two adjacent wythes together.

Heartwood: portion of the tree from the pith (center) to the sapwood.

Heat-absorbing glass: glass containing additives that absorb light in order to reduce glare, brightness, and solar heat gain.

Heat capacity: quantity of heat required to raise the temperature of 1 cubic foot of a material 1 F°.

Heat Mirror: trade name for plastic film treated with selective coating.

Heat pump: mechanical device that transfers heat from a cooler to a warmer medium.

Hip: convex intersection of two roof planes, running from eaves to ridge.

Hip shingle: shingle covering a hip.

Hot wire: current-carrying wire that is not connected to the ground.

Humidifier: appliance for adding water vapor to the air.

Hydrated lime: quicklime and water combined. Also called slaked lime.

Hydronic: method of distributing heat by hot water.

I-beam: steel beam whose section resembles the letter *I*.

Ice dam: ridge of ice at roof eaves.

Incandescent: heated to the point of giving off light.

Infiltration: incursion of outdoor air through cracks, holes, and joints.

Infrared radiation: radiation of a wavelength longer than that of red light. Also known as heat radiation.

Inside sill: window stool.

Insulating glass: factory-sealed double or triple glazing.

Insulation: material with high resistance to heat flow.

Insulation board: wood fiber board available in 1/2-inch and 25/32-inch thicknesses.

Jamb: top and sides of a door or window.

Jamb depth: width of a window frame.

Jamb extender: same as an extension jamb.

Joint compound: material used to finish joints in gypsum drywall. Also known as mud.

Joist: repetitive narrow beam supporting the floor load.

Kiln-dried: lumber dried in a kiln at elevated temperatures. The process removes cellular water.

Kilowatt: 1,000 watts. Abbreviated *kw*.

Kilowatt-hour: unit of electrical energy consumed. One thousand watts of power for a 1-hour duration. Abbreviated *kwhr*.

Knot: section of the base of a branch enclosed in the stem from which it arises. Found in lumber such as pine.

Landing: platform between or at the ends of stairways.

Latex paint: water-based paint.

Lath: perforated base for application of plaster. Formerly wood, now usually metal.

Lattice: framework of crossed strips of wood, plastic, or metal.

Leader: horizontal section of downspout.

Ledger strip: strip of wood forming a ledge on a girder or sill for supporting the bottoms of joists.

Let-in brace: corner brace of 1x4 lumber cut into studs.

Lime: calcium carbonate. When heated, it becomes quicklime.

Linseed oil: oil from the seed of the flax plant.

Lintel: solid member above a door or window that carries the load above.

Live load: temporary load imposed on a building by occupancy and the environment.

Lookout: wood member supporting the end of an overhanging rafter.

Louver: slanted slat of wood, plastic, or metal. Used to admit air but block rain and visibility.

Lumen: measure of total light output. A wax candle gives off about 13 lumens, a 100-watt incandescent bulb about 1,200 lumens.

Masonry: construction consisting of stone, brick, or concrete block.

Masonry cement: cement to which water and sand must be added.

Masonry primer: asphalt primer for bonding asphalt-based products to masonry.

Mastic: thick-bodied adhesive or sealant.

Membrane roof: roofing consisting of a single waterproof sheet.

Metal lath: sheet metal slit and formed into a mesh for use as a plaster base.

Mil: one-thousandth of an inch.

Millwork: building components manufactured in a woodworking plant.

Mineral spirits: petroleum-based solvent for oil-based paints and varnishes.

Miter: to cut at an angle other than 90°.

Miter joint: joint where each member is mitered at equal angles.

Module: repeated dimension.

Modulus of elasticity: (see *elastic*).

Moisture barrier: material or surface with the purpose of blocking the diffusion of water vapor. The same as a vapor barrier.

Moisture content: amount of moisture in a material, expressed as the percentage of dry weight.

Mortar: plastic mixture of cement, sand, and water.

Mortise: hole into which a tenon (tongue) fits.

Movable window insulation: shutter or shade for insulating a window against heat loss.

Natural finish: wood finish that does not greatly alter the unfinished color.

Natural ventilation: air movement in a building due only to natural pressure differences caused by air temperatures.

Neoprene: synthetic rubber.

Neutral wire: grounded wire.

Nonbearing wall: wall or partition that does not carry a load from above.

Nosing: projection of a stair tread beyond the riser; the amount by which the actual tread is wider than the mathematical tread.

Ohm: measure of resistance to electric current.

On-center (oc): framing measurement from the center of one member to the center of the other.

Open valley: roof valley where shingles do not cross the valley intersection; flashing does.

Orientation: placement relative to the sun, wind, view, and so forth.

Overhang: portion of a roof extending beyond the wall line.

Overload: excessive electric current in a conductor. The danger is from overheating. Circuit breakers interrupt circuits upon detecting overloads.

Pane: piece of glass that, when installed in a window, becomes a light.

Panel: thin, flat piece.

Parget: a surface coat of cement over masonry. Also known as parge.

Parquet: thin strips of wood applied in geometric patterns on floors and furniture.

Partition wall: nonbearing wall.

Party wall: common wall that separates two properties.

Passive solar collector: system for collecting solar energy without use of mechanical devices such as fans or pumps.

Penetrating finish: finish that sinks into wood grain and does not leave a hard skin.

Penny: formerly the price in pennies of 100 nails of a certain size; now a measure of length. A 6-penny (6d) nail is approximately 2 inches long.

Perlite: expanded volcanic glass. Used as an insulator and as a lightweight additive to concrete.

Perm: 1 grain of water vapor per square foot per hour per inch of mercury difference in water vapor pressure.

Permeability: ability to transmit water vapor, measured in perms.

Picture window: large fixed window.

Pier: isolated masonry column.

Pigment: powdered dye added to stain or paint.

Pitch: ratio of rise in feet to span in feet of a roof.

Pith: soft core of a tree that represents the original shoot.

Plaster: mortar-like material that hardens after application. Stucco is simply exterior plaster.

Plasterboard: term for gypsum drywall.

Plate: horizontal member at the top or bottom of a wall. The top plate supports the rafter ends. The bottom or sill plate supports studs and posts.

Plenum: ductwork chamber that serves as a distribution point.

Plumb: vertical.

Plywood: wood panel made of three or more veneers of wood alternating in direction of grain.

Pocket door: door that slides into a wall.

Pores: large-diameter wood cells that open to the surface.

Portland cement: strong, water-resistant cement consisting of silica, lime, and alumina.

Preservative: water repellent liquid containing fungicide.

Pressure-treated wood: wood that has been injected with preservative under pressure.

Primer: first coating, applied prior to regular paint.

Psychrometric chart: graph showing the properties of water vapor in air.

Purlin: a horizontal member perpendicular to, and supporting, rafters.

Quarter round: molding whose section is that of a quarter of a circle.

Rabbet: a rectangular shape consisting of two surfaces cut along the edge or end of a board.

Racking: distortion of a building surface from the rectangular in its plane.

Radiant heating: method of heating whereby much of the heat transfer is accomplished by radiation through space from warm building surfaces such as floors, walls, or ceilings.

Rafter: roof beam running in the direction of the slope.

Rail: horizontal member of a door or window sash. Also the top member of a balustrade.

Rebar: abbreviation for *reinforcing bar*. Usually applied to steel bars used in concrete.

Receptacle: electrical device into which a plug may be connected.

Reflectance: decimal fraction of light incident on a surface that is reflected and not absorbed. Absorptance equals 1 minus reflectance.

Reflective glass: glass treated to reflect a fraction (the reflectance) of incident light.

Reflectivity: (see *reflectance*).

Register: grill or grate covering the outlet of a duct.

Reinforcement: (see *rebar*).

Relative humidity: amount of water vapor in air compared with the maximum amount possible, expressed as a percentage.

Resorcinol glue: plastic resin glue that withstands water.

Retrofit: to upgrade a structure using modern materials.

Return: general term for a right-angle turn.

Reverberation time: measure of the length of time a sound wave will bounce around a space before being absorbed.

Ribbon: horizontal strip (usually 1x4) let into studs to support joist ends. Also called a ribband.

Ridge: junction of the top of opposing roof planes.

Ridge board: vertical board between the upper ends of rafters.

Ridge vent: continuous, prefabricated outlet ventilator placed over an opening at the ridge.

Rise: vertical increase in one step of a stair. Also the total vertical span of a stairway from landing to landing, or any vertical change.

Riser: vertical board between stair treads. Also a vertical pipe.

Roll roofing: low-cost asphalt roofing in roll form.

Roof overhang: horizontal projection of the roof beyond the wall.

Run: horizontal span of a flight of stairs.

R-value: measure of resistance to heat flow.

Saddle: pitched section of roof behind a chimney or between a roof and the wall toward which it slopes. Its purpose is to avoid trapped water.

Safety glass: one of a number of types of glass that have been strengthened or reinforced for safety.

Sapwood: wood between the heartwood and the bark, in which the sap runs.

Sash: frame holding the panes of glass in a window or door.

Saturated felt: felt impregnated with asphalt in order to make it water resistant.

Scratch coat: first coat of plaster. It is scratched to provide better bonding with the next coat.

Screen molding: thin wood molding for covering the edge of screening.

Sealant: compressible material used to seal building joints, etc.

Sealed glass: panes of glass with a sealed air space between.

Sealer: liquid applied to unfinished wood to seal the surface.

Selvage: portion of roll roofing meant to be overlapped by the succeeding course.

Sensible heat: heat required to raise the temperature of a material without changing its form.

Service drop: wiring from the utility pole to the service entrance conductors leading to the meter.

Service entrance box: box housing the electrical panel containing the main breaker and branch circuit breakers.

Shading coefficient: ratio of solar gain to the solar gain through a single layer of clear, double-strength glass.

Shake: wood shingle formed by splitting rather than sawing. Also a lumber defect in which the growth rings separate.

Shear: the effect of opposing forces acting in the same plane of a material.

Sheathing: layer of boards over the framing but under the finish.

Shellac: resinous secretion of the lac bug, dissolved in alcohol.

Shelter belt: band of trees and shrubs planted to reduce wind speed.

Shim: thin, tapered piece of wood used to level or plumb.

Shingle: small, thin piece of material, often tapered, for laying in overlapping rows as in roofing or siding.

Shiplap: rabbeted wood joint used in siding.

Siding: exterior finish for a wall.

Sill: lowest horizontal member in a frame. Also the bottom piece of the window rough opening.

Single-phase wiring: wiring in which the voltage exists only as a single sine wave. This is the type used in residences.

Single-strength glass: glass of thickness 0.085 to 0.100 inches.

Skylight: window set into a roof.

Slab-on-grade: concrete slab resting directly on the ground at near-grade level.

Sleeper: wood strip set into or on concrete as a fastener for flooring.

Slider: window that slides horizontally. Also called a sliding window.

Soffit: underside of a roof overhang, cornice, or stairway.

Soffit vent: inlet vent in the soffit. It may be individual or continuous.

Softwood: wood from coniferous, mostly evergreen, trees.

Soil line or pipe: pipe carrying human waste.

Solar radiation: total electro-magnetic radiation from the sun.

Solar-tempered heating system: system deriving a significant fraction of the heating requirement from the sun.

Solder: metal alloy with a low melting point used in joining pipes, electrical wiring, and sheet metal.

Span: distance between supports.

Specific heat: ratio of the heat storage capacity of a material to that of an equal weight of water.

Square: 100 square feet of coverage. Also a carpentry tool for measuring and laying out.

Stack effect: buoyancy of warm gases within a chimney.

Standing-seam: metal roofing technique of folding the upturned edges of adjacent sheets to form a weatherproof seam.

Standing wave: sound wave in a space of a dimension equal to a multiple of the sound's wavelength.

Stile: vertical outside frame member in a door or window sash.

Stool: interior horizontal, flat molding at the bottom of a window.

Stop (molding): thin molding for stopping doors on closure or holding window sash in place.

Storm door or window: removable, extra door or window for reducing winter heat loss.

Story: space between two floors. Also spelled *storey*.

Stringer: side member into which stair treads and risers are set.

Strip flooring: flooring of narrow strips with matched edges and ends.

Stucco: plaster applied to the exterior.

Stud: vertical framing member to which wall sheathing and siding are attached.

Subfloor: first floor laid over the floor joists. The subfloor may also serve as the finish floor.

Supply plumbing: pipes supplying water to a building.

Suspended ceiling: modular ceiling panels supported in a hanging frame.

Tab: exposed portion of an asphalt shingle between the cutouts.

Taping: finishing gypsum drywall joints with paper tape and joint compound.

Tempered glass: glass that has been cooled rapidly to produce surface tension. The result is a stronger-than-normal glass that shatters into relatively harmless cubical fragments when broken.

Tenon: beam-end projection fitting into a mating tenon or hole in a second beam.

Tension: pulling apart; the opposite of compression.

Termiticide: chemical for poisoning termites.

Terne metal: sheet metal coated with a lead-and-tin alloy. Used in roofing.

Therm: 100,000 Btu.

Thermal mass: measure of the ability of a material to store heat (for later release).

Thimble: protective device installed in a combustible wall through which a stovepipe passes.

Threshold: beveled wood strip used as a door sill.

Tie beam: beam placed between mating rafters to form a triangle and prevent spreading. Also known as a collar tie.

Tilt angle: angle of a collector or window from the horizontal.

Timber: lumber that is 5 or more inches thick.

Toenailing: nailing a butt joint at an angle.

Tongue and groove: flooring and sheathing joint in which the tongue of one piece meets a groove in a mating piece.

Touch-sanding: sizing structural wood panels to a uniform thickness by means of light surface sanding.

Transit: surveying instrument, usually mounted on a tripod, for measuring horizontal and vertical angles.

Trap: section of plumbing pipe designed to retain water and block the flow of sewer gas into a building.

Tread: horizontal part of a step. The nosing is physically part of the tread but doesn't count from the design standpoint.

Trim: decorative building elements often used to conceal joints.

Truss: framing structure for spanning great distances, in which every member is purely in tension or in compression.

Tung oil: fast-drying oil from the seed of the Chinese tung tree. Used as a penetrating wood finish.

Ultraviolet radiation: radiation of wavelengths shorter than those of visible radiation.

Underlayment: sheet material or wood providing a smooth, sound base for a finish.

U-value: inverse of R-value.

Valley: intersection of two pitched roofs that form an internal angle.

Vapor barrier: material or surface designed to block diffusion of water vapor.

Varnish: mixture of drying oil and resin without pigment. With pigment added, it becomes enamel.

Veneer: thin sheet of wood formed by slicing a log around the growth rings.

Vent: pipe or duct allowing inlet or exhaust of air.

Vermiculite: mica that has been expanded to form an inert insulation.

Wall: any vertical structure whose height exceeds three times its thickness.

Wane: area of missing wood in lumber due to misjudgment of the log during sawing.

Water hammer: sound made by supply pipes when water is suddenly stopped by the quick closing of a valve.

Water softener: appliance that removes calcium ions from water.

Water table: level of water saturation in soil. Also a setback at foundation level in a masonry wall.

Watt: unit of electrical power. Watts equal volts across the circuit times amps flowing through it.

Weather strip: thin, linear material placed between a door or window and its jambs to prevent air leakage.

Weep hole: hole purposely built into a wall to allow drainage of trapped water.

Wind brace: T-section of metal strip let in and nailed diagonally to studs to provide racking resistance to a wall.

Wythe: single thickness of masonry in a wall.

Sources and Credits

Chapter 1: Design

"Human Dimensions" adapted from *Building Construction Illustrated*, by Francis Ching (NY: Van Nostrand Reinhold, 1975).

"Kitchen Dimensions" adapted from *Low Cost Wood Homes for Rural America*, by L. O. Anderson (Washington, DC: Department of Agriculture, 1969).

"Installing Kitchen Cabinets" adapted from *Install Kitchen Cabinets— Do It Yourself Guide* (84 Lumber and Home Centers).

"Bathroom Dimensions" adapted from *Bathroom Design* (Washington, DC: National Association of Home Builders, 1978).

"Kitchen Plans" adapted from *Kitchen Planning Principles. Equipment. Appliances*, (Urbana - Champaign: Small Homes Council - Building Research Council, University of Illinois, 1984).

"Bathroom Plans" adapted from *Bathroom Design* (Washington, DC: National Association of Home Builders, 1978).

"Stair Design" adapted from *Dwelling Construction Under the Uniform Building Code*, © 1985 ed (Whittier, Calif: with permission of the publishers, the International Conference of Building Officials, 1985).

"Access" adapted from *28 CFR Part 36 Appendix A of the Code of Federal Regulations* (Washington, DC: Department of Justice, 1994).

"Checklist of Code Requirements" adapted from *One and Two Family Dwelling Code* (Falls Church, Va: the Council of American Building Officials, 1995)

Chapter 2: Site

"Magnetic Compass Variation" adapted from *Geodetic Survey*, 1975 (Washington, DC: Department of the Interior, 1975).

"Laying Out Driveways" adapted from *Wood Frame House Construction* (Washington, DC: Department of Agriculture, 1975).

"Typical Soil Conservation Map" adapted from *Soil Survey of Cumberland County, Maine* (Washington, DC: Department of Agriculture, 1974).

"Radon Producing Areas" adapted from "The Radon Report,"*New Shelter Magazine*, January, 1986.

Chapter 3: Masonry

"Brick Positions" and "Brick Sizes and Coursing" adapted from *How Big is a Brick?* (Reston, Va: Brick Institute of America).

"Brick Wall Types" adapted from *Brick Walls for Housing* (Reston, Va: Brick Institute of America).

"Estimating Brick and Mortar" adapted from *Estimating Brick Masonry* (Reston, Va: Brick Institute of America).

"Brick Pavement" adapted from *Exterior Brick Paving* (Reston, Va: Brick Institute of America).

"Block Sizes and Coursing" and "Concrete Pavers"adapted from *Passive Solar Construction Handbook* (Herndon, Va: National Concrete Masonry Association, 1984).

"Checklist of Code Requirements" adapted from *One and Two Family Dwelling Code* (Falls Church, Va: the Council of American Building Officials, 1995)

Chapter 4: Foundations

All materials except "Piers with Floor Insulation" adapted from *Building Foundation Design Handbook* (Oak Ridge, Tenn: Oak Ridge National Laboratory, 1988).

"Checklist of Code Requirements" adapted from *One and Two Family Dwelling Code* (Falls Church, Va: the Council of American Building Officials, 1995)

Chapter 5: Wood

"The Nature of Wood" adapted from *Wood Handbook* (Washington, DC: Department of Agriculture, 1987).

"Lumber Defects" adapted from *Wood Handbook* (Washington, DC: Department of Agriculture, 1987).

"Lumber Grade Stamps" adapted from *Product Use Manual* (Portland, Oreg: Western Wood Products Association, 1986). Stamp used with permission of West Coast Lumber Inspection Bureau.

"Pressure Treated Wood" adapted from *Building Code Requirements for Pressure Treated Wood* (Arlington, Va: Society of American Wood Preservers, 1983).

Chapter 6: Framing

"Stress Value Adjustment Factors" and "Span Tables for S4S Lumber" adapted from *The U.S. Span Book for Major Lumber Species* (Ottawa, Ont: Canadian Wood Council, 1996).

"Wood Trusses" adapted from *Metal Plate Connected Wood Truss Handbook* (Madison, Wis: Wood Truss Council of America, 1990).

"Framing Details" adapted from *Manual for Wood Frame Construction* (Washington, DC: American Forest & Paper Association, 1988).

"Checklist of Code Requirements" adapted from *One and Two Family Dwelling Code* (Falls Church, Va: the Council of American Building Officials, 1995)

Chapter 7: Sheathing

All material adapted from *Design/Construction Guide: Residential and Commercial* (Tacoma, Wash: APA—The Engineered Wood Association, 1996).

Chapter 8: Siding

"Vinyl Siding" adapted from *Installation Instructions for Horizontal and Vertical Solid Vinyl Siding* (Valley Forge, Pa: Certainteed Corporation, 1984).

"Hardboard Lap Siding" adapted from *Hardboard Siding Application and Storage Instructions* (Palatine, Ill: American Hardboard Association, 1983).

"Vertical Wood Siding" adapted from *Redwood Siding Patterns and Application* (Mill Valley, Calif: California Redwood Association).

"Plywood Siding" adapted from *Design/Construction Guide: Residential and Commercial* (Tacoma, Wash: APA—The Engineered Wood Association, 1996).

"Stucco" adapted from "Sticking with Stucco," *Professional Builder*, September 1987.

"Checklist of Code Requirements" adapted from *One and Two Family Dwelling Code* (Falls Church, Va: the Council of American Building Officials, 1995)

Chapter 9: Roofing

"Exposed-Nail Roll Roofing", "Concealed-Nail Roll Roofing", "Double-Coverage Roll Roofing", and "Asphalt Shingles" adapted from *Residential Asphalt Roofing Manual* (Rockville, Md: Asphalt Roofing Manufacturers Association, 1984).

"Cedar Shingles" adapted from *Exterior and Interior Product Glossary* (Bellevue, Wash: Red Cedar Shingle & Handsplit Shake Bureau, 1980).

"Cedar Shakes" adapted from *Exterior and Interior Product Glossary* (Bellevue, Wa: Red Cedar Shingle & Handsplit Shake Bureau, 1980).

"Checklist of Code Requirements" adapted from *One and Two Family Dwelling Code* (Falls Church, Va: the Council of American Building Officials, 1995)

Chapter 10: Windows and Doors

"Skylights" adapted from *Complete Guide* (Greenwood, SC: Velux America, 1997).

"Site-Built Windows" details from Dale McCormick, Cornerstones Energy Group, Brunswick, ME.

"Classic Wood Door Installation" adapted from *Wood Frame House Construction* (Washington, DC: Department of Agriculture).

"Modern Prehung Door Installation" adapted from *Technical Data - Insulated Windows and Doors* (Norcross, Ga: Peachtree Doors, 1987).

"Bulkhead Doors" adapted from *Planning for a Wooden Door Replacement* (New Haven, Conn: The Bilco Company, 1985).

"Checklist of Code Requirements" adapted from *One and Two Family Dwelling Code* (Falls Church, Va: the Council of American Building Officials, 1995)

Chapter 11: Plumbing

"Drain, Waste, and Vent" adapted from *Uniform Plumbing Code Illustrated Training Manual* (Walnut, Calif: International Association of Plumbing and Mechanical Officials, 1985).

"Venting Rules" adapted from *Uniform Plumbing Code Illustrated Training Manual* (Walnut, Calif: International Association of Plumbing and Mechanical Officials, 1985).

"Water Treatment" adapted from *Planning for an Individual Water System*, 3rd ed (Athens, Ga: American Association for Vocational Instructional Materials, 1973), and from *Sears Catalogue* (Chicago: Sears, 1987).

"Checklist of Code Requirements" adapted from *One and Two Family Dwelling Code* (Falls Church, Va: the Council of American Building Officials, 1995)

Chapter 12: Wiring

"Typical Appliance Wattages" adapted from *Home Wind Power* (Washington, DC: Department of Energy, 1981).

"Service Drops and Entrances" adapted from *Standard Requirements: Electric Service and Meter Installations* (Augusta, Me: Central Maine Power Company, 1981).

"Wiring Switches, Receptacles and Lights" adapted from *Simplified Electrical Wiring* (Chicago: Sears, 1974).

"Checklist of Code Requirements" adapted from *One and Two Family Dwelling Code* (Falls Church, Va: the Council of American Building Officials, 1995)

Chapter 13: Insulation

"R-Values of Compressed Batts" data from Owens-Corning Fiberglas.

"R-Values of Walls, Roofs, and Floors" adapted from *Major Conservation Retrofits* (Washington, DC: Department of Energy, 1984).

"Caulks and Weather Strips" adapted from *Air Sealing Homes for Energy Conservation* (Ottawa: Energy, Mines and Resources Canada, 1984).

"Heat Leaks in the Home" adapted from *Cataloguing Air Leakage Components in Houses* (David Harrje, Gerald Gborn, Washington, DC: American Council for an Energy Efficient Economy, 1982).

Chapter 14: Floors, Walls and Ceilings

"Ceramic Tile" adapted from *American National Standard Specifications for Ceramic Tile* (Princeton, NJ: Tile Council of America, 1981).

"Installation" adapted from *Oak Flooring Institute literature* (Memphis, Tenn: Oak Flooring Institute, 1986).

"Gypsum Wallboard" adapted from *Using Gypsum Board for Walls and Ceilings* (Evanston, Ill: Gypsum Association, 1985).

"Achieving Fire Ratings" adapted from *Gypsum Wallboard Construction* (Charlotte, NC: Gold Bond Building Products, 1982).

"Wood Paneling" adapted from *Miracle Worker's Guide to Real Wood Interiors* (Portland, Oreg: Western Wood Products Association, 1986).

"Suspended Ceilings" adapted from *Install Suspended Ceiling – Do It Yourself Guide* (84 Lumber and Home Centers).

"Wood Moldings" adapted from *WM/Series Wood Moulding Patterns* (Portland, Oreg: Wood Moulding and Millwork Association, 1986).

Chapter 15: Heating

"Fireplaces" adapted from *Residential Fireplace Design* (Reston, Va: Brick Institute of America).

"Wood Stove Installation" adapted from *Recommended Standards for the Installation of Woodburning Stoves* (Augusta, Maine: Maine State Fire Marshal's Office, 1979).

"Metal Prefabricated Chimneys" adapted from *Model SSII Metalbestos Chimney Systems by Selkirk* (Nampa, Idaho: Selkirk Metalbestos Company, 1986).

"Checklist of Code Requirements" adapted from *One and Two Family Dwelling Code* (Falls Church, Va: the Council of American Building Officials, 1995)

Chapter 16: Cooling

"Air Conditioning Form" adapted from "Air-Conditioning Guide," *New Shelter*, July 1984.

"Human Comfort Zone" adapted from Passive Cooling and Human Comfort (Cape Canaveral: Florida Solar Energy Center, 1981).

"Attic Radiant Barriers" adapted from *Designing and Installing Radiant Barrier Systems* (Cape Canaveral: Florida Solar Energy Center, 1984).

"Window Shading" adapted from *Comparison of Window Shading Strategies for Heat Gain Prevention* (Cape Canaveral: Florida Solar Energy Center, 1984).

Chapter 17: Passive Solar Heating

"Passive Solar System Types" adapted from *Passive Solar Construction Handbook* (Herndon, Va: National Concrete Masonry Association, 1984).

"Factors for South-Facing Overhangs" adapted from *Passive Solar Construction Handbook* (Herndon, Va: National Concrete Masonry Association, 1984).

"Solar Absorptance" adapted from *Passive Solar Design Handbook*, vol 3 (Washington, DC: Department of Energy, 1980).

"Heat Capacities of Materials" adapted from *Passive Solar Design Handbook*, vol 3 (Washington, DC: Department of Energy, 1980).

"The Mass Patterns" adapted from *Adding Thermal Mass to Passive Designs: Rules of Thumb for Where and When* (Cambridge, Mass: Northeast Solar Energy Center, 1981).

Chapter 18: Lighting

"Lamps" adapted from *Large Lamp Ordering Guide* (Danvers, Mass: Sylvania Lighting Center, 1986).

Chapter 19: Sound

"Sound Transmission" adapted from *Noise-Rated Systems* (Tacoma, Wash: American Plywood Association, 1981) and *Minimum Property Standards*, vol 1 & 2 (Washington, DC: Department of Housing and Urban Development).

Chapter 20: Fasteners

"Metal Framing Aids" adapted from *Full Line Catalog—Construction Hardware* (Montgomery, Minn: United Steel Products Company, 1996).

Index

Page references in *italic* indicate illustrations.
Boldface references indicate tables.